Current Practice of
Gas Chromatography–
Mass Spectrometry

CHROMATOGRAPHIC SCIENCE SERIES

A Series of Textbooks and Reference Books

Editor: JACK CAZES

1. Dynamics of Chromatography, *J. Calvin Giddings*
2. Gas Chromatographic Analysis of Drugs and Pesticides, *Benjamin J. Gudzinowicz*
3. Principles of Adsorption Chromatography: The Separation of Nonionic Organic Compounds, *Lloyd R. Snyder*
4. Multicomponent Chromatography: Theory of Interference, *Friedrich Helfferich and Gerhard Klein*
5. Quantitative Analysis by Gas Chromatography, *Josef Novák*
6. High-Speed Liquid Chromatography, *Peter M. Rajcsanyi and Elisabeth Rajcsanyi*
7. Fundamentals of Integrated GC-MS (in three parts), *Benjamin J. Gudzinowicz, Michael J. Gudzinowicz, and Horace F. Martin*
8. Liquid Chromatography of Polymers and Related Materials, *Jack Cazes*
9. GLC and HPLC Determination of Therapeutic Agents (in three parts), *Part 1 edited by Kiyoshi Tsuji and Walter Morozowich, Parts 2 and 3 edited by Kiyoshi Tsuji*
10. Biological/Biomedical Applications of Liquid Chromatography, *edited by Gerald L. Hawk*
11. Chromatography in Petroleum Analysis, *edited by Klaus H. Altgelt and T. H. Gouw*
12. Biological/Biomedical Applications of Liquid Chromatography II, *edited by Gerald L. Hawk*
13. Liquid Chromatography of Polymers and Related Materials II, *edited by Jack Cazes and Xavier Delamare*
14. Introduction to Analytical Gas Chromatography: History, Principles, and Practice, *John A. Perry*
15. Applications of Glass Capillary Gas Chromatography, *edited by Walter G. Jennings*
16. Steroid Analysis by HPLC: Recent Applications, *edited by Marie P. Kautsky*
17. Thin-Layer Chromatography: Techniques and Applications, *Bernard Fried and Joseph Sherma*
18. Biological/Biomedical Applications of Liquid Chromatography III, *edited by Gerald L. Hawk*
19. Liquid Chromatography of Polymers and Related Materials III, *edited by Jack Cazes*
20. Biological/Biomedical Applications of Liquid Chromatography, *edited by Gerald L. Hawk*
21. Chromatographic Separation and Extraction with Foamed Plastics and Rubbers, *G. J. Moody and J. D. R. Thomas*
22. Analytical Pyrolysis: A Comprehensive Guide, *William J. Irwin*
23. Liquid Chromatography Detectors, *edited by Thomas M. Vickrey*
24. High-Performance Liquid Chromatography in Forensic Chemistry, *edited by Ira S. Lurie and John D. Wittwer, Jr.*
25. Steric Exclusion Liquid Chromatography of Polymers, *edited by Josef Janca*
26. HPLC Analysis of Biological Compounds: A Laboratory Guide, *William S. Hancock and James T. Sparrow*

27. Affinity Chromatography: Template Chromatography of Nucleic Acids and Proteins, *Herbert Schott*
28. HPLC in Nucleic Acid Research: Methods and Applications, *edited by Phyllis R. Brown*
29. Pyrolysis and GC in Polymer Analysis, *edited by S. A. Liebman and E. J. Levy*
30. Modern Chromatographic Analysis of the Vitamins, *edited by André P. De Leenheer, Willy E. Lambert, and Marcel G. M. De Ruyter*
31. Ion-Pair Chromatography, *edited by Milton T. W. Hearn*
32. Therapeutic Drug Monitoring and Toxicology by Liquid Chromatography, *edited by Steven H. Y. Wong*
33. Affinity Chromatography: Practical and Theoretical Aspects, *Peter Mohr and Klaus Pommerening*
34. Reaction Detection in Liquid Chromatography, *edited by Ira S. Krull*
35. Thin-Layer Chromatography: Techniques and Applications. Second Edition, Revised and Expanded, *Bernard Fried and Joseph Sherma*
36. Quantitative Thin-Layer Chromatography and Its Industrial Applications, *edited by Laszlo R. Treiber*
37. Ion Chromatography, *edited by James G. Tarter*
38. Chromatographic Theory and Basic Principles, *edited by Jan Åke Jönsson*
39. Field-Flow Fractionation: Analysis of Macromolecules and Particles, *Josef Janca*
40. Chromatographic Chiral Separations, *edited by Morris Zief and Laura J. Crane*
41. Quantitative Analysis by Gas Chromatography: Second Edition, Revised and Expanded, *Josef Novák*
42. Flow Perturbation Gas Chromatography, *N. A. Katsanos*
43. Ion-Exchange Chromatography of Proteins, *Shuichi Yamamoto, Kazuhiro Nakanishi, and Ryuichi Matsuno*
44. Countercurrent Chromatography: Theory and Practice, *edited by N. Bhushan Mandava and Yoichiro Ito*
45. Microbore Column Chromatography: A Unified Approach to Chromatography, *edited by Frank J. Yang*
46. Preparative-Scale Chromatography, *edited by Eli Grushka*
47. Packings and Stationary Phases in Chromatographic Techniques, *edited by Klaus K. Unger*
48. Detection-Oriented Derivatization Techniques in Liquid Chromatography, *edited by Henk Lingeman and Willy J. M. Underberg*
49. Chromatographic Analysis of Pharmaceuticals, *edited by John A. Adamovics*
50. Multidimensional Chromatography: Techniques and Applications, *edited by Hernan Cortes*
51. HPLC of Biological Macromolecules: Methods and Applications, *edited by Karen M. Gooding and Fred E. Regnier*
52. Modern Thin-Layer Chromatography, *edited by Nelu Grinberg*
53. Chromatographic Analysis of Alkaloids, *Milan Popl, Jan Fähnrich, and Vlastimil Tatar*
54. HPLC in Clinical Chemistry, *I. N. Papadoyannis*
55. Handbook of Thin-Layer Chromatography, *edited by Joseph Sherma and Bernard Fried*
56. Gas–Liquid–Solid Chromatography, *V. G. Berezkin*
57. Complexation Chromatography, *edited by D. Cagniant*
58. Liquid Chromatography–Mass Spectrometry, *W. M. A. Niessen and Jan van der Greef*
59. Trace Analysis with Microcolumn Liquid Chromatography, *Milos Krejcl*

60. Modern Chromatographic Analysis of Vitamins: Second Edition, *edited by André P. De Leenheer, Willy E. Lambert, and Hans J. Nelis*
61. Preparative and Production Scale Chromatography, *edited by G. Ganetsos and P. E. Barker*
62. Diode Array Detection in HPLC, *edited by Ludwig Huber and Stephan A. George*
63. Handbook of Affinity Chromatography, *edited by Toni Kline*
64. Capillary Electrophoresis Technology, *edited by Norberto A. Guzman*
65. Lipid Chromatographic Analysis, *edited by Takayuki Shibamoto*
66. Thin-Layer Chromatography: Techniques and Applications, Third Edition, Revised and Expanded, *Bernard Fried and Joseph Sherma*
67. Liquid Chromatography for the Analyst, *Raymond P. W. Scott*
68. Centrifugal Partition Chromatography, *edited by Alain P. Foucault*
69. Handbook of Size Exclusion Chromatography, *edited by Chi-san Wu*
70. Techniques and Practice of Chromatography, *Raymond P. W. Scott*
71. Handbook of Thin-Layer Chromatography: Second Edition, Revised and Expanded, *edited by Joseph Sherma and Bernard Fried*
72. Liquid Chromatography of Oligomers, *Constantin V. Uglea*
73. Chromatographic Detectors: Design, Function, and Operation, *Raymond P. W. Scott*
74. Chromatographic Analysis of Pharmaceuticals: Second Edition, Revised and Expanded, *edited by John A. Adamovics*
75. Supercritical Fluid Chromatography with Packed Columns: Techniques and Applications, *edited by Klaus Anton and Claire Berger*
76. Introduction to Analytical Gas Chromatography: Second Edition, Revised and Expanded, *Raymond P. W. Scott*
77. Chromatographic Analysis of Environmental and Food Toxicants, *edited by Takayuki Shibamoto*
78. Handbook of HPLC, *edited by Elena Katz, Roy Eksteen, Peter Schoenmakers, and Neil Miller*
79. Liquid Chromatography–Mass Spectrometry: Second Edition, Revised and Expanded, *W. M. A. Niessen*
80. Capillary Electrophoresis of Proteins, *Tim Wehr, Roberto Rodríguez-Díaz, and Mingde Zhu*
81. Thin-Layer Chromatography: Fourth Edition, Revised and Expanded, *Bernard Fried and Joseph Sherma*
82. Countercurrent Chromatography, *edited by Jean-Michel Menet and Didier Thiébaut*
83. Micellar Liquid Chromatography, *Alain Berthod and Celia García-Alvarez-Coque*
84. Modern Chromatographic Analysis of Vitamins, Third Edition, Revised and Expanded, *edited by André P. De Leenheer, Willy E. Lambert, and Jan F. Van Bocxlaer*
85. Quantitative Chromatographic Analysis, *Thomas E. Beesley, Benjamin Buglio, and Raymond P. W. Scott*
86. Current Practice of Gas Chromatography–Mass Spectrometry, *edited by W. M. A. Niessen*

ADDITIONAL VOLUMES IN PREPARATION

Current Practice of Gas Chromatography– Mass Spectrometry

edited by
W. M. A. Niessen
hyphen MassSpec Consultancy
Leiden, The Netherlands

CRC Press
Taylor & Francis Group
Boca Raton London New York

CRC Press is an imprint of the
Taylor & Francis Group, an **informa** business

CRC Press
Taylor & Francis Group
6000 Broken Sound Parkway NW, Suite 300
Boca Raton, FL 33487-2742

First issued in paperback 2019

ISBN-13: 978-0-367-39742-5

**Visit the Taylor & Francis Web site at
http://www.taylorandfrancis.com**

**and the CRC Press Web site at
http://www.crcpress.com**

Preface

Combined gas chromatography–mass spectrometry (GC–MS) is a powerful tool for the quantitative and qualitative analysis of a wide variety of relatively volatile compounds. It is a mature technique, showing excellent perspectives in a variety of application areas. The purpose of this book is to give insight into the actual practice of GC–MS in a number of these application areas. Specialists in these fields were asked to contribute a chapter, briefly describing the state of the art of the application of GC–MS in their field and to illustrate this with results from their own research. This book, therefore, provides a useful guide to the current practice of GC–MS as well as a good perspective on how GC–MS is actually used by researchers in a wide variety of application areas. In this way, this book differs from other books published on GC–MS.

The 20 chapters of this book can be classified into five parts. The text starts with principles and instrumentation for GC–MS, paying attention to general aspects (Ch. 1) and surface ionization for GC–MS (Ch. 2). Industrial and environmental applications of GC–MS are dealt with in the following six chapters. Upstream and downstream applications of GC–MS in the petroleum industry are described in Chapter 3. The identification of chlorinated compounds in the environment is described in two chapters, applying either quadrupole ion-trap technology (Ch. 4) or high-resolution sector instruments (Ch. 5). The use of large-volume injection and on-line solid-phase extraction combined with GC–MS for the analysis of microcontaminants in water samples is described in Chapter 6. Two chapters deal with the use of GC–MS in occupational and environmental health assessment—Chapter 7 in relation to the biological monitoring of 1-nitropyrene, and Chapter 8 with the principles and use of solid-phase microextraction in studies related to occupational exposure to chemicals.

The next five chapters deal with pharmaceutical and clinical applications of GC–MS, paying attention to automated sample pretreatment procedures for rapid GC–MS of drugs (Ch. 9), multidimensional detection strategies in the analysis of anesthetics (Ch. 10), the use of stable isotope-ratio GC–MS in clinical applications (Ch. 11), steroid profiling in relation to a variety of diseases

(Ch. 12), and screening of inborn errors of metabolism (Ch. 13). Toxicological and forensic applications of GC–MS are discussed in three chapters, dealing with clinical and forensic toxicology (Ch. 14), drugs of abuse (Ch. 15), and explosives (Ch. 16). The last four chapters are devoted to food-related applications, paying attention to the analysis of flavors and fragrances (Ch. 17), principles and applications of residue analysis of veterinary hormones (Chs. 18 and 19), and the identification of monoterpenes and sesquiterpenes (Ch. 20).

Current Practice of Gas Chromatography–Mass Spectrometry covers a wide variety of application areas and discusses problems in relation to these applications (e.g., concerning sample pretreatment, analyte derivatization, gas chromatographic separation, and various mass spectrometric approaches to obtain the best possible results). The various contributors succeeded in providing a full perspective on the practice of GC–MS. It was a pleasure to work with them on this project and I would like to thank them for their efforts.

I hope that the reader will benefit from this collection of chapters. For those working in a particular application area, the book provides an up-to-date overview. Also, other chapters give additional ideas to solve analytical problems. Readers will get a good perspective on how GC–MS is used in various application areas and what type of results can be expected.

W. M. A. Niessen

Contents

Preface   *iii*

Contributors   *ix*

Part I: Principles and Instrumentation of Gas Chromatography–Mass Spectrometry

1.   Principles and Instrumentation of Gas Chromatography–Mass
Spectrometry   1
W. M. A. Niessen

2.   Principles and Applications of Surface Ionization in Gas
Chromatography–Mass Spectrometry   31
Toshihiro Fujii

Part II: Industrial and Environmental Applications

3.   Gas Chromatography–Mass Spectrometry in the Petroleum
Industry   55
C. S. Hsu and D. Drinkwater

4.   Analysis of Dioxins and Polychlorinated Biphenyls by
Quadrupole Ion-Trap Gas Chromatography–Mass Spectrometry   95
J. B. Plomley, M. Lauševic, and R. E. March

5.   Gas Chromatography–Mass Spectrometry Analysis of
Chlorinated Organic Compounds   117
M. T. Galceran and F. J. Santos

v

6. On-Line Solid-Phase Extraction–Capillary Gas
   Chromatography–Mass Spectrometry for Water Analysis   155
   Thomas Hankemeier and Udo A. Th. Brinkman

7. Gas Chromatography–Mass Spectrometry in Occupational
   and Environmental Health Risk Assessment with Some
   Applications Related to Environmental and Biological
   Monitoring of 1-Nitropyrene   199
   P. T. J. Scheepers, R. Anzion, and R. P. Bos

8. Application of Solid-Phase Microextraction–Gas
   Chromatography–Mass Spectrometry in Quantitative
   Bioanalysis   229
   Paola Manini and Roberta Andreoli

Part III: Pharmaceutical and Clinical Applications

9. Gas Chromatography–Mass Spectrometry of Drugs in
   Biological Fluids After Automated Sample Pretreatment   247
   M. Valcárcel, M. Gallego, and S. Cárdenas

10. Gas Chromatography–Mass Spectrometry Analysis of
    Anesthetics and Metabolites Using Multidimensional Detection
    Strategies   267
    M. Désage and J. Guitton

11. Gas Chromatography–Mass Spectrometry in Clinical Stable
    Isotope Studies: Possibilities and Limitations   285
    F. Stellaard

12. Clinical Steroid Analysis by Gas Chromatography–Mass
    Spectrometry   309
    Stefan A. Wudy, Janos Homoki, and Walter M. Teller

13. Gas Chromatography–Mass Spectrometry for Selective
    Screening for Inborn Errors of Metabolism   341
    Jörn Oliver Sass and Adrian C. Sewell

Contents

Part IV. Toxicological and Forensic Applications

14. Applications of Gas Chromatography–Mass Spectrometry in
    Clinical and Forensic Toxicology and Doping Control   355
    Hans H. Maurer

15. Detection of Drugs of Abuse by Gas Chromatography–Mass
    Spectrometry   369
    Jennifer S. Brodbelt, Michelle Reyzer, and Mary Satterfield

16. Gas Chromatography–Mass Spectrometry Analysis of
    Explosives   387
    Shmuel Zitrin

Part V. Food-Related Applications

17. Gas Chromatography–Mass Spectrometry Analysis of Flavors
    and Fragrances   409
    M. Careri and A. Mangia

18. Gas Chromatography–Mass Spectrometry for Residue Analysis:
    Some Basic Concepts   441
    *H. F. De Brabander, K. De Wasch, S. Impens, R. Schilt, and
    M. S. Leloux*

19. Applications of Gas Chromatography–Mass Spectrometry in
    Residue Analysis of Veterinary Hormonal Substances and
    Endocrine Disruptors   455
    *R. Schilt, K. De Wasch, S. Impens, H. F. De Brabander, and
    M. S. Leloux*

20. Identification of Terpenes by Gas Chromatography–Mass
    Spectrometry   483
    A. Orav

Appendix   *495*
Index   *499*

Contributors

Roberta Andreoli University of Parma, Parma, Italy

R. Anzion University of Nijmegen, Nijmegen, The Netherlands

R. P. Bos University of Nijmegen, Nijmegen, The Netherlands

Udo A. Th. Brinkman Free University, Amsterdam, The Netherlands

Jennifer S. Brodbelt University of Texas, Austin, Texas

S. Cárdenas University of Córdoba, Córdoba, Spain

M. Careri University of Parma, Parma, Italy

H. F. De Brabander University of Gent, Merelbeke, Belgium

M. Désage Université Claude Bernard, Lyon, France

K. De Wasch University of Gent, Merelbeke, Belgium

D. Drinkwater BASF Corporation, Princeton, New Jersey

Toshihiro Fujii National Institute for Environmental Studies, Onogawa, Tsukuba, Japan

M. T. Galceran University of Barcelona, Barcelona, Spain

M. Gallego University of Córdoba, Córdoba, Spain

J. Guitton   Université Claude Bernard, Lyon, France

Thomas Hankemeier   TNO Nutrition and Food Research, Zeist, The Netherlands

Janos Homoki   University of Ulm, Ulm/Donau, Germany

C. S. Hsu   ExxonMobil Research and Engineering Company, Annandale, New Jersey

S. Impens   University of Gent, Merelbeke, Belgium

M. Lauševic   Technolosko-Metalurski Fakultet, Belgrade, Yugoslavia

M. S. Leloux   State Institute for Quality Control of Agricultural Products, Wageningen, The Netherlands

A. Mangia   University of Parma, Parma, Italy

Paola Manini   University of Parma, Parma, Italy

R. E. March   Trent University, Peterborough, Ontario, Canada

Hans H. Maurer   University of Saarland, Homburg/Saar, Germany

W. M. A. Niessen   hyphen MassSpec Consultancy, Leiden, The Netherlands

A. Orav   Tallinn Technical University, Tallinn, Estonia

J. B. Plomley   MDS SCIEX, Concord, Ontario, Canada

Michelle Reyzer   University of Texas, Austin, Texas

F. J. Santos   University of Barcelona, Barcelona, Spain

Jörn Oliver Sass   Leopold Franzens University Innsbruck, Innsbruck, Austria

Mary Satterfield   University of Texas, Austin, Texas

P. T. J. Scheepers   University of Nijmegen, Nijmegen, The Netherlands

R. Schilt   TNO Nutrition and Food Research, Zeist, The Netherlands

Adrian C. Sewell   Johann Wolfgang Goethe University, Frankfurt, Germany

F. Stellaard   University Hospital Groningen, Groningen, The Netherlands

Walter M. Teller*   University of Ulm, Ulm/Donau, Germany

M. Valcárcel   University of Córdoba, Córdoba, Spain

Stefan A. Wudy   Children's Hospital of the University of Giessen, Giessen, Germany

Shmuel Zitrin   Division of Identification and Forensic Science, Jerusalem, Israel

---

* Deceased.

Current Practice of Gas Chromatography– Mass Spectrometry

1

Principles and Instrumentation of Gas Chromatography–Mass Spectrometry

W. M. A. Niessen
hyphen MassSpec Consultancy, Leiden, The Netherlands

1. INTRODUCTION

Gas chromatography–mass spectrometry (GC–MS) is a combination of two powerful analytical tools: gas chromatography for the highly efficient gas-phase separation of components in complex mixtures, and mass spectrometry for the confirmation of identity of these components as well as for the identification of unknowns. In this introductory chapter, the principles of GC, MS, and GC–MS are described.

1.1. History of Gas Chromatography–Mass Spectrometry

Gas chromatography (GC) was first described in 1952 by James and Martin [1] with the separation of a mixture of small carboxylic acids. Initially, packed-column GC columns were applied. The power of GC was substantially enlarged by the introduction of open capillary columns in 1958 by Golay [2]. In most cases, fragile glass capillary columns were applied. As such, the introduction of the fused-silica capillary column in 1976 by Dandeneau and Zerenner [3] can be considered as a major breakthrough in the development of GC.

    The history of mass spectrometry (MS) started in 1912, when Thomson [4] obtained the mass spectra of compounds such as O_2, N_2, CO, CO_2 and $COCl_2$. These findings were based on the earlier discovery of positive ions by Goldstein (1886) and the deflection of ions in a magnetic field by Wien (1898). Subsequently, Dempster [5] and Aston [6] described the first magnetic sector instruments. The first commercial instruments dedicated to organic analysis were built during World War II. The major application area at that time was petrochemistry.

In 1958, the first on-line GC–MS instruments were introduced. By means of the jet separator, introduced in 1964 by Ryhage [7], the practical aspects of coupling GC and MS were greatly simplified. A major breakthrough in GC–MS was realized in 1968 with the introduction of the first data-processing units [8]. In 1975, the first commercial GC–MS instruments with capillary columns were introduced. In 1981, fused-silica capillary columns were applied for GC–MS [9].

1.2. Scope

This chapter provides a brief and general introduction to gas chromatography, to mass spectrometry, and to the techniques and applications of GC–MS.

2. GAS CHROMATOGRAPHY

Gas chromatography is a physical separation method in which the components in a mixture are selectively distributed between the mobile phase, which is an inert carrier gas, and a stationary phase, which is present as a coating of either column packing particles or the inner column wall. The chromatographic process occurs as a result of repeated sorption/desorption steps during the movement of the analytes along the stationary phase by the carrier gas. The separation is due to the differences in distribution coefficients of the individual components in the mixture. Being a gas-phase separation method, GC requires the analytes to be volatilized prior to their separation. As such, the application of GC is limited to components with sufficient volatility and thermal stability.

Although GC separation can be performed in packed columns as well as open capillary columns, this chapter focuses on the use of capillary columns, as these are most widely applied in GC–MS.

2.1. Instrumental Aspects

The instrumentation for GC consists of a gas control unit, a sample introduction system or injector, a column housed in the temperature-programmable column oven, and a detector or transfer line and/or interface to the mass spectrometer.

The gas control unit performs flow-rate or pressure control of the gas flows through the injector, the column, and the detector of carrier gas and, if required, auxiliary gases. The carrier gas (hydrogen, helium, or nitrogen) typically is applied at a pressure below 0.3 MPa. The flow rate is approximately 20 ml/min for a packed column and 1 ml/min for an open capillary column. In GC–MS applications, helium is most frequently applied as carrier gas. Prior to use, the carrier gas should be cleaned by means of a moisture-and-oxygen trap in order to remove oxygen, water, and hydrocarbons. The presence of oxygen in the carrier gas has a detrimental effect on the stationary phase in the GC column.

The sample introduction is a very critical step in the operation of GC. The aim is to introduce the complete sample in a narrow band at the top of the column, i.e., without thermal degradation and/or component discrimination due to differences in volatility. The most widely applied injection techniques are split injection, splitless injection, and on-column injection (Table 1). Due to the sample splitting in the split injector (typically at a ratio in the range of 1:10 to 1:100, with the higher split ratio used for the smaller internal diameter column), a relatively large sample volume, e.g., 1 µl, can be injected. In splitless and on-column injection, the sample volume is smaller. In splitless injection, the splitter vent is closed for a specified time, typically 50 to 120 s, while the sample flows onto the head of the column. After that, the splitter vent is opened to purge remaining sample and solvent from the injector. The initial column temperature in splitless injection is a critical parameter that depends on the boiling point of the solvent used for dissolving the analytes. Splitless injection provides sample concentration at the top of the column. A comparison of important features of the three injection

Table 1 Comparison of Important Features of Sample Injection Techniques for Capillary GC

	Split	Splitless	On-column
Application	Major components	Trace and major components	Trace and major components
Maximum injectable amount	Depends on split ratio	50 ng per component	100 ng per component
Injection precision	Poor	Good	Excellent
Injector temperature	250–320°C	200–280°C	Column temperature
Initial column temperature	Any	20–40°C below the boiling point of the most volatile sample component	Close to the boiling point of the solvent
Advantages	Variation in split ratio prevents overloading	Direct quantification	Less sample discrimination; direct quantification
Disadvantages	Sample loss; sample discrimination; poor in trace analysis and direct quantification	Limited choice of applicable solvent	Experimentally difficult; risk of column contamination

techniques is given in Table 1. Large-volume injection approaches for capillary GC are discussed in somewhat more detail in section 4.2.

The GC column (see section 2.2) is placed in a temperature-programmable column oven. During chromatography, the oven temperature is linearly increased at a rate of typically 4 to 20°C/min. As a result, components with higher boiling points and/or a stronger retention to the stationary phase are successively released. The maximum temperature depends on the type of stationary phase applied (Table 2). Operating the column near the maximum operating temperature generally results in more severe column bleeding, which in turn leads to a more rapid contamination of the MS ion source.

Compounds with insufficient volatility in the working range of the GC column are not amenable to GC analysis, unless they can be derivatized to more volatile derivatives (see section 4.3).

Instead of MS, other types of detection can be used in GC. The most widely used are thermal conductivity detection and flame ionization detection. In addition, various more specific types are available, e.g., electron-capture detection, thermionic detection, and flame photometric detection. A comparison of some characteristics is given in Table 3. On-line combinations with spectrometric detection, other than MS, are available, e.g., Fourier-transform infrared (FT-IR) and atomic emission detection (AED).

2.2. Gas Chromatography Columns

As indicated before, open capillary columns are applied most frequently in current GC–MS applications. A typical open capillary column is made of fused-silica with an external polyimide coating.

Typical column length is 10 to 100 m, depending on the application. Shorter columns are applied for fast analysis, e.g., for heat-sensitive and high-boiling compounds. Longer columns are applied in high-resolution separations.

Typical column internal diameters (IDs) are 0.25 to 0.53 mm, with a stationary phase film thickness of 0.1 to 2 μm. In GC–MS, mostly 0.25-mm ID columns are applied, as the gas flow at the optimum linear gas velocity of these column (approximately 300 mm/s for helium, corresponding to approximately 1 ml/min) ideally fits the pumping capacity of most benchtop GC–MS systems. The wide-bore 0.53-mm ID columns, showing better inertness and/or higher sample capacity can be used in combination with a jet separator (see section 4.1). In order to reduce column bleeding, usually a chemically bonded or immobilized stationary phase is applied. A thinner film is useful for high-boiling and heat-sensitive compounds, while a thicker film is better for low-boiling compounds.

The stationary phase can be applied at the column wall in different ways. In a wall-coated open tubular (WCOT) column, the stationary phase is applied as a thin (immobilized) liquid film. This column type is applied most frequently.

Table 2 Chemically Bonded General-Purpose Stationary Phases for Open Capillary GC

CP index	Common phases[a]	Composition	Maximum operating temperature (°C)
2	CP-Sil 2	Similar to squalane	200
5	CP-Sil 5 DB-1 HP-1 SPB-1	Dimethyl polysiloxane	325
8	CP-Sil 8 DB-5 HP-5 SPB-5	5% Phenyl, 95% dimethyl polysiloxane	325
19	CP-Sil 19 DB-1701 HP-1701 SPB-1701	14% Cyano-propyl-phenyl, 86% dimethyl polysiloxane	275
24	CP-Sil 24 DB-17 HP-50 SPB-50	50% Phenyl, 50% dimethyl polysiloxane	280
43	CP-Sil 43 DB-225 HP-225	25% Cyano-propyl, 25% phenyl, 50% dimethyl polysiloxane	200
52	CP-Wax 52 DB-Wax HP-INNOWax Supelcowax	Polyethylene glycol	250
58	CP-Wax 58 DB-FFAP HP-FFAB Nukol	Polyethylene glycol nitrotere-phthalic acid ester	250

[a] Phases from Chrompack (CP-Sil), J&W (DB), Agilent Technologies (formerly Hewlett-Packard, HP), Supelco (SPB). Similar phases are also available from a number of other manufacturers.

In addition, support-coated open tubular (SCOT), porous-layer open tubular (PLOT), and wall-coated superior capacity open tubular (WSCOT) columns are applied. In a SCOT column, the column wall is coated with a thin layer of small inert particles that serve as a support for the stationary-phase liquid. The SCOT column provides a higher sample loadability at the expense of separation effi-

Table 3 Characteristics of Detectors for GC

Detector	Specificity	LOD	Linear range
Thermal conductivity (TCD)	Nonspecific	10^{-8} g/ml	10^4
Flame ionization (FID)	CH groups	10^{-13} g/s	10^7
Electron capture (ECD)	Electronegative groups	5×10^{-14} g/s	5×10^4
Thermionic (TID)	P and N	10^{-15} g/s (P) 10^{-14} g/s (N)	10^5
Flame photometric (FPD)	P and S	10^{-13} g/s (P) 10^{-11} g/s (S)	10^5

ciency. In the WSCOT columns, a thicker stationary-phase film is applied, featuring higher sample loadability and enhanced deactivation of the column wall. In a PLOT column, the wall is coated with small adsorptive particles in order to enable gas-solid chromatography.

The polarity of the stationary-phase liquid can be characterized by a number of parameters. For this purpose, Rohrschneider in 1966 and subsequently McReynolds [10] in 1970 proposed a number of test components, representing specific interactions between groups of analytes and the stationary phases. The Kovats retention indices (see section 2.4) of the model compounds benzene, 1-butanol, 2-pentanone, nitropropane, and pyridine on different stationary phases are used to determine the McReynolds constants on these stationary phases. Based on the McReynolds constants, the GC column manufacturer Chrompack introduced the CP index in order to characterize the polarity of stationary phases. The CP index has a value of zero for the highly nonpolar phase squalane and a value of 100 for the very polar phase OV-275. The CP index facilitates the comparison of stationary phases from different manufacturers. A number of general-purpose stationary phases are given in Table 2. Other classification systems for GC stationary phases have recently been reviewed by Abraham et al. [11].

A more polar stationary phase is applied for the analysis of more polar compounds. As the more polar stationary phase generally is more prone to column bleeding, the least polar column applicable is selected for most applications. CP-Sil 5 and CP-Sil 8 (or equivalents) are most widely applied. When these column types do not give adequate resolution, the more polar stationary phases have to be applied. Low-bleed columns especially produced for GC–MS applications are offered as well by several manufacturers.

In addition to these general-purpose stationary phases, several stationary phases have been developed for specific applications, e.g., pesticide analysis or separation of enantiomers (chemically bonded β-cyclodextrin columns, for instance).

2.3. Chromatographic Parameters

A GC separation is performed in the elution mode. The sample is introduced at the top of the column, and with the carrier gas the analyte bands move with different rates through the column and elute one after another from the column. The average rate at which an analyte migrates depends on the fraction of the time spent in the stationary phase, and thus on the affinity of the analyte to the stationary phase. At the detector, a chromatogram with the various analyte peaks appears (Fig. 1).

Several important parameters to characterize the separation and the efficiency can be directly deduced from the chromatogram. The retention time $t_{r,i}$ of the component i is the residence time of the component i and can be measured directly from the chromatogram as the time distance between the sample injection and the top of the peak due to component i (see Fig. 1). From this, a number of other parameters can be calculated (Table 4). The capacity ratio or mass distribution ratio k_i' (Eq. 1) is defined as the ratio of the amount of i in the stationary and the mobile phase, respectively. The capacity ratio is the product of the distribution coefficient and the phase ratio. It follows that the analyte molecules spend an average time fraction of $1/(k_i' + 1)$ in the mobile phase and of $k_i'/(k_i' + 1)$ in the stationary phase. Analytes with different retention times spend different periods of time in the stationary phase. The separation factor $\alpha_{j,i}$ of the phase system for the components i and j (Eq. 2) is another important parameter characterizing the separation.

Besides its retention time, the width of a chromatographic peak is an important parameter. Since it is theoretically assumed that chromatographic peaks are symmetric Gaussian peaks, the width of the peak is measured as the peak standard

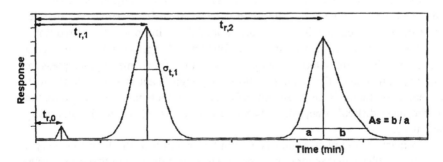

Figure 1 Chromatogram with important parameters indicated.

Table 4 Chromatographic Parameters from the Chromatogram

Parameter	Equation	Equation no.
Capacity ratio	$k_i' = \dfrac{t_i - t_m}{t_m}$	(1)
Separation factor	$\alpha_{j,i} = \dfrac{k_j'}{k_i'} \geq 1$	(2)
Resolution	$R_{ij} = \dfrac{2(t_{r,j} - t_{r,i})}{w_j + w_i}$	(3)
Plate number	$N_i = \dfrac{t_{r,i}^2}{\sigma_{t,i}^2}$	(4)
Plate height	$H_i = \dfrac{L}{N} = \dfrac{L\,\sigma_{t,i}^2}{t_{r,i}^2}$	(5)

deviation $\sigma_{t,i}$ from the half width at 0.607 of the peak height ($w_{t,i} = 4\,\sigma_{t,i}$ for a Gaussian peak, where $w_{t,i}$ is the peak width at the base) (see Fig. 1). From retention time and peak width, the resolution R_s between the two peaks i and j (Eq. 3) can be calculated.

Retention time and peak standard deviation are also used in the characterization of the column efficiency, i.e., in the calculation of the plate number N (Eq. 4) and the plate height H (HETP = height equivalent to a theoretical plate) (Eq. 5). By combining these equations, the resolution can be written as

$$R_{j,i} = (\alpha_{j,i} - 1)\left(\frac{k_j'}{k_j' + 1}\right)\sqrt{\frac{L}{H_i}} \qquad (6)$$

Thus, the resolution depends on the plate height, the column length, the separation factor, and the capacity ratio. The plate height should be as small as possible. The resolution increases with the square root of the column length. A two times better resolution can be achieved with a four times longer column, and thus at the cost of a four times longer analysis time and a four times higher pressure drop. The influence of the capacity ratio is only significant at k' values below 3. The separation factor is the most powerful tool in influencing the resolution, but also the most difficult one, since its value is directly related to the stationary phase used.

In practice, the plate height is a complex function of the linear velocity u of the mobile phase and several other parameters, as indicated in first approximation by the Van Deemter equation:

$$H = A + \frac{B}{u} + (C_m + C_s)u \tag{7}$$

where A represents the contribution from eddy diffusion, B that of longitudinal diffusion, and C_s and C_m that of resistance to mass transfer in stationary and mobile phases, respectively. The eddy diffusion term is zero in an open capillary column. According to the theoretical work by Golay [12], the Van Deemter equation for a WCOT column can be written as

$$H = \frac{2D_m}{u} + \left[\frac{1 + 6k' + 11k'^2}{96(1 + k')^2}\right]\frac{d_c^2 u}{D_m} + \left[\frac{k'}{6(1 + k')^2}\right]\frac{d_f u}{D_s} \tag{8}$$

where D_m and D_s are the diffusion coefficients of component i in mobile and stationary phases, respectively; d_c is the column inner diameter; and d_f is the stationary-phase film thickness.

2.4. Retention Indices

To some extent, the retention time in a standardized GC analysis can be used for positive identification of the compounds analyzed. The most widely applied approach in this respect is the use of the Kovats retention index (RI) [13]. The RI value of a sample peak is determined by comparing its retention time to retention times of closely eluting alkane standards, i.e., the alkanes that elute just before and just after the sample peak. In this way, the RI is quite insensitive to small changes in experimental conditions. In the Kovats RI system, normal alkanes are assigned an RI value of $100\times$ the carbon number, e.g., n-pentane has an RI of 500. In an isothermal separation, the RI of component i with retention time $t_{r,i}$ (and net retention time of $t'_{r,i} = t_{r,i} - t_{r,0}$ is thus calculated from

$$RI = 100n + 100\frac{\log(t'_{r,i}) - \log(t'_{r,n})}{\log(t'_{r,n+1}) - \log(t'_{r,n})} \tag{9}$$

For temperature gradient operation, the RI equation simplifies to:

$$RI = 100n + 100\frac{t'_{r,i} - t'_{r,n}}{t'_{r,n+1} - t'_{r,n}} \tag{10}$$

Retention time indices are available for many substances as data collections for a variety of stationary phases [14–15]. Computerized retention index libraries are available. Alternative RI systems have recently been reviewed by Castello [16].

3. MASS SPECTROMETRY

The principle of MS is the production of gas-phase ions that are subsequently separated according to their mass-to-charge (m/z) ratio and detected. The resulting mass spectrum is a plot of the (relative) abundance of the generated ions as a function of the m/z ratio. Extreme selectivity can be obtained, which is of utmost importance in quantitative trace analysis.

3.1. Instrumental Aspects

The mass spectrometer nowadays is a highly sophisticated and computerized instrument. It consists of five parts: sample introduction, ionization, mass analysis, ion detection, and data handling. Modern mass spectrometers have total computer control over the various parts.

In GC–MS systems, sample introduction is performed from the open capillary chromatographic column, either directly or via an open split coupling (see section 4.1). The ionization of the analytes is generally performed by either electron ionization or chemical ionization (see section 3.2). After the production of ions, these are separated according to their m/z ratio in the mass analyzer. Although linear quadrupole analyzers are most widely applied, other analyzer types, i.e., (magnetic) sector, quadrupole ion trap, and time-of-flight, are applied as well (see section 3.3). The detection of ions is mostly performed by means of an electron multiplier.

In GC–MS systems, analyte ionization, mass analysis, and ion detection take place in a high-vacuum system. In most benchtop GC–MS systems, the vacuum system consists of one pumped chamber, evacuated by means of a small turbomolecular pump, backed by a mechanical fore pump. Large systems as well as modern research-grade mass spectrometers generally contain two differentially pumped vacuum chambers, separated by means of a baffle containing a slit, i.e., the ion source housing and the analyzer region.

Efficient means to collect and handle the enormous amounts of data that are generated in the operation of a mass spectrometer, especially in on-line combination with chromatography, are of utmost importance. Highly advanced computer programs are currently available for use in handling, interpretation, and reporting the data (see section 4.4).

3.2. Analyte Ionization

A wide variety of ionization techniques are available for organic mass spectrometry. In this section the most important techniques in relation to GC–MS are discussed.

3.2.1. Electron Ionization

The oldest and most frequently applied ionization technique is electron ionization (EI). In EI, the analyte vapor is subjected to bombardment by energetic electrons (typically 70 eV). Most electrons are elastically scattered, others cause electron excitation of the analyte molecules upon interaction, while a few excitations cause the complete removal of an electron from the molecule. The last type of interaction generates a radical cation, generally denoted as $M^{+\bullet}$, and two electrons:

$$M + e^- \rightarrow M^{+\bullet} + 2e^- \tag{11}$$

The $M^{+\bullet}$ ion is called the molecular ion. Its m/z ratio corresponds to the molecular mass M_r of the analyte. The ions generated in EI are characterized by a distribution of internal energies, centered around 2 to 6 eV. The excess internal energy of the molecular ion can give rise to unimolecular dissociation reactions resulting in structure-dependent fragment ions. Typical fragmentation reactions of a molecule M upon electron ionization result in the formation of an ionized fragment accompanied by the loss of either a radical $R^\bullet$ or a neutral N:

$$M^{+\bullet} \rightarrow F_1^+ + R^\bullet \tag{12}$$
$$M^{+\bullet} \rightarrow F_2^{+\bullet} + N \tag{13}$$
$$F_2^{+\bullet} \rightarrow F_3^+ + R^\bullet \tag{14}$$

The fragmentation reactions are discussed in somewhat more detail in section 5.2. Electron ionization is performed in a high-vacuum ion source (typical pressure $\leq 10^{-2}$ Pa), excluding intermolecular collisions. Electron ionization spectra are highly reproducible. Extensive computer-searchable libraries of EI mass spectra are available to assist in the identification of unknowns (see section. 5.1).

3.2.2. Chemical Ionization

Chemical ionization (CI) is based on gas-phase chemical reactions and can be considered as a versatile ionization technique [17]. Chemical ionization is primarily based on ion–molecule reactions between reagent gas ions and the analyte molecules, e.g., proton transfer (addition), charge exchange, electrophilic addition, and anion abstraction in positive-ion CI (PCI) and proton transfer (abstraction) in negative-ion CI (NCI). Next to these, electron capture is a NCI method for the formation of negative ions that is not based on ion–molecule reactions.

In most GC–MS applications, CI is performed at ion source pressures between 1 and 100 Pa. Reagent gas ions are produced by bombardment of the reagent gas, e.g., methane, isobutane, or ammonia, by energetic electrons (100 to 400 eV), i.e., by EI, followed by a series of ion–molecule reactions. Due to the relatively high source pressure, frequent intermolecular and ion–molecule collisions occur in the source. Upon EI of ammonia in a CI source, protonated ammonia NH_4^+ and $(NH_3)NH_4^+$ are the most abundant ions formed. They can react with an analyte molecule M in a proton-transfer reaction, resulting in a product ion with low internal energy. The m/z of this protonated molecule corresponds to $M_r + 1$ and can thus be used to determine the molecular mass of the analyte.

For a proton transfer reaction to proceed, the proton affinity (PA) (Table 5) of the analyte molecule M must exceed that of the reagent gas; thus, the reaction between NH_4^+ and pyridine will yield the protonated pyridine, while no proton-transfer reaction will occur between NH_4^+ and water. The reaction products show little internal energy (depending on the PA difference between analyte and reagent gas) and a narrow internal energy distribution. As a result, generally little fragmentation is observed.

Chemical ionization can also be used to produce negative ions either by ion–molecule reactions or by electron capture. In the former case, proton transfer or abstraction takes place when the gas-phase acidity of the reagent gas exceeds that of the analyte. In electron-capture NCI, ionization takes place by capture of slow, "thermal" electrons with energies between 0 and 1 eV by the analyte molecules resulting in the generation of radical anions. The process must be performed in a high-pressure ion source, since the high pressure slows down the electrons and at the same time removes the excess energy from the radical anion formed upon electron attachment. The latter is important because the electron affinity of most organic molecules does not exceed 2.5 eV. The formation of negative ions by electron capture can occur by either an associative or a dissociative mechanism:

$$AB + e^- \rightarrow AB^{-\bullet} \tag{15}$$
$$AB + e^- \rightarrow A^- + B^\bullet \tag{16}$$

Table 5 Proton Affinity of Some Frequently Used Reagent Gases for CI

Reagent gas	Pa (kJ/mol)
H_2	422
CH_4	536
H_2O	723
i-C_4H_{10}	823
NH_3	857
Pyridine	921

Electron capture is a highly selective ionization method, as only a limited number of analytes are prone to efficient electron capture. Fluorinated compounds are extremely sensitive to electron capture. Chemical derivatization is often applied to label analytes with, for example, pentafluorobenzyl groups, as is frequently done in gas chromatography with electron-capture detection (see section 4.3).

3.2.3. Field Ionization

In field ionization (FI), the sample vapor is guided along an emitter that is especially activated to provide microneedles at the surface. A high electric field (10^7 to 10^8 V/m) is applied at the emitter. High local electric fields at the microneedles enable electron tunneling from the sample to the emitter: radical cations are generated with little fragmentation. The technique of FI recently gained interest for its use in combination with time-of-flight mass analyzers.

3.3. Mass Analysis

Mass analysis, i.e., the separation of ions according to their m/z ratio in either time or space, can be achieved in a number of ways. Under GC–MS conditions, mostly singly charged ions are generated. The basic principles of the four most important types of mass analyzers are briefly discussed in this section. An excellent and more elaborate discussion on the various mass analyzers and their advantages and limitations is given by Brunnee [18].

3.3.1. Sector Instruments

In a single-focusing sector instrument, the ions with mass m and z elementary charges and a particular kinetic energy are introduced into a magnetic field B. The kinetic energy of the ions is determined by the voltage V with which the ions are accelerated toward the source exit slit (typically 5 to 8 kV). When the magnetic force is counterbalanced by the centrifugal force, ions are transmitted to the detector. This leads to the following equation:

$$m/z = \frac{B^2 r^2 e}{2V} \tag{17}$$

where e is the elementary charge and r is the radius of curvature of the path through the magnetic field. This equation indicates that by variation of the radius of curvature, ions differing in m/z value are separated in space. By variation of either B or V, ions of different m/z values can be detected by a detector at a fixed position behind a slit as being separated in time. The most common way is by performing an exponential magnet scan.

Upon ionization, ions with a distribution of ion kinetic energies are generated. The resolution of the mass analysis by a sector instrument can be improved by means of an electrostatic analyzer (ESA). In principle, the ESA can be applied in various geometries, e.g., the Nier–Johnson geometry (EB) and the reversed geometry (BE) (Fig. 2). In most cases, the instrument is designed in such a way that velocity focusing takes place: ions with one particular m/z value but different kinetic energies are deflected toward one focal point. Instruments with both a

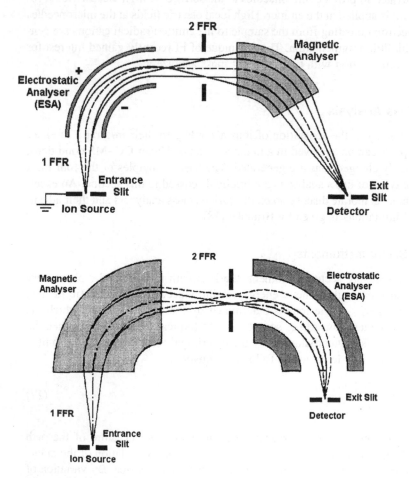

Figure 2 Geometries of double-focusing sector instrument: Nier–Johnson (EB) geometry and reversed geometry.

magnetic and an electrostatic sector, usually called double-focusing instruments, are capable of high-resolution mass determination with high ion transmittance. Some aspects and features of a double-focusing sector instrument are

- Ions from the ion source enter the magnetic sector or ESA from a slit rather than a point. This affects the resolution, which in principle can be increased by narrowing the slits at the source and/or detector side. Obviously, narrowing the slit has an adverse effect on the ion transmission. In practice, one often has to strike a compromise between sensitivity and resolution.
- A double-focusing sector instrument allows collision-activated ion dissociation (CID) for tandem mass spectrometry (MS–MS) in the field-free regions, e.g., between the entrance slit and the first sector, or in between the two sectors. Special linked scanning procedures of the instrument, such B/E linked scans, allow the acquisition of product-ion mass spectra.
- A double-focusing sector instrument allows high-resolution measurements, enabling (1) determination of accurate m/z of ions (at 1 ppm or better) and (2) improvement of selectivity in the analysis of compounds in complex matrices.

Double-focusing sector instrument are still widely used in GC–MS for a number of applications, e.g., in the analysis of polychlorodibenzodioxins and related compounds.

3.3.2. Linear Quadrupole Instruments

The linear quadrupole mass analyzer is a mass filter. It consists of four hyperbolic rods that are placed parallel in a radial array. Opposite rods are charged by a positive or negative DC potential U at which an oscillating radiofrequency voltage $V_0 \cos \omega t$ is superimposed. The latter successively reinforces and overwhelms the DC field. Ions are introduced into the quadrupole field by means of a low accelerating potential, typically only a few volts. The ions start to oscillate in a plane perpendicular to the rod length as they traverse through the quadrupole filter. The trajectories of the ions of one particular m/z are stable. These ions are transmitted toward the detector. Ions with other m/z have unstable trajectories and do not pass the mass filter because the amplitude of their oscillations becomes infinite. Ions of different m/z can consecutively be transmitted by the linear quadrupole filter toward the detector when the DC and AC potentials are swept, while their ratios and oscillation frequencies are kept constant.

The quadrupole analyzer thus acts as a band pass filter, the resolution of which depends on the ratio of DC and AC potentials. Generally, "unit-mass" resolution is achieved, which indicates that, for instance, m/z = 100 and m/z =

101 can be distinguished, but m/z 100 and m/z 100.1 cannot. The quadrupole mass filter is suitable for the determination of the nominal masses of a compound and its fragment ions. The quadrupole is easy to use. The electric voltages can be rapidly varied under computer control, enabling fast scanning. The quadrupole mass filter is the most widely applied mass analyzer in GC–MS.

3.3.3. Quadrupole Ion-Trap Instruments

An important development in quadrupole technology is the three-dimensional ion trap [19]. A quadrupole ion trap consists of a cylindrical ring electrode to which the quadrupole field is applied, and two end-cap electrodes (Fig. 3). The top end cap contains holes for introducing ions or electrons into the trap, while the bottom end cap contains holes for ions ejected toward the electron multiplier. Ions that are generated either inside the trap itself or in an external ion source are stored in the trap. A relatively high pressure of helium bath gas (0.1 Pa) is present in the ion trap in order to stabilize the ion trajectories.

    The quadrupole ion trap differs from a linear quadrupole filter because (1) with an ion trap, the mass analysis process is a discontinuous, pulsed process, while with a linear quadrupole, the process is continuous and (2) with an ion trap, scanning of various m/z is based on m/z-selective instability, while with the linear quadrupole, scanning is based on m/z-selective stability. The scanning process of an ion trap consists of a number of consecutive steps:

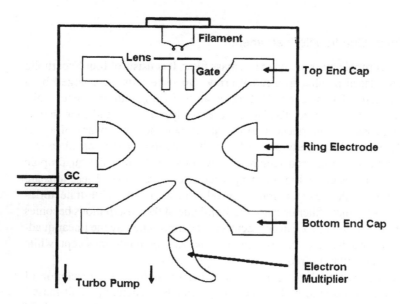

Figure 3   Schematic diagram of an quadrupole ion-trap mass spectrometer.

- During a preionization step, a short ionization pulse is generated and the ions are removed from the ion trap. From the ion current measured, the ionization time is determined: long when no compound elutes and short at the top of a peak. The ionization time must be varied in order to avoid overloading the trap with ions; the resulting space charge effects result in loss of resolution and quantitation accuracy.
- After setting an appropriate radiofrequency storage voltage at the ring electrode, ions are generated by EI either inside the trap or injected from the external ion source and stored.
- In full-scan mode, the ions of different m/z are consecutively ejected by ramping the radiofrequency voltage at the ring electrode. The ejected ions are detected by an electron multiplier outside the trap.

The ion-trap instrument is a versatile tool in GC–MS.

3.3.4. Time-of-Flight Instruments

In a time-of-flight (TOF) instrument [20], a package of ions is accelerated by a potential V into a field-free linear flight tube; the time t_{flight} needed to reach a detector placed at a distance d is measured. That time is related to the mass-to-charge ratio:

$$ t_{flight}^2 = \frac{md^2}{2zeV} = \frac{m}{z}\left[\frac{d^2}{2eV}\right] \tag{18} $$

The ion source must be pulsed in order to avoid the simultaneous arrival of ions of different m/z. The use of TOF instruments in GC–MS has recently gained considerable interest for two reasons: scan speed and mass accuracy. The TOF–MS is an integrating rather than a scanning detector. This means that the acquisition rate is limited by the ion pulse frequency and the spectrum storage speed. With current computer technology, the number of mass spectra that can be acquired can be on the order of 500 spectra per second. This enables the use of TOF-MS for ultrafast GC analysis. An example is shown in Figure 4, where a mixture of laboratory solvents is separated on a 1 m × 0.1 mm ID fused-silica open capillary column; total analysis time is approximately 7 sec.

Due to the energy dispersion of the ions leaving the ion source, the resolution achievable in a TOF instrument is generally limited. The resolution can be improved by the use of an electrostatic reflectron. A reflectron consists of a series of lens plates with different voltages, forming a retarding field. Ions with higher kinetic energies will penetrate the retarding field more deeply, will spend more time turning around, and will catch up with the ions with lower kinetic energy and reach the detector simultaneously. A further improvement of the resolution can be achieved from the use of delayed extraction or time-lag focusing, in which

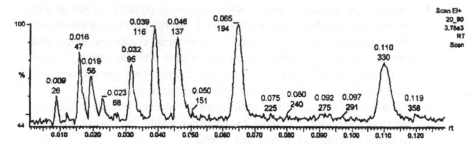

Figure 4   Analysis of laboratory solvent mixture by fast GC–TOF-MS. Column 1 m ×
0.1 mm ID (0.1 μm HP1 phase). (From Ref. 28.)

ions are trapped between two grids before being accelerated into the flight tube,
or by orthogonal acceleration of the ions. In this way, sufficient resolution can
be achieved to perform accurate mass determination (at 5 ppm). As such, a GC–
TOF-MS instrument can be an interesting alternative to a double-focusing sector
instrument because it is easier to operate.

   The relative merits of the various mass analyzers are discussed in detail
by Brunnee [18]. There is not one ideal mass analyzer. The choice depends on
the application. A practical comparison of mass analyzers is provided in Table 6.

4.   PRACTICAL ASPECTS OF GAS CHROMATOGRAPHY–
   MASS SPECTROMETRY

4.1.   Gas Chromatography–Mass Spectrometry Interfacing

For use in combination with packed-column GC, a variety of interfaces for GC–
MS were developed. The aim of these devices was to achieve analyte enrichment,

Table 6   Comparison of Some Features of Various Mass Analyses

Feature	Double-focusing sector	Linear quadrupole	Ion trap	TOF
Scanning		Fast	Fast	Super fast
Vacuum	Critical		Bath gas	
Mass range	10,000	4000	2000	No limits
Acquisition	Full-scan, SIM	Full-scan, SIM	Full-scan, SIM	Full-scan
Resolution	High	Unit-mass	Unit-mass	High
Mass accuracy	Accurate (1 ppm)	Nominal	Nominal	Accurate (5 ppm)

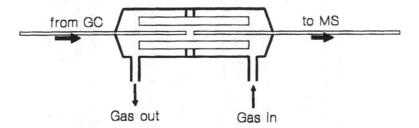

from GC   to MS

Gas out   Gas In

Figure 5 Schematic diagram of the open split coupling for GC–MS.

i.e., a better ratio between analyte and carrier gas. The jet separator, the Watson-Biemann or effusion interface, and the membrane interface are the most frequently applied devices of the type.

With the introduction of open capillary columns for GC and GC–MS, an analyte enrichment interface is no longer required as the optimum flow rate of such a column is readily amenable to the vacuum system of a benchtop GC–MS system. At present, two types of GC–MS coupling are applied, i.e., the direct coupling and the open split interface. In the direct coupling, the column effluent of the GC column is directly introduced into the ion source of the mass spectrometer. While this approach is very simple, it has some disadvantages that are avoided by the use of an open split coupling (Fig. 5). In a direct-coupled GC–MS, the column outlet is at high vacuum, resulting in changes in the chromatogram similar to those obtained from a GC–FID. As the complete output of the GC column is introduced into the source, the risk of source detuning and contamination is higher, e.g., due to the solvent pulse, flow-rate changes during temperature programming, and sample contaminants. While in direct coupling, the vacuum system must be switched off for changing the GC column; this is not required with the open split coupling.

4.2. Sample Pretreatment and Large-Volume Injection

A major disadvantage of the small column internal diameters used in open capillary GC is the limited injection volume. This limitation can partly be reduced by the use of preconcentrating sample pretreatment, such liquid-liquid extraction (LLE), solid-phase extraction (SPE), and solid-phase microextraction (SPME). These sample pretreatment techniques are widely applied throughout the daily practice of GC–MS, as exemplified by the sample pretreatment procedures described in the various chapters of this book.

In addition, large-volume sample introduction techniques have been introduced for open capillary GC, e.g., on-column injection, loop-type injection, and

Table 7 Large-volume Introduction Techniques for Capillary GC

ON-COLUMN INJECTION

Two types of evaporation techniques can be applied:

Conventional retention gap technique

Sample is injected at a temperature below the solvent boiling point. If the retention gap can be wetted by the solvent, a flooded zone is formed. The solvent film evaporates from the rear to the front and volatile analytes are reconcentrated by the solvent trapping effect. In addition, phase soaking effects reconcentration of the analytes due to the increased retention power of the thicker stationary phase. Less volatile components remain spread over the retention gap and are reconcentrated by the phase-ratio-focusing effect.

Partially concurrent solvent evaporation (PCSE)

Sample is injected into the GC under conditions that cause the major part of the solvent to evaporate while the remaining solvent floods the retention gap; that is, the solvent introduction rate is higher than the evaporation rate. In this way, about 90% of the introduced solvent can be evaporated during introduction. Volatile analytes are reconcentrated due to phase soaking and solvent trapping in the remaining solvent film. Less volatile components remain spread over the retention gap and are reconcentrated by the phase-ratio-focusing effect.

LOOP-TYPE INJECTION

The sample in the loop is pushed into the GC by the carrier gas. Two types of solvent evaporation technique can be applied:

Fully concurrent solvent evaporation technique (FCSE)

Sample is injected at a temperature above the solvent boiling point. The sample is completely evaporated during injection. No flooded zone is formed. Volatile analytes co-evaporate with the solvent. Less volatile components remain spread over the retention gap and are reconcentrated by the phase-ratio-focusing effect.

Co-solvent trapping

A small amount of a higher boiling co-solvent (e.g. octadecane) is added to the main solvent to create a layer of condensed liquid ahead of the main evaporation site. The main solvent evaporates concurrently, and part of the co-solvent evaporates together with the main solvent. Boiling point and amount of co-solvent must be adjusted such that some co-solvent is left behind as a liquid and spreads into the retention gap. Volatile analytes are reconcentrated due to solvent trapping in the co-solvent. Less volatile components remain spread over the retention gap and are reconcentrated by the phase-ratio-focusing effect.

PTV INJECTION

A programmed temperature vaporizer injector (PTV) basically is a split-splitless injector with temperature control, i.e., the vaporizer chamber can be heated or cooled rapidly. Three types of large-volume introduction techniques can be distinguished:

PTV solvent split injection

The sample is injected in a packed liner with an open split exit at an injector temperature below the solvent boiling point. Volatile compounds co-evaporating with the solvent are lost. After solvent evaporation, the analytes retained in the liner are transferred to the GC column. The maximum introduction volume which can be injected "at once" mainly depends on the liner dimensions: a 1 mm i.d. liner can hold 20–30 μl of liquid, 3–4 mm i.d. liners can hold up to 150 μl of liquid. Higher sample volumes have to be introduced in a speed-controlled manner where the speed is adjusted to the evaporation rate.

PTV large-volume splitless injection

The sample is introduced in a packed liner at a temperature below or close to the solvent boiling point. The split exit is kept closed, i.e., the flow rate through the liner is equal to the column flow rate. The evaporating solvent is vented via the GC column. Volatile components co-evaporating with the solvent are trapped in the swollen stationary phase of the GC column.

PTV vapor overflow

The sample is rapidly injected into a packed liner at a temperature far above the boiling point of the solvent. During solvent evaporation, the split exit is closed but the septum purge is wide open; the evaporating solvent escapes through the purge exit. After solvent evaporation, the injector is heated to effect transfer of the analytes to the GC column. The technique has also been carried out in a conventional split/splitless injector.

Source: Ref. 21.

programmed temperature vaporizer (PTV) injection. These techniques enable injection of volumes as large as 100 µl onto the GC column. These techniques have recently been reviewed by Hankemeier [21] and are briefly described in Table 7. In on-column and loop-type large-volume injection systems, a retention gap is used. This is an uncoated deactivated fused-silica injection column that enables the reconcentration of broadened bands. The performance of both on-column and loop-type injections can be further enhanced by the use of a solvent vapor exit (SVE). The SVE is a solvent release system that helps to protect the GC detector from vapor and to accelerate solvent evaporation. The SVE is positioned prior to the GC column [21].

4.3. Analyte Derivatization

Compounds that not amenable to GC analysis, either because of limited thermal stability or insufficient volatility, can sometimes be made amenable by means of derivatization. A general aim in this type of derivatization is the reduction of analyte polarity by chemical substitution of active protons in the analyte. In MS, derivatization may result in additional effects, e.g., enhancement of the intensity of the molecular ion, changes in the fragmentation directing functionality, and/or improvement of the ionization efficiency. A clear example of the last is the introduction by derivatization of fluorine groups in a molecule, enhancing its amenability to electron-capture detection. A wide variety of derivatization reagents are available [22]. An overview of frequently applied derivatization agents for various compound classes is given in Table 8.

Table 8 Common Derivatization Reagents for GC–MS

Compound class (functional group)	Reagent	Group replacing functional group	Increase in M_r per derivatized functional group
Alcohols (—OH)	TMS	—OSi(CH$_3$)$_3$	72
	TBDMS	—OSi(CH$_3$)$_2$C(CH$_3$)$_3$	114
	Acetylation	—OC(=O)CH$_3$	42
Carboxylic Acids	Methylation	—OCH$_3$	14
(—OH)	TMS	—OSi(CH$_3$)$_3$	72
Amines and Amides	Acetylation	—NRC(=O)CH$_3$	42
(—NRH)	TMS	—NRSi(CH$_3$)$_3$	72
	TFA	—NRC(=O)CF$_3$	96
Carbonyl Compounds	Methoxime	—C=N—OCH$_3$	29
(C=O)	Oxime/TMS	—C=N—OSi(CH$_3$)$_3$	87
	Phenylhydrazone	—C=H—NH—C$_6$H$_5$	102

Derivatization obviously complicates the analytical method, as a (series of) time-consuming step(s) must be included. Sometimes, the derivatization generates additional problems due to artifact formation. Routine derivatization in trace quantitative analysis is often difficult to perform.

4.4. Data Acquisition and Processing

Two general modes of data acquisition are available in MS: full-scan acquisition and selective ion monitoring (SIM). In full-scan analysis, a continuous series of mass spectra is acquired during the chromatographic run. For high-efficiency open capillary GC columns, sufficiently fast scanning is required in order to acquire a sufficient number of data points (typically 10 to 20) to adequately describe the chromatographic peak profile. However, in routine quantitative analysis of a limited number of components, better results in terms of lower detection limits are achieved by the use of SIM, in which the intensity of a (number of) ion(s) is monitored. The choice between full-scan and SIM acquisition in a particular application depends on the required detection limit and information content.

As a result of data acquisition, a three-dimensional data array along the axes time, m/z, and ion intensity is generated in the data system. This array can be processed in a number of ways. In the total-ion chromatogram (TIC), the total ion intensity per spectrum is plotted as a function of time. This provides more-or-less universal detection, with a chromatogram comparable to a FID chromatogram. From the peaks in the TIC, a mass spectrum may be obtained. In order to minimize concentration effects on the spectrum quality in narrow GC peaks, an averaged mass spectrum is often obtained. Background subtraction can also be applied to enhance the spectrum quality. The mass spectrum may be computer searched against a library of mass spectra to enable provisional identification (see section 5.1). An alternative to the TIC is the mass chromatogram or extracted ion current (XIC) chromatogram, where the ion intensity of a selected m/z is plotted as a function of time.

In addition to these general modes of data processing, a variety of more specialized procedures are available, e.g., for quantitative analysis, including fully automated peak integration and calibration by linear regression.

5. ANALYTICAL STRATEGIES USING GAS CHROMATOGRAPHY–MASS SPECTROMETRY

In this section, a number of important aspects in the practice of GC–MS are discussed. Attention is paid to identification and confirmation of identity by library search, structure elucidation and qualitative analysis, quantitative analysis and use of isotopically labeled internal standards, the potential of high-resolution MS, as well as multidimensional detection approaches.

5.1. Library Search for Identification of Unknowns

A powerful tool in the interpretation of data is the computer search through mass spectral libraries. Large mass spectral libraries are commercially available, such as the NIST Library containing about 130,000 spectra, and the Wiley Library containing about 310,000 mass spectra. Various search algorithms are available, e.g., the probability-based matching (PBM) algorithm delivered by Palisade Corporation in combination with the Wiley Library. The libraries can be implemented to be directly searched from the main data acquisition and data processing software platforms used in GC–MS.

Although the library search is a powerful tool in the identification of unknowns, it can also be a tool given too much credit, especially by unexperienced users. The library search algorithms apply statistical tools to come to their conclusions. The results of this process are easily overemphasized when the best hit from the search is assumed to indicate the identity of the unknown. In such cases, the cast is lost out of sight: the "stupidity" of the computer with its ability to very rapidly search through a large database and to perform statistical calculations to indicate goodness of fit, and the "intelligence" of the operator with the ability to critically evaluate the search results and to make decisions based on the outcome of the computer search routine. In the evaluation of the search algorithms, an answer is considered to be correct if the retrieved mass spectrum is the correct compound or one of its stereoisomers. Positional isomers are often difficult to discriminate by MS; the GC retention time of reference compounds often yields a better result, or the use of an alternative technique such as FT-IR or NMR.

The library search will possibly only lead to a correct identification if the spectrum of the unknown is actually present in the library and the GC separation has been sufficiently efficient to obtain a sufficiently clean mass spectrum. When the unknown is not present in the library, the search procedure also yields valuable information in pointing out certain structural elements present in the unknown as well as structural similarities with known compounds. However, this information is only useful in combination with a proper interpretation of the mass spectrum. A fast check of the mass spectra found by the search routine against the supposed structure is advisable as well, in order to eliminate possible errors in the library. The Wiley Library, for example, contains almost identical mass spectra for 2-butanol and 2-methyl-1-propanol, while theoretically and practically, these mass spectra are distinctly different. It must also be taken into account that the vast majority of the spectra available in the library are of compounds having molecular masses between 150 and 300 Da. The number of spectra of compounds with molecular masses above 400 Da is limited, although the number of possible compounds obviously increases with increasing molecular mass. Amenability with GC–MS and wide availability of such compounds limit their appearance in the mass spectral library.

The ultimate proof of identity is in buying or synthesizing the appropriate reference compound, matching GC retention characteristics and mass spectra, and evaluation in both GC and MS of possible isomers.

5.2. Structure Elucidation in Qualitative Analysis

Next to the library search, the interpretation of mass spectra is an important aspect of identification of unknowns. A huge amount of information on the fragmentation reactions of molecular ions and the resulting fragments is available [23]. Helped by the knowledge of possible functionalities and directions from the library search, the identity of an unknown may be determined by interpretation of the spectrum. Sometimes, this can be quite difficult. The description of the various fragmentation reactions is beyond the scope of this book.

The basic interpretation procedure of a mass spectrum includes the following steps. Study all available information on the unknown compound and on the sample in which it was found. Assess the general appearance of the mass spectrum: pay attention to the intensity of the supposed molecular ion and the extent of fragmentation. Then apply the computer library search and carefully study the characteristics of the various hits. At this stage, the actual interpretation of the spectrum, start by first determining the molecular ion by applying the fact that the molecular ion must be an odd-electron ion, showing possibly a number of logical neutral losses (loss of 1, 2, 15 or 18 Da) but no losses of 3 to 14 or 21 to 25 Da. Eventually, a soft ionization technique such as chemical ionization can be applied to confirm the molecular mass. When a clear isotopic pattern of the molecular ion is observed, this can be applied to determine the elemental composition, provided that the isotopic pattern was measured with sufficient accuracy and the molecular mass of the compound is not higher than 300 to 400 Da. The procedure continues in searching for important and informative ions, e.g., peaks due to other odd-electron ions, peaks with high relative abundance and at high m/z. Possible structure assignments may follow from the low-mass ion series, the primary neutral losses from the molecular ion, and important characteristic ions. Finally, all information must be combined in the postulation of possible molecular structures. The spectrum of the unknown may be compared with reference spectra, either from the library or from the analysis of reference compounds.

5.3. Quantitative Analysis and Isotope Dilution

Next to structure elucidation of unknowns and confirmation of identity, quantitative analysis is an important application area of GC–MS. The power of GC–MS in quantitative analysis can be attributed to the excellent selectivity of the detection and the good limits of detection (LOD) achievable. Selective ion monitoring using a number of characteristic ions for a particular compound can be applied for the best results.

The power of GC–MS in quantitative analysis is even further enhanced because of the ability to apply the ideal internal standard, which is the isotopically labeled internal standard, showing (almost) identical characteristics to the compound of interest during sample pretreatment, GC separation, and MS ionization. The small mass difference enables the discrimination between the compound of interest and its internal standard. The use of isotopically labeled internal standards is frequently demonstrated throughout the various applications collected in this book. Appropriate algorithms for quantitative analysis using isotopically labeled internal standards are available as part of MS data processing software.

The way SIM is applied in quantitative analysis strongly depends on the application area and on the complexity of the samples analyzed. While in some pharmaceutical applications SIM can be performed at only one m/z of a characteristic ion, in the areas of residue analysis several ions must be applied. Criteria are available for confirmation of identity in such applications, based on the coelution of peak profiles of the various ions selected within a particular small retention time window. The peak areas of the various ions must agree with predetermined area ratios, within an area window of, for example, ±10% of the area ratios in the reference mass spectrum.

5.4. High-Resolution Mass Spectrometry

Although in most applications of GC–MS a linear quadrupole or quadrupole ion trap is used for mass analysis, yielding unit-mass resolution only, higher mass resolution can be achieved as well, especially by means of a double-focusing sector instrument. Such instruments have been for many years, and still are, routinely used in various application areas of GC–MS, e.g., analysis of polychlorinated compounds such as polychlorodibenzodioxins and polychlorobiphenyls. Highly sophisticated and dedicated instruments are available for this purpose.

In principle, there are two types of applications of high-resolution MS. High-resolution MS enables a more accurate determination of the m/z of any ion in the mass spectrum. When an accuracy better than 5 ppm can be achieved, the accurate mass determined can be used to predict possible elemental compositions for a particular ion, which may help in identification of the compound of interest. The predictive power obviously increases at higher resolution and better mass accuracy, but in general, working at higher resolution becomes increasingly more difficult. Recently, an alternative instrument was introduced for this type of application, i.e., an orthogonal-acceleration reflectron time-of-flight (oaReTOF) instrument (e.g., Micromass GCT). With such an instrument, mass accuracy better than 5 ppm can be routinely achieved, while it is more user friendly in operation than a high-resolution sector instrument.

Another application of high-resolution MS in GC–MS is based on the ability to enhance the selectivity of the MS as a detector by increasing the resolution. This is, for example, applied in the analysis of polychlorinated compounds. Due to the negative mass defect of the chlorine atom, the m/z of a polychlorinated compound will have a negative mass defect as well. These compounds preferentially accumulate in the fat tissue, containing only components with a positive mass defect. Using SIM at a mass resolution in the range of 10,000, the excess of fat-related compounds will no longer interfere in measuring the signals due to the polychlorinated compounds. While the actual signal due to the polychlorinated compounds will decrease as a result of the increase in mass spectral resolution, the detection limit will improve, because the noise will decrease even further. It must be pointed out that a double-focusing sector instrument is required for this type of application. An oaReTOF instrument cannot be applied with a similar gain in signal-to-noise ratio, since it cannot be operated in SIM mode but only as an integrating detector.

5.5. Multidimensional Detection Approaches

Although GC–MS is a powerful tool in solving many analytical problems, it is not the general solution to all problems. For several applications, the use of multidetection systems can be extremely useful. While multidetection systems containing both FID and MS are readily available, a more powerful multidetection system combines MS with another spectrometric detection system, e.g., Fourier-transform infrared (FT-IR) or atomic emission detection (AED).

The on-line combination of GC with FT-IR and MS is commercially available from Agilent Technologies [24]. This on-line combination can be realized either in a parallel configuration, applying a postcolumn splitter (MS:IR between 1:1 and 1:10), or in a serial configuration, i.e., IR flow cell followed by MS interface. In many structure elucidation problems, FT-IR can provide useful complementary information to the GC–MS information. Examples are characterization of positions of aromatic substitution, and *cis/trans*-isomerism in fatty acid esters.

The on-line combination of GC–MS and AED is not (yet) commercially available [25–27]. In most cases, a parallel configuration is applied. The AED allows selective detection of any element except helium, with detection limits in the pg/s range for most elements, which is almost independent of the structure of the analyte. As such, AED allows the determination of the elemental composition of an unknown compound. The power of GC–AED in combination with GC–MS is readily demonstrated in Figure 6. By combined use of GC–MS and GC–AED, the presence of elements such as sulfur, phosphorus, chlorine, and

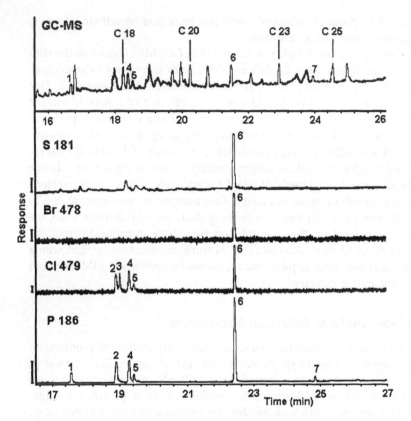

Figure 6 Element specific detection by GC–AED assisting in the interpretation of a complex GC–MS chromatogram from the analysis of a wastewater sample. Peaks: 1, tributylphosphoric acid; 2, tris(2-chloroethyl)phosphate; 3, hexachlorocyclohexane; 4, tris(2-chloro-1-methyl-ethyl) phosphate; 5, bis(2-chloro-1-methyl-ethyl)(2-chloropropyl)phosphate; 5, bromophos-ethyl; and 7, tris(2-butoxyethyl) phosphate. (From Ref. 26.)

bromine is indicated in peaks in the complex TIC from GC–MS obtained by the analysis of a wastewater sample.

REFERENCES

1. A.T. James and A.J. Martin, Biochem. J. Proc., 50 (1952).
2. M.J.E. Golay, in V.J. Coates, H.J. Noebels, and I.S. Fagerson, eds., Gas Chromatography, 1958, Academic Pr., p. 1.
3. R.D. Dandeneau and E.H. Zerenner, J. High Resolut. Chromatogr., 2 (1979) 351.

4. J.J. Thomson, Rays of Positive Electricity and Their Application to Chemical Analysis, 1913, Longmans Green, London.
5. A.J. Demster, Phys. Rev., 11 (1918) 316.
6. F.W. Aston, Philos. Mag., 38 (1919) 707.
7. R. Ryhage, Anal. Chem., 36 (1964) 759.
8. R.E. Finnigan, D.W. Hoyt, and D.E. Smith, Environ. Sci. Technol., 13 (1979) 534.
9. F. Friedli et al., J. High Resolut. Chromatogr., 4 (1981) 495.
10. W.O. McReynolds, J. Chromatogr. Sci., 8 (1970) 685.
11. M.H. Abraham, C.F. Poole, and S.K. Poole, J. Chromatogr. A, 842 (1999) 79.
12. M.J.E. Golay, in D.H. Desty, ed., Gas Chromatography, 1959, Butterworth, London, p. 36.
13. E. Kováts, Helv. Chim. Acta, 41 (1958) 1915.
14. Sadtler Standard Gas Chromatography Retention Index Library, 1984, Sadtler Research Laboratories, Philadelphia, PA.
15. R.A. de Zeeuw, J.P. Franke, H.H. Maurer, and K.P. Fleger, Gas Chromatographic Rention Indices of Toxicologically-Relevant Substances on Packed or Capillary Columns with Dimethylsilicone Stationary Phases, 1992, VCH, Weinheim, Germany.
16. G. Castello, J. Chromatogr. A, 842 (1999) 51.
17. A.G. Harrison, Chemical Ionization Mass Spectrometry, 2d ed., 1992, CRC Pr., Boca Raton, FL.
18. C. Brunnee, Int. J. Mass Spectrom. Ion Proc., 76 (1987) 125.
19. R.E. March, J. Mass Spectrom., 32 (1997) 351.
20. E.W. Schlag, ed., Time-of-Flight Mass Spectrometry and Its Applications, 1994, Elsevier, Amsterdam.
21. Th. Hankemeier, Automated Sample Preparation and Large-Volume Injection for Gas Chromatography with Spectrometric Detection, Ph.D. thesis, 2000, Free University Amsterdam, Netherlands.
22. F.G. Kitson, B.S. Larsen, and C.N. McEwen, Gas Chromatography and Mass Spectrometry, a Practical Guide, 1996, Academic Pr., San Diego, CA.
23. F.W. McLafferty and F. Tureček, Interpretation of Mass Spectra, 1993, University Science Books, Mill Valley, CA.
24. R.J. Leibrand, ed., Basics of GC–IRD and GC–IRD–MS, 1993, Hewlett-Packard, Palo Alto, CA.
25. S. Pedersen-Bjergaard and T. Geibrokk, J. High Resol. Chromatogr., 19 (1996) 597.
26. Th. Hankemeier, J. Rozenbrand, M. Abhadur, J.J. Vreuls, and U.A.Th. Brinkman, Chromatographia, 48 (1998) 273.
27. H.G.J. Mol, Th. Hankemeier, and U.A.Th. Brinkman, LC–GC Int., 12 (1999) 108.
28. S.C. Davis, A.A. Makarov, and J.D. Hughes, Rapid Comm. Mass Spectrom. 13 (1999) 237.

2
Principles and Applications of Surface Ionization in Gas Chromatography–Mass Spectrometry

Toshihiro Fujii
National Institute for Environmental Studies, Onogawa, Tsukuba, Japan

1. INTRODUCTION

The term "surface ionization" is frequently used to describe the process in which ionization takes place within some critical distance from the surface. A number of surface-dependent ionization methods, such as field desorption, fast atom bombardment, and laser desorption, are described as surface ionization methods. Another use of this term is more restricted; surface ionization (SI) is defined as the ionization process, which can be interpreted by use of the Saha–Langmuir equation.

In 1923, Kingdom and Langmuir [1] first observed SI (Fig. 1). This phenomenon consisted of desorption of cesium (Cs) atoms in the form of positive ions from the surface of a heated tungsten (W) filament. Subsequently, numerous studies of positive-ion SI have been conducted, since this effect opens interesting possibilities for analysis of chemical species with low ionization energy (IE), for ion production, and for the detection of molecular and atomic beams. After the Saha–Langmuir equation [2] was established and well defined, further studies have [3,4] been directed to the atoms of virtually all elements, which have a low ionization energy. Especially, SI was studied for suitability as a source of ions for precise isotope-ratio measurements [5] and isotope-dilution techniques. It is apparent, however, that its use has, until recently, been restricted to metals.

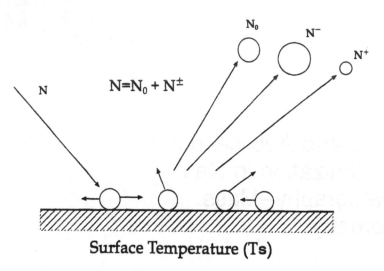

Surface Temperature (Ts)

Figure 1 Fluxes (N) of molecules from a molecular beam impinge on surface and then scatter partly as neutral particles (N_0), positive ions (N^+), and negative ions (N^-). Under steady state conditions, the flux of particles desorbing from the surface ($N_0 + N^+$) or ($N_0 + N^-$) is equal to that of particles (N) absorbing on the surface from the gas phase.

The SI of organic compounds has its origin in the discovery [6,7] that many residual organic gases in a mass spectrometer with a thermionic ion source are capable of forming ions on a heated metal surface. A surface-ionization organic mass spectrometry (SIOMS) study of individual organic compounds including alkyl-substituted amines and hydrazine was first performed by Zandberg et al. [8] in the 1960s using oxidized W. It is now established [9–11] that organic compounds with heteroatoms ionize on metal emitters, especially refractory rhenium (Re) metal oxides. In other words, when organic molecules interact with a hot surface, thermionic emission takes place, resulting in the emission of organic ions (see Fig. 1). Nitrogen-containing compounds are ionized most effectively; extremely high ionization efficiency of amines on tungsten oxide emitters has been demonstrated.

Since then, new developments related to the chemistry of organic compounds have followed. Davis [12] has constructed a magnetic sector mass spectrometer with a direct air sample inlet system and thermionic rhenium emitter. He demonstrated that many organic compounds in normal laboratory air were efficiently ionized on an oxidized rhenium surface. Along with these results, it was suggested that the SI phenomenon could be exploited in gas chromatography (GC) and mass spectrometry (MS).

This chapter reports on and discusses several aspects of SI in GC–MS: (1) the principles of SI, (2) the instrumental and operational description of the GC–MS system, (3) the most important performance and response characteristics, and (4) some applications [13] in chemical analysis to demonstrate that SI can have a unique role in GC–MS.

2. PRINCIPLES

2.1. Saha–Langmuir Model

Surface ionization is conventionally interpreted by the use of the Saha–Langmuir equation (Eq. 1), which is based upon the assumption that thermal and charge equilibria are established between the species on the surface material and the surface material itself [2]. This equation describes the temperature dependence of the degree of ionization, α,

$$\alpha = N^+/N_0 = g^+/g_0 \exp[(\psi - IE)/kT] \tag{1}$$

where N^+ is the number of ions leaving the surface per unit area; N_0 is the number of neutral species emitted from the same surface in this time; ψ is the work function of the surface at which ionization occurs at temperature T; k is the Boltzmann constant; IE is the ionization energy of the emitting chemical species; and g^+/g_0 is the ratio of the statistical weights of the ions and the neutral species.

Surface-ionization mass spectrometry (SI-MS) studies of organic compounds show that in most cases the compounds decompose into radicals, which have a lower ionization energy (IE) than the corresponding molecules, and are ionized more efficiently. For a given type of secondary species, s, formed on the surface, the resulting positive thermionic emission currents (J) and their dependence on surface temperature (T) are described by the ionization efficiency, $\beta_s(T)$, and by $Y_s(T)$, the yield of chemical reactions on the surface, such that

$$J_s(T) = NY_s(T) \cdot \beta_s(T) \tag{2}$$

when a stationary beam of organic molecules, N, impinges from the gas phase on the surface. $\beta_s(T)$ is expressed as follows using the Saha–Langmuir equation (see Eq. 1):

$$\beta_s(T) = \{1 + g^+/g_0 \exp[(IE - \psi)/kT]\}^{-1} \tag{3}$$

if equilibrium thermal ionization can be assumed. The combination of Eqs. (2) and (3) tells us that the emitter surface must have good pyrolytic properties [$Y_s(T)$ should be high] and the work function of the emitter must be high. Also, we see that the SI process is specific, since it is strongly dependent on the *IE* of the species.

2.2. Work Function of Surface

The work function, ψ, of a solid surface is defined as the minimum energy necessary to extract a conduction electron in the solid into the outside space of the surface. This is equal to the energy gap between the vacuum level and the Fermi level. Thus, this property is easily understood to think that the value of ψ is the electron affinity of the solid.

One of the techniques for measuring the work function of surfaces uses the thermal emission of electrons from the heated surface under study [4]. According to the Richardson-Schottky equation, the electron current density, I, depends on the surface temperature, T, and the work function, ψ:

$$I = AT^2 \exp(-\psi/kT) \tag{4}$$

where I is the saturation electron emission current density (ampere/cm^2). The electron current density, I, collected on the anode is plotted for different values of the anode voltage, V_a. For each temperature, a plot is made of log I versus V_a and the straight-line (Schottky) portion is extrapolated to zero field, yielding the zero-field current density $I_o(T)$. Then, a plot for $\log(I_o/T^2)$ versus $1/T$ gives a slope $(-\psi/K)$ and intercept, a, from which the so-called Richardson work function, ψ, and the preexponential factor, A, may be determined. In theory, A should be a constant for all surfaces, equal to 120 A/cm$^2 K^2$, and the slope of the Richardson plot should be constant as well.

The work functions of metal surfaces show a marked dependence on the crystal planes that constitute the surface. For polycrystal metal surfaces, the experimentally determined value of the work function is an average of the work functions of all the crystal planes that are present on the surface. It is made up of the contributions from the individual crystal planes. If ψ_e, ψ^+ and ψ^- are defined as the mean effective work functions for thermionic electron emission, positive surface ionization, and negative surface ionization, respectively, it is generally evident that $\psi^+ > \psi^-$, and $\psi^- = \psi_e$.

2.3. Ion Formation Mode of Surface Ionization of Organic Compounds

In 1981, a study of SIOMS commenced in the author's laboratory under the experimental conditions that organic molecules are supplied from the gas phase onto a heated emitter SI ion source [13]. Interaction of organic molecules with the hot emitter surface leads to the thermionic emission. The SI mass spectra of the more than 600 organic compounds studied indicate that, depending on their nature, one of three basically different processes occurs [14,15]. For a molecule, M, these processes are classified as shown in the following sections [16].

2.3.1. Molecular Surface Ionization (MSI)

$M^{+\cdot}$ ions are observed only for M having not too high IE. Aniline is the typical example. Molecular surface ionization (MSI) is easier not only in the case of a lower *IE*, but also in the case where a dissociative surface reaction takes place very slowly since the ionization process competes with a pyrolytic surface reaction. The $M^{+\cdot}$ intensity varies significantly with the work function of the emitter surface. According to this behavior, MSI is a process conveniently described by the Saha–Langmuir equation, no matter whether the MSI process occurs in an equilibrium or in a nonequilibrium.

2.3.2. Dissociative Surface Ionization (DSI)

$[M - H]^+$, $[M - CH_3]^+$, or $[M - OH]^+$ ions are often observed for many organic compounds such as toluene. One of the possible mechanisms is the concerted process:

$$M \rightarrow (M - H)^+ + H^{\cdot} + e \tag{5}$$

$$AP[(M - H)^+] = IP(M - H) + D[(M - H) - H] \tag{6}$$

According to the calculations, appearance potential (AP) should be 11.3 eV for toluene. Even if the effect of adsorption energy of atomic hydrogen is taken into account, these values are too high to allow the formation of the ions on the surface.

　　The consecutive process is more likely: the formation of surface dissociative reaction product with a low ionization potential (IP) followed by the Saha–Langmuir model SI on the hot surface emitter with a high work function. In this mechanism, the ionization currents for $[M - H]$ radical are determined not only by the ionization efficiency, given by the Saha–Langmuir equation, but also by the efficiency of the formation of this $[M - X]$ product on the emitter surface [16]. The large $[M - H]^+$ ion peak from toluene can be explained by this process since the ionization potential of benzyl radical, 7.63 eV, is sufficiently low to be ionized [17].

2.3.3. Associative Surface Ionization (ASI)

Some basic compounds, such as the pyridine molecule [15], generate the protonated (associated) molecule on the hot emitter surface. A bimolecular surface reaction of proton transfer complex is the most probable mechanism of formation. The process is shown in Eq. (7), in which two molecules on the surface produce the ion of interest:

$$2M \rightarrow (M + H)^+ + (M - H)^{\cdot} + e \tag{7}$$

2.3.4. Reactive Surface Ionization (RSI)

Diphenylamine (DPA) proved to be an interesting compound for SIOMS [18]. The SI mass spectrum of DPA shows that, in addition to DSI and MSI ions, there are ions formulated as $[(C_6H_5)_2N=CH_2]^+$. These ions are apparently different from those of ASI, in which an association process takes place. We speculate that DPA may act as a nucleophile in the reaction with the CH_3 radicals on the emitter surface. The radicals are probably produced due to a thermal decomposition on the hot emitter surface, which leads to the formation of ammonium ions according to the proposed scheme. The proposed rearrangement reactions lead to the formation of new molecules from amines or their radicals, which then ionize by SI according to the same general rule for ion formation.

$$(C_6H_5)_2NH^{+\cdot} + CH_3^{\cdot} \rightarrow (C_6H_5)_2N^+ = CH_2 + H_2 \tag{8}$$

It should be noted that ASI is a single instance of RSI for which the exothermic reaction on the surface is responsible, e.g., the reaction of halogen gases with alkali oxides [19].

In experiments aiming at a search for emitters, which surface-ionize organic compounds more efficiently [20] (based on the understanding that the work function of the surface is greatly increased by the chemisorption of electronegative gases [21]), a considerable increase in alkali metal ion currents is always observed, when chlorine (or fluorine) is introduced into a mass spectrometer with a thermal ionization source. This phenomenon is well interpreted as the so-called "stimulated surface ionization" [22] in which the exothermic reaction on the surface is responsible for the ionization. This is another interesting example of RSI.

2.3.5. Surface-Ionization Mass Spectra

Table 1 shows some principal schemes for the SI mass spectra of six compounds, which are chosen from the organic compounds with the different functional group. For most compounds, the $[M - H]^+$ intensity is highest, and the electrochemical properties of the radical substituent in the molecule affect the ionization efficiency of $[M - H]^+$ ions. $M^{+\cdot}$ ions are formed when the molecular IE is small. $[M + H]^+$ ions with great intensity can be formed by a protonation process in cases where the AP of $[M + H]^+$ ions is small.

Table 1 Surface Ionization Mass Spectrum and Sensitivity of Selected Organic Compounds

Compound (M,IE)	Most intense peaks (m/z)	Relative sensitivity (A torr^{-1})	SI/EI (n)[a]	Remarks[b]
Trimethylamine (59, 7.8)	58, [M − H]$^+$	5.6 × 10	650	
	42	2 × 10^{-3}	(58)	
	30	2.5 × 10^{-2}		
Piperidine (85, 8.7)	86, [M + H]$^+$	3.3 × 10^{-2}	367	AP(8.0)
	84, [M − H]$^+$	45.8	(84)	
	82, [M − 3H]	7.4		
	80, [M − 5H]$^+$	6.2		
Benzaldehyde (106, 9.51)	105, [M − H]$^+$	6 × 10^{-1}	1.8	
			(106)	
Aluminium acetyl aceto-nate[c] (324, ?)	27, Al$^+$	100	1.6	T$_c$ = 125°C
	225, [M-acac]$^+$	11.1	(225)	
	141	7.8		
	143	4.4		
	226	2.1		
Proline (115, 9.36)	70, [M − COOH]$^+$	100	0.70	T$_c$ = 200°C
	98, [M − OH]$^+$	31	(70)	
	68	20		
	96	15		
	112, [M − 3H]$^+$	1.5		
	110, [M − 5H]$^+$	1.4		
	114, [M − H]$^+$	0.8		
Nicotene[c] (162, ?)	84	100	3.2	T$_c$ = 100°C
	130	25	(84)	
	161, [M − H]	12	16	
	159, [M − 3H]$^+$	1.4	(161)	

The admission of liquid sample was controlled by using an ionization gauge, calibrated by a capacitance manometer and a variable leak valve. Solid samples were introduced either through the gas chromatograph or through the solid probe with a heated quartz ampule.
[a] The (n) in the SI/EI column indicates the mass number, n, of the base peak in the EI spectrum for comparison.
[b] T$_c$ is the quartz ampule temperature.
[c] The spectrum of the compound is expressed as the pattern coefficient in the sensitivity column.

Most DSI ions have been postulated in a purely empirical way. However, the composition of some ions, which are not commonly observed species, has not been elucidated thus far. Further investigation of the SI ions of compounds with labeled functional groups is a plausible approach to this problem. There is no doubt that SIOMS with the labeling technique should provide a unique method for studying the ion structure of organic ions. The sensitivity and SI/EI ratio in Table 1 is detailed in later sections.

3. INSTRUMENTATION

Systematic studies of SI of individual organic compounds suggested that the SI phenomenon could be exploited [23] in GC–MS. That is, a sensitive and selective detection of specific compounds, such as amines, eluted from a gas chromatograph may be possible by utilizing the SI of the organic compounds. The instrument should have (1) the mass spectrometer with a high current sensitivity and an adequate resolution with respect to m/z, (2) the most effective emitters with a high work function, which are stable with time, (3) methods for measuring the emitter temperature (T_s) and work function (ψ) under working conditions, and (4) a capillary column GC equipped with a multiple sample inlet system.

3.1. Gas Chromatograph–Surface-Ionization Organic Mass Spectrometer

A schematic view of the GC–SIOMS apparatus is shown in Figure 2. The design of the SI probe assembly is such that the Re ribbon filament (for instance, 6 × 0.75 × 0.025 mm) can be placed in the center of the conventional electron ionization (EI) ion source chamber. This combined ion source allows an easily inter-

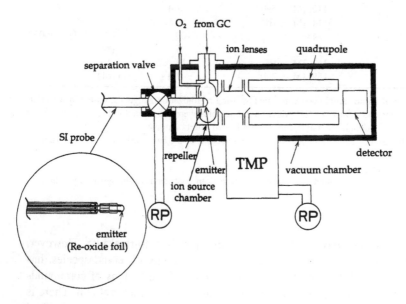

Figure 2 Diagram of the GC–MS apparatus. RP, rotary pump; TMP, turbomolecular pump. The SI probe is inserted through the separation valve. The inset shows details of the SI probe with the detachable Re-oxide emitter.

changeable operation in both the SI and the EI modes. The details are discussed in section 4.3 below. The Re ribbon filament emitter was used in the presence of oxygen (O_2) at a controlled pressure of 1×10^{-5} torr, which was obtained by adjusting the variable leak valve. A continuous admission of O_2 allows an increase in work function, which is essential for the high SI efficiency. Operating with the continuous oxygen flow caused no observable deterioration in the performance of the mass spectrometer.

3.2. Instrumental Details

3.2.1. Surface Ionization Probe

A SI probe (Shimadzu Corp., Japan) is shown in the insert of Figure 2. The probe has almost the same configuration as the solid inlet probe except that the solid sample holder is replaced by the Re filament assembly consisting of a spotwelded Re filament across the posts on the Kovar seal. The filament, which is maintained at the same potential as the accelerating voltage, is heated using a power supply. The filament emitter temperature can be measured by an optical pyrometer, which measures the luminescence of the hot filament through a glass window on the analyzer manifold.

3.2.2. Emitter Material

According to Eq. (1), the thermionic emission capability depends on a maximum value of work function. In other words, the emitter should have the highest possible work function. Further requirements include high stability with respect to the vapor to be ionized. Under SIOMS conditions, the emitter is likely to change its surface nature, which has an inevitable effect on the work function, and the high-temperature durability.

Only a few studies [24–26] in the search for the high work function materials have been reported until now. The first attempts were made by Zandberg et al. [27] regarding an efficient emitter for organic compounds. Oxygen was supplied to a W wire. The work function of this oxidized W wire was estimated to be 6.5 eV. They were also able to detect many nitrogen (N)-containing organic compounds with the tungsten oxide emitter. Unfortunately, the oxide coating evaporates at the prevailing high temperature. According to Davis [12], oxygen in approximately 10^{-5} torr air increases the work function of a Re filament to about 7.2 eV. Greaves and Stickney [28] suggested that the combination Re-O_2 would be a better choice than either W-O_2 or molybdenum (Mo)-O_2 if a high work function were desired.

Known refractory materials exhibiting higher work functions than W, such as Re [26] and iridium (Ir) [24], or semiconductors such as silicon (Si) [25] have been investigated. The conclusion is that oxidized Re seems to be the best emitter material for detecting organic substances for practical use in SIOMS.

3.2.3.　Mass Spectrometer

The quadrupole mass spectrometer has become one of the most useful of mass spectrometers and has enjoyed widespread application in SIOMS, too, because of the following characteristics: (1) small size and weight since there is no magnetic field, (2) rapid scanning, (3) easy operation at relatively high pressure (10^{-4} torr) and low ion energy (<10 eV), (4) wide dynamic range, and (5) electrical adjustment of the resolving power.

　　A time-of-flight (TOF) mass spectrometer has several distinct advantages over quadrupoles, namely, high ion transmission efficiencies at all mass ranges; very high data rates; simplicity of design, construction and repair; and adaptability to most pulse-type ionization sources. A TOF instrument was also modified for use with a SI source [29]. The author claimed that formation of a potential well within the ionization region to trap ions prior to acceleration improved sensitivity without any apparent loss in resolution.

3.2.4.　Gas Chromatograph Inlet

Most organic samples are admitted to the mass spectrometer through the gas chromatograph with multiple inlet systems. For solids, which are not amenable to GC analysis, the sample can be introduced using a solid probe with a heated quartz ampule.

3.2.5.　Thermionic Diode

It is desirable to build the thermionic emission diode designed by Eckstrein and Forman [30] in the mass spectrometer, since it enables a quantitative measurement of work function variation with the adsorbed O_2 on the Re surface and the discovery of an even higher work function surface (see section 2.2).

3.3.　Operational Remarks

The effect of the emitter surface temperature, electric field and gaseous environment around the emitter on the analyte response should be considered. Figure 3 shows the curve of $i(T)$ for ions with mass $[M + H]^+$ of pyridine in ASI, with $[M - H]^+$ of xylene in DSI, and with $M^{+\cdot}$ of aniline in MSI [11]. An ionization threshold temperature (T_o) was observed around 400°C, which is compound dependent. The T_o value for all the compounds is less than T (475°C) for the appearance of sodium ions (Na^+) and potassium ions (K^+) from Na and K impurity atoms in Re. The i versus T curve for aniline resembles that for the ionization

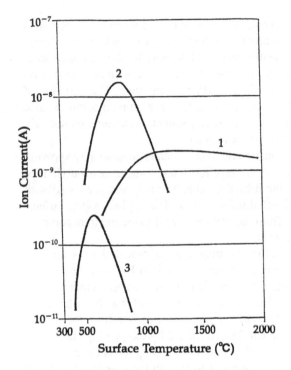

Figure 3 Surface temperature dependence of the resulting ion current of surface ionization for (1) aniline $M^{+\cdot}$, (2) xylene $[M - H]^+$, and (3) pyridine $[M + H]^+$.

of atoms, while the i versus I curves were bell-shaped with the maximum for
both DSI and ASI cases. The maximum, which occurs at filament temperature
of 520 to 800°C varies from sample to sample. The observation that ion currents
decrease with filament temperature after reaching the maximum is mainly attributable to the decrease in the work function [31] of the Re filament with evaporation of the oxides from the Re surface at higher temperatures. Generally, the
emitter temperature, which provides the maximum intensity for the ASI process,
is 360°C lower than that for the DSI process.

For the interpretation of these temperature characteristics, the temperature
variation of the ion formation and ion desorption rate should be taken into account. The rate law for the desorption process, the energy of which is ΔE, is
described as follows:

$$V = V_o \exp(-\Delta E/kT) \tag{9}$$

where V represents the rate of desorption at T, and Vo is the preexponential factor.
The initial intensity increase with the temperature may be interpreted partly by

this equation. The temperature dependency is complicated when $[M + H]^+$ ions are formed through the adsorption process of the proton. The dissociative chemisorption of the proton is the essential step, which must be the activated step. However, molecular association such as the proton-transfer-complex formation is reduced by the thermal energy at higher surface temperature. The variation of the lifetime of the pyridine and proton on the surface at high temperature is also rate-determining for the ion formation. All these parameters explain qualitatively the shape of the $i(T)$ curve in a complex manner.

To determine the effects of an electric field, which is generated by applying the same voltage to the emitter and the repeller, a test sample of tributylamine (TBA) was introduced through the diffusion tube. The response increased with increasing applied voltage and leveled off at ~10 V. The applied voltage affects the desorption of the ion species from the surface as well as the ion transportation to the inlet of the mass spectrometer.

The ionization behavior over the operating time varies with the gas environment. It was known previously that, during Si on oxidized Re in vacuum under steady conditions (p = constant, T = constant), the ion current gradually decreases in time. There is a poisoning of the emitters, which has been linked to carbonization of the surface [13,32]. Only removing the carbon from the surface and reoxidizing the emitter, which can be established by oxygen injections, can restore the ionizing ability. Thus, in order to avoid poisoning the emitter, continuous oxygen introduction should be employed to sustain the constant ionization efficiency.

4. RESPONSE AND PERFORMANCE CHARACTERISTICS

Response and performance characteristics were assessed using (1) calibration curve and linear dynamic range, (2) sensitivity, (3) selectivity, and (4) noise and minimum detectable amount.

4.1. Calibration Curve

The relationship between the ion intensity of the most intense peak, $[M - H]^+$, and the amount of sample was tested using TBA. It was shown that the linear portion of the calibration curve covers nearly four orders of magnitude with an uncertainty of ±3.0%. The linearity is limited at higher TBA concentrations by the decrease in the work function of the Re filament due to carbonization of the surface [32]. At extreme amounts of TBA, no resulting ion current was observed.

The relationship of the intensity (I) of the protonated molecule to the amount of sample (P) was studied by using pyridine-d_5. Experimental data at a constant temperature of 550°C show a straight line in the plot of the square root of I (C_5D_5ND) versus the sample partial pressure. This indicates that the process is bimolecular, which means that the $[M + H]^+$ ions are formed by reaction either between two pyridine molecules or between a pyridine molecule and its dissociated species on the hot surface.

4.2. Minimum Detectable Amount, Sensitivity, and Selectivity

For a given organic compound, the sensitivity (see Table 1) may be defined as the maximum response of the base ion emitted per a given amount of compound. Here, sensitivity (A/torr) is expressed in amperes of the resulting ion current of the base peak in the SI spectrum per torr of sample. The sensitivity can be varied, depending on operating conditions such as the emitter temperature (T_s). Both sensitivity and noise increase as emitter temperature is increased. Thus, an optimum must be determined in order to find the minimum detectable amount.

Under the optimal conditions, the sensitivity measurements were made using the electron multiplier with a gain of 5×10^3 in the quadrupole mass spectrometer in the author's laboratory. The sample pressure was adjusted to approximately 5×10^{-7} torr, which is low enough to be within the linear range of the response curve. The sensitivity for different organic compounds, which is easily determined by comparison of the signal currents, varies in a wide range from compound to compound. For instance, sensitivity was 1.58 A/torr for tributylamine (TBA) and 2.98×10^{-6} A/torr for dodecane. The selectivity between two compounds, which is defined as the ratio of sensitivities of these two compounds, was

$$S(\text{TBA})/S(\text{dodecane}) = 5.3 \times 10^{-5} \tag{10}$$

4.3. Comparison Between Surface Ionization and Electron Ionization

A combination-type of ion source readily allows the comparison between the sensitivities under SI and EI conditions. The ratio of the base peak intensities obtained in the SI mode and in the EI mode (SI/EI) is shown Table 1, presented previously. If the detection limit in the EI mode is known for a given instrument, the SI/EI ratio reported in Table 1 can be employed to calculate the detection limit achievable in the SI mode.

The results of SI/EI are particularly interesting. Alkylamine and piperidine have significantly higher sensitivities in SI than in EI. Although it is not widely known, the SI of organic compounds provides a highly valuable analytical method.

5. GAS CHROMATOGRAPHY–MASS SPECTROMETRY APPLICATIONS

The SI mass spectra for more than 600 organic substances have been obtained so far in the systematic study on the SIOMS. This study will be extended to various classes of organic compounds: (1) N-containing compounds [16,18], (2) hydrocarbon compounds and halogenated hydrocarbon compounds [33], (3) O-containing compounds [34], (4) organometallic compounds [35–39], and (5) biologically important compounds [40]. The mass spectral report should indicate the predominant ions in the SI spectrum, the possible ion structures, the sensitivity (A/torr), and so on. Available IE and AP data from the literature [17] are also desirable (cf. Table 1). The following subsection describes briefly the SI mass spectral features of the various classes of organic compounds. This is useful for the interpretation of SI mass spectra from the GC–MS. Examples of three applications are then presented to provide a broader understanding of the potential of SI in GC–MS.

5.1. Mass Spectral Features

5.1.1. Nitrogen-Containing Compounds

SIOMS is (particularly) well suited to the analysis of alkylamine, aminoalcohols, quaternary ammonium salts, and hydrazines, which exhibit an intense molecular ion, together with several diagnostic ions for structural determination. In particular, alkylamine and hydrazines show sensitivities in SI that are significantly higher than for EI. The appearance of the ion in the spectrum of tetramethylammonium chloride provides evidence for the evaporation of intact salt molecules. Aminoalcohols provide a characteristic set of peaks similar to that for methylamines. The spectra show very intense $[M - H]^+$ ions and many diagnostic ions, such as $[M - 3H]^+$, and $[M + H]^+$. This is in contrast to the EI spectra [41], which show relatively weak (or no) molecular ions and very extensive fragmentation. Several 5-membered and 6-membered ring N,N-heterocyclic amine compounds were studied. Pyrrole, indole, and carbazole exclusively provide an intense molecular ion, while pyridine, quinoline, and acridine invariably exhibit $[M + H]^+$ ions with great intensity.

5.1.2. Hydrocarbons and Halogenated Hydrocarbons

Most of the alkane, alkene, and alkyne compounds tested do not have SI sensitivities, except a few species such as cyclohexene and 1,3,5-cycloheptatriene. Aromatic-aliphatic hydrocarbons generate intense DSI ions with a small number of peaks, whereas polycyclic aromatic hydrocarbons (PAHs) give dominant $M^{+\cdot}$

ions. Other suitable compounds are terpenoids, which form many peaks of significant intensity.

5.1.3. Oxygen-Containing Compounds

In general, many oxygen-containing organic compounds show lower SI sensitivities than analogous nitrogen-containing compounds, but the SI mode gives greater output currents for three compounds (benzyl alcohol, 4-methoxytoluene, and benzaldehyde) than does the EI mode.

5.1.4. Organometallic Compounds

Organometallic compounds are virtually complexes and are an interesting class of compounds. The relatively high thermal stability and small IE values of some of these compounds suggest the possible formation of ions in SI mode. Metallocenes of five transition metals (iron, nickel, cobalt, zirconium, and titanium) are efficiently surface ionized [35,38] in either the molecular or radical form. The carbonyls [36] can provide an efficient method for producing ligand-free metal ion in the gas phase, while [M-acac]$^+$ types of ions have been produced by all metal-acetylacetonate (acac) complexes examined [36]. Surface ionization has tremendous potential for elucidating certain aspects of complex organometallic chemistry and, hence, SIOMS should become important in this field.

5.1.5. Biologically Important Compounds

The SI techniques are apparently useful for amino acids, purine bases, and alkaloids, but not for steroids and carbohydrates. Some of the SI ions observed can simply be assigned and used for structure determination, while a very intense species can be valuable for trace analysis. However, some of the biomolecules give many more unidentified fragment peaks than other compounds.

5.1.6. Tabular Correlation of Surface-Ionization Ionic Species

Since the reference mass spectra of known compounds have been run previously for a number of years, correlations of SI mass spectra with structure can be made for many of the common classes of organic compounds. Most of these correlations emphasize the spectral pattern or simple decomposition pathways to be expected for a particular molecular structure. This has led to the tabulation of mass spectral correlations to provide empirical and structural formulas of ions that might be found at a particular m/z in a mass spectrum, plus an indication of how each such ion might have arisen. Such a table has been reported elsewhere [10].

5.2. Determination of Trimethylamine in Air

The occurrence and determination of aliphatic amines have received a great deal of attention in recent years [42–44]. These foul-smelling compounds have been found in a number of ambient environments [45–47] and become a source of serious social and psychological problems. They are also involved in nitrosamine synthesis in air [48], because methylamines react with nitrogen oxides (NO_x) and O_2.

The necessity of determining these compounds at low levels in complex matrices has stimulated a study [49] on the applicability of GC–SIOMS, based upon the findings of a surprisingly high sensitivity for trimethylamine (TMA) in SI. This compound yields much higher ionization efficiency in the SI mode than in the conventional EI mode.

The analytical procedure [49] employs the direct gas sample inlet without complicated sample preparation and selective ion monitoring (SIM). It saves time and minimizes the risk of sample contamination and sample loss during manipulation. Ambient air analyses were performed as follows. A 1-ml sample was injected into a gas chromatograph with a 1-ml glass syringe. The SI spectra of

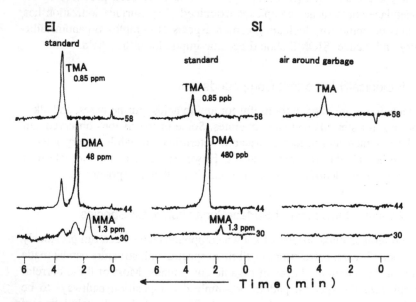

Figure 4 SIM chromatograms of aliphatic amines in a 1-ml air sample: spectrum of a 1-ml standard gas sample in EI mode (left), spectrum of a 1-ml standard gas sample in SI mode (center), and spectrum of an air sample from garbage site in the SI mode (right). All measuring ranges of ion currents are 1×10^{-10} A (full scale), except that of ion currents at m/z 58 in the SI mode, which are 5×10^{-9} A (full scale).

monomethylamine (MMA), dimethylamine (DMA), and TMA are dominated by only a few peaks [20], showing that the $[M - H]^+$ intensity is highest for these compounds. The mass spectrometer was set to monitor m/z 30 for MMA and TMA, m/z 44 for DMA, and m/z 58 for TMA. The effectiveness of the technique is demonstrated by a comparison of single ion chromatograms obtained with SI and EI of a known mixture of MMA, DMA, and TMA (Fig. 4). The much greater sensitivity of SI for TMA than that of EI is clearly demonstrated. The sensitivities of SI and EI for MMA are comparable. Other experiments with TMA gave a peak with a signal-to-noise ratio (S/N) of 6 for a 225-pg/L (0.8 ppb by volume) sample, which could not be detected in a single ion trace using EI. These results are consistent with previous observations of a very high sensitivity for TMA with a SI detector [50].

A 333-fold improvement in the detection limit was achieved by the use of the SI ion source. Estimated from recovery study, the relative standard deviation (RSD) of peak area of TMA was 6.8% at a concentration of 225 pg/L. The method offers a few advantages over the flame ionization detector, which is commonly used. Because no preconcentration is needed, the analysis time can be less than 10 minutes, which is much shorter than a GC method including a preconcentration procedure. A second advantage is a high degree of specificity due to the characteristics of SI. Therefore, the method may be useful in routine analysis for a large number of air samples.

5.3. Determination of Lidocaine in Serum

Lidocaine is used extensively to treat ventricular cardiac arrhythmias, especially in patients who have had cardiac surgery or sustained an acute myocardial infarction [51]. Among many methods for its determination, the use of capillary GC–EI-MS in SIM mode may be the best [52]. It is applied in many clinical studies.

Gas chromatography–surface-ionization organic mass spectrometry also seems to be a promising method for the detection of various drugs, such as lidocaine. This drug was selected because it can be expected that the amine radical moiety has the potential to be surface ionized efficiently [53] and the previous SI detection studies [50] showed promising results.

Figure 5 shows the mass spectra of lidocaine obtained by SI and EI along with total ion chromatograms (TICs) [54]. The samples were 10-ml drug-free serum extracts spiked with lidocaine in concentrations of 40 mg/ml and 400 mg/ml. Interestingly, the TIC in the SI mode shows no interference from the compounds associated with the blood. The lidocaine peaks in both TIC traces yield easily identified mass spectra, which are, as expected, almost identical to that obtained from the direct introduction of the pure lidocaine sample.

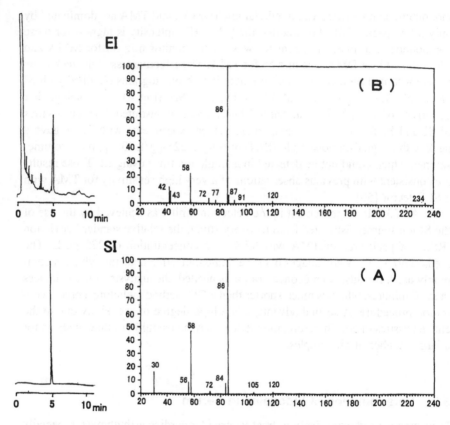

Figure 5 Comparison of TICs (left) and mass spectra (right) of lidocaine in a spiked serum sample obtained by (A) using the SI mode and (B) using the conventional EI mode. On the assumption that the extraction efficiency is 100%, the lidocaine peak of the TIC profile in the SI mode corresponds to 0.8 mg, while the peak in the EI mode corresponds to 8 mg. Mass spectral intensity is plotted as the normalized percentage scale.

Comparison with the EI mode reveals that GC–MS in SI mode does not give rise to peak broadening, tailing, and baseline drift in the TIC profiles, owing to its fast response characteristics. Thus, the SI is compatible with capillary column techniques.

Another important aspect of SIOMS is its high sensitivity, which is demonstrated by a comparison of the TIC traces obtained with the SI and EI of a spiked serum sample at different concentrations. If a spiked sample of 500-µl serum was extracted and a 10-µl injection of the extract was made, the lidocaine detection limit in serum was 42 ng/ml (S/N 3) from TIC profiles obtained with SI. Certainly, a SIM procedure for the m/z 86 base peak would allow the determination of much lower concentrations, i.e., at or below the clinically important concentrations.

The accuracy and precision of this assay were determined by measuring five different plasma samples spiked with lidocaine. The RSD was 2.5%. The good reproducibility of the method is due to the reliability of the GC–MS operational mode and to the minimal handling of samples. The recovery from five analyses ranged 101.0 ± 27%.

Hence, GC–SIOMS appears to have great potential for use in the routine biomedical analysis of drugs, many of which possess nitrogen heteroatoms. A preliminary study showed that a routine determination of the antidepressant drug imipramine in serum at concentrations as low as 1.5 ng/ml (S/N 3) can be reliably made.

Gas chromatography–surface-ionization organic mass spectrometry can be also used successfully for sensitive and selective detection of imipramine in serum [55]. In this respect, the GC–SIOMS method can result in new opportunities in the field of pharmacology.

5.4. Determination of s-Triazine Herbicides in Water

s-Triazine derivatives are important compounds in agriculture and industry because of their herbicidal properties. Triazine herbicides are some of the most widely applied pesticides in the United States and Europe. The most common member of this class, atrazine, was the most heavily used pesticide in recent years. The triazines represent a class of environmental chemicals that are continuously causing severe poisoning [56]. Analysis is often needed for low-level concentrations in the environmental media. Recent studies indicate a relationship between water levels of these pesticides and environmental response [57]. Water levels can be very low, even in fatal impact.

Some research was performed for s-triazines in the aqueous environments [58]. Figure 6 shows the GC–SIOMS results obtained by TIC and by SIM (m/z 213). For the TIC profile, a 10-µl river water extract was injected (concentrated 100-fold, spiked with the model compounds [see caption] at concentrations of 0.8 mg/ml each). The peaks yield easily identified mass spectra that are almost identical to those obtained from the direct introduction of the pure s-triazines. Interestingly, both chromatograms in the SI mode show little interference from other compounds in the river water.

Certainly, the SIM detection, using for instance the m/z 213 ion of simetryne, allows the determination of much lower concentrations. If a spiked sample of water was treated (again concentrated 100-fold) and a 10-µl injection of the

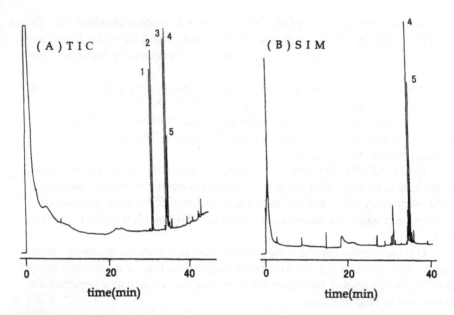

Figure 6 Gas chromatograms of a spiked river water sample obtained by full scan (left) and SIM (right). Peaks: (1) atrazine, (2) simazine, (3) prometryne, (4) simetryne, and (5) ametryne. On the assumption that the extraction efficiency is 100%, each peak of the profiles in the TIC corresponds to 0.8 mg, while the simetryne peak (4) in the SIM corresponds to 28 ng.

water extract was made, the simetryne detection limit in the water was calculated to be 22 pg/ml (S/N 3). This value is at or below that typically needed in the environmental analysis.

This is the first study on GC–SIOMS for the characterization of some pesticides as an alternative, or at least as complementary, to conventional EI-MS. It yielded a better understanding of the possibilities and limitations of the SI technique.

6. SUMMARY

The principal features of GC–SIOMS are:

1. The simplicity, specificity, and sensitivity of this method provided by detecting abundant ions formed upon SI on a refractory metal surface. This enables sensitive and selective analysis of organic compounds at low concentration levels in various media. Because of the high speci-

ficity, the matrix interferences are greatly reduced, making cleanup procedures usually not necessary. Owing to the increased sensitivity of the detector, a better detection capability is achieved.

2. GC–SIOMS allows identification of many organic compounds with certainty complementary to conventional EI.

3. GC–SIOMS may detect a large number of drug or pesticide metabolites under the conditions that an extensive SI mass spectral library can be established.

4. To fully exploit advantages of GC–SIOMS, an SI mass spectral library is also essential.

5. At present, the GC–SIOMS response is fairly unpredictable for a particular compound because not all IE values are available.

This chapter demonstrates successful examples of GC–SIOMS. It is certain that this method can be extended to analysis of organic compounds in many fields that can be ionized efficiently on the SI emitter. These compounds include alkylamines, diamines, amino alcohols, hydrazines, nitrosamines, 5-membered and 6-membered ring N,N-heterocyclic amines, polyaromatic hydrocarbons, terpenoids, organometallic compounds, amino acids, purine bases, alkaloids, and so on.

Systematic research on SIOMS is important for the interpretation of SI mass spectra and for the development and application of GC–SIOMS. Applications to liquid chromatography and supercritical fluid chromatography may well be possible. The resulting establishment of SIOMS will also lead to a better understanding of the substantial advantages of the SI method in the MS analysis of organic compounds. In addition, it will also provide information on the stability and the structure of charged organic compounds with complex composition and on the heterogeneous catalysis on a solid.

Among the many future prospects and developments in GC–SIOMS, hyperthermal (or superthermal) surface ionization (HSI), as described by Danon and Amirav [59], seems to be one of the most interesting topics. Their experimental setup employs scattering techniques involving 1 to 20 eV supersonic molecular beams [60] and various kinds of surfaces, showing that a single collision of fast molecules with a surface can produce highly efficient ionization. They demonstrated very efficient ionization upon high-energy impact of molecules on a solid, e.g., anthracene and N,N-diethylamine impinging on diamond. This research can be regarded as second-generation SI studies. A comparative study of SI and HSI was made [61] on some drug compounds in the thermal and hyperthermal energy ranges. Many features of HSI also stimulate the application of SI to the GC–MS field [62,63]. The concern may be focused on the fast GC–MS with a supersonic molecular beam. The interesting field of HSI has been excellently summarized by Amirav [64].

ACKNOWLEDGMENTS

This chapter embodies efforts with T. Kitai, H. Suzuki, M. Obuchi, M. Ogura, H. Gimba, H. Ishii, K. Kakizaki, K. Shouji, K. Inagaki, Y. Kurihara, and K. Hatanaka. Among others whose contributions are greatly acknowledged are Drs. H. Kishi, H. Arimoto, and Y. Mitsutsuka, and M. Kareev for helpful discussion, and M. Isobe and M. Kanaya for the manuscript preparation.

This work was supported by grants 09307008 and 09554052 from the Ministry of Education, Science and Culture of Japan and by grant 50 from the Japanese STA under the framework of Japan–Brazil cooperative scientific agreements.

REFERENCES

1. K.H. Kingdom and I. Langmuir, Phys. Rev., 21 (1923) 380.
2. I. Langmuir and K.H. Kingdom, Proc. Roy. Soc. A, 107 (1925) 61.
3. E.Ya. Zanderg and N.I. Ionov, Surface Ionization, 1971, Israel Program for Scientific Translations, Jerusalem, p. 354.
4. M. Kaminskey, Atomic and Ionic Impact Phenomena of Metal Surfaces. 1965. Springer-Verlag, New York, p. 402.
5. M.G. Ingram and R.J. Hayden, Handbook on Mass Spectrometry. Nuclear Science Series, Report No. 14, 1954, NRC, USA.
6. J. Koch, Z. Phys., 100 (1936) 669.
7. (a) G.H. Palmer, J. Nucl. Energy, 7 (1958) 1.
    (b) E.Ya. Zandberg and N.I. Ionov, Dokl. Akad. Nauk., 141 (1961) 139.
8. (a) E.Ya. Zandberg, U.Kh. Rasulev, and B.N. Schstrov., Dokl. Akad. Nauk., 172 (1967) 885.
    (b) E.Ya. Zandberg and U.Kh. Rasulev, Zh. Eksp. Teor. Khim., 7 (1971) 363.
9. (a) E.Ya. Zanderg and U.Kh. Rasulev, Russ. Chem. Rev., 51 (1982) 819.
    (b) U.Kh. Rasulev and E.Ya. Zanderg, Prog. Surf. Sci., 28 (1988) 181.
10. (a) T. Fujii, Eur. Mass Spectrom., 2 (1996) 91.
    (b) T. Fujii, Eur. Mass Spectrom., 2 (1996) 263.
11. T. Fujii and H. Arimoto, Am. Lab., August (1987) 54.
12. (a) W.E. Davis, Environ. Sci. Technol., 11 (1977) 587.
    (b) W.E. Davis, Environ. Sci. Technol., 11 (1977) 593.
13. T. Fujii, Int. J. Mass Spectrom. Ion Proc., 57 (1984) 63.
14. T. Fujii, J. Phys. Chem., 88 (1984) 5228.
15. T. Fujii, H. Suzuki, and M. Obuchi, J. Phys. Chem., 89 (1985) 4687.
16. T. Fujii and T. Kitai, Int. J. Mass Spectrom. Ion Process., 71 (1986) 129.
17. (a) R. Walder and J.L. Franklin, Int. J. Mass Spectrom, Ion Phys., 36 (1980) 85.
    (b) H.M. Rosenstock, K. Draxl, B.W. Steiner, and J.T. Herron, J. Phys. Chem. Ref. Data, 6 (1977) 783.
18. T. Fujii and H. Jimba, Int. J. Mass Spectrom. Ion Process., 79 (1987) 221.
19. T. Fujii, J. Chem. Phys., 87 (1987) 2321.

20. T. Fujii, Int. J. Mass Spectrom. Ion Process., 87 (1989) 51.
21. (a) M.L. Shaw and N.P. Carleton, J. Chem. Phys., 44 (1966) 3387.
    (b) D.L. Fehrs and R.E. Stickney, Surf. Sci., 17 (1969) 298.
22. (a) R.K.B. Helbing, Y.F. Hsieh, and V. Pol, J. Chem. Phys., 63 (1975) 5058.
    (b) Y.F. Hsieh, R.K.B. Helbing, and V. Pol, Can. J. Phys., 55 (1977) 2166.
    (c) R.J. Gordon, D.S.Y. Hsu, Y.T. Lee, and D.R. Hershbach, J. Chem. Phys., 63 (1975) 5056.
23. T. Fujii and H. Arimoto, Anal. Chem., 57 (1985) 2625.
24. R.G. Wilson, J. Appl. Phys., 44 (1973) 2130.
25. J. Pelletier, D. Gervais, and C. Pomot, J. Appl. Phys., 55 (1984) 994.
26. B. Rasser, D.I.C. Pearson, and M. Remy, Rev. Sci. Instrum., 51 (1980) 474.
27. E.Ya. Zandberg, U.Kh. Rasulev, and Sh.M. Khalikov, Soc. Phys. Tech. Phys., 21 (1976) 483.
28. W. Greaves and R.E. Stickney, Surf. Sci., 11 (1968) 395.
29. D.A. Chatfield and M. Ajami, Int. J. Mass Spectrom. Ion Process., 77 (1987) 241.
30. (a) B.H. Eckstrein and R. Forman, J. Appl. Phys., 32 (1962) 82.
    (b) H. Kishi and T. Fujii, Chem. Phys., 192 (1995) 387.
31. B. Weber and A. Cassuto, Surf. Sci., 36 (1973) 105.
32. A. Persky, E.F. Greene, and A. Kuppermann, J. Chem. Phys., 49 (1968) 2347.
33. T. Fujii, H. Ishii, and H. Jimba, Int. J. Mass Spectrom. Ion Process., 93 (1989) 79.
34. T. Fujii, K. Kakizaki, and Y. Mitsutsuka, Int. J. Mass Spectrom. Ion Process., 104 (1991) 129.
35. T. Fujii and H. Ishii, Chem. Phys. Lett., 163 (1989) 69.
36. T. Fujii, K. Kakizaki, and H. Ishii, Chem. Phys., 163 (1992) 413.
37. T. Fujii, K. Kakizaki, and H. Ishii, Chem. Phys. 147 (1990) 213.
38. T. Fujii, K. Kakizaki, and H. Ishii, J. Organometallic Chem., 426 (1992) 361.
39. T. Fujii and H. Ishii, J. Organometallic Chem., 391 (1990) 147.
40. T. Fujii, Y. Inagaki, and Y. Mitsutsuka, Int. J. Mass Spectrom. Ion Process., 124 (1993) 45.
41. F.W. Mclafferty and D.B. Stauffer, The Wiley/NBS Registry of Mass Spectral Data, 1988, J. Wiley, New York.
42. S. Fuselli, C. Cerquiglini, and E. Chiacchierini, Chim. Ind., 60 (1978) 711.
43. R.E. Brubaker, H.J. Muranko, D.B.Smith, G.J. Beck, and G.J. Scovel, J. Occup. Med. 21 (1979) 688.
44. K. Kuwata, E. Akiyama, Y. Yamagali, H. Yamagali, and Y. Kuge, Anal. Chem., 55 (1983) 2199.
45. S. Fuselli, G. Benedetti, and R. Mastrangei, Atoms. Environ., 16 (1982) 2943.
46. A.R. Mosier, C.E. Andre, and F.G. Viets, Environ. Sci. Technol., 7 (1973) 642.
47. H.R. Van Langenhove, F.A. Van Wassenhove, J.K. Coppin, M.R. Van Acker, and W.M. Schamp, Environ. Sci. Technol., 16 (1982) 883.
48. D.H. Fine, D.P. Rounbehler, E. Sawicki, and K. Krost, Environ. Sci. Technol., 11 (1977) 577.
49. T. Fujii and T. Kitai, Anal. Chem., 59 (1987) 379.
50. T. Fujii and H. Arimoto, in: Herbert H. Hill, ed., Surface Ionisation Detector, Detectors for Capillary Chromatography (Chemical Analysis Series), 1993, J. Wiley, New York, pp. 169–191.

51. C.E. Hignite, C. Tschanz, J. Steiner, D.H. Huffman, and D.L. Azarnoff, J. Chromatogr., 161 (1978) 243.

52. H. Heusler, J. Chromatogr. 340 (1985) 273.

53. T. Fujii, H. Jimba, and H. Arimoto, Anal. Chem., 62 (1990) 107.

54. T. Fujii, Y. Kurihara, H. Arimoto, and Y. Mitsutsuka, Anal. Chem., 66 (1994) 1884.

55. T. Fujii, Y. Kurihara, H. Arimoto, and Y. Mitsutsuka, Anal. Chem., 66 (1994) 1884.

56. U.S. Department of Agriculture (USDA), National Agricultural Statistics Service, Agricultural Chemical Usage 1992, Field Crops Summary, 1993, USDA, Washington, DC.

57. U.S. Environmental Protection Agency (EPA), National Survey of Pesticides in Drinking Water Wells, Phase II Report, EPA 570/9-91-020, 1992, EPA, Springfield, VA.

58. T. Fujii, K. Hatanaka, H. Arimoto, and Y. Mitsutsuka, J. Mass Spectrom., 32 (1997) 408.

59. (a) A. Danon and A. Amirav, J. Chem. Phys., 86 (1987) 4708.
    (b) A. Danon and A. Amirav, J. Phys. Chem., 93 (1989) 5549.

60. (a) J.B. Anderson, R.P. Andres, and J.B. Fenn, Adv. Chem. Phys., 10 (1965) 275.
    (b) R. Campargue, J. Phys. Chem., 88 (1984) 4466.
    (c) A.P. Miller, in: G. Scoles, ed., Atomic and Molecular Beam Methods, vol. 1, 1988, Oxford Univ. Pr., New York, pp. 14–53.

61. S. Dagan, A. Amirav, and T. Fujii, Int. J. Mass Spectrom. Ion Process., 151 (1995) 159.

62. S. Dagan and A. Amirav, Int. J. Mass Spectrom. Ion Process., 133 (1994) 187.

63. H. Kishi and T. Fujii, J. Phys. Chem., 99 (1995) 11153.

64. A. Amirav, Org. Mass Spectrom., 26 (1991) 1.

3

Gas Chromatography–Mass Spectrometry in the Petroleum Industry

C. S. Hsu
ExxonMobil Research and Engineering Company, Annandale, New Jersey

D. Drinkwater
BASF Corporation, Princeton, New Jersey

1. INTRODUCTION

Although the basic principle of mass spectrometry (MS) was discovered as early as 1910 by Sir J. J. Thomson, it was not until the end of World War II that MS was first developed for analyses of gas and hydrocarbon mixtures [1]. Almost all of the mass spectrometers at that time were made by researchers or specialists. The first commercial mass spectrometer manufactured by Consolidated Engineering Corporation (CEC) was delivered to the Atlantic Refining Company in 1946 for analysis of hydrocarbon fractions in gasoline boiling range [2]. Since then, the petroleum industry has pioneered the use of MS in chemical research. Many advances in MS were driven by the needs of the petroleum industry for analyzing components in complex mixtures.

In early MS analysis, the whole sample, either a single pure component or a complex mixture, was introduced into the ion source of the mass spectrometer without prior chromatographic separations. For mixtures, the spectra can be very complex. Since the spectrum is a linear superposition of the mass spectra of constituent pure components, very accurate mixture analysis is possible. In fact, for two decades the main application of MS was quantitative analysis of light hydrocarbon mixtures, often with accuracy of $\pm 1\%$ absolute [3]. The components present are commonly grouped into a predetermined number of compound types,

55

with their distributions determined by mathematical deconvolution of the characteristic ions of each type using matrix algebra [4]. However, due to the presence of isobaric ions that would interfere with the measurement, fractionation into saturate and aromatic fractions prior to MS analysis was usually necessary.

Double-focusing high resolution MS was later introduced to provide accurate mass measurement of isobaric ions from saturates and aromatics. With accurate mass measurement, elemental composition of individual ions can be determined. Hence, the distributions of compound types and carbon number within each compound type can be obtained without the need of fractionation into saturates and aromatics. However, this method cannot be used to differentiate isomers with identical chemical formula. Since many of these isomers have different gas chromatographic retention times, coupling MS with gas chromatography (GC) would become a logical option for improving mixture analysis through additional dimensions [5].

Early GC–MS applications utilized magnetic sector and linear time-of-flight (TOF) mass spectrometers [6–8]. Both types of mass spectrometers had limitations. A magnetic sector instrument needed to be scanned rather slowly to avoid hysteresis of the magnet. Although early TOF instruments could achieve much faster scans, they had low mass range and poor resolution. Hence, it is not surprising that GC–MS was later dominated by quadrupole mass spectrometers.

A quadrupole mass spectrometer has much more tolerance to pressure variation than a sector mass spectrometer because a much lower voltage (20 to 100 V) is needed to extract ions out of the ion source, compared with 6 to 8 kV needed to accelerate ions in a sector instrument. Sensitivity is also increased due to large ion flux through the quadrupole analyzer with no slits used in the ion path. In a quadrupole mass spectrometer, the mass is a linear function of scan time by scanning direct-current (DC) voltage and keeping radio-frequency (RF) voltage constant, while in a sector or a TOF mass spectrometer, the mass is not linear to scan or ion-arrival time. The linear mass scale allows a data system to acquire and analyze data more easily. Great advances in GC–MS were made possible by computerization to improve instrument control and data acquisition and to enable rapid data analysis [9].

A combined GC–MS is composed of a gas chromatograph, a mass spectrometer, and an interface [10,11]. Each of them can have many different variations. Almost all of the columns used in the GC analysis can be used in GC–MS. In early GC–MS applications, glass GC columns packed with various solid supports coated with liquid phase were used. The effluent from the column was introduced through a jet or a membrane separator to the mass spectrometer ion source under vacuum (in the range of 10^{-5} to 10^{-6} torr). Capillary open tubular columns, first with stainless steel then with fused silica, were developed to improve resolution of GC separation. With a capillary column, less vapor is introduced into the mass spectrometer source, thus eliminating the need of an interface to divert a large volume of carrier gas. The end of a capillary column can be inserted inside the ion source near or under the electron beam for electron ionization (EI) or chemical ionization (CI). Typical mass spectrometers are quadrupole, ion trap, magnetic sector types, although Fourier-transform ion-cyclotron resonance (FT-ICR) and TOF have also been used. The use of TOF-MS is expected to increase in the future due to recent improvements in mass range and resolution using a reflectron. EI and CI are commonly used ionization methods, although field ionization (FI) has also been used.

Petroleum fractions and products may contain thousands of components [12]. The simplest fraction is natural gas that contains mainly methane. Depending on its source, natural gas can also contain carbon dioxide, hydrogen sulfide, and other light hydrocarbons (ethane, propane, and butanes). Other fractions with increasing boiling points are naphtha, middle distillates, gas oils, and residua. These fractions can be further refined or blended to produce gasolines, diesel and jet fuels, kerosene, lubricant oils, heating oils, and asphalt. Some of these fractions are converted into basic chemicals, such as solvents, olefins, aromatics, etc. Light olefins, such as ethylene, propylene, butenes, butadiene, and isoprene, are important feedstocks for synthetic polymers. The fuels and lubricant oils are normally additized with small amounts of additives to improve stability and performance. The heaviest fractions are cracked thermally or catalytically into more valuable, lower-boiling products and cokes, or used as low-value fuels and paving material (asphalts).

Gas chromatography–mass spectrometry is suitable for analyzing fractions from natural gas to vacuum gas oils with boiling points up to 650°C. For analyzing whole crude oil, preasphaltening prior to GC–MS analysis can avoid or reduce contamination at the GC injector. Due to extreme complexity in molecular distribution in middle distillates and gas oils, liquid chromatographic separations according to the polarity of the molecules would facilitate the subsequent GC–MS analysis. Nonvolatile fractions, such as residua, can be analyzed by pyrolysis GC–MS, where samples are pyrolyzed at temperatures greater than 450°C, usually between 600 and 800°C, for subsequent GC–MS analysis.

Gas chromatography–mass spectrometry provides information of compounds by retention time from GC and mass spectral patterns from MS. Retention time is usually expressed by retention index that relates the retention time to a homologous series of reference compounds. Combined use of retention indices and mass spectra allows compound identification with high specificity. Overlapping GC compounds can be differentiated by the difference in their mass spectra. On the other hand, isomers with similar or identical MS can be differentiated by the difference in GC retention time. Thus, GC and MS complement each other for mixture analysis.

2. CHALLENGES IN THE ANALYSIS OF PETROLEUM AND ITS DERIVATIVES

Crude oil is an enormously complex material, with boiling and mass ranges from methane all the way up to carbon particles. Since GC and MS each offer the greatest separating power available to the analytical chemist, the GC–MS combination is an obvious choice for these very complex materials. However, some students might confuse mass spectrometry with some sort of spectroscopic technique because most GC–MS performed in industry and academia today use very powerful GC separation to provide almost pure components to the mass spectrometer. Then, qualitative or quantitative information is sought for that one GC peak, in much the same way GC–IR might be used to get the infrared spectrum of various individual eluting analytes. Such idealities are far from possible in the field of petroleum analysis, where the GC, even with the highest resolution capillary columns available, cannot baseline separate petroleum components, except for the lightest gases. Fortunately, the mass spectrometer is a powerful separation tool in its own right. Even a large number of components that might be coeluting under a given GC peak can be separated further for qualitative and/or quantitative analysis of the particular compounds or compound classes of interest.

Petroleum products produced in a refinery from crude oil suffer from the added complication that each is far more structurally diverse than the parent crude oil while having a much narrower boiling distribution. Since the highest number of theoretical plates are available from nonpolar GC columns, which separate largely by boiling point, the mass spectrometer takes on an increasing amount of the separating work as the boiling range of the sample decreases. In many cases "classical" GC–MS methodology is insufficient to characterize these products, and it is necessary to do derivatization or preseparation by liquid chromatography (LC), supercritical fluid chromatography (SFC), supercritical fluid extraction (SFE), or liquid-liquid extraction (LLE) prior to GC–MS analysis. Another possibility is to inject the total sample on the GC–MS and to use high-resolution mass spectrometry (HRMS), mass spectrometry–mass spectrometry (MS–MS, or tandem mass spectrometry), or selective ionization methodology to gain additional selectivity beyond "classical" high-resolution GC with low-resolution electron-ionization MS (HRGC–LR-EI-MS). These methods of precolumn or postcolumn selectivity of specific compound classes are discussed in a later section of this chapter.

In the petroleum industry, the gasses and liquids recovered from nature are typically referred to as "upstream," while the refinery streams within the refinery and resulting petroleum and petrochemical products from the refinery are typically referred to as "downstream." This terminology will be used through the rest of this chapter, which includes separate sections on upstream- and downstream-specific GC–MS applications.

Upstream analysis is primarily concerned with differences among various crude oils, and what can be learned about the source biomaterial, its geographical distribution, and the transformation processes it went through to turn bio-organisms into kerogen, bitumen, oil, and gas. Petroleum geochemistry relies almost totally on the material's characterization data from techniques such as GC–MS to characterize the oil fields, their history, and their future.

Downstream analysis is usually performed for totally different reasons than upstream analysis and is often done by different laboratories or scientists than the upstream analyses for this reason. The line between petroleum products and petrochemical products has become fuzzy since the time long ago when petroleum products were merely distillation cuts of crude oil. A modern refinery has become far more complex with less gasoline, for example, coming directly from distillation of crude oil, more from numerous catalytic processing of desirable molecules from less-useful fractions of the crude oil. All of these streams in the refinery must be analyzed and monitored in order to ensure an economical refining process. Several refinery streams are blended to produce petroleum products and petrochemicals according to specifications. Products leaving the refinery are often subjected to compliance specification testing. The last major challenge to the downstream GC–MS laboratory is the analysis of petroleum product constituents in the environment, which entails soil, water, and air analyses.

Gas chromatography–mass spectrometry has been applied in almost all phases of the petroleum industry, including upstream exploration and production, downstream refining processes and products, petrochemicals, and environmental compliance. The following sections describe some of the GC–MS applications in each petroleum area.

3. GAS CHROMATOGRAPHY–MASS SPECTROMETRY IN PETROLEUM UPSTREAM RESEARCH

The success of petroleum exploration depends on the effectiveness of sedimentary basin assessment that involves many different geophysical and geochemical disciplines, including seismology, sedimentology, petrology, etc. Petroleum geochemistry applies chemical principles to the study of the origin, migration, accumulation, and alteration of petroleum, and uses this knowledge in exploration and recovery of oil, gas, and related bitumens.

3.1. Biomarker Analysis

Gas chromatography–mass spectrometry has been extensively used in petroleum organic geochemistry. In fact, the rapid development of petroleum geochemistry and its applications to oil potential assessment in petroleum exploration since the

1960s has largely been due to the development of GC–MS. Extracts from geological samples such as crude oils, source rocks, coals, sediments, or shales are complex mixtures of organic compounds. Some of these compounds are biomarkers that can be used to relate the origin and depositional environment of petroleum and coal [13]. Biomarkers are hydrocarbon molecules that retain basic carbon skeletons of deeply buried, ancient, living organisms that underwent physicochemical transformation (metagenesis, diagenesis, and maturation) during the geological history of the sedimentary basin. Hence, they carry important information on the sources and the fate of biological materials that are ultimately transformed into petroleum and other fossil fuels. The distribution of these molecules in source rocks and oils can provide information about sources of organic matter and depositional environments of sedimentary basins, as well as the level of maturity, age, migration, and alteration (such as biodegradation) of the petroleum [14]. The distributions of biomarkers have also been extensively used as "fingerprints" for oil/oil and oil/source rock correlation [15,16].

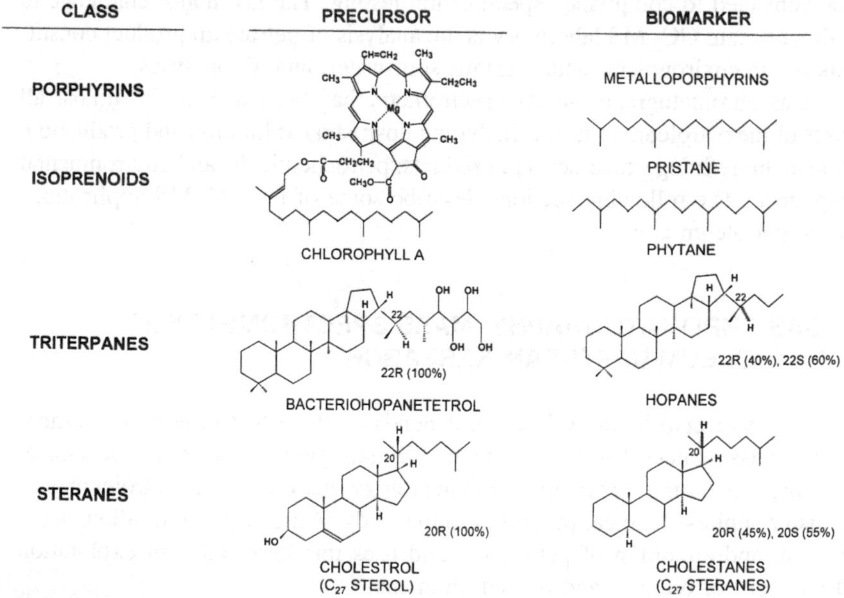

Figure 1   Typical biomarker compounds and their precursors. Chlorophyll A undergoes physicochemical transformation to form metalloporphyrins (mainly nickel and vanadyl), pristane, and phytane. The R-configuration of bacteriohopanetetrol at C-22 is converted to hopanes with 40% R- and 60% S-configurations in mature oils. Similarly, the R-configuration of cholesterol at C-20 reaches equilibrium values of 40% R- and 60% S-configurations in mature oils.

Biomarkers include compounds such as isoprenoids, triterpanes, steranes, aromatized steranes, sulfur-containing steranes, and porphyrins [17,18]. They are distributed in the saturate, aromatic, and NSO (polar) fractions of oils and rock extracts. Figure 1 illustrates a few typical biomarker compounds and their precursors. This product/precursor relationship is the foundation of biomarker applications in geochemistry. For example, steranes are derived from sterols that are widespread and abundant in eukaryotic organisms: animals, plants, fungi, and algae. In general, sediments derived from marine sources contain high concentrations of C_{27} sterols and those from terrigenous (terrestrial) sources contain high concentrations of C_{29} sterols. Thus, the relative abundance of C_{27} and C_{29} steranes gives some indication of whether the oils originate from sources enriched in marine or terrigenous organic matter. Diagenesis and maturation of the precursors lead to defunctionization and formation of thermodynamically more stable isomers. For example, the R-configuration of bacteriohopanetetrol at C-22 is converted to hopanes with 40% R- and 60% S-configurations in mature oils. Similarly, the R-configuration of cholesterol at C-20 reaches equilibrium values of 45% R- and 55% S-configurations in mature oils.

Mackenzie et al. [19] described the conversion of sterols to various steranes via the formation of sterenes. In general, sterols are biosynthesized with stereo-

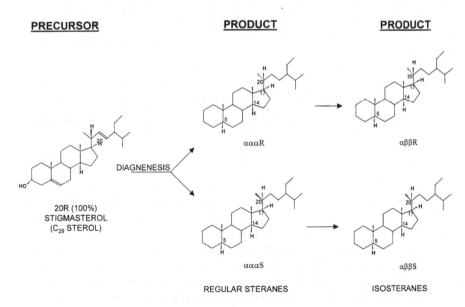

Figure 2 Conversion of sterols under thermal stress. With increasing thermal stress, steranes possessing the immature βααR (not shown) and αααR configurations are thought to be isomerized to αααS as well as αββS and αββR epimers.

chemical configurations of 5β(H),14α(H),17α(H),20R and 5α(H),14α(H), 17α(H),20R (abbreviated as βααR and αααR, respectively). With increasing thermal stress steranes possessing these immature βααR and αααR configurations are thought to be isomerized to αααS as well as αββS and αββR epimers, as shown in Figure 2 [20]. In depositional environments rich in detrital clays, the rearrangement of methyl groups from the C-10 and C-13 positions to the C-5 and C-14 positions is thought to occur on clay mineral surfaces, yielding 13α(H),17β(H) (abbreviated as αβ) and 13β(H),17α(H) (or βα) isomers. Steranes with ααα configurations are referred to as regular steranes, αββ as isosteranes, and αβ and βα as diasteranes. The conversion of sterane isomers constitutes the chemical basis for using the relative abundance of these isomers as maturity parameters in the GC–MS measurement. Further dehydrogenation of steranes yields aromatized steranes.

3.2. Gas Chromatography–Mass Spectrometry–Mass Spectrometry Parent Analysis of Steranes

Other than n-alkanes, pristane, phytane, and other isoprenoid alkanes that are present in abundance in unaltered oils and easily recognizable by GC and GC–MS analyses, most of the biomarker components are present in trace amounts. Among these, naphthenic hydrocarbons, including tricyclic terpanes, pentacyclic triterpanes, and steranes, have been studied most extensively because they carry more specific information on source and depositional environment than normal and isoprenoid alkanes [21].

Figure 3 is a GC–MS reconstructed ion chromatogram (RIC) that is equivalent to GC flame-ionization trace of a typical crude oil. It illustrates that triterpanes and steranes cannot be detected by GC alone due to low abundance and interference of coeluting components. Through the use of mass chromatograms of characteristic fragment ions at 191 Da for triterpanes and 217 Da for steranes from a full-scan GC–MS analysis, triterpanes and steranes can be found to elute between n-C_{22} and n-C_{36} on a nonpolar methyl silicone column. However, in order to enhance the GC–MS detection of these components, selective ion monitoring (SIM) technique for only the ions of interest can be used. Figure 4 shows the increase of the signal-to-noise ratio when only the 191- and 217-Da ions are monitored for triterpanes and steranes, respectively.

Since triterpanes (hopanes) are prevalent in the swamp/peat environment due to bacterial decay of high plant material and steranes are mostly generated by algae, the relative amounts of triterpanes and steranes have been used for estimating the organic matter type and depositional environment [22]. Figure 4 also illustrates the difficulties of measuring distributions of steranes by GC–MS. While most triterpanes are well resolved in the 191-Da mass chromatogram,

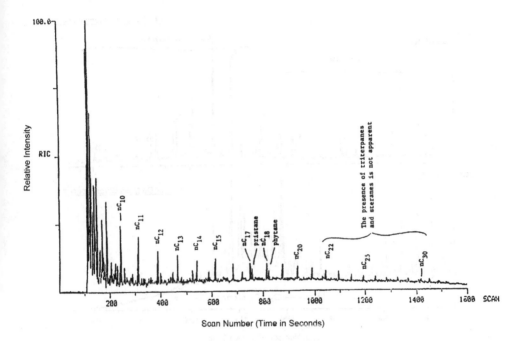

Figure 3 Gas chromatography–mass spectrometric reconstructed ion chromatogram (RIC) of a typical crude oil. A RIC is equivalent to a GC flame ionization detection trace. The presence of triterpanes and steranes eluting between n-C_{20} and n-C_{35} are not evident due to low abundance and interference of coeluting compounds. Conditions: Finnigan (San Jose, CA), 9610 gas chromatograph, Supelco (Bellefonte, PA), 30 m × 0.25 mm ID (0.1-μm film thickness) DB-1 fused silica capillary column, temperature programmed from 120 to 310°C at 8°C/min, 1-μl injection at 50:1 split. Mass spectrometer: Finnigan TSQ-46B, 70-eV electron ionization, full-scan mode from 35 to 500 Da in 1 sec cycle time.

many of the isomeric steranes coelute and interfere with each other. This necessitates compromise when steranes are measured by GC–MS.

These coelution problems can be largely resolved by coupling MS–MS with GC, that is, GC–MS–MS. The MS–MS can be in a form of two mass analyzers in tandem that monitor precursor ions and fragment ions in respective analyzers. Other forms of MS–MS include metastable monitoring performed on a double-focusing sector mass spectrometer using specific scanning techniques [23,24]. Most of the GC–MS–MS applications for biomarkers utilize precursor (or parent) scans where the precursor ions of selected characteristic fragment ions are monitored. Tandem triple-stage quadrupole (TSQ) mass spectrometers are most suitable for this task. In a TSQ mass spectrometer, the first quadrupole is

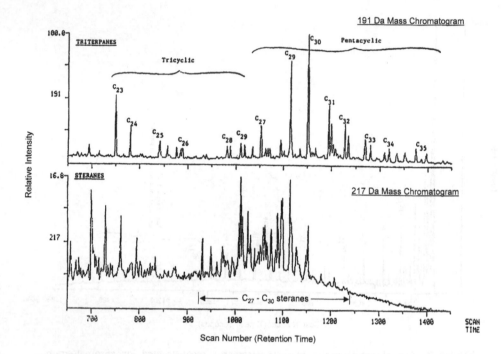

Figure 4 The 191- and 217-Da mass chromatograms from SIM GC–MS analysis. Gas chromatography–mass spectrometry conditions are the same as Figure 3 except the acquisition being in a SIM mode rather than a full-scan mode; residence time per mass, 100 msec.

used as the first stage of the mass analyzer (MS-1) to scan or focus precursor (parent) ions, while the third quadrupole is used as the second stage of the mass analyzer (MS-2) to scan or focus product (daughter) ions. Only radio-frequency (RF) voltage is applied on the second-stage quadrupole for collimating ions. The second-stage quadrupole is often used as a collision region to enhance fragmentation of precursor ions. A TSQ mass spectrometer can perform precursor (parent), product (daughter), and neutral scans [25]. Ion trap and FT-ICR mass spectrometers, on the other hand, can only be used for product (daughter) scans.

Although daughter scans are most commonly used in the MS–MS applications, parent scans yield the most interference-free biomarker distributions. In a parent scan mode, second-stage MS is used as a filter to allow selected ions characteristic to biomarkers to fly through while first-stage MS is scanned to record the molecular ions that produce the selected ions. Only biomarker compounds that can yield both the molecular and characteristic fragment ions are detected while other coeluting nonbiomarker compounds are filtered out. Thus, GC–MS–MS provides interference-free measurement of biomarker compounds, resulting in baseline resolution of the chromatographic peaks.

The advantage of GC–MS–MS lies in the ability to distinguish among homologous series of steranes by linking their molecular ions with their characteristic fragment ion at 217 Da. Figure 5 shows typical GC–MS–MS chromatograms for steranes. The top four traces (A through D) are selected GC–MS–MS measurements between the molecular ions and the 217-Da ion for the C_{27} through C_{30} steranes, respectively. The bottom trace (E) is the integration of traces A through D, which would be similar to the 217-Da mass chromatogram from the GC–MS analysis, as shown in Figure 5. Thus, the GC–MS–MS technique can determine distributions of diasteranes, isosteranes, and regular steranes for each carbon homologue without any interference due to coelution. For example, Figure 5 shows that the two C_{27} $\alpha\beta\beta$ steranes ((20R)-5α(H),14β(H),17β(H)-cholestane and (20S)-5α(H),14β(H),17β(H)-cholestane) actually coelute with the (20R)-

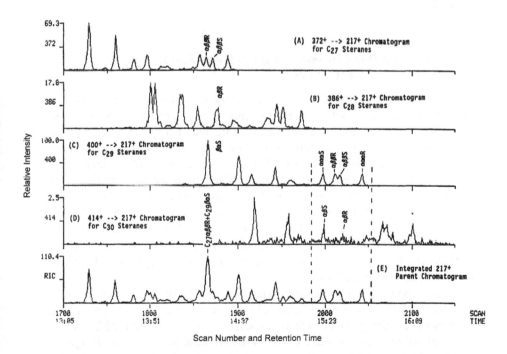

Figure 5 Gas chromatography–mass spectrometry–mass spectrometric chromatograms of steranes in a mature oil. Conditions are the same as Figure 3 except the acquisition being in parent scan mode rather than a full-scan mode; residence time per parent mass, 100 msec.

24-methyl-13α(H),17β(H)-diacholestane (C_{28} $\alpha\beta$R) and (20S)-24-ethyl-13β(H), 17α(H)-diacholestane (C_{29} $\beta\alpha$S). For the (20S)-24-ethyl-5α(H),14α(H),17α(H)-cholestane (C_{29} $\alpha\alpha\alpha$) and (20S)-24-ethyl-5α(H),14β(H),17β(H)-cholestane (C_{29} $\alpha\beta\beta$), the only interference is the coelution of the C_{30} $\alpha\beta$ steranes, (20S)-24-propyl-13α(H),17β(H)-diacholestane and (20R)-24-propyl-13α(H),17β(H)-diacholestane, that are usually present in trace amounts [28].

Steranes are routinely employed in a variety of geochemical applications. They are used to evaluate thermal maturity of oils and source rocks from the abundance of $\alpha\alpha\alpha$S relative to $\alpha\alpha\alpha$R, to estimate organic matter inputs (marine versus terrigenous) from the ratio of C_{27}/C_{29} steranes, and to assess the geological age of the source rock from the ratio of C_{28}/C_{29} steranes in the oil generated [14,26,27]. The C_{30} steranes, shown in the $414^+ \rightarrow 217^+$ GC–MS–MS chromatograms of Figure 5, are unambiguous indicators of contributions from marine-derived organic matter [29]. Their presence in oils and source rocks cannot be unambiguously determined by GC–MS due to severe interference from C_{28} and C_{29} steranes. Hence, GC–MS–MS is the only means to accurately measure this valuable sterane marker.

3.3. Pyrolysis Gas Chromatography–Mass Spectrometry for Kerogens and Coals

Source rocks, coals, shales, and other sedimentary rocks contain solvent-insoluble and nonvolatile macromolecular constituents called kerogens [30]. Kerogen, which represents the most abundant form of organic deposit (accounts for over 90% total organic carbon) on earth, is the principal precursor of petroleum and natural gas [31,32]. Flash pyrolysis GC–MS has been used for the analysis of kerogens. Prior to analysis, kerogens are isolated from the associated mineral matrix and solvent-soluble organic matter by hydrochloric acid or hydrofluoric acid (HCl/HF) digestion and ultrasonic extraction with methylene chloride (CH_2Cl_2) and methanol (CH_3OH), as suggested by Eglingon and Douglas [33]. In an early application, Gallegos [34] applied flash pyrolysis (rapid thermal decomposition) GC–MS to studies of pyrolytic release of terpanes and steranes from kerogen of Green River shale.

Curie point pyrolysis GC–MS has been applied to study the differences between pyrolysates of different macerol types with the same coal rank and the differences between pyrolysates from the same macerol types with different coal rank [35]. Among the coal macerols in Yorkshire Coal Basin (United Kingdom) of Upper-Carboniferous age, the distribution patterns of benzenes and naphthalenes in vitrinites of higher rank are very similar to those in inertinites of lower rank and exinites. A decrease of phenolic compound as a function of increasing percentage of carbon (increasing rank) is observed.

3.4. Geochemical Applications of Gas Chromatography– Isotope-Ratio Mass Spectrometry (GC–IRMS)

Another aspect of organic geochemistry important to petroleum exploration is the use of carbon, hydrogen, and sulfur isotopic compositions for helping in the determination of the origin of individual oil and gas accumulations and their genetic relationship to other accumulations. The isotopic analyses are done on individual compounds and also on bulk fractions. The isotopic compositions of individual compounds can be used for tracing the origin of individual compounds in petroleum as well as for practical applications of correlating one geological occurrence of petroleum with another. Interpretation of the isotopic data is normally done in conjunction with additional analyses including GC, GC–MS, and GC–MS–MS.

The carbon isotopic composition of individual compounds in gases, oils, condensates, and source rock extracts is measured by GC–combustion–isotope-ratio MS (GC–C–IRMS) or isotope-ratio-monitoring GC–MS (irm-GC–MS). This is a specifically designed capillary GC–MS system in which hydrocarbon components from the GC effluent are combusted to carbon dioxide (CO_2) and water (H_2O). The CO_2 of combustion is continuously analyzed on-line by a triple-collector isotope-ratio mass spectrometer to provide precise $^{13}C/^{12}C$ ratio measurements of the individual components [36–38]. The entire combustion, water removal, and isotope ratio measurements are made while maintaining GC resolution. By convention, the $^{13}C/^{12}C$ ratio of the components is reported as $\delta^{13}C$ in parts per thousand or per mil (‰) according to the following equation:

$$\delta^{13}C \ (‰) = \{[(^{13}C/^{12}C)_{sample}/(^{13}C/^{12}C)_{standard}] - 1\} \times 1000 \tag{1}$$

where the standard is an international reference (PDB, or Pee Dee Belemnite) [39].

Some examples of petroleum applications of GC–IRMS include:

1. Recognition of naturally occurring mixtures of reservoired oils and condensates from multiple geological sources by comparison of the $\delta^{13}C$ composition of individual light hydrocarbon components for defining petroleum systems [40,41].
2. Comparison and correlation of low-molecular-weight condensates from gas with oils [42]. Individual compounds in condensates are found to exhibit a wide range of isotopic compositions. This wide range provides a new dimension for interpreting the origin of the compounds, beyond what can be done from compositions alone.
3. $\delta^{13}C$ analysis of individual gas components for determining the origin of gases and the genetic relationship of one accumulation to another [43,44]. The $\delta^{13}C$ values of the individual gas components are found

to vary significantly and systematically with the $\delta^{13}C$ in an increasing order of $C_1 < C_2 < C_3 < n\text{-}C_4$.

4. Pyrolysis GC–IRMS of sedimentary kerogens and natural biopolymers to establish the relationship between the isotopic signature of pyrolysis products and their parent macromolecules [45,46].

5. Tracing the origin of individual compounds in crude oils that have distinctive $\delta^{13}C$ values [47,48].

4. GAS CHROMATOGRAPHY–MASS SPECTROMETRY FOR DOWNSTREAM REFINERY STREAMS AND PRODUCTS

Most petroleum upstream applications for GC–MS are for analyzing whole oils, while most downstream applications are for analyzing distillation fractions, refining streams, and products in the marketplace. In the downstream applications, knowledge of the composition of petroleum and its fractions allows the refiner to optimize conversion of raw petroleum into high-value products. It is important to understand the chemical transformations that occur in a refinery, and the terminology of petroleum products.

Initially, petroleum is distilled into fractions up to 350°C (650°F) under atmospheric pressure into gas, naphtha, middle distillates, gas oils, and residua. The atmospheric residua can be further distilled up to 350°C under vacuum to produce vacuum gas oils, lubricant oils, and vacuum residua. A modern refinery is no longer just a big distillation column that sells various boiling fractions to different consumer markets. While distillation is usually the first step, it is important to understand that a refinery processes 100,000 to 500,000 barrels of oil a day, and must turn every barrel of that oil into something that can be sold economically in the marketplace. For this reason, many of the crude oil fractions (''streams'') undergo catalytic transformation into more valuable streams and then are blended to produce petroleum products or petrochemicals that provide the highest value to the final end use.

Modern refineries use a sophisticated combination of heat, catalyst, and hydrogen. The gas oils and residua from vacuum distillation are further processed by cracking, either thermally or catalytically (such as fluid catalytic cracking, or FCC), and coking to break large molecules into lighter compounds, and produce highly olefinic streams [12]. This is in sharp contrast to crude oil, which has almost no olefinic content. Hydrotreating catalysts reduce olefins and aromatics and remove heteroatom-containing compounds to produce environmentally acceptable products and to avoid poisoning catalysts downstream. Isomerization and reforming catalysts rearrange molecules to those having higher value, e.g., gasolines of high octane number and high energy content. These refinery streams are blended into products, such as gasoline, diesel and jet fuels, lubricant oils, etc. Additives may be added into these products to improve their stability and performance. All together, the catalysts that turn less valuable streams in refineries into more valuable streams in refineries are a $2.1 billion/year business worldwide [49].

For gaseous mixtures, GC alone can sometimes provide sufficient information on composition. Gas chromatography–mass spectrometry is therefore normally applied to semi-volatile liquid streams and products: in the range of naphtha to vacuum oils in refinery streams and gasolines to lubricant oils in the petroleum products. For heavier and higher-boiling fractions than vacuum gas oils, pyrolysis GC–MS can be used for analysis. Most refinery streams and products are largely free of heteroatoms or metals, to usually the tens of ppm total level, which means individual heteroatom-containing molecules are generally present below GC–MS detection limits. To analyze these molecules, enrichment through sophisticated chromatographic procedures is usually needed.

4.1. Fuel Analysis

The simplest liquid mixtures are naphtha or gasolines. Even for these fractions, accurate analysis by GC alone can be difficult due to coelution of components even when a very high-resolution GC method is used. Gasoline, being one of the most important high-value products and the most thoroughly analyzed, is mostly a synthetic product with very strict compositional parameters. Gas chromatography–mass spectrometry has been used to determine compound type, including paraffins, isoparaffins, olefins, naphthenes, and aromatics (PIONA), and carbon number distributions of a gasoline sample [50]. Gas chromatography–mass spectrometry mapping is used to resolve coeluting compounds with different mass spectral patterns. The coeluting compounds are then identified by the "difference spectra." Gas chromatography–mass spectrometry has also been used to determine the efficiency of a GC column for separating naphtha components [51].

The largest current application of GC–MS to refinery streams and products is due to recent government regulations. For example, the U.S. Environmental Protection Agency (EPA) specified in the *Federal Register* in 1994 that GC–MS must be used to determine the total aromatics in gasoline [52]. The method was later implemented as American Society for Testing and Materials (ASTM) method D5769, which uses deuterated surrogates and 23-component QC mix with accuracy of each component in the QC mix to be within 5% relative to the true, weighted concentration [53]. As part of the new rules for reformulated gasoline (RFG), the U.S. government has mandated strict limits on various compositional parameters of gasoline to be used in parts of the United States with air quality problems. These parameters are to be determined by specified methods, and the method specified for total aromatics is GC–MS. There are 50 or more significant aromatic species in gasoline, and one of the best ways to quantify all of them individually is by GC–MS. The prior method for total aromatics was by open column chromatography with yardstick measurement of colored bands, which suffered from serious precision problems and did not provide quantitation for individual species. Figure 6 shows a portion of the D5769 chromatogram of a gasoline sample between 6 and 6.5 minutes. The distribution of alkanes is displayed by the 57-Da ion current trace, while the distribution of C-3 benzenes is displayed by the 120-Da ion current trace. The selectivity of aromatic compounds

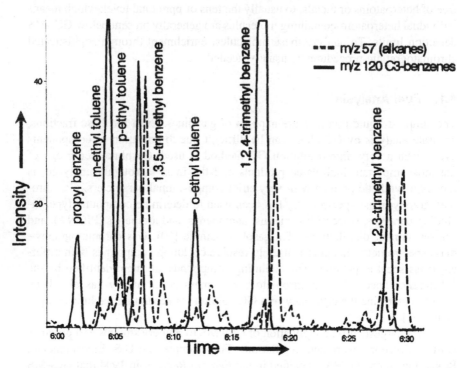

Figure 6 Partial selected ion chromatogram of ASTM D5769 data (see text for reference) for a reformulated gasoline. Solid line is m/z 120 ion current for various C-3 alkylbenzenes. Dotted line is m/z 57 ion current, base peak for saturate gasoline components (unidentified), showing the selectivity provided by GC–MS for badly coeluting gasoline components. Cycloparafins and olefins are also present but not displayed. Conditions: gas chromatograph: Varian 3400 (Varian, Palo Alto, CA) running 20 m × 0.10 mm ID DB-1 phase (0.01-μm film thickness) at 1000:1 split of 0.1 ul injection of gasoline dosed with mixture of isotopically labeled aromatic standards as per method. Mass spectrometer: Finnigan (Finnigan, San Jose, CA) TSQ710 in 70-eV EI, full-scan mode, 20 scans/sec.

using fast chromatography is demonstrated in this example in which the analysis of whole sample is completed in 9 minutes. Every batch of gasoline sold as RFG (approximately 40% of the U.S. gasoline market) must be analyzed by this method. Briefly, the sample is first dosed by weight with a mixture of deuterated aromatic compounds as internal standards (IS) for quantitation. The compounds of interest are quantified separately versus a relevant deuterated IS component. This method has the added advantage of being able to quantify individual target species such as benzene and methyl *tert*-butyl ether (MTBE), among others [53].

Retention indices are very useful for compound identification by GC–MS, especially for isomers that yield similar or identical mass spectra. Lai and Song [54] have determined retention indices of over 150 compounds with temperature programming of moderately polar and slightly polar GC columns for the comparison of jet fuels JP-8 derived from petroleum and coal. They found that aliphatic compounds give nearly constant retention indices at different heating rates, while aromatic compounds show relatively large temperature dependence. With different polarity of the GC columns, the difference in retention indices is relatively small with aliphatic compounds but becomes larger with polycyclic aromatic and polar compounds. Petroleum-derived jet fuel saturates are mainly *n*-alkanes and isoalkanes. Coal-derived jet fuel saturates are mainly cycloparaffins with 1 to 3 rings, with some *n*-alkanes. Similarly, petroleum-derived jet fuel aromatics are dominated by alkyl benzenes and alkyl naphthalenes, while coal-derived jet fuel aromatics are rich in hydroaromatics [54].

4.2. Lubricant and Mineral Oil Base Stocks

Vacuum gas oils (VGOs) are typically used for lubricant and mineral oil base stock. These materials vary in carbon number from 12 to 30 or more, and generally present a totally unresolved "hump" for the total ion chromatogram, and almost every nominal mass is present at almost every retention time. In this case, the GC–MS analyst is limited to either target compound analysis of higher concentration components, or more selective analysis using high-resolution GC–high-resolution MS or high-resolution GC–MS–MS. Both are discussed below in section 6 on special techniques providing extra selectivity for specific compound classes. Important exceptions to this generalization about lubricant base stocks are: (1) waxy oils that contain primarily *n*-alkanes that are well resolved chromatographically, and (2) synthetic and "semisynthetic" base stocks that are oils of the same boiling range that have been replaced with or doped with poly-alpha-olefin (PAO) or polyether petrochemical streams to increase various lubricant performance properties. Such samples may be analyzed either by conventional GC–MS using high-temperature GC columns or by pyrolysis GC–MS where the sample is flash-heated to between 600 and 1000°C prior to the injection port, and the ensuing olefinic pyrolysis fragments are cryotrapped on the GC column.

Figure 7 shows the GC–MS from a semisynthetic oil showing the PAO distribution along with the mineral oil "hump" characteristic of conventional vacuum gas oils [55]. Pyrolysis GC–MS is also commonly applied to finished (additized) motor oils, which may contain polymeric additives with molar mass up to 500,000 Da.

4.3. Process Gas Chromatography–Mass Spectrometry

Traditionally, refineries have made great use of gas chromatography for process stream characterization. It is not unusual to walk through a refinery and see dozens of process GC instruments stuck to walls and bulkheads. They generally contain very long capillary columns (100 m × 0.25 mm DB-1 columns are very common) and take automated quantitative flame ionization detection (FID) measurements of various constituents. The data are then fed to the computerized plant control room. The slow analysis time and limitation to only the largest GC peaks causes some amount of efficiency loss. These units are slowly being replaced by process GC–MS instrumentation using shorter, faster columns, as well as process near-infrared (NIR), nuclear magnetic resonance (NMR), and 2-dimensional GC (GC–GC) instrumentation in that order.

Typically, fuels products are blended under computer control. There are constraints against which the blend recipe is constantly checked. The constraints are typically product specifications (cetane, octane, cloud point, distillation points, etc.) and blend stock quality and availability. The quality of the on-going blend is monitored using key quality analyzers that feed back to the blend control computer the current quality status. The blend recipe is then adjusted to optimize the quality (specifications) using the currently available blending components.

A refinery-compatible GC–MS instrument has been developed recently for use in on-line blending and process control applications [55]. For rapid analysis, a 10-m microbore GC column coated with 0.17-μm film thickness of DB-5 (methyl silicone) is used. The column temperature is programmed for gasolines from 35 to 230°C and for diesel fuels from 110 to 295°C, both at 30°C/min. Typical results of this rapid on-line GC–MS analysis for gasolines, jet fuels, and diesel fuels are shown in Figure 8.

Partial least square (PLS) modeling is used to develop property predictions for the streams of interest [56]. The procedure requires the use of a training or reference sample set that has known values for the properties being modeled. The reference samples are analyzed by GC–MS after which the data generated are treated by multivariate correlation methods. The resulting inferential models are then used with subsequent GC–MS analysis of an unknown mixture to produce a predicted value for the property or groups of properties of interest. In the on-line case, the training set is developed from actual blender samples with on-line GC–MS data coupled with standard laboratory analyses over a long period

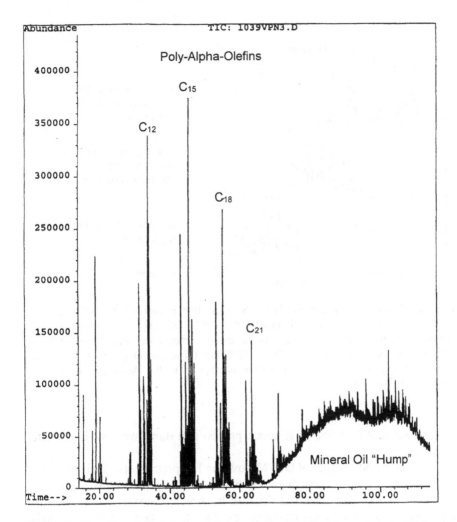

Figure 7 Gas chromatography–mass spectrometry total ion chromatogram (TIC) of a semisynthetic motor oil base stock. The peaks at lower retention time (RT) are poly-alpha-olefin oligomer clusters separated by three carbons. The wide "hump" at high RT is the mineral oil component, with insufficient chromatographic resolution to distinguish individual species without mass spectral separation. Conditions: Hewlett Packard (Palo Alto, CA) 5890 gas chromatograph with MSD detector in 70-eV EI mode, 30 m × 0.25 mm ID DB-5MS phase (0.25-μm film thickness) with temperature ramp from 35 to 350°C at 2°C/min.

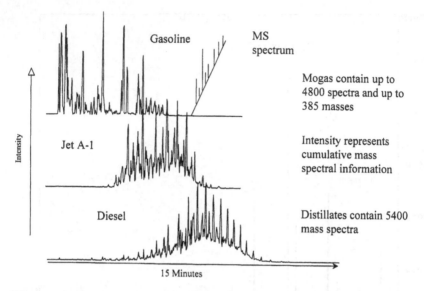

Figure 8 Typical results of on-line GC–MS for gasoline, jet fuel, and diesel fuel. Conditions: GC–MS: Hewlett-Packard (Palo Alto, CA) 5971/2A MSD, 10-m microbore GC column coated with DB-5 (0.17-μm film thickness), temperature ramp 35 to 230°C at 30°C/min for gasoline and 110 to 295°C at 30°C/min for jet fuel and diesel fuel, sample size 0.5 μl injection with 250:1 split, mass scan range 10 to 300 Da for gasoline and 10 to 400 for distillate fuels.

of time to account for refinery variations, process unit shutdowns, and crude slate changes.

One of the advantages of GC–MS over an IR spectroscopic analyzer is the ability to measure distillation characteristics as well as predict other properties. There are several other materials that can be directly measured and reported. These include benzene, total aromatics, oxygenates, certain sulfur compounds and additives. The properties that can be predicted include (among others) cetane number and index, research and motor octanes, refractive index, distillation properties, aniline point, cloud point, pour point, volatility, flash point, density, conductivity, and viscosity [57].

5. ENVIRONMENTAL ANALYSIS FOR TARGET COMPOUNDS BY GAS CHROMATOGRAPHY–MASS SPECTROMETRY

In recent years, environmental analysis of petroleum and its products as adulterants in air, water, or soil has become increasingly important for regulatory, health, and legal reasons. Because these matrices are so complicated it is not possible to obtain the necessary compositional profile by physical or spectroscopic means (such as the "total petroleum hydrocarbon" analyses performed with IR or GC on soil), GC–MS has been the analysis method of choice for these sorts of studies. Gas chromatography–mass spectrometry analyses of environmental samples are almost always target compound analyses, where the analyst needs to quantify one or many petroleum product species. While this analysis may seem simple, accurate quantitation in these samples can be exceeding difficult. Once an adequate individual species compositional profile has been determined for a sample, the data manipulation to determine the actual product identity and mixture analysis can get quite complicated due to the constant background of non-point-source petroleum products in any given vicinity.

5.1. "Modified 8270 Analysis"

The widely applied GC–MS method for complex samples was written many years ago as part of the EPA SW-846 protocol for analysis of solid waste [58]. In this protocol, an array of extraction and cleanup procedures (EPA methods 3500s and 3600s) are specified for the separation of nonpolar target analytes from polluted water or solid waste, where the actual cleanup method to be applied is dependent on the complexity of the sample.

The most common procedure for soil or water is ultrasonic extraction of the analytes into methylene chloride following careful quantitative addition of stable isotope–labeled internal standard surrogate species to determine the recovery and efficacy of the extraction. This solid waste procedure is commonly applied to soil, sludge, body tissues, plant matter, and even polar solid petroleum products such as asphalt. SW-846 then specifies the use of GC–MS (EPA method 8270) to quantify for a list of about 80 enumerated target analytes, which include a wide range of common industrial pollutants and pesticides. This list does include a few compounds present in petroleum products, but the 8270 methodology is in practice able to determine the amount of any nonpolar GC–MS analyte. Thus, the general analysis of environmental samples for species present in petroleum products but not specifically listed in the EPA method is often referred to by the catchall term "modified 8270 analysis." This method is sufficiently rigorous in its use of internal standards and surrogates. This is in contrast to other methods, which use external or non–isotopically labeled standards and do not take into account the difference in compound affinities for the soil or water in question, thus leading to possible misleading results. Other advantages of the "8270" procedure are: (1) it is run by numerous contract laboratories, which can do the extractions and/or GC–MS injections cheaply in an automated manner and provide the petroleum analyst with the raw data files for further data reduction, and (2) it allows for the use of extra cleanup steps or selectivity methodologies (such as pre-LC or MS–MS) as long as the standards and controls come out with the right answers.

The modified 8270 method is used extensively for the analysis of poly-nuclear aromatic hydrocarbons (PNAs or PAHs). Soot is believed to be the first known human carcinogen, and sootlike molecules, *nonalkylated* PAHs, are gener-ally recognized as some of the most chronically toxic hydrocarbons. Eighteen nonalkylated PAHs are on the EPA priority pollutant list, and are included in the classical 8270 list. While petroleum products can contain upwards of 50 mass % alkylated PAHs, it is very rare for a petroleum product to contain any signifi-cant amounts of nonalkylated PAHs that come primarily from combustion pro-cesses (pyrogenic rather than petrogenic). Samples from the environment, how-ever, often contain nonalkylated PAHs, which derive from used motor oil runoff on roadways, vehicle and power plant particulate (often coated with nonalkylated PAHs), or simple soot from any of a number of sources. Fortunately, these non-alkylated PAHs are easy to analyze by GC–MS since they are easy to ionize with little fragmentation. A high-temperature GC column must be used in some instances to elute the high-boiling species.

5.2. Fugitive Compound Analysis

The most complicated problems for the environmental petroleum analyst are those products, either crude oil or refined and additized products, that somewhere in transport or storage get released unintentionally to the environment. These "spills" are referred to as fugitive products, and can often be characterized by modified 8270 procedures to get the distribution of major petrogenic species. These species distributions are then compared with the distribution of those spe-cies in various products to determine the age, source, or concentration of a given product in the sample. This can be a very complicated procedure, since, as men-tioned above, hydrocarbons are present everywhere in the environment.

Figure 9 shows the "fingerprint" of the C-3 dibenzothiophenes and C-3 phenanthrenes in a possible fugitive product. These fingerprints are very different among different sources, and are a powerful tool to determine the identity of a fraction in a mixture attributable to a given product. Perhaps the most famous example of this was when these same compound fingerprints shown in Figure 9 were used in analysis of Prince William Sound sediments to show that in some areas a large fraction of the oil was from petrogenic hydrocarbons from seeps and not fugitive product at all [59]. For urban samples, where hydrocarbon content is always a mixture of ambient pollution from nonpoint sources, it is usually neces-sary to do a quantitative analysis, where 10 to 100 compounds are measured in the various possible sources and then compared statistically with the distribution measured in the sample. In this case, the various possible concentration vectors are fed into a multiple least squares analysis to determine the loading scores of

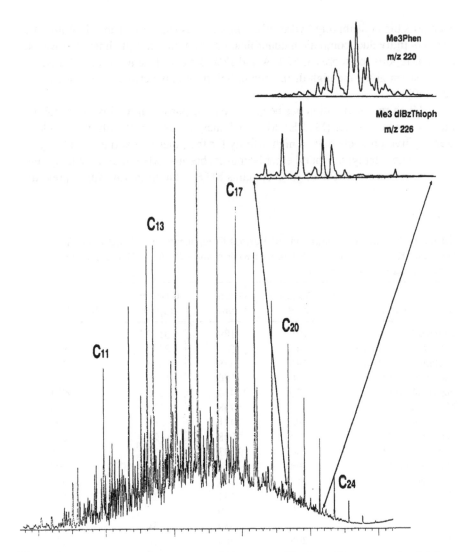

Figure 9 Gas chromatography–mass spectrometry TIC of possible source material for a fugitive product. Inserts show the "fingerprint" of C-3 phenanthrenes and C-3 dibenzothiophenes, respectively. These profiles are very different from one petroleum product to another and their profile and/or total abundance can be used for quantitative identification of fugitive product mixtures. Conditions: Varian 3400 gas chromatograph (Varian, Palo Alto, CA) with Finnigan (San Jose, CA) SSQ170 EI mass spectrometer running a temperature gradient on a 100 m × 0.25 mm ID DB-5MS (0.25-μm film thickness) fused silica capillary GC column (special order from J&W Scientific, Folsom, CA).

each possible contributing hydrocarbon mixture, as shown in Table 1. While the errors of individual component quantification are high, multiple linear regression analysis reports the scores as 83% A and 17% B, very close to the actual mixture composition given the high difference in boiling points between these two products [60].

Much recent attention has been cast on one particular additive to gasoline, methyl *tert*-butyl ether (MTBE). Most polymer and detergent additives in additized fugitive products have a high affinity for the polar soil surfaces and stay at the spill site. Methyl *tert*-butyl ether is unusual because when present in a fugitive gasoline product (almost all RFG contains MTBE due to its low vapor pressure

Table 1   Relative Concentrations of Various Components in a Diesel Fuel (A), a Gasoline (B), and a Synthetic Mixture of 87% A and 13% B, as Determined by Semiquantitative GC–MS analysis[a]

Species measured	Mass used	Diesel (ppm)	Gasoline (ppm)	Mixture 83% A/17% B
Naphthalene	128	189	3119	694
2-Methyl naphthalene	142	504	2347	876
1-Methyl naphthalene	142	348	955	481
Naphthalene C2	156	4550	1337	4285
Naphthalene C3	170	10433	386	9981
Naphthalene C4	184	5748	112	8215
Naphthalene C5	198	2471	30	3901
Naphthalene C6	212	1008	21	1612
Dibenzothio-C0	184	33	0	25
Dibenzothio-C1	198	210	2	199
Dibenzothio-C2	212	318	1	335
Phenanthrene	178	144	14	120
Anthracene	178	0	3	1
Phen + anth-C1	192	559	54	515
Phen + anth-C2	206	465	50	545
Phen + anth-C3	220	188	16	296
Pyrene-C0	202	13	3	10
Pyrene-C1	216	25	10	31
Pyrene-C2	230	25	10	34
Chrysene-C0	228	0	1	1
Acenaphthene	154	44	6	48

[a] While the errors of individual component quantification are high, multiple linear regression analysis reports the scores as 83% A and 17% B, very close to the actual mixture composition given the high difference in boiling points between these two products. Regression results: 87% A/13% B $r^2 = 0.9479$ (column normalized).

and high octane rating and oxygen content) it moves underground in the water table ahead of the gasoline, giving a noticeable taste and odor to the water before the gasoline arrives. Many underground gasoline spills went unnoticed for years while the groundwater polluted by gasoline remains tasteless and odorless until the addition of odoriferous MTBE caused the public attention to this serious threat to drinking water quality. Methyl *tert*-butyl ether, along with acetone, is one of the few common compounds for which the GC–MS analyst cannot use the EPA methylene chloride extraction methods, such as 3510B or 3520B, due to the poor recoveries of these components. These materials may be analyzed by headspace GC–MS, assuming an isotopically labeled internal standard is used to account for sample differences in vapor pressure due to fuel and other cosolvents in the aqueous matrix.

5.3. Diesel Emissions

The combustion of diesel fuels produces mainly carbon dioxide and water. The emitted exhausts also contain small amounts of carbon monoxide, nitrogen oxides (NO_x, including NO, NO_2, N_2O, etc.), sulfur oxides (SO_2/SO_3), unburned hydrocarbons, and particulates. These minor components are pollutants and legally regulated. When exhaust gas is cooled to collect the condensates, the organic fractions are largely condensed on the particulate soots [61,62].

Some diesel particulates are of respirable size, and the organic species attached to them may constitute an inhalation health hazard to the human population [63,64]. Organic compounds on diesel particulates have been extracted and studied [62,65]. In the soluble organic fractions (SOFs), much attention has been paid to polycyclic aromatic hydrocarbons (PAHs), nitrated PAHs, and oxygenated PAHs, particularly for those showing mutagenic activities.

Uncombusted PAH from diesel fuels can react with oxides of nitrogen in the exhaust to form nitrated PAH. Some of them, for example, 1-nitropyrene and dinitropyrenes, have been shown to be potent mutagens in Ames assays [64]. Due to this environmental health concern, nitrated PAHs have been studied by various GC and MS techniques [66,67]. Oxygenated PAHs found in diesel particulates include polycyclic ketones and quinones, carboxaldehydes, and hydroxy-PAH [68].

6. SPECIAL TECHNIQUES FOR SELECTIVE COMPOUND CLASSES

Most of the aforementioned GC–MS techniques can be performed on a commercial instrument. However, there are many special techniques that are developed at various laboratories for the effective determination of specific or targeted compound classes of interest.

6.1. Gas Chromatography–Low-Energy Ionization Mass Spectrometry

Petroleum and petroleum distillate cuts often consist of hundreds, if not thousands, of individual components. Except for the lowest-boiling fractions, it is virtually impossible to separate, identify, and/or quantitate all of the significant components in a given sample. In many cases, petroleum chemists and engineers are able to work with simplified characterizations that express a given sample in terms of hydrocarbon groups or "types," such as "paraffins," "naphthenes," "benzenes," etc. Several mass spectrometric methods have been developed that characterize petroleum samples into such hydrocarbon types. These methods make use of mass spectral ions that are characteristic of the various groups of hydrocarbons. Ion intensity values for these characteristic ions can be summed to give a quantitative measure of the amounts of each of the hydrocarbon groups [53]. In most cases, however, these MS methods do not provide carbon number distributions of each hydrocarbon group.

6.1.1. Gas Chromatography–Low-Voltage Electron-Ionization Mass Spectrometry

One method that has been used in the petroleum industry to provide carbon number distributions of aromatic compounds makes use of low-voltage electron ionization (LV-EI). At ionizing voltages on the order of about 10 eV, saturated molecules, such as paraffins and naphthenes, do not ionize to any great extent. Aromatic molecules, having lower ionization potentials, still do ionize. The amount of energy transferred to the aromatic molecules under such conditions, however, is such that most of the molecular ions formed do not break apart, and thus retain the molecular weight information. This makes it possible to obtain carbon number distributions of the various aromatic types. These "types" are generally thought of in terms of "z-series," where the "z" comes from the empirical formula C_nH_{2n+z} [69]. Using low-resolution mass spectrometry, it is possible to obtain carbon number distributions for aromatics in the "-6", "-8", "-10", "-12", "-14", "-16", and "-18" z-series. Aromatic sulfur compounds and more-condensed aromatic hydrocarbons overlap with these z-series, but are often present at low enough levels as to cause only minor problems.

Gas chromatography–mass spectrometry is a technique used frequently to analyze naphtha and distillate cuts of petroleum. Even with high-resolution GC columns, it is usually impossible to completely separate components in such samples. In some cases, coeluting components can be identified and quantified by careful analysis of the mass spectra produced by such mixtures. This process becomes more difficult for higher-boiling petroleum fractions. The top trace of Figure 10 shows a total ion chromatogram (TIC) for a typical diesel fuel sample, using a 70-eV electron beam as ionization means. A short HP-5 fused silica capillary column (10 m × 0.10 mm ID) was used for this analysis. The GC oven temperature was ramped at a high rate in order to elute the sample quickly. As can be seen from the chromatogram, the sample is very complex. The top spectrum of Figure 11 shows a summed 70-eV mass spectrum for the entire sample. This ionization energy is 5 to 10 times the amount necessary to ionize most hydrocarbon molecules. As a result, a considerable amount of the molecular ions formed fragment into smaller pieces. It is this 70-eV mix of parent and fragment ions that is typically used as a "fingerprint" of a given compound. While this often works well for completely separated components, it also makes spectral interpretation more difficult in cases where components are not separated in the gas chromatograph. In the case of aromatic compounds, low ionizing voltage offers a partial solution to this problem. A low ionizing voltage TIC for the same diesel fuel sample is shown in the bottom trace of Figure 10. The summed mass spectrum for the entire sample is given in the bottom spectrum of Figure 11. The spectrum consists mainly of molecular ions, with little fragmentation. The saturates, which gave large ion signals in the low mass region at 70 eV, do not contribute much to the overall signal under the low eV conditions. By determining response factors for the aromatic compound types, it is now possible to obtain quantitative carbon number distributions for each z-series. In addition, it is possible to obtain isomer distributions at the lower carbon number ranges of each z-series by making use of extracted ion chromatograms (mass chromatograms). Thus, low ionizing voltage provides a means to simplify the characterization of aromatics in complex petroleum samples, yet still provides a level of detail allowing chemists and engineers to determine differences among various samples.

6.1.2. Gas Chromatography–Charge-Exchange Mass Spectrometry

Alternatively, the sample can be analyzed under low-energy ionization conditions using benzene [70] or carbon disulfide (CS_2) [71] as a charge exchange reagent gas. In low-voltage EI, molecular ions are formed with a distribution of internal energies. In contrast, for charge exchange the internal energy (IE) of the molecular ion is well defined and equal to the difference between the recombination energy (RE) of the reagent gas (benzene or CS_2) and the ionization potential (IP) of the sample molecule. For example, in CS_2 charge exchange (CS_2-CE)

$$CS_2^{+\cdot} + M \rightarrow CS_2 + M^{+\cdot} \tag{2}$$

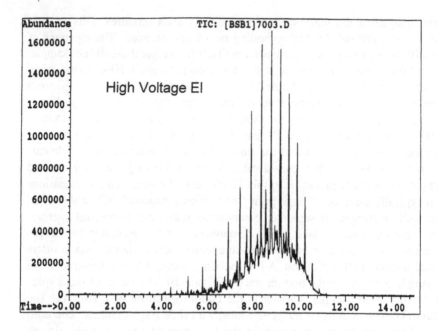

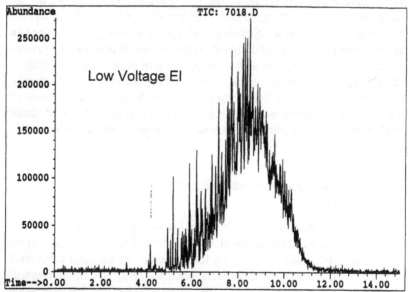

Figure 10 Comparison of GC–MS TICs of a typical diesel fuel between high-voltage and low-voltage EI runs. The top trace shows high-voltage (70 V) EI results. Normal paraffins are shown as sharp peaks above the envelope (hump). At low-voltage (10 V) EI, the normal paraffin peaks disappear with the low-ionization-potential aromatic components predominant in the chromatogram, shown as the bottom trace.

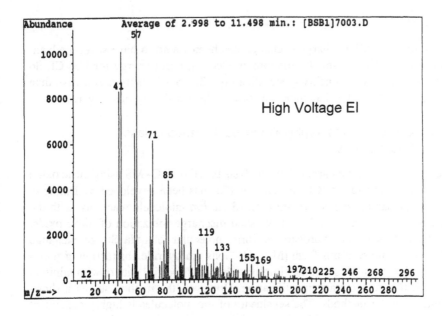

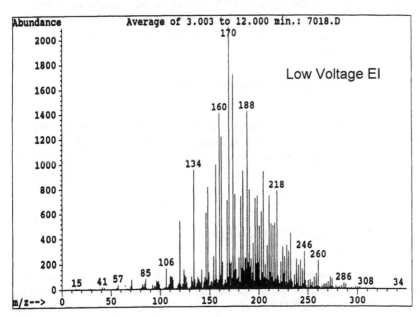

Figure 11 Comparison of high-voltage and low-voltage EI spectra of the diesel fuel. In the high-voltage EI spectrum (top), significant amounts of fragment ions, shown as odd masses, are present. In the low-voltage EI spectrum (bottom), essentially molecular ions, shown as even masses, are present.

$$IE(M^{+\cdot}) = RE(CS_2^{+\cdot}) - IP(M) \qquad\qquad (3)$$

Carbon disulfide charge exchange has been found to have several advantages over LV-EI: (1) small variations in electron beam energy for CS_2-CE do not affect the relative sensitivity significantly, (2) CS_2-CE is more sensitive than LV-EI, and (3) CS_2-CE has more uniform molar sensitivity than LV-EI.

6.1.3. Gas Chromatography–Chemical-Ionization Mass Spectrometry

A comprehensive hydrocarbon-type analysis based on GC–MS using nitric oxide as a chemical ionization (CI) reagent gas [72] has been developed by Dzidic et al. [73] The sample molecules are ionized via ion–molecule reactions, with the reactant NO^+ ion generated by Townsend discharge ionization of nitric oxide. For example, aromatic hydrocarbons form molecular ions via charge exchange with NO^+, while paraffins form $[M - H]^+$ ions via hydride abstraction of molecules by NO^+ ions. This method simplifies the mass spectral pattern with minimal fragmentation to facilitate quantification. However, the method shares a shortcoming of all CI methods. The sensitivity of the molecules is highly dependent on ion source pressure, making it difficult to reproduce the quantification results. In addition, the coproduction of other pseudo-molecular ions, such as $[M + NO]^+$ ions for aromatic compounds, would complicate the interpretation and quantification of the components.

6.1.4. Gas Chromatography–Field-Ionization Mass Spectrometry

Another emerging method is the coupling of GC with field-ionization MS (GC–FI-MS) [74]. Field-ionization MS yields essentially molecular ions for all hydrocarbons, except isoparaffins, with uniform sensitivities within each compound class [75]. This would greatly simplify calibration since each class would require only one calibration compound.

6.2. Gas Chromatography–High-Resolution Mass Spectrometry

Gallegos [76,77] pioneered the use of GC–medium- or high-resolution MS (HRMS) for the analysis of sulfur and nitrogen compounds. With a resolving power of 3400, CHS^+ and CH_2N^+, characteristic fragment ions for sulfur and nitrogen compounds, respectively, are well resolved from other isobaric ions. Monitoring these characteristic ions during entire GC–MS runs can be used as sulfur- and nitrogen-specific detection for the sulfur and nitrogen compounds.

When a double-focusing sector mass spectrometer is used for GC–HRMS of the ionic species in entire spectra, the mass spectrometer has to be scanned

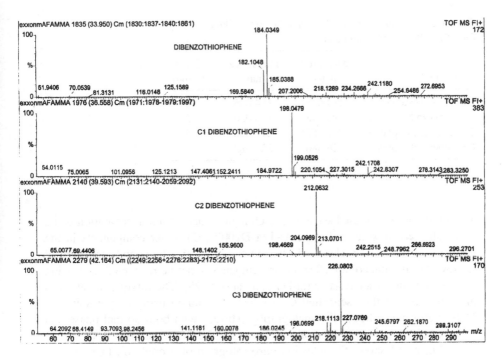

Figure 12   Field-ionization mass spectra of dibenzothiophenes (DBTs) in a diesel fuel analyzed by GC–TOF-MS in FI mode. Error in accurate mass measurement for DBTs is less than 3 mmu. Note that there are no significant fragment ions present.

in a slow speed. In addition, the mass calibration compound has to be bled into the ion source for accurate mass measurement. These would severely lower the chromatographic resolution and affect the sensitivity of the measurement. Although ultrahigh-resolution FT-ICR-MS can be coupled with GC for GC–HRMS, a huge amount of data points collected during each GC–MS run limits its applications. Recent advances in TOF-MS using delayed extraction and a reflecting analyzer (reflectron) have greatly improved the accuracy of mass measurement. Figure 12 shows the mass spectra of the dibenzothiophene (DBT) series in a diesel fuel analyzed by a GC–TOF-MS using field ionization (FI). As shown in Table 2, even under FI conditions that in general yield less accurate mass measurement results than EI, the accuracy of the mass measurement is within 3 mmu.

6.3.   Sample Pretreatment Prior to Gas Chromatography– Mass Spectrometry

For very complex mixtures such as petroleum and its fractions, isolation of various compound classes by LC prior to GC–MS analysis is a common procedure.

Table 2 Demonstration of the Mass Accuracy Obtained in GC–FI-TOF-MS Measurements

Compound	Measured mass	Theoretical mass	Error (mmu)
Dibenzothiophene	184.035	184.035	0
C1-Dibenzothiophene	198.048	198.050	2
C2-Dibenzothiophene	212.063	212.066	3
C3-Dibenzothiophene	226.080	226.082	2

The most commonly used separation technique is open column separation using clay followed by silica gel, e.g., ASTM D2007 [52]. Polar components in the mixture are retained on clay, and saturate and aromatic compounds are allowed to flow through the column. The polar components can be recovered later using a polar solvent for analysis or further separations [78]. The mixture of saturates and aromatics is then separated on a silica gel column using proper solvents. High-performance liquid chromatography (HPLC) with both normal phase and reversed phase procedures can also be used for the same purposes [58].

Figure 13 shows the total ion chromatogram of a crude oil that has had 95% of the nonaromatic compounds removed by solid-phase extraction (SPE). Oil is put onto silica stationary phase with saturates eluting under pentane and the aromatics coming off with ethyl ether. This 2-minute procedure simplifies the chromatogram significantly as aromatics have fewer possible isomers than the corresponding saturate species that may have isobaric masses. The aromatic sulfur compounds are not separated on the silica from the other aromatic hydrocarbons, but isobaric species of these series have significantly different boiling points and can usually be identified by retention time.

Petroleum fractions have also be separated on a nonaqueous ion exchange column into acid, base, and neutral fractions [79].

Selective derivatization can be used to identify a specific class of compounds of interest in a complex mixture. For example, after the removal of thiophenes on an alumina column, sulfides can be distinguished from thiols in light fuel distillates by selective derivatization of thiols with pentafluorobenzoyl chloride [80]. The sulfur compounds are then identified by GC–MS.

For polar molecules that cannot or are difficult to flow through a GC column, derivatization into less-polar compounds is a common practice. For example, carboxylic acids in petroleum were esterified with fluoroalcohols to reduce polarity and GC retention time [81]. Another advantage of this approach is that significantly higher molecular weights of the esters move molecular ions and most fragment ions of interest out of mass regions of potential interfering ions.

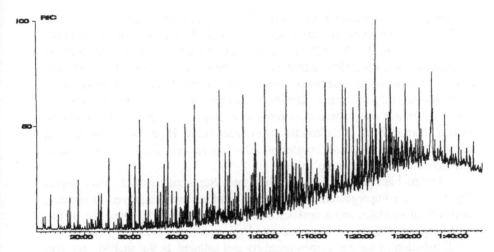

Figure 13 Gas chromatography–mass spectrometry TIC of the aromatic fraction of an Arabian crude oil. While the TIC of the whole oil yields a chromatographically unresolved "hump," this sample, which has had approximately 95% of the saturates removed by a fast silica SPE cleanup procedure, shows good resolution of the aromatic components, due to the fact that there are far more isomers at a given carbon number of saturates than aromatics. Conditions: 100 mg oil deposited on 1-g silica SPE cartridge; 2 ml pentane, followed by 2 ml ethyl ether. The ether fraction was partially evaporated and injected splitless into a Varian 3400 gas chromatograph (Varian, Palo Alto, CA) using a 100 m × 0.25 mm ID DB5-MS (0.1-μm film thickness) capillary GC column (special order from J&W Scientific, Folsom, CA). Column temperature ramp 35 to 350°C at 2°C/min. Mass spectrometer: Finnigan (San Jose, CA) SSQ710 in 70-eV EI full-scan mode.

Hofmann et al. [82] used pentafluorophenyldiazoalkanes as derivatization reagents to eliminate the problems of decomposition when using pentafluorobenzyl bromide. Gas chromatography–negative-ion mass spectrometry was used to identify and quantify the acids. Phenolic components in fuels have been converted into corresponding trifluoroacetate esters prior to GC–MS analysis [83]. Primary, secondary, and tertiary amines in petroleum have been differentiated by exhaustive trifluoroacetylation [84]. Most primary amines add two, secondary amines add one, and tertiary amines add no trifluoroacetyl group. However, the differentiation of amine types can be conveniently performed using deuterated ammonia (ND_3) CI [78].

6.4. Strategies for Olefin Analysis

Analysis of olefins is important in the petroleum industry due to their possible adverse effects on the quality of petroleum products. However, the complicated mixture containing olefins produced from catalytic and coking processes such as FCC (discussed above) presents a serious analytical challenge for any method, including "classical" GC–MS. These materials span the same boiling range as paraffins and cycloparaffins (naphthenes) of the same carbon number and have an exponentially larger number of isomers than the paraffins, which makes them difficult to separate by even the highest-resolution GC columns. In addition, olefins have similar mass spectral patterns as cycloparaffins under EI conditions, making them difficult to differentiate from each other. High-resolution MS will not work because the exact masses of these species are the same as those of corresponding cyclic saturate species.

Olefins bind preferentially to silver ions. Supercritical fluid chromatography with silver-impregnated columns has therefore been employed to separate olefins from saturates and aromatics in the same sample [85]. Once the olefins have been isolated, they can be analyzed by conventional GC–MS to determine the distribution of various carbon numbers and isomers. In the analysis, the aromatics, olefins, and saturates are resolved by SFC as three separate peaks. With careful switching between the two SFC columns to elute one fraction at a time through the septum into an unmodified GC–MS injection port, the fraction is trapped cryogenically on the GC column head. The trapped fraction is analyzed by GC–MS prior to subsequent fractions being introduced onto the injector [86]. Figure 14 shows the GC–MS profiles of the three different SFC fractions. All analyses were performed on one 10-μl injection and data taken into the same data file with all three instruments under the control of the MS data station.

Olefins do exhibit selective chemistry, and many derivatization procedures have been attempted on these either prior to analysis or in the ionization chamber to achieve olefin selectivity. Acetone CI gives preferential acetylation at the double bond, which can be used to differentiate olefins from cycloparaffins [87]. Classically, olefins are derivatized prior to analysis as part of the sample preparation procedure. Some of these methods are possible in a complex mixture such as gasoline, most notably the bis-alkylthiolation of the double bond using a reagent such as dimethyldisulfide [88,89]. This interesting technique has the advantage of adding two methyl sulfide groups into a molecule, possibly pushing the GC retention time out of that observed for the rest of the sample. The fragmentation pattern is indicative of starting double bond position, an advantage over hydrogenation. Hydrogenation of olefins can be easily carried out by exposing the gasoline to hydrogen gas in the presence of a palladium (Pd) catalyst. Figure 15 shows the results from an experiment where 20 mg 1% Pd on C powder was put onto the frit in the injection port of a GC, and then hydrogen used as the carrier gas to hydrogenate the olefins on-line. With on-line hydrogenation, the olefinic components in the 112-Da ion current trace disappear, leaving cycloparaffins unchanged. The alkane distributions shown in the bottom two 114-Da ion

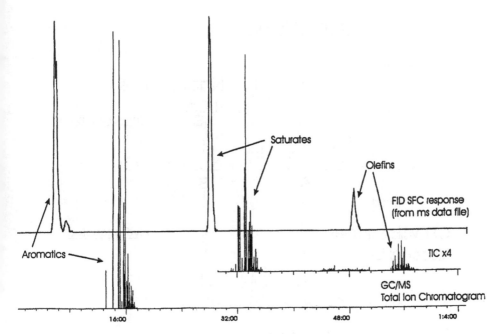

Figure 14 Supercritical fluid chromatography–gas chromatography–mass spectroscopy chromatograms. Top chromatogram shows the SFC-FID chromatogram, bottom chromatogram shows the GC–MS TIC. Gasoline, 10 μl is injected into a column-switching supercritical fluid chromatograph [85] that elutes the aromatic, saturate, and olefin fractions as three separate peaks under computer control. The SFC effluent is split between the FID detector and a capillary, which pierces the septum of an unmodified GC–MS injection port. The injection port is set at 5000:1 split to release the 100% CO_2 SFC mobile phase. During a SFC peak, the split is turned off, and the sample cryogenically trapped on the GC column at −40°C (the CO_2 passes through the GC column). When the SFC peak is over, the split is turned back on so helium flows through the GC column, and the GC temperature ramped to acquire the GC–MS data on that specific SFC cut. When the fast GC–MS run is over and the gas chromatograph cool, the supercritical fluid chromatograph is allowed to elute the next peak. All data are acquired in the same run into the same data file, with the SFC, GC and MS all controlled by the MS data system. Conditions: Dionex (Sunnyvale, CA) supercritical fluid chromatograph, Varian 3400 gas chromatograph (Varian, Palo Alto, CA) with 20 m × 0.10 mm ID DB-1 column (0.10-μm film thickness), Finnigan SSQ710 mass spectrometer in 70-eV EI mode scanning 20 scans/sec in full-scan mode.

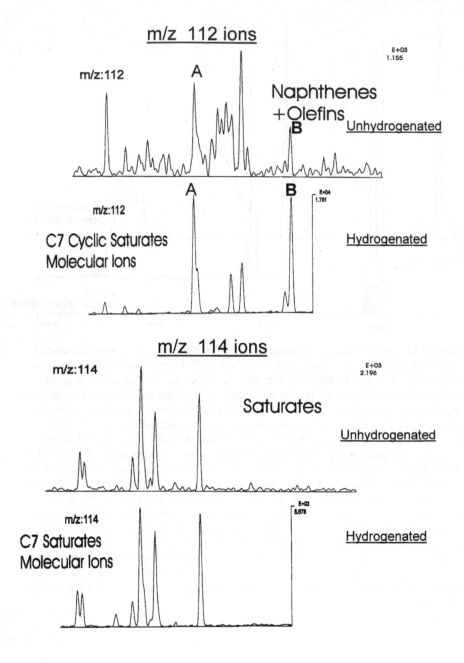

m/z 112 ions

m/z:112   A

Naphthenes
+Olefins Unhydrogenated
B

E+03
1.155

A   B   E+04
1.781
m/z:112

C7 Cyclic Saturates
Molecular Ions   Hydrogenated

m/z 114 ions

m/z:114

Saturates

Unhydrogenated

E+03
2.196

m/z:114   E+03
6.578

C7 Saturates
Molecular Ions   Hydrogenated

current traces are the same with and without hydrogenation. This procedure serves to identify olefin GC peaks by observing their disappearance, but destroys all information about the location of the double bond.

7. CONCLUSIONS

Gas chromatography–mass spectrometry has been an indispensable tool in all phases of the petroleum industry, including research, development, and business support. Diverse techniques have been developed to meet the needs of obtaining compositional information on very complex mixtures derived from petroleum and fossil fuels. Many specialized techniques also have been developed for the analysis of target compounds of concern to processes and the environment.

The analytical needs of the petroleum industry continue to drive the ever-increasing selectivity and sensitivity of GC–MS instrumentation and provide analysts in other industries with new analytical techniques that they may apply as their samples become more complex. New innovations in GC–MS to gain ever-greater speed and resolution in GC and MS, such as fast GC–MS, GC–GC–MS, and GC–TOF-MS, will open new avenues of analysis and problem solving for tomorrow's upstream and downstream analysts.

ACKNOWLEDGMENTS

The authors wish to thank Alan James of Exxon Production Research Company for his discussions in GC–IRMS, Steve Colgrove of Exxon Development and Research Laboratories for discussions in GC–low-voltage EI-MS, and Terry Ashe of Imperial Oil for discussions in process GC–MS.

Figure 15 Four selected ion chromatograms of a gasoline sample with and without hydrogenation at the gas chromatograph injection port. The top two chromatograms show 112-Da ions (molecular ions of C_8 cycloparafins and olefins), with and without, respectively, 10 mg 1% Pd on C catalyst in the gas chromatograph injection port running hydrogen carrier gas. The bottom two chromatograms show 114-Da ions (molecular ions of C_8 acyclic alkanes), with and without, respectively, injection port hydrogenation. Olefins, but not the isomeric cycloparaffins, are hydrogenated, resulting in the disappearance of peaks in the second chromatogram and the increase of the major paraffin peaks in the fourth chromatogram.

REFERENCES

1. A.K. Brewer and V.H. Dibeler, J. Res. Natl. Bur. Standards, 35 (1945) 125–139.
2. R.A. Brown, R.C. Taylor, F.W. Melpolder, and W.S. Young, Anal. Chem., 20 (1948) 5–9.
3. F.W. McLafferty and F. Turecek, Interpretation of Mass Spectra, 4th ed., 1993, University Science Books, Mill Valley, CA, p. 12.
4. T.R. Ashe and S.G. Colgrove, Energy Fuels, 5 (1991) 356–360.
5. J.C. Holmes and F.A. Morrell, Appl. Spectroc., 11 (1957) 86.
6. R.S. Golke, Anal. Chem., 31 (1959) 535–541.
7. A.A. Ebert, Jr., Anal. Chem., 33 (1961) 1865–1870.
8. D.O. Miller, Anal. Chem., 35 (1963) 2033–2035
9. D.H. Smith, M. Achenbach, W.J. Yeager, P.J. Anderson, W.L. Fitch, and T.C. Rindfleisch, Anal. Chem., 49 (1977) 1623–1632.
10. G.A. Junk, Int. J. Mass Spectrom. Ion Phys., 8 (1972) 1–71.
11. W.H. McFadden, Techniques of Combined Gas Chromatography/Mass Spectrometry: Applications in Organic Analysis, 1973, J. Wiley, New York.
12. W.K. Robbins and C.S. Hsu, Kirk-Othmer Encylopedia of Chemical Technology, vol. 18, 4th ed., 1996, J. Wiley, pp. 352–370.
13. W.K. Seifert, J.M. Moldowan, and R.W. Jones, Proceedings of the 10th World Petroleum Congress, Bucharest, Romania, 1979, paper SP8, Heyden, London, pp. 425–440.
14. K.E. Peters and J.M. Moldowan, The Biomarker Guide, 1992, Prentice-Hall, Englewood Cliffs, NJ, pp. 196–197.
15. W.K. Seifert and J.M. Moldowan, Geochim. Cosmochim. Acta, 43 (1979) 111–126.
16. J.-Y. Shi, A.S. Mackenzie, R. Alexander, G. Eglington, A.P. Gower, G.A. Wolff, and J.R. Maxwell, Chem. Geol., 35 (1982) 1–31.
17. J.M. Moldowan, P. Albrecht, and R.P. Philp, Biological Markers in Sediments and Petroleum, 1989, Prentice-Hall, Englewood Cliffs, NJ.
18. R.P. Philp and J.-N. Oung, Anal. Chem., 60 (1988) 887A–896A.
19. A.S. Mackenzie, S.C. Brassell, G. Eglington, and J.R. Maxwell, Science, 217 (1982) 491.
20. A.S. Mackenzie, Applications of biological markers in petroleum geochemistry, in J. Brook and D. Welte, eds., Advances in Petroleum Geochemistry, vol. 1, 1984, Academic Pr., London, pp. 115–213.
21. R.P. Philp, Mass Spectrom. Rev., 4 (1985) 1–54.
22. G.H. Isaksen, Abstract, 15th Meeting, EAOG, Manchester, U.K., 1991, pp. 361–363.
23. E.J. Gallegos, Anal. Chem., 48 (1976) 1348–1351.
24. G.A. Warburton and J.E. Zumberge, Anal. Chem., 55 (1983) 123–126.
25. F.W. McLafferty, ed., Tandem Mass Spectrometry, 1983, J. Wiley, New York.
26. A.S. Mackenzie, Applications of biological markers in petroleum geochemistry, in J. Brooks and D. Welte, eds., Advances in Petroleum Geochemistry, vol. 1, 1984, Academic Pr., London, pp. 115–213.
27. P.J. Grantham and L.L. Wakefield, Org. Geochem., 12 (1988) 61–73.
28. C.S. Hsu, A.G. Requejo, A. Lorber, S.C. Blum, G. Isaksen, and G. Hieshima, Proceedings of the 41st ASMS Conference on Mass Spectrometry and Allied Topics, San Francisco, CA, May 30–June 4, 1993, ASMS, Santa Fe, New Mexico, p. 144.

29. J.M. Moldowan, W.K. Seifert, and E.J. Gallegos, AAPG Bull., 69 (1985) 1255–1269.
30. C.S. Hsu, Encyclopedia of Analytical Science, (1995), Academic Pr., New York, pp. 2028–2034.
31. B. Durand, ed. Kerogen—Insoluble Oragnic Mater for Sedimentary Rocks, 1980, Edition Techniq, Paris.
32. B.P. Tissot and D.H. Welte, Petroleum Formation and Occurrence, 2nd ed., 1984, Springer-Verlag, New York.
33. T.I. Eglington and A.G. Douglas, Energy Fuels, 2 (1988) 81–88.
34. E.J. Gallegos, Anal. Chem., 47 (1975) 1524–1528.
35. M. Nip, J.W. de Leeuw, and P.A. Schenck, Geochim. Coschim. Acta, 52 (1988) 637–648.
36. D.E. Matthew and J.M. Hayes, Anal. Chem., 50 (1978) 1465–1473.
37. K.H. Freeman, The Carbon Isotope Compositions of Individual Compounds from Ancient and Modern Depositional Environments, 1991, Ph.D. dissertation, Indiana University, Bloomington, Indiana.
38. D.A. Merritt, W.A. Brand, and J.M. Hayes, Org. Geochem., 21 (1994) 573–583.
39. H. Craig, Geochim. Cosmochim. Acta, 12 (1957) 133–149.
40. H.M. Chung, C.C. Walters, S. Buck, and G. Bingham, Org. Geochem., 29 (1998) 381–396.
41. M. Rooney, A.A.K. Valetich, and C.E. Griffith, Org. Geochem., 29 (1998) 241–254.
42. M. Bjory, P.B. Hall, and R.P. Moe, Org. Geochem., 21 (1994) 761–776.
43. A.T. James, AAPG Bull., 74 (1990) 1441–1458.
44. M. Schoell, AAPG Bull., 67 (1983) 2235–2238.
45. G.P.A. Muscio, B. Horsfield, and D.H. Welte, Org. Geochem., 22 (1994) 461–476.
46. M.A. Goni and T.I. Eglington, J. High Resol. Chromtogr., 17 (1994) 476–488.
47. M. Schoell, M.A. McCaffrey, F.J. Fago, and J.M. Moldowan, Geochim. Cosmochim. Acta, 56 (1992) 1391–1399.
48. W.A. Hartgers, S.J.S. Damste, A.G. Requejo, J. Allan, J.M. Hayes, X. Ling, T.-M. Xie, J. Primack, and J.W. de Leeuw, Org. Geochem., 22 (1994) 703–725.
49. C&E News, Am. Chem. Soc., Sept. 20, 1999.
50. S. Teng, A.D. Williams, K. Urdal, J. High Resol. Chromatogr., 17 (1994) 469–475.
51. D. Estel, M. Mohnke, F. Biermans, H. Rotzsche, J. High Resol. Chromatogr., 18 (1995) 403–412.
52. U.S. Environmental Protection Agency (EPA), EPA 40 CFR Part 80, Fed. Reg., 59 (1994) 7826–7828.
53. American Society for Testing and Materials (ASTM), Annual Book of ASTM Standards, 1999, American Society for Testing and Materials, West Conshohocken, PA.
54. W.-C. Lai and C. Song, Fuel, 74 (1990) 1436–1451.
55. V.P. Nero and D. Drinkwater, Proceedings of the 46th ASMS Conference on Mass Spectrometry and Allied Topics, Orlando, FL, May 30–June 4, 1998, ASMS, Santa Fe, New Mexico, p. 813.
56. T.R. Ashe, R.J. Falkiner, and T.C. Lau, U.S. Patent 5,600,134, 1997.
57. T.R. Ashe, J.D. Kelly, T.C. Lau, and R.T.-K. Pho, U.S. Patent 5,602,755, 1997.
58. U.S. Environmental Protection Agency (EPA), Test Methods for Evaluating Solid Wastes: Physical and Chemical Methods, SW-846, vol. 1A, 1995, EPA, Washington, DC.

59. A.E. Bence, K.A. Kvenvolden, and M.C. Kennicutt II, Org. Geochem., 24 (1996) 7–42.
60. V.P. Nero and D. Drinkwater, Encyclopedia of Environmental Analysis and Remediation, vol. 6, 1998, J. Wiley, New York, pp. 3668–3686.
61. T. Ogawa, M. Araga, M. Okada, and Y. Fujimoto, SAE 952351, (1998) 21–41.
62. T. Ogawa, K. Nakakita, M. Yamanoto, M. Okada, and Y. Fujimoto, SAE 971605, (1997) 33–46.
63. A. Seaton, W. MacNee, K. Donaldson, and D. Godden, Lancet, 345 (1995) 176–178.
64. World Health Organization (WHO), Diesel Fuel and Exhaust Emission, Environmental Health Criteria 171, WHO, 1996, Geneva, Switzerland.
65. P.T. Williams, K.D. Bartle, and G.E. Andrews, Fuel, 65 (1986) 1150–1158.
66. M.C. Paputa-Peck, R.S. Marano, D. Schuetzle, T.L. Riley, C.V. Hampton, T.J. Prater, L.M. Skewes, T.E. Jensen, P.H. Ruehle, L.C. Bosch, and W.P. Duncan, Anal. Chem., 55 (1983) 1946–1954.
67. T.R. Henderson, J.D. Sun, R.E. Royer, C.R. Clark, A.P. Li, T.M. Harvey, D.H. Hunt, J.E. Fulford, A.M. Lovette, and W.R. Davidson, Environ. Sci. Technol., 17 (1983) 443–449.
68. D.R. Chaudhury, Environ. Sci. Technol., 16 (1982) 102–108.
69. C.S. Hsu, K. Qian, and Y.C. Chen, Anal. Chim. Acta, 264 (1992) 79–89.
70. S.C. Subba Rao and C. Fenselau, Anal. Chem., 50 (1978) 511–515.
71. C.S. Hsu and K. Qian, Anal. Chem., 65 (1993) 767–771.
72. D.F. Hunt and D.F. Harvey, Anal. Chem., 47 (1975) 2136–2141.
73. I. Dzidic, H.A. Peterson, P.A. Wadsworth, and H.V. Hart, Anal. Chem., 64 (1992) 2227–2232.
74. R. Malhotra, M.J. Coggiola, S. Young, C.S. Hsu, G.J. Dechert, P.M. Rahimi, and Y. Briker, Am. Chem. Soc. Div. Petrol. Chem. Preprints, 43 (1998) 507–509.
75. Z. Liang and C.S. Hsu, Energy Fuels, 12 (1998) 637–643.
76. E.J. Gallegos, Anal. Chem., 47 (1975) 1150–1154.
77. E.J. Gallegos, Anal. Chem., 53 (1981) 187–189.
78. C.S. Hsu, K. Qian, and W.K. Robbins, High Resol. Chromatogr., 17 (1994) 271–276.
79. J.B. Green, R.J. Hoff, P.W. Woodward, and L.L. Stevens, Fuel, 63 (1984) 1290–1301.
80. J.S. Thomson, J.B. Green, and T.B. McWilliams, Energy Fuels, 11 (1997) 909–914.
81. J.B. Green, S.K.-T. Yu, and R.P. Varna, J. High Resol. Chromatogr., 17 (1994) 427–438.
82. U. Hoffmann, S. Holzer, C.O. Meese, J. Chromatogr., 508 (1990) 349–356.
83. J. Green, S.K.-T. Yu, R.P. Vrana, J. High Resol. Chromatogr., 17 (1994) 439–451.
84. J.S. Thomson, J.B. Green, T.B. McWilliams, and S.K.-T. Yu, J. High Resol. Chromatogr., 17 (1994) 415–426.
85. E.N. Chen, D.E. Drinkwater, and J.M. McCann, J. Chrom. Sci., 33 (1995) 353–359.
86. D. Drinkwater, E.N. Chen, and V.P. Nero, Proceedings of the 45th ASMS Conference on Mass Spectrometry and Allied Topics, Palm Springs, CA, June 1–5, 1997, p. 178.
87. S.G. Roussis and J.W. Fedora, Anal. Chem., 69 (1997) 1550–1556.
88. G.W. Francis and K. Veland, J. Chromatogr., 219 (1981) 379–384.
89. B.D. Bhatt, S. Ali and J.V. Prasad, J. Chrom. Sci., 31 (1993) 113–119.

4

Analysis of Dioxins and Polychlorinated Biphenyls by Quadrupole Ion-Trap Gas Chromatography–Mass Spectrometry

J. B. Plomley
MDS SCIEX, Concord, Ontario, Canada

M. Lauševic
Tehnolosko-Metalurski Fakultet, Belgrade, Yugoslavia

R. E. March
Trent University, Peterborough, Ontario, Canada

1. INTRODUCTION

The experimental identification and measurement of dioxins and polychlorinated biphenyls (PCBs) became pressing issues once it was confirmed that some dioxin and PCB congeners were toxic to animals. Endeavors were initiated to determine the toxicity of each dioxin and PCB congener and to determine mass spectrometric protocols for the identification and measurement of each congener as it elutes from a gas chromatographic column. We are concerned here with the second endeavor and, in particular, with the application of the quadrupole ion trap as a tandem mass spectrometer of high detection specificity and sensitivity for dioxins and PCBs. Since dioxins are chemically very similar to furans (dioxins having two rather than one bridging oxygen atoms between the two benzo-moieties), they are produced concurrently and occur together in the environment. Hence any discussion of the determination of dioxins involves determination of furans also. Here, an understanding of quadrupole ion trap theory is assumed and reference is made to texts to assist the reader through the subsequent discussion of the determination of dioxins, furans, and PCBs. The section on dioxins and furans deals with the tandem mass spectrometric determination of tetra- to octa-chloro-substituted compounds with emphasis on the determination of coeluting compounds. The section on PCBs focuses on the use of chemical ionization (CI) for the determination of coeluting compounds, and the application of mass spectrometry (MS) for distinguishing toxic from nontoxic congeners.

2. THE QUADRUPOLE ION TRAP

The basic theory of operation of quadrupole devices was enunciated almost 100 years before the quadrupole ion trap was invented by Paul and Steinwedel [1]. The pioneering work of the inventors was recognized by the award of a 1989 Nobel Prize in Physics to Wolfgang Paul [2]. The quadrupole ion trap functions both as an ion store for gaseous ions and as a mass spectrometer. The mass-selective axial instability mode of ion trap operation developed by Stafford et al. [3] made possible the commercialization of the quadrupole ion trap. It should be recalled that the trapping parameter q_z is given by

$$q_z = \frac{8eV}{m(r_0^2 + 2z_0^2)\Omega^2} \tag{1}$$

where e is the electronic charge, V is the zero-to-peak amplitude of the RF potential of radial frequency Ω applied to the ring electrode, m is the mass of an ion, $2z_0$ is the separation of the end-cap electrodes, and r_0 is the inner radius of the ring electrode. The secular frequencies of ion motion are given by $\omega_{z,n}$

$$\omega_{z,n} = \left(n + \frac{1}{2}\beta_z\right)\Omega, \, 0 \leq n < \infty \tag{2}$$

where

$$\beta_z \approx \frac{q_z}{\sqrt{2}} \tag{3}$$

There are detailed accounts by Dawson and Whetten [4] and by Dawson [5] of the early development of quadrupole devices; a full account of ion trap theory by March et al. [6]; reviews by Todd [7], Cooks et al. [8], Glish and McLuckey [9], and March [10]; and three volumes entitled *Practical Aspects of Ion Trap Mass Spectrometry* [11] that contain accounts from 30 laboratories engaged in ion trap research.

3. TANDEM MASS SPECTROMETRIC DETERMINATION OF DIOXINS AND FURANS

3.1. PCDDs and PCDFs

Polychlorodibenzo-p-dioxins (PCDDs) and polychlorodibenzofurans (PCDFs) are two classes of compounds that are of environmental concern because of the high toxicity of those isomers with 2,3,7,8-tetrachloro-substitution [12]. The PCDDs/PCDFs are monitored in air, rain, effluents, soil, and biota matrices. Usually, the total concentration of all PCDD/PCDF congener groups (i.e., total tetrachlorinated dioxins, total pentachlorinated dioxins, etc.) and the concentrations of each of the seventeen 2,3,7,8-substituted toxic isomers are monitored. Such determinations require extensive sample preparation, sixteen expensive $^{13}C_{12}$-2,3,7,8-substituted internal standards [13], and mass spectrometers of high capital cost. High-resolution MS [14,15] or triple-stage quadrupole MS (TSQMS) [16–18] is necessary in order to differentiate between (1) PCDDs/PCDFs and interferences, such as polychlorinated biphenyls [19], and (2) $^{13}C_{12}$-PCDFs and native PCDDs.

3.1.1. Tandem Mass Spectrometry

In 1994, a rapid screening technique using the quadrupole ion trap was demonstrated for the detection and quantitation of 2,3,7,8-TCDD. This technique involved operation of the ion trap in a tandem mass spectrometric (MS–MS) mode and monitoring of the predominant transition $[M + 2]^{+\bullet} \rightarrow [M + 2 - COCl^{\bullet}]^{+}$ [20,21]. The initial sensitivity of the ion-trap MS–MS technique was 500 fg/μl with a signal-to-noise ratio (S/N) of 5:1. In ion-trap MS–MS mode, it is now possible to deconvolute mass spectra generated from analytes that coelute [22,23] and monitor multiple product ions, from tetra- to octa-PCDDs/PCDFs, in a single chromatographic acquisition with a detection limit of less than 100 ppt [24]. Up to eight scan functions can be applied in a chromatographic retention-time window. Collision-induced dissociation (CID) can be affected in four modes: (1) single-frequency irradiation (SFI), (2) multifrequency irradiation (MFI), (3) secular-frequency modulation (SFM), and (4) nonresonant excitation [25]. Single-frequency irradiation is performed at a fixed value of q_z by the application of a dipolar supplementary AC signal across the end-cap electrodes. Secular-frequency modulation involves modulating the RF drive potential amplitude applied to the ring electrode such that ions move into and out of resonance with an applied single-frequency waveform. Multifrequency irradiation involves the application of a waveform consisting of several frequency components at a constant value of q_z. Tandem mass spectrometry offers high specificity for the determination of PCDDs/PCDFs [17], minimizes the likelihood of isobaric interferences, and reduces significantly chemical noise with concomitant enhancement of the S/N ratio. The specificity of MS–MS has permitted utilization of

TSQMS instruments of relatively low mass resolution for the determination of PCDDs/PCDFs [16,18,26]. Compared with the TSQMS in MS–MS mode, the ion trap offers three advantages. First, the ion trap accumulates ions mass-selectively over time, such that a selected target number of ions will be stored over a wide range of eluant concentrations. Second, CID in the ion trap is wrought by many (approximately 200) collisions of mass-selected ions with helium buffer gas atoms wherein the energy transferred in a single collision is rarely greater than that of a vibrational quantum such that the lowest energy dissociative channels are accessed exclusively. Third, the mass-selected ions are dissociated completely and ≈90% of fragment ions are confined subsequently.

3.1.2. Resonance Excitation Modes

In a previous comparison [27] of resonant excitation by SFI with SFM and MFI with bandwidths between 1 and 2 kHz, it was found that MFI was the most appropriate mode for an analytical protocol since it exhibited high CID efficiency during a brief "tickling" period of approximately 10 msec and obviated the strict requirement for waveform calibration.

3.1.3. Isolation of Molecular Ions

In quadrupole ion-trap MS–MS [28], the two most intense ions in the molecular ion cluster are selected for isolation, that is, $[M]^{+\cdot}$ and $[M + 2]^{+\cdot}$ (where $[M]^{+\cdot}$ is the molecular ion containing only $^{35}Cl^{\cdot}$ atoms and $[M + 2]^{+\cdot}$ contains one $^{37}Cl^{\cdot}$ atom) for T_4CDF, P_5CDF, H_6CDF, H_7CDF, T_4CDD, P_5CDD, H_6CDD, H_7CDD, and their $^{13}C_{12}$-labeled analogues, and $[M + 2]^{+\cdot}$ and $[M + 4]^{+\cdot}$ (where $[M + 4]^{+\cdot}$ is the molecular ion containing two $^{37}Cl^{\cdot}$ atoms) for O_8CDF, O_8CDD, and $^{13}C_{12}$-O_8CDD. T_4CDD/F is an accepted abbreviation for tetrachlorodibenzo-*p*-dioxin/tetrachlorodibenzofuran; thus P_5 indicates a pentachlorinated species and H_6 a hexachlorinated species, etc.

Optimization of the ion isolation process entails construction of part of a scan function that will effect complete ejection of all other ions while minimizing parent ion losses. The CID stage entails the completion of the same scan function such that the isolated molecular ion species, $[M]^{+\cdot}$ and $[M + 2]^{+\cdot}$, are excited resonantly and dissociate. Optimization of this process entails selection of the value for q_z and selection of the frequency (or band of frequencies), amplitude, and duration of application such that all of the irradiated ions are dissociated and the retention of fragment ions is maximized. This process is simplified by selecting a common value for q_z, say $q_z = 0.4$, for all scan functions.

3.1.4. Instrumentation

The experiments were performed on a Varian (Walnut Creek, CA) Saturn 3D bench-top gas chromatograph–tandem mass spectrometer equipped with a Varian 8200 autosampler. Saturn 5.2 was used for data acquisition that is compatible with the multiple-scan-function software Ion Trap Toolkit™ for MS–MS 1.0. Varian Toolkit software permits the determination of approximately 200 scan functions that can then be recalled in a specified time sequence. The injector was maintained at 300°C, the manifold heater at 220°C, and the transfer line at 260°C. Helium pressure at the head of the column was maintained at 26 psi. The 60 m × 0.25 mm ID × 0.25 μm DB-5 (diphenyldimethylpolysiloxane) fused-silica capillary column was obtained from J&W Scientific (Folsom, CA). The analytical ramp was scanned at 5555 Da/sec; axial modulation was carried out with an amplitude of 3 $V_{(0-p)}$ and at a frequency of 485 kHz. The electron multiplier was biased at a voltage of approximately 1800 V to provide an ion signal gain of 10^5.

3.1.5. Scan Functions

A scan function is a graphical representation of the temporal variation of potentials applied to ion trap electrodes throughout an experiment. The scan function employed for the determination of T₄CDDs is shown in Figure 1, where the abscissa represents time and the ordinate represents the amplitudes of the voltages. For a given trap magnitude, geometry, and drive frequency, the amplitude

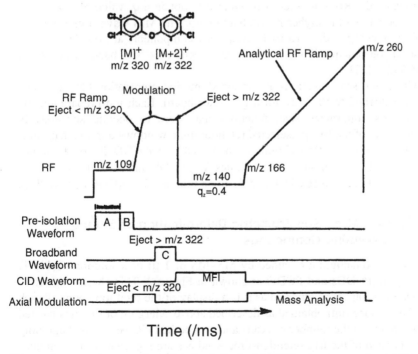

Figure 1   Scan function for MS/MS of dioxin (T₄CDD); the prescan for AGC is not shown.

of the RF voltage determines the low-mass cutoff (LMCO) value, which is the lowest mass/charge ratio for ions confined in the ion trap. The LMCO value during the ionization period (A on Fig. 1) was set at m/z 160. A total ion target of 35,000 counts was set for the automatic gain control (AGC) algorithm. The filament emission current was 50 μA. The supplementary alternating waveforms were applied to the end-cap electrodes in dipolar fashion; such waveforms are employed for ion isolation, ion excitation, and axial modulation.

A coarse-isolation waveform for the ejection of all ions except those selected covered the range 3.7 to 513.5 kHz with a 1-kHz notch corresponding to the secular frequency of the selected molecular ions; this waveform of amplitude some 20 $V_{(0-p)}$ was imposed during ionization (A on Fig. 1) and prolonged after ionization during period B. For T_4CDD shown in Figure 1, the notch is centered at 174.5 kHz for m/z 320 and m/z 322 with a q_z value of about 0.45. Fine isolation was achieved by ramping the RF amplitude until the LMCO was just less than m/z 320 so as to eject ions of lower mass/charge ratio; ion ejection was facilitated by concurrent application of axial modulation with an amplitude of 3 $V_{(0-p)}$. The RF amplitude was then decreased so that ions with m/z >322 were ejected upon application of a broadband waveform having an amplitude of 30 $V_{(0-p)}$ and lasting for 5 msec (C on Fig. 1).

Finally, the RF amplitude was reduced to obtain a q_z value of 0.4 for the selected ion having the higher mass/charge ratio. The MFI waveform employed for CID was composed of 13 to 17 frequency components at 500-Hz intervals and was of constant amplitude. In each scan function the MFI band was centered on 153.6 kHz [27].

Optimized scan functions are arranged into ion-preparation files and each file is downloaded to the waveform generator board. Each ion-preparation file included those optimized scan functions that were to operate successively throughout one of the five preselected retention-time windows during a gas chromatography (GC) run: 20 to 27 minutes for T_4CDF and T_4CDD, 27 to 34 minutes for P_5CDF and P_5CDD, 34 to 42.5 minutes for H_6CDF and H_6CDD, 42.5 to 51 minutes for H_7CDF and H_7CDD, and 51 to 58 minutes for O_8CDF and O_8CDD.

3.2. Tandem Mass Spectrometric Determination of Eluting and Coeluting Compounds

A total ion chromatogram obtained by MS–MS of 1 μl of a solution (Table 1) containing 35 PCDDs and PCDFs is shown in Figure 2. This chromatogram is a merged total ion chromatogram as it is a display of the total ion count of each merged mass spectrum obtained as described above. Merging of the files has led to a contraction in the number of scans and a smoothed GC peak. The beginning and end of each of the five retention-time windows are represented by diamond-shaped indicators on the abscissa; the appropriate ion-preparation file was acti-

Table 1 PCDD and PCDF Components of the 18 Peaks in the Total Ion Chromatogram Shown in Figure 2, and Their Concentrations

Peak number[a]	Compound	Conc. (pg/μl)	Compound	Conc. (pg/μl)	Compound	Conc. (pg/μl)
1	2,3,7,8-T_4CDF	200	$^{13}C_{12}$-2,3,7,8-T_4CDF	100		
2	$^{13}C_{12}$-1,2,3,4-T_4CDD	100				
3	2,3,7,8-T_4CDD	200	$^{37}Cl_4$-2,3,7,8-T_4CDD	200	$^{13}C_{12}$-2,3,7,8-T_4CDD	100
4	1,2,3,7,8-P_5CDF	1000	$^{13}C_{12}$-1,2,3,7,8-P_5CDF	100		
5	2,3,4,7,8-P_5CDF	1000	$^{13}C_{12}$-2,3,4,7,8-P_5CDF	100		
6	1,2,3,7,8-P_5CDD	1000	$^{13}C_{12}$-1,2,3,7,8-P_5CDD	100		
7	1,2,3,4,7,8-H_6CDF	1000	$^{13}C_{12}$-1,2,3,4,7,8-H_6CDF	100		
8	1,2,3,6,7,8-H_6CDF	1000	$^{13}C_{12}$-1,2,3,6,7,8-H_6CDF	100		
9	2,3,4,6,7,8-H_6CDF	1000	$^{13}C_{12}$-2,3,4,6,7,8-H_6CDF	100		
10	1,2,3,4,7,8-H_6CDD	1000	$^{13}C_{12}$-1,2,3,4,7,8-H_6CDD	100		
11	1,2,3,6,7,8-H_6CDD	1000	$^{13}C_{12}$-1,2,3,6,7,8-H_6CDD	100		
12	1,2,3,7,8,9-H_6CDD	1000	$^{13}C_{12}$-1,2,3,7,8,9-H_6CDD	100		
13	1,2,3,7,8,9-H_6CDF	1000	$^{13}C_{12}$-1,2,3,7,8,9-H_6CDF	100		
14	1,2,3,4,6,7,8-H_7CDF	1000	$^{13}C_{12}$-1,2,3,4,6,7,8-H_7CDF	100		
15	1,2,3,4,6,7,8-H_7CDD	1000	$^{13}C_{12}$-1,2,3,4,6,7,8-H_7CDD	100		
16	1,2,3,4,7,8,9-H_7CDF	1000	$^{13}C_{12}$-1,2,3,4,7,8,9-H_7CDF	100		
17	O_8CDD	2000	$^{13}C_{12}$-O_8CDD	200		
18	O_8CDF	2000				

[a] Peak number corresponds to the elution order as shown in Figure 2.

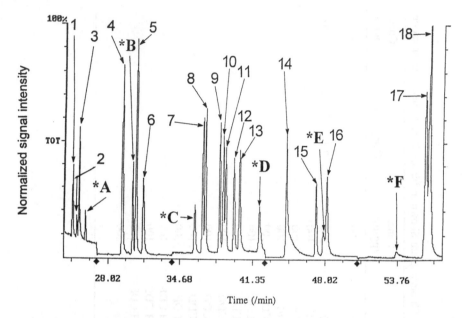

Figure 2 Total ion chromatogram obtained after the mass spectrum merging procedure for dioxins and furans contained in 1 μl of solution described in Table 1. The diamond shapes reported on the abscissa indicate the retention-time windows during which successive degrees of chlorinated compounds were eluted.

vated during each window. The peaks indicated with asterisks and identified as A through F in Figure 2 are due to impurities present in the sample.

Consider peaks 1, 2, and 3 observed during the first retention-time window of Figure 2. Peak 1 is due to 2,3,7,8-T₄CDF and its coeluting ¹³C-labeled isotopomer, peak 2 is due solely to ¹³C₁₂-1,2,3,4-T₄CDD, while peak 3 is a composite not only of the native and ¹³C-labeled compounds but of ³⁷Cl₄-2,3,7,8-T₄CDD also, as reported in Table 1. For 2,3,7,8-T₄CDF, the molecular ions M⁺˙ of m/z 304 and [M + 2]⁺˙ of m/z 306 were isolated simultaneously then dissociated using MFI with the first scan function of the first ion-preparation file. Fragment ion signal intensities arising from the loss of COCl˙ to yield [M − COCl˙]⁺ of m/z 241 and m/z 243 were recorded. The fragment ions of m/z 241 and 243 arise as follows:

$$[C_{12}H_4Cl_4O]^{+\bullet} \rightarrow [C_{11}H_4Cl_3]^+ + COCl^\bullet \tag{4}$$
$$\text{(m/z 304)} \qquad \text{(m/z 241)}$$

$$[C_{12}H_4Cl_3{}^{37}ClO]^{+\bullet} \rightarrow [C_{11}H_4Cl_3]^+ + CO^{37}Cl^\bullet \tag{5}$$
$$\text{(m/z 306)} \qquad \text{(m/z 241)}$$

$$\rightarrow [C_{11}H_4Cl_2{}^{37}Cl]^+ + COCl^\bullet \tag{6}$$
$$\text{(m/z 243)}$$

Now as 2,3,7,8-T$_4$CDF coelutes with the labeled isotopomer ^{13}C$_{12}$-2,3,7,8-T$_4$CDF, the second scan function of the first ion-preparation file is used for the ionization of ^{13}C$_{12}$-2,3,7,8-T$_4$CDF and simultaneous isolation of the molecular ions M$^{+\cdot}$ of m/z 316 and [M + 2]$^{+\cdot}$ of m/z 318; these isolated ion species are then dissociated using MFI. The fragment ion signal intensities arising from the loss of ^{13}COCl$^{\cdot}$ to yield [M$-^{13}$COCl$^{\cdot}$]$^+$ of m/z 252 and m/z 254 were recorded; these fragment ions arise as follows:

$$[^{13}C_{12}H_4Cl_4O]^{+\cdot} \rightarrow [^{13}C_{11}H_4Cl_3]^+ + {}^{13}COCl^{\cdot} \tag{7}$$
$$\underset{(m/z\,316)}{} \qquad \underset{(m/z\,252)}{}$$

$$[^{13}C_{12}H_4Cl_3{}^{37}ClO]^{+\cdot} \rightarrow [^{13}C_{11}H_4Cl_3]^+ + {}^{13}CO^{37}Cl^{\cdot} \tag{8}$$
$$\underset{(m/z\,318)}{} \qquad \underset{(m/z\,252)}{}$$

$$\rightarrow [^{13}C_{11}H_4Cl_2{}^{37}Cl]^+ + {}^{13}COCl^{\cdot} \tag{9}$$
$$\underset{(m/z\,254)}{}$$

Peak 2 of Figure 2 corresponds to MS–MS of labeled ^{13}C$_{12}$-1,2,3,4-T$_4$CDD for which a third scan function is required in order to control ionization and isolation of the molecular ions M$^{+\cdot}$ of m/z 332 and [M + 2]$^{+\cdot}$ of m/z 334. These isolated ion species were then dissociated and the fragment ion signal intensities arising from the loss of ^{13}COCl$^{\cdot}$ to yield [M-^{13}COCl$^{\cdot}$]$^+$ of m/z 268 and m/z 270 were recorded. The fragment ions of m/z 268 and 270 arise as follows:

$$[^{13}C_{12}H_4Cl_4O_2]^{+\cdot} \rightarrow [^{13}C_{11}H_4Cl_3O]^+ + {}^{13}COCl^{\cdot} \tag{10}$$
$$\underset{(m/z\,332)}{} \qquad \underset{(m/z\,268)}{}$$

$$[^{13}C_{12}H_4Cl_3{}^{37}ClO_2]^{+\cdot} \rightarrow [^{13}C_{11}H_4Cl_3O]^+ + {}^{13}CO^{37}Cl^{\cdot} \tag{11}$$
$$\underset{(m/z\,334)}{} \qquad \underset{(m/z\,268)}{}$$

$$\rightarrow [^{13}C_{11}H_4Cl_2{}^{37}ClO]^+ + {}^{13}COCl^{\cdot} \tag{12}$$
$$\underset{(m/z\,270)}{}$$

Peak 3 of Figure 2 is composed of the fragment ion counts from three compounds, thus two additional scan functions were used in sequence for the MS–MS determination of the coeluting 2,3,7,8-T$_4$CDD, ^{13}C$_{12}$-2,3,7,8-T$_4$CDD, and ^{37}Cl$_4$-2,3,7,8-T$_4$CDD. The scan function for ^{13}C$_{12}$-2,3,7,8-T$_4$CDD is the same as that used for ^{13}C$_{12}$-1,2,3,4-T$_4$CDD of peak 2. For 2,3,7,8-T$_4$CDD, the molecular ions M$^{+\cdot}$ of m/z 320 and [M + 2]$^{+\cdot}$ of m/z 322 were isolated then dissociated and the fragment ion signal intensities arising from the loss of COCl$^{\cdot}$ to yield [M $-$ COCl$^{\cdot}$]$^+$ of m/z 257 and m/z 259 were recorded. The fragment ions of m/z 257 and 259 arise as follows:

$$[C_{12}H_4Cl_4O_2]^{+\cdot} \rightarrow [C_{11}H_4Cl_3O]^+ + COCl^{\cdot} \tag{13}$$
$$\underset{(m/z\,320)}{} \qquad \underset{(m/z\,257)}{}$$

$$[C_{12}H_4Cl_3{}^{37}ClO_2]^{+\cdot} \rightarrow [C_{11}H_4Cl_3O]^+ + CO^{37}Cl^{\cdot} \tag{14}$$
$$\underset{(m/z\,322)}{} \qquad \underset{(m/z\,257)}{}$$

$$\rightarrow [C_{11}H_4Cl_2{}^{37}ClO]^+ + COCl^{\cdot} \tag{15}$$
$$\underset{(m/z\,259)}{}$$

For $^{37}Cl_4$-2,3,7,8-T_4CDD, the molecular ion $M^{+\cdot}$ of m/z 328 was isolated then dissociated and the fragment ion signal intensity arising from the loss of $CO^{37}Cl^{\cdot}$ to yield $[M-CO^{37}Cl^{\cdot}]^+$ of m/z 263 was recorded. The fragment ions of m/z 263 arise as follows:

$$[C_{12}H_4{}^{37}Cl_4O_2]^{+\cdot} \rightarrow [C_{11}H_4{}^{37}Cl_3O]^+ + CO^{37}Cl^{\cdot} \qquad (16)$$
$$\text{(m/z 328)} \qquad\qquad \text{(m/z 263)}$$

In the above discussion of peaks 1, 2, and 3 of Figure 2 where five different scan functions were required for MS–MS of six different tetrachlorinated dioxins and furans, only a single fragmentation channel was considered, that of the loss of $COCl^{\cdot}$. While $COCl^{\cdot}$ is the major fragmentation channel, it is not the sole loss channel. For the dioxins $[C_{12}H_{8-x}Cl_xO_2]^{+\cdot}$, where x = 4 to 8, the other fragmentation channels involve the loss of $Cl^{\cdot}$, $2COCl^{\cdot}$, and $(CO)_2Cl^{\cdot}$, while for the furans, where again x = 4 to 8, the other fragmentation channels involve the loss of $Cl^{\cdot}$, $COCl_2$, and $COCl_3^{\cdot}$.

Let us examine the three selected ion chromatograms shown in the unmerged file format in Figure 3(a). The top trace shows the product-ion chromatogram for O_8CDF, the middle trace for O_8CDD, and the bottom for $^{13}C_{12}-O_8CDD$. O_8CDD (middle) and $^{13}C_{12}-O_8CDD$ (bottom) coelute, while elution of the furan, O_8CDF (top), overlaps the other two compounds. Yet the mass selectivity is not lost as is seen from the accompanying product-ion mass spectrum for each compound. Each product-ion mass spectrum shown in Figure 3(a) can be compared with the corresponding mass spectrum taken from the three selected ion chromatograms (Fig. 3(b)) obtained after ion isolation and prior to CID. The mass spectra of Figure 3(b) show the isolated molecular $[M + 2]^{+\cdot}$ and $[M + 4]^{+\cdot}$ peaks. It should be noted that the abscissa of each trace in Figure 3(a) (CID) and Figure 3(b) (isolation) indicates both number of mass spectra and retention time. The retention-time windows illustrated in parts (a) and (b) are very similar, however the number of mass spectra/unit time in the former is twice that in the latter. The acquisition rate during CID (Fig. 3(a)) is twice that during isolation (Fig. 3(b)) due to the smaller mass range scanned in the former.

Figure 3 Unmerged ion chromatograms for three coeluting octachlorinated compounds. Each chromatogram is composed from the signal intensities of the ion species identified on the ordinate. (a) The fragment ion signal intensities were obtained from CID mass spectra and one such mass spectrum is displayed for each compound. (b) The isolated molecular ion signal intensities were obtained prior to CID and a mass spectrum showing the isolated molecular ions is displayed for each compound.

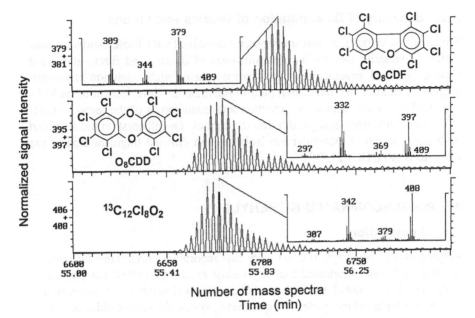

(a)

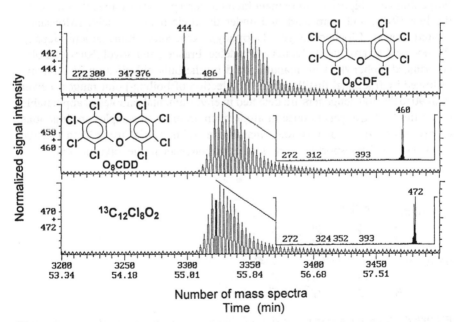

(b)

3.3. Summary of Determination of Dioxins and Furans

The quadrupole ion-trap mass spectrometer combined with Toolkit software constitutes a versatile tool for the determination of dioxins and furans that have similar chemical structure and may coelute during high-resolution chromatographic separation. We have attempted to illustrate the development of an MS–MS method for the ultra-trace detection and quantitation of the tetra- to octa-PCDDs/PCDFs using isotopic dilution techniques. This illustration has involved both optimization of molecular ion isolation and directed fragmentation of the tetra- to octa-PCDDs/PCDFs.

4. POLYCHLORINATED BIPHENYLS

4.1. Introduction

Polychlorinated biphenyls (PCBs) were first reported in environmental samples in 1966 [29]. Polychlorinated biphenyls exist as 209 individual compounds or congeners that exhibit all the variations of position and number of chlorine substitutions in a biphenyl molecule. The numbering system for sites of chlorine substitution on the biphenyl molecule is shown in Figure 4. Ballschmiter and Zell [30] proposed a system wherein each congener is assigned a number from 1 to 209. Polychlorinated biphenyls were manufactured for a period of some 50 years, until the late 1970s, and were marketed under the trade names Aroclor (Monsanto, United States), Clophen (Bayer, Germany), Kanechlor (Kanegafuchi, Japan), Phenoclor (Caffaro, Italy), Pyralene (Prodelec, France), and Sovol (Russia). Polychlorinated biphenyls were manufactured by the bulk chlorination of biphenyl followed by fractional distillation and collection of boiling-point ranges. Fewer than 90 to the 209 congeners are detected in environmental samples in appreciable quantities [31]. Congener concentrations in environmental samples are functions of original product compositions, magnitudes of industrial usage, and relative persistence and transport properties of each congener in the environment.

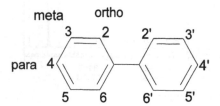

Figure 4 Numbering system for sites of chlorine substitution on PCBs, and positions of *ortho*, *meta*, and *para* chlorines on the biphenyl molecule.

The early quantification of PCBs in environmental samples was accomplished by comparing chromatographic profiles with Aroclor mixtures; PCB analysis data were reported in terms of Aroclor equivalents. This simple approach was clearly unsatisfactory. Congener 153 (2,2',4,4',5,5'-hexachlorobiphenyl) is a dominant constituent in most environmental samples; it is a major component of both Aroclor 1254 (6.2%) and Aroclor 1260 (10.6%). Due to the high resistance of this congener to metabolic and microbial breakdown [32], it is readily bioaccumulated and appears in marine and freshwater biota from remote locations [33]. However, congener 153 is not very toxic in that its toxicity is 5000 times less than that of the most toxic PCB congener, congener 169 [34].

A congener-specific approach to PCB analysis was required for two principal reasons [35]. First, there are significant differences in congener relative concentrations in environmental samples and commercial mixtures. Second, the enormous differences in the toxicities of PCB congeners necessitate congener-specific analysis in order to calculate the total toxic potential of a mixture of PCBs. For the more toxic congeners, information is required on their incidence, environmental weathering, bioaccumulation, and metabolism.

The toxicity of PCB mixtures is due principally to a small group of non-*ortho*- and mono-*ortho*-substituted congeners [37] such as congener 169 (3,3',4,4',5,5'-hexachlorobiphenyl). Congeners with partial planarity, due to substitution at only one *ortho* position (mono-*ortho*-PCBs) exhibit enhanced toxicity relative to nonplanar PCBs that have two or more chlorines in *ortho* positions. The similarity in structure between these compounds and 2,3,7,8-tetrachlorodibenzo-*p*-dioxin (Fig. 5) may account for the comparable biochemical and toxic responses in humans, laboratory animals, and mammalian cells in culture [34]. It has been suggested that a small number of readily resolvable congeners be selected for quantitation during routine environmental analysis [36]. Interest in the toxicological effects of PCBs has been focused on a small group of planar PCB congeners referred to as coplanar PCBs [37].

4.2. Toxicity Equivalent Factors

The development of toxicity equivalent factors (TEFs) for halogenated aromatic compounds, as proposed by Safe [34], is based on the observed common receptor-mediated mechanism of action of toxic halogenated aromatics. The relative toxicity of 2,3,7,8-TCDD is taken as unity on the TEF scale, and individual TEF values have been derived for the other stereoisomers including PCBs [38]. Recently, a thermodynamic model for calculating the TEF values for PCBs was reported by Kafafi et al. [39]; their model permitted calculation of the dissociation constants of complexes formed between the aryl hydrocarbon receptor and a PCB molecule. McFarland and Clarke [40] have argued that, as the toxicity of individual congeners varies over a range of 10^{12}, the determination of total PCB content is of

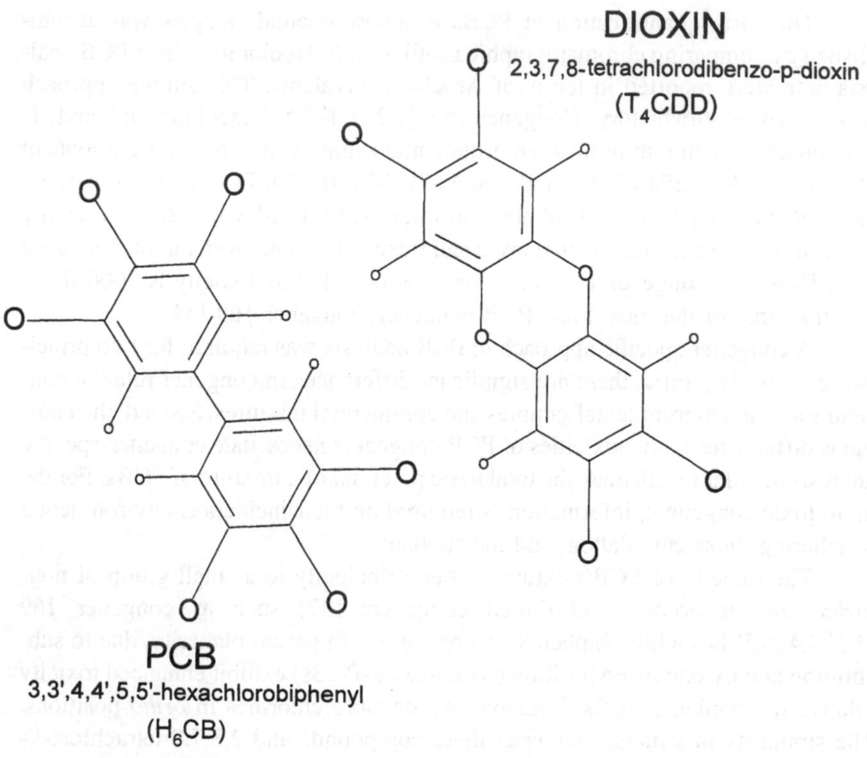

DIOXIN
2,3,7,8-tetrachlorodibenzo-p-dioxin
(T_4CDD)

PCB
3,3',4,4',5,5'-hexachlorobiphenyl
(H_6CB)

Figure 5 Comparison of the molecular configuration and positions of chlorine substitution for 2,3,7,8-tetrachlorodibenzo-*p*-dioxin and 3,3',4,4',5,5'-hexachlorobiphenyl (congener 169).

limited value; thus, an attempt has been made to recognize and to distinguish the highly toxic PCB congeners from those congeners that are less toxic. Based on the TEF approach [38], where the toxicity of a given PCB congener is expressed relative to that of TCDD having unit value, a toxicity index is then given as a single number equal to the sum of the product of concentration and TEF for each congener present. A single TEF value can express an assessment of the total toxicity of a sample and is more informative than total PCB content.

As experimental toxicity data for most PCBs are not known, the toxicities of PCBs as determined by theoretical calculation carried out by Kafafi et al. [39] have been considered. In Table 2, the results of Kafafi et al. [39] are reviewed as a function of the total number of chlorine atoms substituted in the PCB molecule and the number of chlorine atoms substituted in *ortho* positions only. The values given in Table 2 are those suggested by Safe [38] and correspond to a

Table 2 TEFs of PCBs as Calculated by Kafafi et al. [39][a]

Isomer	Non-*ortho*	Mono-*ortho*	Di-*ortho*	Tri-*ortho*	Tetra-*ortho*
Cl	10^{-6}–10^{-7}	10^{-9}	—	—	—
	(2)	(1)			
2Cl	10^{-4}–10^{-6}	10^{-7}–10^{-8}	10^{-10}	—	—
	(5)	(5)	(2)		
3Cl	10^{-4}–10^{-6}	10^{-6}	10^{-7}–10^{-10}	10^{-12}	—
	(5)	(11)	(9)	(1)	
4Cl	10^{-3}–10^{-4}	10^{-5}–10^{-7}	10^{-7}–10^{-9}	10^{-10}–10^{-11}	10^{-13}
	(5)	(14)	(17)	(5)	(1)
5Cl	10^{-2}–10^{-3}	10^{-4}–10^{-6}	10^{-5}–10^{-8}	10^{-8}–10^{-10}	10^{-11}–10^{-13}
	(2)	(11)	(18)	(11)	(2)
6Cl	10^{-1}	10^{-4}–10^{-5}	10^{-5}–10^{-7}	10^{-8}–10^{-10}	10^{-10}–10^{-12}
	(1)	(5)	(17)	(14)	(5)
7Cl	—	10^{-3}	10^{-5}–10^{-6}	10^{-7}–10^{-9}	10^{-9}–10^{-11}
		(1)	(7)	(11)	(5)
8Cl	—	—	10^{-4}	10^{-6}–10^{-7}	10^{-8}–10^{-10}
			(2)	(5)	(5)
9Cl	—	—	—	10^{-6}	10^{-8}
				(5)	(2)
10Cl	—	—	—	—	10^{-7}
					(1)

[a] TEFs were calculated as a function of the total number of chlorine atoms in the PCB molecule and the number of *ortho*-substituted chlorine atoms. The numbers in parentheses are the total number of isomers for which the range of TEF values is given. The most toxic congeners are in the box.

range of toxicity for isomers with a given number of *ortho*-substituted chlorine atoms while the number in parentheses give the number of isomers in each group.

The toxicity increases regularly across Table 2 from right to left as the number of *ortho*-substituents decreases in an isomer group (in each row) and down Table 2 as the degree of chlorine substitution increases (in each column). Consequently, the more toxic PCBs are non- and mono-*ortho*-substituted congeners containing four to seven chlorine atoms in the molecule; the locus of these most toxic congeners is the boxed area in Table 2 that contains 39 congeners. In a hypothetical mixture of equal amounts of all 209 PCBs, the 39 most toxic congeners account for 99.5% of total toxicity [41].

A similar approach to a congener-specific detection procedure using a limited number of congeners [42–46] has been adopted for use in Germany and the Netherlands; for this procedure, seven congeners are prescribed, that is, 28, 52,

101, 118, 138, 153, and 180; however, the choice of congeners is not related to toxicity.

Recently, 36 congeners have been identified as being the most environmentally threatening of the PCBs based on their reported frequency of occurrence in environmental samples, their relative abundance in animal tissue, and their potential for toxicity [40]. This list includes all save one, #28, of the congeners selected by Germany and the Netherlands, but only 13 of the 39 toxic congeners identified in Table 2.

4.3. Variation of Elution Order with Molecular Structure and Toxicity

In general, PCB molecules elute from a GC column in order of increasing chlorine number, while the elution among isomers depends upon the number of *ortho*-substituted chlorines. The retention time is shortest for tetra-*ortho*-substituted isomers and increases as the number of *ortho*-substituted chlorine atoms decreases.

Since the retention times of mono-*ortho*- and non-*ortho*-substituted PCBs are usually greater than those of other isomers, coplanar PCBs can coelute with nonplanar PCBs of higher chlorine number. The marked overlap of isomer groups with respect to elution is illustrated by the chromatograms of a mixture of 13

Table 3 Mixture of 13 PCB Congeners Containing Non-*ortho*-, Mono-*ortho*-, and Di-*ortho*-Substituted Tetra-, Penta-, and Hexachlorobiphenyl Isomers

Chlorine number	Positional substitution	*ortho*	IUPAC number	Elution order number	Relative retention time[a]
Tetra	2,2′,5,5′-	di	52	1	0.757
	2,2′,4,4′-	di	47	2	0.768
	2,3,4,4′-	mono	66	3	0.840
	3,4,4′,5-	non	81	6	0.910
	3,3′,4,4′-	non	77	7	0.926
Penta	2,2′,4,5,5′-	di	101	4	0.870
	2,3′,4,4′,6-	di	119	5	0.888
	2′,3′,4,4′,5-	mono	118	8	0.964
	2,3,3′,4,4′-	mono	105	10	1.011
	3,3′,4,4′,5-	non	126	11	1.075
Hexa	2,2′,4,4′,5,5′-	di	153	9	1.000
	2,3,3′,4,4′,5-	mono	156	12	1.176
	3,3′,4,4′,5,5′-	non	169	13	1.282

[a] Retention time for congener 153 is taken as unity.

PCBs subjected to electron ionization (EI). The mixture contains the non-*ortho*-, mono-*ortho*-, and di-*ortho*-substituted Cl_4, Cl_5, and Cl_6 isomers listed in Table 3 together with their elution order. In Figure 6, the top trace presents the total ion chromatogram of the mixture, and the 13 PCBs are marked in their elution order (1 to 13). The succeeding traces show the selected ion chromatograms for the isomers containing four, five, and six chlorine atoms, respectively, where the intensity sum of the $M^{+\cdot}$, $[M + 2]^{+\cdot}$, and $[M + 4]^{+\cdot}$ ions has been plotted in each case. The 13 PCBs have been identified in the selected ion chromatograms by their IUPAC numbers. While compounds eluted from the column in the order di-*ortho*-substituted before mono-*ortho*- before non-*ortho*- in each isomer group, the isomer groups overlap in retention time.

4.4. Identification of Non- and Mono-*ortho*-Substituted Polychlorinated Biphenyls

Positive identification of non- and mono-*ortho*-substituted congeners can be accomplished with EI. The fragmentation pattern observed in the EI mass spectra of all PCBs shows a variety of chlorine atom losses from the molecular ion. The

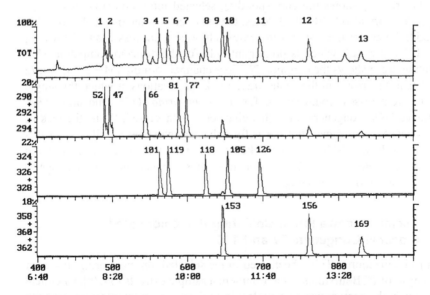

Figure 6 Chromatograms for the mixture of 13 PCBs listed in Table 3. The top trace shows the total ion chromatogram with the peaks numbered in order of relative retention time. The succeeding traces show the selected ion chromatograms of the intensity sum of the $M^{+\cdot}$, $[M + 2]^{+\cdot}$ and $[M + 4]^{+\cdot}$ ions from isomers containing four, five and six chlorine atoms, respectively. Each peak in the selected ion chromatograms is identified by its IUPAC number.

EI mass spectra for the non- and mono-*ortho*-substituted congeners are virtually indistinguishable yet differ significantly from those of the di-, tri- and tetra-*ortho*-substituted congeners, which are also virtually indistinguishable. Single chlorine atom loss is specific to the di-, tri- and tetra-*ortho*-substituted congeners and is pronounced. As the 39 more-toxic congeners (see Table 2) do not exhibit loss of a chlorine atom, the selective ion monitoring can facilitate clear differentiation between toxic and nontoxic congeners on the basis of their fragmentation behavior.

Single chlorine atom loss from the molecular ions of di-*ortho*-substituted tetrachlorinated and hexachlorinated PCBs is shown in Figure 7. Figure 7(a) shows selected ion chromatograms of the five tetrachlorinated PCBs of the mixture of 13 PCBs identified by their IUPAC number and listed in Table 3; the top trace is of the molecular ion cluster represented as the sum of $M^{+\cdot}$, $[M + 2]^{+\cdot}$, and $[M + 4]^{+\cdot}$; the center trace shows the single chlorine atom loss channel as a sum of m/z 255, 257, and 259, which is prominent only for the di-*ortho* compounds 52 and 47; the bottom trace shows the dominant 2Cl-loss channel. The arrows in the center trace mark the locations at which the single chlorine atom losses would appear for the mono-*ortho*- (66) and non-*ortho*-substituted congeners (81 and 77).

Figure 7(b) shows the corresponding selected ion chromatograms of the three hexachlorinated PCBs of 13 PCBs; the top trace is of the molecular ion $M^{+\cdot}$, m/z 358; the center trace shows the single chlorine atom loss channel forming $[M-^{35}Cl]^{+}$, m/z 323, which is prominent only for the di-*ortho* compound 153; the bottom trace shows the common 2Cl-loss channel forming $[M-2^{35}Cl]^{+\cdot}$, m/z 288. Again the arrows in the center trace mark the locations at which the single chlorine atom losses would appear for the mono-*ortho*- (156) and non-*ortho*-substituted (169) congeners, were that channel to be accessed. While the lack of a fragmentary feature may not normally enhance analytical specificity, in this case such a distinguishing feature, taken together with retention time, selected molecular ion, and selected 2Cl-loss channel does enhance the analytical specificity for these more toxic congeners.

4.5. Chemical Ionization of Coeluting Polychlorinated Biphenyl Congeners 77 and 110

Chemical ionization (CI) has been used in the selected ion monitoring mode for the analysis of PCB mixtures in environmental sample extracts [34,38]. Chemical ionization is the protonation of a molecule to form a pseudomolecular ion with little accompanying fragmentation. However, CI reagent ion species are formed concurrently in ion/molecule reactions with other ion species; thus, in a given CI system, several reagent ion species can be found. These species usually differ with respect to electron affinity [47] and the propensity with which the proton

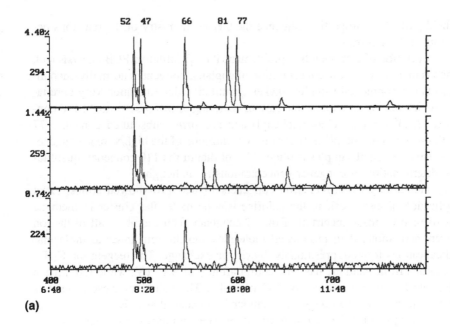

(a)

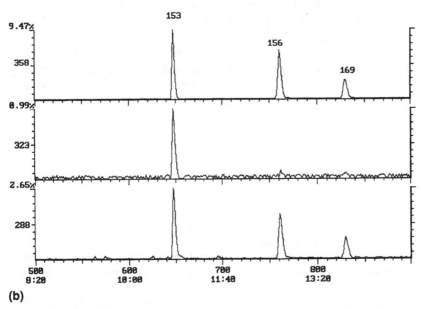

(b)

Figure 7 Selected ion chromatograms for the molecular ion $M^{+\cdot}$, fragment ion [M − Cl$^\cdot$]$^+$, and fragment ion [M − 2Cl$^\cdot$]$^{+\cdot}$: (a) five tetrachlorobiphenyl isomers, (b) three hexachlorobiphenyl isomers.

is held [48]. Both properties influence the thermochemistry of the reactions occurring in the ion trap.

Coelution of compounds is problematical to quantitative PCB analysis and,
when combined with low concentration of coplanar congeners normally encountered in environmental samples, makes quantitative determination very unreliable. An interesting example is the case of coeluting di-*ortho*-substituted 110 congener (2,3,3',4',6-pentachlorobiphenyl) and non-*ortho*-substituted congener 77
(3,3',4,4',-tetrachlorobiphenyl). As the concentration of the highly toxic congener
77 in environmental samples is often <1% of that of the 110 congener, quantitative determination of congener concentrations is challenging.

Preliminary experiments on congeners 77 and 110 showed almost exact
duplication of each peak in the chlorine isotope molecular cluster in methane
positive-ion CI mass spectrum of the 77 congener. Thus, almost half of the ion
current corresponded to protonated molecules and the remainder to molecular
radical cations formed, presumably, by charge exchange. The reagent ion $C_2H_5^+$
as isolated from methane has an electron affinity of 8.13 eV, while C_2H_4 has a
relatively low proton affinity of 680 kJ mol^{-1}. The values of these properties
relative to those of the congeners examined indicated that $C_2H_5^+$ had potential
analytical application for PCB determination. The dominant ion species observed
in the mass spectrum obtained with mass-selected $C_2H_5^+$ were the $[M + H]^+$
ions of each congener. The isotopic ratios for each cluster of protonated congeners were in good agreement with those expected for pure CI. The mass spectrum
was entirely free of fragment ions below $[M_{77} + H]^+$. If the behavior of each of
congeners 77 and 110 is characteristic of that of coplanar and planar PCB molecules, respectively, then CI of PCBs using selected $C_2H_5^+$ ions from methane
offers a promising method for the determination of PCBs.

REFERENCES

1. W. Paul and H. Steinwedel, Apparatus for separating charged particles of different
   specific charges. German Patent 944,900, 1056; U.S. Patent 2,939,952, 7 June 1960.
2. W. Paul, Angew. Chem., 29 (1990) 739–748.
3. G.C. Stafford, Jr., P.E. Kelley, J.E.P. Syka, W.E. Reynolds, and J.F.J. Todd, Int. J.
   Mass Spectrom. Ion Process., 60 (1984) 85–98.
4. P.H. Dawson and N.R. Whetten, Adv. Electr. Electron. Phys., 27 (1969) 59–185.
5. P.H. Dawson, Quadrupole Mass Spectrometry and Its Applications, 1976, Elsevier,
   Amsterdam, Netherlands.
6. R.E. March, R.J. Hughes, and J.F.J. Todd, Quadrupole Storage Mass Spectrometry,
   1989, Wiley Interscience, New York.
7. J.F.J. Todd, Mass Spectrom. Rev., 10 (1991) 3–52.
8. (a) R.G. Cooks and R.E. Kaiser, Jr., Accounts Chem. Res., 23 (1990) 213–219;
   (b) B.D. Nourse and R.G. Cooks, Anal. Chem. Acta., 228 (1990) 1–21.

9. G.L. Glish and S.A. McLuckey, Int. J. Mass Spectrom. Ion Process., 106 (1991) 1.
10. R.E. March, Int. J. Mass Spectrom. Ion Process., 118/119 (1992) 71–135.
11. R.E. March and J.F.J. Todd, eds., Practical Aspects of Ion Trap Mass Spectrometry, vols. 1–3, Modern Mass Spectrometry Ion Trap Series, 1995, CRC Pr., Boca Raton, FL.
12. R.E. Clement, Anal. Chem., 63 (1991) 1130A–1139A.
13. U.S. Environmental Protection Agency (EPA), Method 1613: Tetra-Through Octa-Chlorinated Dioxins and Furans by Isotope Dilution HRGC/HRMS, Revision A. 1990, EPA, Washington, DC.
14. R.E. Clement and H.M. Tosine, Mass Spectrom. Rev., 7 (1988) 593.
15. V.Y. Taguchi, E.J. Reiner, D.T. Wang, O. Meresz, and B. Hallas, Anal. Chem., 60 (1988) 1429–1433.
16. Y. Tondeur, W.N. Niederhut, J.E. Campana, and S.R. Missler, Biol. Environ. Mass Spectrom., 14 (1987) 449.
17. E.K. Chess and M.L. Gross, Anal. Chem., 52 (1980) 2057.
18. E.J. Reiner, D.H. Shellenberg, and V.Y. Taguchi, Environ. Sci. Technol., 25 (1991) 110.
19. Ontario Ministry of the Environment and Energy, Method for the Determination of Polychlorinated Dibenzo-p-Dioxins and Polychlorinated Dibenzofurans in Fish. 1993, Ontario Ministry of the Environment and Energy, Etobicoke, Ontario, Canada.
20. J.B. Plomlcy, C.J. Koester, and R.E. March, Org. Mass Spectrom., 29 (1994) 372–381.
21. J.B. Plomley, C.J. Koester, and R.E. March, Proceedings of the 42nd ASMS Conference on Mass Spectrometry and Allied Topics, Chicago, IL, 1994, p. 718.
22. J.B. Plomley, R.S. Mercer, and R.E. March, Proceedings of the 43rd ASMS Conference on Mass Spectrometry and Allied Topics, Atlanta, GA, 1995, p. 230.
23. G. Hamelin, C. Brochu, and S. Moore, Organohalogen Compd., 23 (1995) 125–130.
24. C.P.R. Jennison, Oral presentation, Annual Western Pesticide Conference, Calgary, Canada, April, 1999.
25. M. Wang, S. Schachterle, and G. Wells, J. Am. Soc. Mass Spectrom., 7 (1996) 668.
26. M.J. Charles, B. Green, Y. Tondeur, and J.R. Hass, Chemosphere, 19 (1989) 51.
27. J.B. Plomley, R.E. March, and R.S. Mercer, Anal. Chem., 68 (1996) 2345–2352.
28. M.J. Charles, W.C. Green, and G.D. Marbury, Environ. Sci. Technol., 29 (1995) 1741.
29. S. Jensen, Report of a new chemical hazard. New Sci., 32 (1966) 612–615.
30. K. Ballschmiter and M. Zell, Fresenius J. Anal. Chem., 302 (1980) 20–31.
31. C.D. Metcalfe, in J.W. Kiceniuk and S. Ray, eds. Analysis of Contaminants in Edible Aquatic Resources, 1994, VCH Publishers., New York pp. 305–338.
32. G. Sundstrom, O. Hutzinger, and S. Safe, Chemosphere, 5 (1976) 267.
33. C.G. Muir, R.J. Nostrom, and M. Simon, Environ. Sci. Technol., 22 (1988) 1071.
34. S. Safe, Crit. Rev. Toxicol., 21 (1990) 51.
35. J.C. Duinker, D.E. Schulz, and G. Petrick, Chemosphere, 23, (1989) 1009–1028.
36. J.C. Duinker, D.E. Schulz, and G. Petrick, Mar. Pollut. Bull., 19 (1988) 19–25.
37. S. Tanabe, Environ. Pollut., 50 (1988) 5–28.
38. S. Safe, Environ. Health Perspect., 100 (1992) 259.

39. S.A. Kafafi, H.Y. Afeefy, A.H. Ali, H.K. Said, and A.G. Kafafi, Environ. Health
     Perspect. 101 (1993) 422–428.
40. V. McFarland and J. Clarke, J. Environ. Health Persp., 81 (1989) 225.
41. M. Lausevic, X. Jiang, C.D. Metcalfe, and R.E. March, Rapid Commun. Mass Spec-
     trom., 9 (1995) 927–936.
42. K. Ballschmiter, R. Bacher, A. Mennel, R. Fischer, U. Riehle, and M. Swerev, J.
     High Resolut. Chromatog., 15 (1992) 260.
43. DIN 51527, Determination of PCBs: Parts 1 and 2, Beut Verlag GmbH, Berlin,
     1987.
44. Rijks Institut voor Volksgezondheid en Milieuhygiene (RIVM), Werkdocument
     rondzending analyses van PCBs in afval olie volgens de IvM methode, 1989,
     RIVM, Bilthoven, Netherlands.
45. P. de Voogt, Chemosphere, 23 (1991) 901.
46. T.J. Munsteren, A.H. Roosand, and W.A. Traag, Int. J. Environ. Anal. Chem., 14
     (1983) 147.
47. S.G. Lias, J.E. Bartmess, J.F. Liebman, J.L. Holmes, R.D. Levin, and W.G. Mallard,
     Gas-Phase Ion and Natural Thermochemistry, 1988, American Institute of Physics,
     New York.
48. S.G. Lias, J.F. Liebman, and R.D. Levin, J. Phys. Chem. Ref. Data, 13 (1984) 695–
     808.

5

Gas Chromatography–Mass Spectrometry Analysis of Chlorinated Organic Compounds

M. T. Galceran and F. J. Santos
University of Barcelona, Barcelona, Spain

1. INTRODUCTION

In the last three decades, there has been considerable public concern about the presence of halogenated anthropogenic compounds in the environment because of their persistence, the bioaccumulation potential, and the health risks that they bring about [1–3]. Examples of members of this family of xenobiotics are polychlorinated biphenyls (PCBs), polychlorinated terphenyls (PCTs) and toxaphene. Polychlorinated biphenyls were described as ubiquitous, environmentally harmful substances more than 30 years ago [4]. In addition, the presence of PCBs and toxaphene in the environment has been extensively documented [5–7]. In contrast, little attention has been paid to polychlorinated terphenyls, which are similar to PCBs in chemical characteristics and applications. In this chapter, we discuss different methods described in the literature for the analysis of PCTs and toxaphene using high-resolution gas chromatography (HRGC) coupled to mass spectrometry (MS).

Polychlorinated terphenyls have been used extensively in sealants, hydraulic fluids, electrical equipment, plasticizers, and paints because of their desirable electrical and flame-retardant properties [8]. The Monsanto Chemical Co. started the production of both PCTs and PCBs in the United States in 1929. Since then, PCTs have also been produced in France, Italy, Germany, and Japan. Information about PCT production is incomplete, but during the period 1955 to 1980, approximately 60,000 metric tons were produced worldwide [8,9], which is 15 to 20 times lower than the total PCB production during the same period [7]. In 1972, the manufacture of PCTs was voluntarily discontinued in the United States, and later in such countries as Germany, Italy, and France, in 1974, 1975, and 1980, respectively.

Polychlorinated terphenyls have been produced as technical mixtures under different commercial names such as Aroclor, Kanechlor C, Leromoll, Clophen, Cloresil, Electrophenyl T-60, Phenoclor, and Terphenyl Chlore T-60. These technical formulations consisted of mixtures of congeners with variable degrees of chlorination. Commercial formulation of PCTs could be expected to be much more complex than those of PCBs because of the presence of an additional aromatic ring, which increases the positions available for chlorination. The theoretical number of PCT congeners with $C_{18}H_{14-x}Cl_x$ ($1 \leq X \leq 14$) is 8557 [10,11], as shown in Table 1.

Little information is available about the distribution, fate, and effects of PCTs in the environment. However, as highly chlorinated aromatic compounds, PCTs should be highly resistant to biodegradation and photodegradation, to the effect of acids, bases, and oxidizing agents, and should easily accumulate in adipose tissues of living organisms. Comparable toxic effects to those of PCBs on animals have been reported for commercial PCT formulations [12]; however, some investigators found that their effects were less severe [8,9].

Table 1 Theoretical Number of PCT Congeners

| Homologue | Terphenyl configuration | | | Total |
	ortho-terphenyl	meta-terphenyl	para-terphenyl	
Mono-CTs	5	6	4	15
Di-CTs	28	28	21	77
Tri-CTs	86	90	58	234
Tetra-CTs	217	217	142	576
Penta-CTs	391	400	244	1035
Hexa-CTs	574	574	356	1504
Hepta-CTs	636	648	388	1672
Octa-CTs	574	574	356	1504
Nona-CTs	391	400	244	1035
Deca-CTs	217	217	142	576
Undeca-CTs	86	90	58	234
Dodeca-CTs	28	28	21	77
Trideca-CTs	5	6	4	15
Tetradeca-CTs	1	1	1	3
Total	3239	3279	2039	8557

Source: adapted from Refs. 10 and 11.

From an environmental point of view, polychlorinated terphenyls have been identified in various matrices such as soil and sediments [13–15], water [15,16], waste oils [17], birds [18], aquatic organisms [8,18–23], wolves and pigs [24], human tissue [25,26], as well as in many other samples [27].

Analysis of PCTs has proved to be difficult because of the complexity of the mixtures, the high boiling points of the heavily chlorinated congeners, and the coelution of the lower chlorinated PCTs with some PCBs, but the use of gas chromatographic capillary columns longer than 50 m improves the separation. Nevertheless, the problem of the chromatographic coelution cannot be solved even using multidimensional gas chromatography (GC). Capillary gas chromatographic columns with high thermal stability and relatively nonpolar stationary phases such as 5% fenil–95% methyl siloxane, have been commonly used for PCT separation. Stationary phases of different polarity [10] do not result in better separations. In addition, the high numbers of possible PCT congeners and the unavailability of individual standards hamper their accurate quantification. Recently, some individual PCT congeners became commercially available from Promochem (Wesel, Germany), but the number is still very small. This problem can be partly solved using commercial mixtures of PCTs, such as Aroclor 5432, 5442 or 5460, as secondary standards for identification and quantification of total PCTs.

High-resolution gas chromatography (HRGC) combined with an electron-capture detector (ECD) [9,12,20], an electrolytic conductivity detector (ELCD) [13,22], and/or coupled to MS with electron ionization (EI) [9,25,28,29] or negative chemical ionization (NCI) [15,21] are commonly used for the analysis of PCTs. In general, HRGC–ECD provides adequate sensitivity and selectivity in the analysis of PCTs, but the presence of other related compounds, such as PCBs of high chlorinated level or other halogenated compounds of high molecular weight such as polychlorinated naphthalenes and organochlorine pesticides, could potentially interfere with the final determination of these compounds. In addition, only a semiquantitative analysis of PCTs, expressed as total content, can be achieved by HRGC–ECD by assuming equal response factors for all congeners and similar composition between samples and commercial standard mixtures. High-resolution gas chromatography–mass spectrometry (HRGC-MS) techniques could be a good choice to solve interference problems in the analysis of PCTs. Several approaches using low-resolution MS (LRMS), high-resolution MS (HRMS), and tandem mass spectrometry (MS–MS) have been proposed; they are discussed in the respective sections later in this chapter.

Toxaphene is considered to be the most extensively used insecticide in the world. It has been widely used in insect control on cotton and other crops, and in the control of the external insects on livestock [30]. It was banned in the United States in the early 1980s and some time later in Canada and some European countries [31] because of its acute and chronic toxicity to humans, environmental persistence, and bioaccumulating capabilities [32,33]. However, toxaphene and other related compounds are still being manufactured and widely used in Central and South America, African countries, India, the former Soviet Union, and some East European countries [34,35]. Its cumulative global usage averaged 450 thousand metric tons, but the total global production between 1950 to 1993 was estimated to approach 1.3 megatons [36], which is a higher value than that of polychlorinated biphenyls.

Toxaphene was first manufactured in the United States, in 1945, and it is mainly composed by polychlorobornanes (CHBs, 76%), polychlorobornenes (18%), polychlorobornadienes (2%), other chlorinated hydrocarbons (1%), and nonchlorinated hydrocarbons (3%) [37], with a total chlorine content of 67 to 69% [38]. The actual composition of this extremely complex mixture is not fully characterized. Vetter [39] estimated that there are 32,768 theoretically possible CHB congeners. The number of possible congeners decreases considerately after applying some restrictions due to unfavorable chlorine substitutions on the ring and bridge carbon atoms. Oddly enough, the number of toxaphene congeners found in environmental samples is lower than those in technical formulations. These dissimilarities are due to photodegradation, selective bioaccumulation, and/or the metabolism of toxaphene congeners in aquatic and terrestrial organisms, including humans [34,40]. Although the total number of congeners decreases, the remaining toxaphene mixture is still complex and requires powerful techniques of separation and analysis.

There is an additional problem of nomenclature. In fact, several systematic nomenclature systems for toxaphene congeners have been proposed [21,41–47], but none of them has been universally adopted until now. At the Workshop on Toxaphene, held in Burlington (Ontario, Canada) in 1993, it was agreed that the term chlorobornanes (CHBs) would be used for toxaphene compounds [48]. In addition, difficulties in formulating the correct systematic names for chlorobornanes were theoretically solved when the International Union of Pure and Applied Chemistry (IUPAC) presented the definitive numbering of the carbon skeleton (Fig. 1). Inadequate use of the IUPAC nomenclature leads to some confusion [49,50], particularly with the C-8 and C-9 positions of the bornane skeleton. As IUPAC names are very long and not very easy to use, several authors proposed more simple nomenclatures. Initially, some toxaphene congeners were numbered by the research group of Parlar [41,42] using their GC retention time on a given stationary phase, but as each number was assigned to a chromatographic peak, congener coelution could not be excluded. Other nomenclatures were based on binary coding systems, such as those proposed by Tribulovich et al. [43] and Nikiforov et al. [44], Oehme and Kallenborn [45], and Andrews and Vetter [46], which were difficult to handle. Recently, Wester et al. [47,51] have proposed a new system that provides structural information. It is applicable to all the congeners and allows discerning enantiomers, and is relatively easy to use. In this

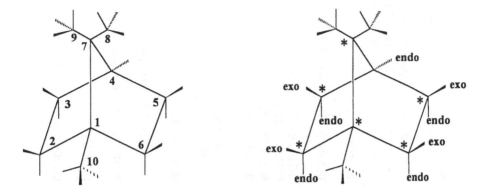

Figure 1 Structure of bornane skeleton with numbering of carbon atoms and endo and exo positions (asterisk indicates chirality carbon atom).

nomenclature, every digit in the code represents the chlorine substitution of the carbon number. The first part of the code reflects the conformation of the six-membered ring (C-2 to C-6) (0, none; 1, endo; 2, exo; 3, both), and the second part gives the number of chlorine atoms attached to C-8, C-9, and C-10, respectively. The 8-digit number is preceded by "B" for bornanes, "E" for bornenes, or "D" for bornadienes. An overview of some of these nomenclature systems is presented in Table 2. The various nomenclature systems are still used by different authors, and it would be necessary to adopt a unique nomenclature system in order to end the current confusing situation.

Toxaphene has been described as a ubiquitous contaminant in various environmental compartments and has a widespread distribution around the world [52–54], even in very remote areas, such as the polar regions, due to aerial transport [55]. Extensive reviews on the chemistry, biochemistry, toxicity, analysis, and environmental fate of toxaphene have been published [32,38,52]. Toxaphene, similar to other pollutants such as DDT, PCBs, and other organochlorine compounds, was found in air [54–56], freshwater [57,58], food [59], soil and sediment [60,61], human milk [53], and even marine biota [41,62–66].

The analysis of toxaphene is currently performed by HRGC [31,52,67]. In general, nonpolar capillary columns with 5% phenyl–95% methyl polysiloxane stationary phases are used. Other columns with semipolar and polar stationary phases, such as cyanopropylphenyl methyl polysiloxane and polyethylene glycol polysiloxane, have also been used [63,68]. However, care is recommended because some compounds may decompose on these polar phases [67]. On the other hand, injector temperatures higher than 240°C produced a substantial decrease in the signal due to decomposition on active sites of the gas chromatograph injector [63]. In general, it is recommended to work with on-column or splitless injec-

Table 2 Correlation of Different Names of Chlorinated Bornane Congeners Appearing in the Literature

Parlar's No. [41,42]	IUPAC Name (Racemate) [49,50]	Proposed by Nikiforov and Tribulovich [43,44]	Proposed by Oehme and Kallenborn [45]	Proposed by Andrews and Vetter [46]	Proposed by Wester et al. [47,51]	Other Names (Vetter et al. [49], Stern et al. [79])
21	2-endo,2-exo,5-endo,5-exo,9,10,10-HeptaCHB	HpCB-6533	99–043	B7-449	B[30030]-(012)	
32	2-endo,2-exo,5-endo,6-exo,8,9,10-HeptaCHB	HpCB-6452	195–241	B7-515	B[30012]-(111)	Toxicant B
	2-exo,3-endo,5-exo,9,9,10,10-HeptaCHB	HpCB-3207	134–113	B7-1453	B[21020]-(022)	TOX 7
	2-exo,3-endo,6-endo,8,9,10,10-HeptaCHB	HpCB-3157	98–033	B7-1462	B[21001]-(112)	
	2-exo,5-endo,5-exo,9,9,10,10-HeptaCHB	HpCB-2439		B7-1715	B[20030]-(022)	
39	2-endo,2-exo,3-exo,5-endo,6-exo,8,9,10-OctaCHB	OCB-6964	299–421	B8-531	B[32012]-(111)	
25	2-endo,2-exo,3-exo,8,9,9,10-OctaCHB	OCB-6686	11–631	B8-623	B[32000]-(221)	
51	2-endo,2-exo,5-endo,5-exo,8,9,10,10-OctaCHB	OCB-6549	99–423	B8-786	B[30030]-(112)	
38	2-endo,2-exo,5-endo,5-exo,9,9,10,10-OctaCHB	OCB-6535	99–063	B8-789	B[30030]-(022)	
42β	2-endo,2-exo,5-endo,6-exo,8,9,9,10-OctaCHB	OCB-6454	291–461	B8-806	B[30012]-(121)	Toxicant A$_2$
42α	2-endo,2-exo,5-endo,6-exo,8,8,9,10-OctaCHB	OCB-6460	291–641	B8-809	B[30012]-(211)	Toxicant A$_1$
	2-endo,2-exo,5-endo,6-exo,8,9,10,10-OctaCHB	OCB-6453	193–243	B8-810	B[30012]-(112)	
26	2-endo,3-exo,5-endo,6-exo,8,8,10,10-OctaCHB	OCB-4921	297–603	B8-1413	B[12012]-(202)	T2, TOX 8

No.	Compound	Code		B-code	Formula	Notes
40	2-endo,3-exo,5-endo,6-exo,8,9,10,10-OctaCHB	OCB-4917	297–243	B8-1414	B[12012]-(112)	
41	2-exo,3-endo,5-exo,8,9,9,10,10-OctaCHB	OCB-3223	70–463	B8-1945	B[21020]-(122)	
44	2-exo,5-endo,5-exo,8,9,9,10,10-OctaCHB	OCB-2455	98–463	B8-2229	B[20030]-(122)	
58	2-endo,2-exo,3-exo,5-endo,5-exo,8,9,10,10-NonaCHB	NCB-7061	107–243	B9-715	B[32030]-(112)	
	2-endo,2-exo,3-exo,5-endo,5-exo,9,9,10,10-NonaCHB	NCB-7047	107–033	B9-718	B[32030]-(022)	
	2-endo,2-exo,3-exo,5-endo,6-exo,8,9,9,10-NonaCHB	NCB-6966	199–043	B9-742	B[32012]-(121)	
	2-endo,2-exo,5-endo,5-exo,6-exo,8,9,9,10-NonaCHB	NCB-6582	227–641	B9-1011	B[30032]-(121)	
62	2-endo,2-exo,5-endo,5-exo,8,8,9,10,10-NonaCHB	NCB-6551	099–643	B9-1025	B[30030]-(122)	
56	2-endo,2-exo,5-endo,6-exo,8,8,9,10,10-NonaCHB	NCB-6461	291–645	B9-1046	B[30012]-(212)	
59	2-endo,2-exo,5-endo,6-exo,8,9,9,10,10-NonaCHB	NCB-6455	291–463	B9-1049	B[30012]-(122)	
50	2-endo,3-exo,5-endo,6-exo,8,8,9,10,10-NonaCHB	NCB-4925	297–643	B9-1679	B[12012]-(212)	T12, Toxicant Ac, TOX 9
63	2-exo,3-endo,5-endo,6-exo,8,8,9,10,10-NonaCHB	NCB-3261	326–643	B9-2206	B[21022]-(212)	
	2-endo,2-exo,3-exo,5-endo,5-exo,8,9,9,10,10-DecaCHB	DCB-7063	103–643	B10-831	B[32030]-(122)	
	2-endo,2-exo,3-exo,5-endo,6-exo,8,8,9,10,10-DecaCHB	DCB-6967	199–643	B10-860	B[32012]-(212)	
69	2-endo,2-exo,5-endo,6-exo,8,9,9,10,10-DecaCHB	NCB-6583	355–463	B10-1110	B[30032]-(122)	

tion at temperatures below 220°C [50] in order to prevent thermal degradation, and to use the shortest possible column to avoid long retention times and high elution temperatures [67].

High-resolution gas chromatography with chiral capillary columns has also been applied to the analysis of some CHB enantiomers [67,69–71]. Enantioseparation has been achieved using chiral stationary phases (CSPs). Several studies have been performed in order to determine the behavior of enantiomers in the biota and the environment. In fact, differences in the enantiomeric ratios have been observed in seals [52,72], monkey fatty tissue [70], cod liver, and fish-oil samples [73]. This fact suggests that the biological transformation of some enantiomer pairs can change depending on the type of biota. Multidimensional GC has also been used for the analysis of toxaphene congeners. The resolution required for CHB separation can be achieved by selecting an adequate combination of columns. De Boer et al. [68,74] proposed the use of Ultra 2/Rtx 2330 to analyze CHB congeners in biological samples. In general, the main drawback of multidimensional GC is the relatively long analysis time compared with the single-column method.

Two commonly used techniques to analyze CHBs in biological or environmental samples at residue levels are HRGC–ECD and HRGC–MS. Although GC–ECD offers high sensitivity at low cost, it is always questionable as a confirmatory method due to the low specificity in the analysis of complex samples, even if extensive cleanup procedures are applied. Interferences from organochlorine pesticides and PCBs and limitations in the quantitative analysis of toxaphene are frequently observed. In contrast, GC–MS is generally recognized as a more powerful analytical tool with adequate sensitivity and specificity. Electron-capture negative-ion MS (ECNI-MS) and EI-MS are the two most popular MS techniques. Nevertheless, quantification of toxaphene is difficult, mainly due to the substantial difference in peak profiles of environmental/biological samples and those of industrial formulation, and to the lack of congener-specific standards of the different chlorobornanes. The research group of Parlar [41,75] succeeded in producing the 22 more important single congeners of toxaphene. Most of them are octa- and nonachlorobornanes, which are commercially available from Ehrenstorfer (Augsburg, Germany] or Promochem (Wesel, Germany). For an accurate quantification of CHBs, it is absolutely essential to prepare isotopically labeled internal standard congeners.

The aim of this chapter is to present and discuss the different GC–MS techniques and the advances achieved in the characterization and analysis of polychlorinated terphenyls and toxaphene. We have paid special attention to the performances of the GC–MS techniques and the advantages and drawbacks of its application as a reliable method for the analysis of these compounds in commercial mixtures and samples. The use of HRGC–ECNI-MS, HRGC–EI-LRMS, HRGC–EI-HRMS, and HRGC–MS–MS is discussed in the following sections.

2. HIGH-RESOLUTION GAS CHROMATOGRAPHY–LOW-RESOLUTION MASS SPECTROMETRY (HRGC–LRMS)

For PCT and toxaphene analysis, cleanup steps are mandatory. They allow the separation of these halogenated compounds from interferences of the sample matrix. Little attention has been paid to the efficiency of the extraction and isolation of these compounds from environmental and biological samples. However, the procedures currently used for the analysis of other organochlorine compounds such as polychlorinated biphenyls and organochlorine pesticides with similar lipophilic and structural characteristics are commonly applied. In general, column chromatography with different sorbents, such as silica gel, silica/sulfuric acid, alumina, Florisil [20,21,23,42,58,64], and graphitized carbon [76], and reversed-phase C-8 and C-18 [77] and preparative gel permeation chromatography [13,25,60,62,78] have been used. However, as none of these procedures provided enough preseparation, HRGC–MS seems to be the preferred technique as opposed to HRGC–ECD for the analysis of PCT and toxaphene mixtures. In fact, HRGC–LRMS has been used for the analysis of these compounds in commercial mixtures and in environmental and biological samples, with EI-MS mainly used for PCTs and ECNI-MS for toxaphene. These methods are discussed in the next sections.

2.1. Electron-Ionization Mass Spectrometry (EI-MS)

Electron ionization has been currently chosen as a reference ionization method in MS and it has been commonly used in the GC–MS analysis of PCTs and toxaphene. In fact, EI is readily available in all mass spectrometers and, in general, is stable, reproducible, and easy to operate on a daily basis.

The coupling of HRGC with LRMS (resolving power 500) using selective ion monitoring (SIM) allows one to obtain both structural information and homologue composition [13,19]. The EI mass spectra of the PCT congeners show an intense molecular ion $M^{+\bullet}$ [28] and relatively intense peaks due to successive loss of two chlorine atoms from the molecular cluster and small fragments due to $[M - Cl]^+$. The homologue profiles for PCT mixtures can be obtained by selecting the most intense ion of the molecular cluster from each homologue group. The traces of Aroclor 5460 are shown in Figure 2 as an example. According to this, coelution of PCT congeners containing fewer chlorines did not interfere with the detection and the measurement of the $M^{+\bullet}$ ions of a homologue group of congeners, but there was interference by the $[M - Cl_2]^+$ fragment ions of homologues containing two additional chlorine atoms [20,37]. Another potential interference may be produced by the $[M - Cl]^+$ fragment ions, although this contribution does not seem to be important because the signals of these ions are very weak. In order to avoid these internal interferences, HRMS with resolving

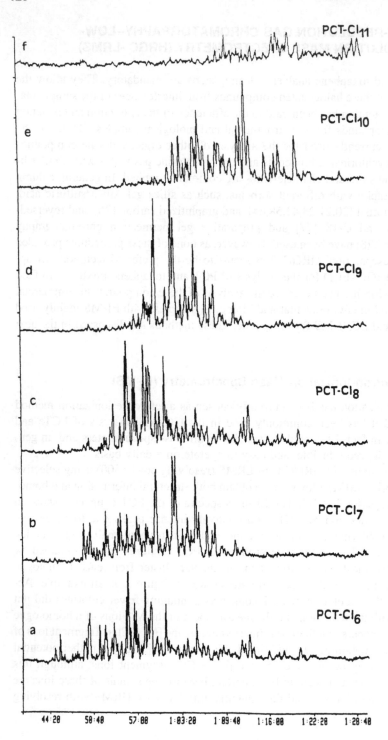

power of 30,000 is required [20]. Nevertheless, the homologue distribution of PCT mixtures in commercial formulations and in environmental samples can be approximately calculated by eliminating from each SIM trace for each homologue the contribution of the $[M - Cl_2]^+$ ions, assuming that virtually no coelution between homologues containing two additional chlorine atoms occurs. This approach was used to determine the approximate homologue composition of the PCT commercial mixtures and for the identification of the source of PCT contamination in biota [20].

Electron ionization has also been used for the analysis of toxaphene. However, the mass spectra of CHBs are characterized by complex fragmentation patterns. In contrast to PCTs, molecular ions are normally absent in CHBs and major fragment ions in the high-mass range are produced by the sequential loss of a combination of Cl, HCl, $CHCl_2$ or CH_2Cl from the molecules. In addition, some characteristic fragment ions in the low-mass range are also observed [31]. Stern et al. [79] studied extensively the fragmentation mechanism and reported three major fragmentation pathways: (1) the loss of Cl and HCl, (2) the retro–Diels-Alder reaction, and (3) the loss of $CHCl_2$ and CH_2Cl.

The ions most commonly chosen for quantitative analysis of CHBs are the $[M - Cl]^+$ ions because they offer molecular mass information and have high specificity. However, these ions are usually very weak in EI-MS, resulting in relatively high detection limits. In addition, at low resolving power, the $[M - HCl - Cl]^+$ ions from each homologue group interfere with the $[M - Cl]^+$ ions of homologues with 1 chlorine atom less. Significant contributions in the profile of hepta-CHBs on the hexa-CHBs, or the octa-CHBs on the hepta-CHBs are observed in the HRGC–EI-MS SIM traces of $[M - Cl]^+$ ions for the homologues of toxaphene. This internal interference is named the "mass leakage" or "crossover" problem by Lau et al. [31]. An additional problem in the use of EI-LRMS is the potential interference of the $[M - Cl]^+$ isotope ions of chlorinated camphenes and monounsaturated bornenes (2 nominal mass units lower than fully saturated CHBs) on the signal of the $[M - HCl - Cl]^+$ ions from the higher CHB homologues. High-resolution mass spectrometry at a resolving power higher than 20,000 is required in order to remove these specific interferences [31,80].

In general, HRGC–EI-LRMS has been used as a complementary technique to HRGC–ECD or HRGC–ECNI-MS in the analysis of toxaphene in the environment and biota samples [81,82], but its use is limited. Nevertheless, this technique in combination with protium-nuclear magnetic resonance (^{1}H-NMR), linked-field

---

Figure 2 EI-LRMS-selected ion monitoring for homologues of Aroclor 5460. (a) hexachloroterphenyls, (b) heptachloroterphenyls, (c) octachloroterphenyls, (d) nonachloroterphenyls, (e) decachloroterphenyls, and (f) undecachloroterphenyls. (From Ref. 20.)

scans, EI-HRMS, ECNI-HRMS, and MS–MS have been often used in the identification, structure elucidation, and characterization of major toxaphene congeners from biota samples and technical toxaphene [79,80,82,83].

2.2. Electron-Capture Negative-Ion Mass Spectrometry (ECNI-MS)

The technique of HRGC–ECNI-MS has also been applied for the analysis of PCTs and toxaphene. Although this technique is the most popular method for the analysis of toxaphene, to our knowledge, it has only been used by Canton and Grimalt [15] and Wester et al. [21] for the analysis of PCTs. In general, the usefulness of ECNI-MS in the analysis of halogenated compounds is due to its higher sensitivity and specificity compared with EI-MS. The mass spectra obtained with ECNI-MS for PCT and toxaphene congeners are characterized by a low fragmentation pattern using methane as a moderating gas. For example, intense signals of the molecular $[M]^{-\cdot}$ cluster ions and weak signals of the $[M - Cl + H]^-$ and $[M - 2Cl + 2H]^-$ cluster ions are obtained for PCT congeners, as can be seen in Figure 3, whereas for toxaphene, ECNI produces intense $[M - Cl]^-$ ions for CHBs with 7 or more chlorine atoms, and both $[M]^{-\cdot}$ and $[M - Cl]^-$ ions for lower-chlorinated homologues [31,62,72]. For the PCT quantitative analysis, Wester et al. [21] studied various methods to obtain unambiguous data. They suggested the use of the total ion current (TIC) chromatogram of $[M]^{-\cdot}$ and $[M + 2]^{-\cdot}$ ions of the molecular cluster for each homologue group from trichloroterphenyls to undecachloroterphenyls. However, low-chlorinated terphenyls (≤ 4 chlorine atoms) show low sensitivity and, in addition, other polychlorinated compounds, such as PCBs, polychlorinated naphthalenes (PCNs), and some organochlorine pesticides, can cause false positives. Additionally, the authors emphasize the importance of the proper selection of the standard mixture, especially because of the large differences in response factors, which give higher errors in HRGC–ECNI-MS than in HRGC–EI-MS or HRGC–ECD. In unfavorable conditions, this error can be as high as 500%. Nevertheless, the sensitivity achieved with HRGC–ECNI-MS is 5- to 10-fold higher than HRGC–ECD or HRGC–ELCD with a repeatability of 15%.

On the other hand, while HRGC–ECNI-MS is highly sensitive for the analysis of toxaphene, some disadvantages have been reported, such as low reproducibility and variations of the response factors with ECNI conditions (ion source temperature, pressure of the moderating gas, and instrument design) [31,62]. In addition, a wide range of response factors in congeners has been reported [36,42,64]. The low responses of some CHB congeners can produce false negatives, and, therefore, the reliability of the quantitative determination of toxaphene cannot be assured. For instance, Lau et al. [84] reported that the nonachlorobornane congener, Parlar No. 62, did not give the expected $[M - Cl]^-$ (m/z 413)

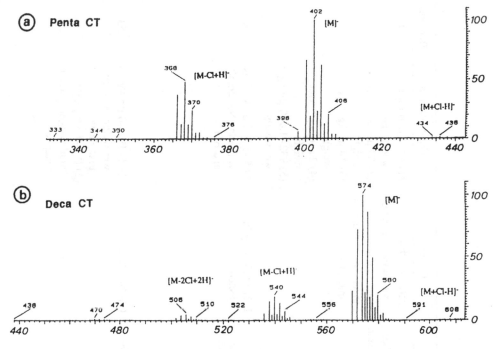

Figure 3 ECNI-mass spectra of (a) a pentachloroterphenyl and (b) a decachloroterphenyl. (From Ref. 21. Copyright 1996 American Chemical Society.)

and only a weak signal of the [M − HCl − Cl]⁻ ion was observed. In order to increase the intensity of the [M − Cl]⁻ ion of Parlar No. 62, Swackhamer et al. [62] recommend a decrease in the ion source temperature to approximately 100°C. Some authors [85] indicate that a stereoisomeric effect is responsible for the different responses while others [31,42,72] suggest that the absence of chlorine atoms in C-3 and C-6 positions of the six-membering ring can explain the poor response of some congeners. Nevertheless, the lack of response of these congeners prevents their identification and determination in environmental and biological samples because low or null response is obtained using ECNI-MS. This is an important problem because Parlar No. 26, 50, and 62 have been proposed to be used to monitor CHBs residues in fish and foodstuffs [78].

Polychlorinated biphenyls are a potential interference in the toxaphene analysis using HRGC–ECNI-LRMS in SIM mode [60,84–87]. The presence of a small amount of oxygen in the ion source produces [M − Cl + O]⁻˙ and [M − H + O]⁻˙ fragment ions from hexa- and heptachlorobiphenyls with the same m/z as hepta- and nonachlorobornanes. This interference can be solved by using HRMS. Although HRGC–ECNI-MS has some drawbacks, it has been exten-

Table 3 Analysis of Toxaphene Using HRGC–ECNI-MS

Sample	GC–MS conditions	Quanfication	Comments	Concentration	Reference
Fish (salmon and cod liver) oil (1) Red fish (2) Trout (3) Halibut (4) Caviar (5)	HRGC–ECNI-MS-SIM MS: Hewlett-Packard 5988A MS ECNI: methane 2 × 10^{-4} Torr 100°C ion source temperature 200 µA emission current	Congener-specific method: (chlorobornane derivatives standards)	LOD: 0.3–7.0 pg injected of pure congener standard Linear over 4 orders of magnitude	(1) 62–5,020 µg kg^{-1} wet weight (2) 53 µg kg^{-1} wet weight (3) 14 µg kg^{-1} wet weight (4) 21 µg kg^{-1} wet weight (5) 932–1,070 µg kg^{-1} wet weight	41
Fish and fish products (caviar, sea eel, cod fish, cod liver, salmon oil)	HRGC–ECNI-MS-SIM MS: Hewlett-Packard 5988A MS ECNI: methane 100°C ion source temperature 200 µA emission current	Congener-specific method: Σ (Parlar No. 26, 50 and 62)	LOD: 0.3–7 pg injected	4.65–1,394 pg kg^{-1} as Σ (Parlar No. 26,50 and 62)	42
Fish (1) Milk (2)	HRGC–ECNI-MS-SIM ECNI: methane, 1 Torr 100°C ion source temperature	Homologue specific method: Internal standard: 1,2,3,4-tetrachloronaphthalene	CV(%): 22%	(1) 1–19,000 µg kg^{-1} wet weight as total toxaphene (2) 0.3–7.6 ng kg^{-1} lipid weight as total toxaphene	53

Matrix	Instrument / Method	Method / Standards	Analytical performance	Concentration	Ref.
Air	HRGC–ECNI-MS-SIM MS: Hewlett-Packard 5989A MS-Engine ECNI: methane 1.2 Torr 150°C ion source temperature 230 eV, 300 mA emission current	Congener-specific method: Standards: 22 congeners, 11 congeners and 5 congeners Internal standards: 1,3,5-tribromobenzene and octachloronaphthalene	High variability of the response factor (CV(%): 177%)	0.9–10.1 pg/m^3	54
Lake trout (1) Whitefish (2)	HRGC–ECNI-MS-SIM MS: Hewlett-Packard 5985B MS ECNI: methane, 0.35 Torr 100°C ion source temperature	Homologue specific method:	CV(%): 28% LOD: 1 ng g^{-1} lipid	(1) 4,400–15,000 ng g^{-1} lipid (2) 4,500–10,000 ng g^{-1} lipid	57
Water (1) Sediment (2) Phytoplankton, net zooplankton (3) Lake trout (4)	HRGC–ECNI-MS-SIM MS: : Hewlett-Packard 5988A MS ECNI: methane 1.0 Torr 100°C ion source temperature	Homologue specific method: Internal standard: PCB-204	CV(%): 19–22%	(1) 0.17–1.12 ng L^{-1} (2) 15 ng g^{-1} dry weight (3) 50–250 ng g^{-1} dry weight (4) 4.3–10.3 µg g^{-1} lipid	58
Fish (troup, whitefish, carp)	HRGC–ECNI-MS-SIM MS: Hewlett-Packard 5985B MS ECNI: methane, 0.35 Torr 100°C ion source temperature 200 µA emission current, 200 eV	Homologue specific method: Σ hepta-, octa-, nona-, deca-CHBs Internal standard: PCB-204	LOD: 75 pg injected Linear response over 4 orders of magnitude	220–510 ng g^{-1} of whole fish	62

Table 3 Continued

Sample	GC–MS conditions	Quanfication	Comments	Concentration	Reference
Fish: (1) herring, (2) mackrel, halibut redfish, sardines (3) salmon	HRGC–ECNI-MS-SIM MS: Finnigan SSQ70 MS ECNI: methane, 0.5 Torr 100°C ion source temperature 90 eV	Congener-specific method: Parlar No. 26,50 and 62		(1) 1–33.6 µg kg^{-1} wet weight (2) 2.3–13.5 µg kg^{-1}wet weight (3) 6.4–19.2 µg kg^{-1} wet weight	63
Fish and fish products	HRGC–ECNI-MS-SIM MS: Hewlett-Packard 5988A MS ECNI: methane 100°C ion source temperature 200 µA emission current	Congener-specific method: Parlar No. 26, 32, 50, 62 and 69		12–755 µg kg^{-1} as Σ (Parlar No. 26, 50,62)	64
Lake Superior food web (planktonic organisms) (1) Lake trout (2)	HRGC–ECNI-MS-SIM	Homologue specific method: Σ (hepta-, octa-, nona-CHBs) Internal standard: PCB-204	LOD: 92 ng of total toxaphene	(1) 21–1,360 ng g^{-1} wet weight (2) 12–24 ng g^{-1} wet weight	59

Sample	Instrument/conditions	Method	Analytical data	Results	Ref.
Soil	HRGC–ECNI-MS-SIM MS: Finigan-MAT 4021; ECNI: methane, 0.4 Torr; 170°C ion source temperature; 70 eV, 500 µA emission current	Homologue specific method: Internal standard: PCB-204	Precision CV% = 10% Reproducibility CV% = 10% Linearity range: 0.1–1 mg kg^{-1}	16–69 µg kg^{-1} of soil	60
Sediment	HRGC–ECNI-MS-SIM MS: Hewlett-Packard 5988A MS; ECNI: methane, 1 Torr; 100°C Ion source temperature	Homologue specific method: Σ (hexa-, hepta-, octa-, nona-CHBs). Internal standard: PCB-204	LOD: 3–6 ng injected Precision, CV% = 22%	33–15 ng g^{-1} of sediment	61
Cod liver oil NIST SRM 1588	HRGC–ECNI-MS MS: Finnigan 4500-MS; ECNI: methane, 0.5 Torr; 100°C ion source temperature, 90 eV	Congener-specific method: Parlar No. 26, 50, 62 and 32	CV(%): 20%	Parlar No. 26: 375 µg kg^{-1} of fat Parlar No. 50: 434 µg kg^{-1} of fat Parlar No. 62: 220 µg kg^{-1} of fat	78
Whole blood	HRGC–ECNI-MS-SIM MS: Hewlett-Packard 5989A MS-Engine; ECNI: methane, 1.9 bar; 120°C ion source temperature, 200 eV	Homologue specific method Congener-specific method: T2 (Parlar No. 26) and T12 (Parlar No. 50) Internal standard: ^{13}C-PCB-153 and PCB-209		Total toxaphene: 162–174 ng L^{-1} T2 and T12 is the 90% of the total toxaphene content	86

Table 3 Continued

Sample	GC–MS conditions	Quanfication	Comments	Concentration	Reference
Fish Herring oil	HRGC–ECNI-MS-SIM MS: Hewlett-Packard 5988A MS ECNI: methane, 1 Torr 120°C ion source temperature	Homologue specific method		0.1–7 mg kg^{-1} lipid	87
Beluga whale blubber (1) Arctic char, whole fish (2) Milk (3)	HRGC–ECNI-MS-SIM MS: VG-7070E-HF double-focusing MS ECNI: methane 175°C ion source temperature 100 eV, 200 μA emission current	Congener-specific method: T2 (a) and T12 (b) congeners		(1a) 602–1300 ng g^{-1} of lipid (1b) 1050–2350 ng g^{-1} of lipid (2a) 43.6–73.9 ng g^{-1} of lipid (2b) 83.2–137.9 ng g^{-1} of lipid (3a) 69.5 ng g^{-1} of lipid (3b) 151 ng g^{-1} of lipid	88

Seal blubber	HRGC–ECNI-MS-SIM MS: Hewlett-Packard 5989B MS ECNI: methane, 1.2 bar 200°C ion source temperature	Congener-specific method: Σ (Parlar No. 26 (TOX8) and 62 (TOX9))	TOX8: 5.3 μg kg^{-1} TOX9: 3.0 μg kg^{-1} Total toxaphene as Σ (TOX8, TOX9): 8.3 μg kg^{-1}	89
Fish	HRGC–ECNI-MS-SIM MS: Finigan TSQ-70 MS ECNI: Methane 9,400 mTorr	Homologue specific method	LOD: < 0.3 ng g^{-1} as total toxaphene 1.6 ng g^{-1} neuston 12.8 ng g^{-1} pollack	91
Soil	HRGC–ECNI-MS-SIM MS: Hewlett-Packard 5989A MS-Engine ECNI: methane, 0.4 Torr 150°C ion source temperature	Homologue specific method: Internal standard: PCB-107	668–1950 mg kg^{-1}	92

sively used for the analysis of toxaphene using SIM mode. The $[M]^{-\cdot}$ ions of hexachlorobornane/bornene (m/z 340 and 342, respectively) and the $[M - Cl]^-$ from hepta- to decachlorobornanes/bornenes (m/z 374 and 376, m/z 408 and 410, m/z 442 and 444, and m/z 476 and 478, respectively) have been commonly used for quantitative analysis. This method was proposed by Swackhamer et al. [57,62] and it has been used by several researchers for the determination of total toxaphene as well as the specific CHB congeners in environmental and biota samples [31,42,86–92]. In addition, some other ions have also been used in order to correct the above-mentioned PCBs interferences. Two approaches have been proposed for quantitative analysis. The first is based on the use of technical toxaphene as standard and provides a rough estimation of the total toxaphene concentration [68] and the homologue distribution of the CHB. The uncertainty in this determination may be higher than using individual compounds, particularly in samples from biota with a high biotransformation rate for toxaphene, such as marine mammals [93] and human milk [53].

The second approach for toxaphene analysis is the congener-specific method. Nevertheless, HRGC, which offers high selectivity, or multidimensional GC [68] is required when coelution with other organochlorine compounds occurs.

A summary of some selected literature data is provided in Table 3, where some HRGC–ECNI-MS methods in SIM mode are given. Detection limits for the individual congeners (Parlar No. 26, 50, and 62] between 0.3 and 7 pg injected are achieved and a wide range of toxaphene concentrations in environmental and biological samples is found. Some authors used the technical toxaphene as standard to estimate total concentration [58,90–92]. Recently, the individual determinations of Parlar No. 26 and 50 for marine mammals [63,88] and Parlar No. 26, 50, and 62 for fish and fish products [42,63,64,70] have been used as indicators of the level of toxaphene contamination.

3. HIGH-RESOLUTION GAS CHROMATOGRAPHY–HIGH-RESOLUTION MASS SPECTROMETRY (HRGC–HRMS)

High-resolution gas chromatography–high-resolution mass spectrometry offers high selectivity with acceptable sensitivity and is an accurate and powerful technique for the quantitative analysis of organic contaminants such as polychlorinated dibenzo-p-dioxins and dibenzofurans in complex environmental matrices [94]. High-resolution mass spectrometry has a very high capacity to separate close masses and allows the removal of the contributions of interfering compounds. The technique of HRGC–HRMS in SIM mode has been used in the analysis of PCTs and toxaphene with both EI and ECNI. The performance of these two ionization modes is described in detail in the next sections.

3.1. Electron-Ionization High-Resolution Mass Spectrometry (EI-HRMS)

High-resolution gas chromatography–electron-ionization high-resolution mass spectrometry (HRGC–EI-HRMS) operated in SIM mode has been used for the analysis of PCTs and toxaphene [23,28,80]. This technique allows high selectivity in the analysis of PCTs by eliminating the internal interferences in the measurement of the molecular cluster ions of each PCT homologue group. For toxaphene analysis, a higher specificity on the CHB compounds than LRMS and unambiguous determination of the individual congeners in environmental and biota samples have been achieved.

The analysis of PCTs using EI-LRMS presents some interferences due to the $[M - Cl_2]^+$ fragment ions of the homologues with two additional chlorine atoms, which can be overcome using HRMS [28]. A scheme of the contribution of the nonachloroterphenyl molecular ion to the selected heptachloroterphenyl molecular ion is given in Figure 4, as example of the high resolution needed. The loss of two chlorine atoms gives the same $[M - 2Cl]^+$ cluster fragment that interferes with the molecular cluster ion of octachloroterphenyls. In order to avoid this interference, a resolution of 25,300 is required. In general, for the complete elimination of all interferences, a resolution higher than 34,600 is needed [28].

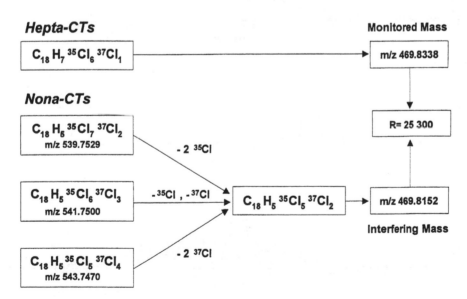

Figure 4 Interferent mass of nonachloroterphenyls on the heptachloroterphenyls monitored molecular mass for Aroclor 5460. (From Ref. 28.)

The HRGC–EI-LRMS SIM profile for the heptachloroterphenyls and the corresponding HRGC–EI-HRMS SIM profiles for the hepta- and nonachloroterphenyls after the loss of two chlorine atoms (m/z 469.8152) are shown in Figure 5. The contribution of the $[M - 2Cl]^+$ ion of nonachloroterphenyls on the molecular ion of heptachloroterphenyls is completely eliminated at a resolving power of 35,000. Under these conditions, the homologue distributions of PCT mixtures can be determined. For example, the values obtained for Aroclor 5460 and Leromoll 141 [28] indicate that Aroclor 5460 is mainly composed of terphenyls with 7 (10.4%), 8 (45.7%), 9 (29.4%), and 10 (11.9%) chlorine atoms with the largest group being octachloroterphenyls, whereas Leromoll 141 is composed by penta- (12.0%), hexa- (30.2%), hepta- (42.6%), and octachloroterphenyls (14.1%). These homologue distributions could be considered as reference values for the determination of the composition of these commercial formulations. The tech-

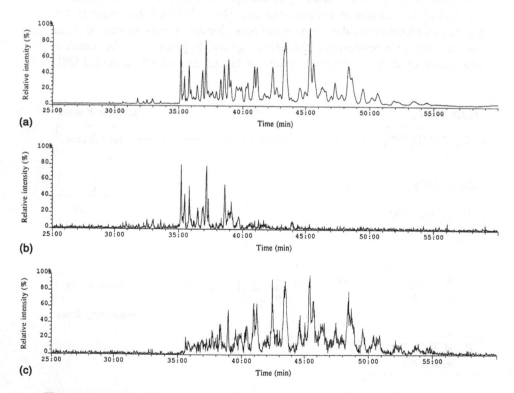

Figure 5 HRGC–EI-selected ion monitoring using (a) LRMS for heptachloroterphenyls, m/z 469.8, (b) HRMS for heptachloroterphenyls, m/z 469.8338, and (c) nonachloroterphenyls after loss of two chlorine atoms, m/z 469.8152. (From Ref. 28.)

nique of HRGC–HRMS has been used in the analysis of PCTs by only a small number of authors [23,28].

Electron-ionization high-resolution mass spectrometry has also been applied for the analysis of toxaphene to prevent the potential interferences of other organochlorine compounds. Three different approaches have been used for the analysis of these compounds. The first is based on the monitoring of the m/z 158.9768 and the isotope peaks at m/z 160.9739 and 162.9709, which correspond to the dichlorotropylium ion structure ($C_7H_5Cl_2^+$) formed through successive dechlorinations and rearrangements of CHBs. The use of these ions in combination with HRMS at a resolving power higher than 10,000 was firstly proposed by Saleh [37] in 1983. It has been used by several laboratories [95–97] with relatively good results. The correct isotope ratios of $100:65:11\%$ for m/z 158.9768, 160.9739, and 162.9709, respectively, give an unambiguous identification of CHBs. This method has been applied in the determination of toxaphene in biota samples [95,96] and limits of detection lower than 10 μg g^{-1} for total toxaphene [97] and 0.2 ng g^{-1} for specific single congeners (2 pg injected) [98,99] have been reported.

The characteristic ion at m/z 159 can be considered as a universal ion for all toxaphene congeners regardless of their degree of unsaturation and chlorination. Therefore, chlorobornenes and chlorobornadienes can be determined in a single injection. Total toxaphene concentration can be calculated by comparing the total areas of samples and technical toxaphene standard. One of the advantages of EI-HRMS is that these ions (159/161) are not affected by the presence of common environmental pollutants such as PCBs and organochlorine pesticides, although other compounds, such as chlorinated pinenes, camphenes, and compounds with a similar structure to toxaphene, can interfere [46]. In addition, since the m/z 159/161 ions are end products from a series of fragmentation pathways, the identity of the original compound is lost and different homologue groups cannot be distinguished.

The second approach proposed in the literature for the analysis of toxaphene by EI-HRMS is based on the determination of the homologue composition. Santos et al. [80], using SIM mode, studied the two most intense ions of the $[M - Cl]^+$, $[M - HCl]^{+\bullet}$, and $[M - Cl - HCl]^+$ clusters for each homologue group (between Cl_6 and Cl_{10}). From these profiles, they concluded that the $[M - Cl]^+$ cluster ions had the maximum intensity. The homologue profiles for toxaphene at different resolving powers are shown in Figure 6 as an example The interferences of homologues with an additional chlorine atom are not observed at a resolving power of 20,000, and at 10,000 most of the interferences are eliminated. By integrating the EI-HRMS SIM profiles for each homologue group using $[M - Cl]^+$ ions (hexa-CHBs, m/z 308.9352; hepta-CHBs, m/z 342.8963; octa-CHBs, m/z 376.8573, nona-CHBs, m/z 410.8173; and deca-CHBs, m/z 444.7793), the homologue distribution of commercial toxaphene can be calculated. Using this approach, it is deduced that toxaphene was mainly composed

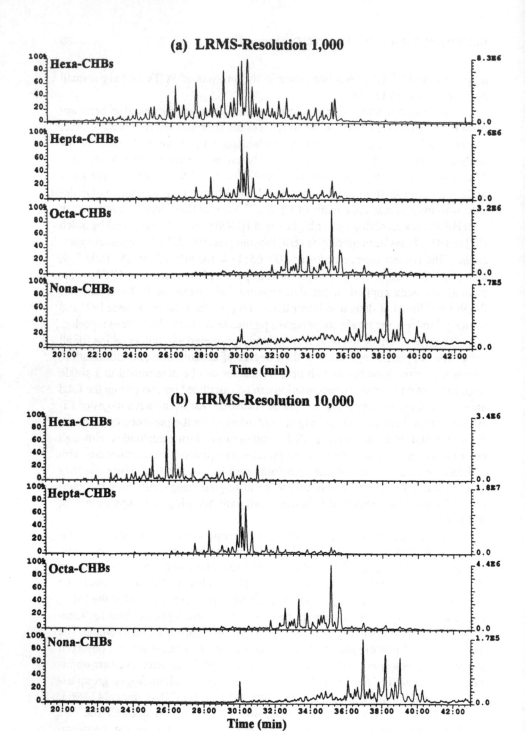

Figure 6 HRGC–EI-MS-selected ion monitoring of $[M - Cl]^+$ ions for the homologues of toxaphene at resolving power of (a) 1,000, (b) 10,000, and (c) 20,000. (From Ref. 80.)

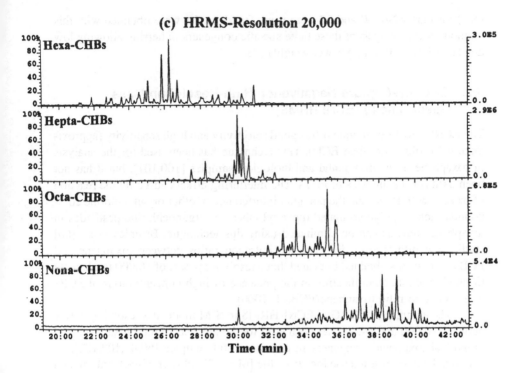

(c) HRMS-Resolution 20,000

Time (min)

of CHBs with 7 chlorine atoms, with smaller amounts of 6 and 8, and very low amounts of 9 and 10. This homologue distribution of toxaphene was distinctly different from the one obtained by HRGC–ECNI-MS [62,91], and these differences might be due to the MS techniques used. The ECNI-MS gives variable responses for CHBs, even for congeners of the same homologue group, and it is highly affected by ECNI conditions. In contrast, the EI-HRMS technique gives very similar responses for each homologue group and eliminates interferences among them. Therefore, this technique can be considered as reference for the determination of toxaphene homologue composition. The HRGC–EI-HRMS in SIM mode has been also applied to the quantification of CHB congeners in commercial toxaphene mixtures, and the detection limits obtained with this technique were between 5 and 9 pg injected [80].

Finally, a third method has been proposed for the analysis of toxaphene using EI-HRMS in SIM mode [98,99]. This method is based on the congener-specific analysis for Parlar No. 26, 50, and 62, which are considered as indicators of toxaphene contamination. Selective ion monitoring at a resolving power of 10,000 of mass fragments at m/z 340.8806 ([M − Cl − HCl]⁺ cluster ion) for Parlar No. 26 and at m/z 338.8649 and 340.8620 ([M − Cl − 2HCl]⁺ cluster ions) for Parlar No. 50 and 62 were used. Good results were obtained with this method for the analysis of these three specific congeners in herring allowing low detection limits (0.2 ng g^{-1} wet weight) [98].

3.2. Electron-Capture Negative-Ion High-Resolution Mass Spectrometry (ECNI-HRMS)

The ECNI-HRMS technique offers good selectivity and high sensitivity (approximately 10-fold better than ECD). This technique has been used for the analysis of toxaphene in environmental and biological samples [100,101], but it has not been used for the analysis of PCTs. The interfering effect caused by the reaction of oxygen with PCBs and the potential interference of other organochlorine compounds such as polychlorinated diphenyl ethers or organochlorine pesticides in toxaphene analysis can be eliminated using this technique. Brumley et al. [60] studied the influence of different concentration ratios between toxaphene and PCBs. The results obtained indicated that a resolving power of 10,000 is adequate to obtain reliable quantification in the presence of high concentrations of PCBs (concentration ratio toxaphene/PCBs 1:1000).

The technique of HRGC–ECNI-HRMS in SIM mode at a resolving power between 10,000 and 14,000 has been used for the analysis of total toxaphene and individual toxaphene congeners in environmental samples [60,82,102,103]. In general, the two most intense ions from the [M − Cl]$^-$ cluster of each chlorinated homologue group (m/z 308.8962 and 310.9323 for hexachlorobornanes, m/z 376.8573 and 378.8543 for heptachlorobornanes, and m/z 410.8183 and 412.3154 for octachlorobornanes) have been selected for identification and quantification of toxaphene. In all cases, methane has been used as moderating gas for ECNI. Using this technique, low detection limits for individual toxaphene congeners ranging from 0.06 to 0.7 µg kg^{-1} for sediments and from 0.02 to 0.2 µg kg^{-1} for fish [102], or between 0.3 and 7.0 pg injected [103], have been obtained. Although the use of individual standards has increased in recent years, the number and the type of congeners that must be analyzed in environmental samples have not yet been established. For example, Kimmel et al. [103] proposed to add Parlar No. 40, 41, and 44 to the commonly used congeners, Parlar No. 26, 50, and 62 [42,63,64] for quantification of toxaphene. The sum of these six congeners seems to be suitable for quantification of toxaphene residues in fish and fish oils and represents at least 80 to 85% of the total toxaphene.

In general, ECNI-HRMS can provide higher sensitivity, whereas higher selectivity can be achieved using the EI-HRMS mode. This can be explained by the broad fragmentation pattern obtained with EI, which allows one to chose a more specific ion to be monitored. Thus, both HRMS techniques may offer greater possibilities than LRMS for the analysis of toxaphene.

4. HIGH-RESOLUTION GAS CHROMATOGRAPHY– TANDEM MASS SPECTROMETRY (HRGC–MS–MS)

As has been pointed out previously, HRMS can be used for the characterization and quantification of PCTs and toxaphene, but its application to routine analysis is difficult because of the high resolving power required, and the high cost of the instrumentation. EI-MS–MS can be an excellent alternative to HRMS because of its high sensitivity and its lower cost.

Tandem MS monitoring selected fragment ions produced in the collision cell of a sector or multiquadrupole instrument or in an ion-trap chamber offers advantages over single-stage MS in terms of selectivity of analysis and specificity in molecular structure determination [104]. Tandem MS techniques have been widely used with good results for the analysis of polyhalogenated compounds, such as isomers of the polychlorinated dibenzo-*p*-dioxins and dibenzofurans [105], PCBs [106], PCNs, and related compounds [107]. However, the application of MS–MS to the analysis of PCTs and toxaphene has been very limited.

For PCT analysis, Santos et al. [29] reported good results in the determination of the homologue distribution of these compounds using HGC–MS–MS. Experiments have been performed using the precursor-ion and product-ion scan modes, which give information about interferences and typical fragmentation reactions, as well as selective reaction monitoring (SRM), which allows one to obtain information about the distribution of PCT homologues. The precursor-ion and product-ion mass spectra of a heptachloroterphenyl from Aroclor 5460 obtained by collision-activated dissociation (CID) with xenon as collision gas are given in Figure 7 as an example [29]. In the precursor-ion spectrum of m/z 470 for heptachloroterphenyls (see Fig. 7a), the ions at m/z 540, 542 and 544, corresponding to the nonacloroterphenyls, and the ions at m/z 505 and 507, corresponding to the nonachloroterphenyls after the loss of 1 chlorine atom, are observed. In this spectrum, it can be seen that the major interferences in the heptachlototerphenyls are the nonachloroterphenyls [28]. In the product-ion mass spectrum of a heptachloroterphenyl (precursor-ion m/z 470) (see Fig. 7b), intense fragment ions are observed corresponding to the loss of 2 and 4 chlorine atoms, $[M - 2Cl]^+$ and $[M - 4Cl]^+$. Using the information of the product-ion and precursor-ion mass spectra, the transitions ($[M]^{+\bullet} \rightarrow [M - 2^{35}Cl]^+$) for each homologue group can be selected and used in SRM experiments. From these experiments, it can be deduced that some interferences of the homologues with two additional chlorine atoms occurred. The inability of the HRGC–MS–MS experiments to completely overcome interferences is mainly due to the low resolution of the first MS stage, which allows some interfering ions access to the CID cell when chromatographic coelution occurs. Although the interferences are not completely eliminated, the distribution of PCT mixtures can be calculated taking the

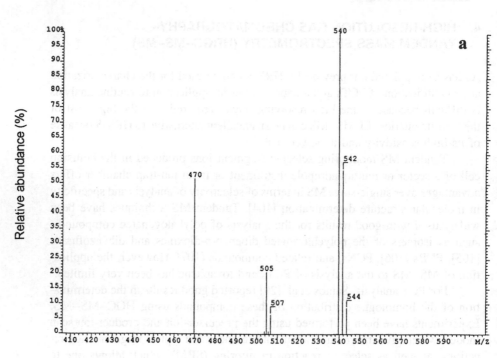

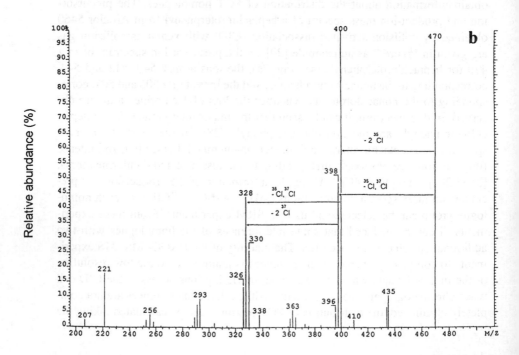

coelution and the effect of the contribution of the $[M - 2Cl]^+$ fragment ions into account, following the same strategy as proposed with LRMS [20]. In order to compare the capabilities of LRMS, HRMS, and MS–MS for the study of homologue distribution of PCTs, the results obtained for two commercial formulations, Aroclor 5460 and Leromoll 141, are given in Table 4. The percentages of the homologue distribution obtained using HRMS and MS–MS are very similar, which ensures that the interference of the $[M - 2Cl]^+$ fragment ion with $[M]^{+\cdot}$ of each homologue is minimized by MS–MS. Consequently, HRGC–MS–MS in SRM mode can be considered as a useful and relatively inexpensive technique for the determination of the homologue distribution of PCTs. This technique has also been successfully applied to the identification and quantification of PCTs in fish samples, in which relatively low levels of PCTs have been detected [28]. Sensitivity greater by a factor of about 3 than that obtained under the best HRMS conditions was achieved.

Electron-ionization tandem mass spectrometry (EI-MS–MS) has also been applied for the analysis of toxaphene using SRM [107,108]. This technique has been used for the identification, characterization, and structural elucidation of polychlorobornanes in technical toxaphene and marine mammals. Buser et al. [107] reported detailed fragmentation pathways for two toxaphene congeners isolated from marine mammals, Parlar No. 26 and 50. These authors concluded that the most important fragmentation pathway of these two CHB congeners involves the elimination of $C_2H_2Cl_2$ and C_2HCl_3 from $[M - HCl]^{+\cdot}$ ions through a retro–Diels-Alder mechanism. The transition corresponding to the loss of $C_2H_2Cl_2$ can be considered specific for these compounds. In Figure 8, the SRM chromatograms of technical toxaphene corresponding to the retro–Diels-Alder transitions (a) $374^+ \rightarrow 244^+$ and (b) $374^+ \rightarrow 278^+$ for octachlorobornanes, and (c) $408^+ \rightarrow 278^+$ and (d) $408^+ \rightarrow 312^+$ for nonachlorobornanes are shown as an example. As can be seen, TOX9 (Parlar No. 50) can be identified as a major component in technical toxaphene, whereas TOX8 (Parlar No. 26) is only present as a minor component. In addition, low responses for these congeners were obtained when the loss of C_2HCl_3 was registered. Different fragmentation pathways for toxaphene have been studied [108]: the loss of Cl and HCl, the retro–Diels-Alder mechanism, and the loss of $CHCl_2$ and CH_2Cl. From these fragmentation reactions, it can be deduced that higher selectivity and sensitivity can be obtained using both the retro–Diels-Alder and the loss of $CHCl_2$ and CH_2Cl for the analy-

---

Figure 7 MS–MS spectra of a heptachloroterphenyl of Aroclor 5460. (a) Precursor-ion spectrum of the molecular ion m/z 470. (b) Product-ion spectrum of the ion m/z 470. (From Ref. 29.)

Table 4 Homologue Distribution for Aroclor 5460 and Leromoll 141 Obtained by HRGC–EI-LRMS-SIM (Resolving Power 1,500), HRGC–EI-HRMS (Resolving Power 35,000) and HRGC–EI-MS–MS-MRM

| | Percentage of Homologues | | | | | |
| | Aroclor 5460 | | | Leromoll 141 | | |
Homologues	LRMS-SIM	HRMS-SIM	MS–MS-MRM	LRMS-SIM	HRMS-SIM	MS–MS-MRM
Penta-CTs	ND[a]	ND	ND	ND	12.0	11
Hexa-CTs	9	0.9	1	20	30.2	32
Hepta-CTs	11	10.4	11	33	42.6	44
Octa-CTs	25	45.7	44	39	14.1	12
Nona-CTs	34	29.4	29	8	0.7	0.6
Deca-CTs	18	11.9	13	0.2	0.4	ND
Undeca-CTs	3	1.2	2	ND	ND	ND
Dodeca-CTs	ND	ND	<0.5	ND	ND	ND

[a] ND: not detected.
Source: Adapted from Refs. 28 and 29.

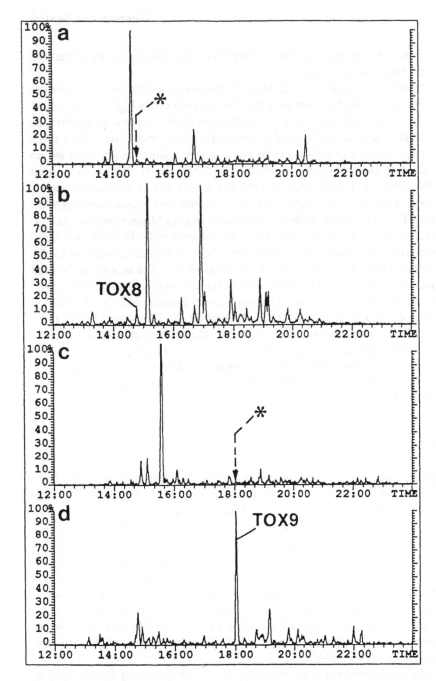

Figure 8 Selected reaction monitoring chromatograms of technical toxaphene. Retro–Diels-Alder transitions from octachlorobornanes (a) $374^+ \rightarrow 244^+$ (loss of C_2HCl_3), (b) $374^+ \rightarrow 278^+$ (loss of $C_2H_2Cl_2$), and from nonachlorobornanes (c) $408^+ \rightarrow 278^+$ (loss of C_2HCl_3), and (d) $408^+ \rightarrow 312^+$ (loss of $C_2H_2Cl_2$). Absence of signal at retention times of TOX 8 (Parlar No. 26) and TOX 9 (Parlar No. 50) is marked by asterisk in chromatograms a and c, respectively. Retention times indicated in min.). (From Ref. 107.)

sis of homologue groups in technical toxaphene, while the specificity of the loss of Cl and HCl was very small.

Recently, ion-trap–based EI-MS–MS has been used for the determination of toxaphene in biological samples [109]. Ion trap MS offers high selectivity and sensitivity at relatively low cost. In contrast with hybrid or triple-stage quadrupole instruments, where the isolation of precursor ions and further dissociation takes place in space, ion-trap MS–MS takes place in time. Consequently, there are no transmission losses and, hence, it provides better sensitivity. For use in ion-trap MS–MS, Chan et al. [109] proposed the ion at m/z 125 as a precursor ion for toxaphene analysis. This corresponds to a chlorinated monochlorotropylium structure. Polychlorinated biphenyls and other organochlorine compounds also produce the fragment ion at m/z 125 in the ion-trap MS–MS mode, but these compounds do not produce the product ion m/z 89, which corresponds to totally dechlorinated toxaphene. Therefore, the transition m/z 125 to m/z 89 can be used for quantification of total toxaphene and individual toxaphene congeners [109]. Using this method, instrumental detection limits between 1 and 3 pg μl^{-1} for the individual congeners Parlar No. 26, 32, 50, and 62 and 280 pg μl^{-1} for the toxaphene technical mixture have been achieved. Coefficients of variation of 14% were obtained. Although ion-trap MS–MS allows good results in terms of accuracy and precision, some interferences from organochlorine pesticides have been detected. Nevertheless, these interferences seem not to be important in environmental and biological samples.

5. CONCLUSIONS

The HRGC–MS technique is a powerful tool for the analysis of PCTs and toxaphene in environmental and biological samples. Both HRGC–EI-MS and HRGC–ECNI-MS in the SIM mode at a low resolving power seem to be suitable for the analysis of these compounds, but some problems have been detected. In order to avoid the interferences produced by the presence of other organochlorine compounds and the contribution of homologues with higher degrees of chlorination, HRMS is required. Neither of these two LRMS techniques can be claimed to be superior to the other. The method based on ECNI-MS in SIM mode has relatively good sensitivity for most of the toxaphene congeners, but false negatives can be produced, whereas the method based on EI-MS in SIM mode has relatively lower sensitivity, although more similar and reproducible responses among congeners are obtained.

High-resolution mass spectrometry offers high selectivity for the analysis of PCTs and toxaphene congeners in environmental and biological samples and avoids the specific internal interferences and the effect of environmental matrices. Both HRGC–EI-MS and HRGC–ECNI-MS in the SIM mode can be used as reference methods because unambiguous determination of these compounds is achieved. In addition, low detection limits and good reproducibility are obtained. Nevertheless, the high cost of the instrumentation makes it difficult to incorporate these techniques in routine laboratories. Tandem MS can be an excellent alternative to HRMS. It is a highly selective and relatively inexpensive technique and can be considered as promising for the analysis of PCTs and toxaphene.

Future development in GC–MS analysis will require the use of more stable GC capillary columns or multidimensional GC coupled to advanced MS techniques such as HRMS and MS–MS with EI and ECNI ionization. The availability of further individual congeners for specific analysis and isotopically labeled standards will assure an accurate quantitative analysis. In addition, the development of more selective analytical methods for the isolation of these compounds from environmental and biological matrices can help to enhance the selectivity of the LRMS techniques. Further research and collaborative studies still need to be done to solve the problems associated with the analysis of PCTs and toxaphene.

REFERENCES

1. R.D. Kimbrough and A.A. Jensen, Halogenated Biphenyls, Terphenyls, Naphthalenes, Dibenzodioxins, and related Compounds, 2d ed., 1989, Elsevier/North-Holand Biomedical Pr., Amsterdam, Netherlands.
2. O. Hutzinger, The Handbook of Environmental Chemistry, vol. 3, Part B, 1982, Springer, Berlin, pp. 89–92.
3. J.E. Hose, J.N. Cross, S.G. Smith, and D. Diehl, Environ. Pollut., 57 (1989) 139–148.
4. S. Jensen, New Sci., 32 (1966) 612–622.
5. M.D. Erickson, Analytical Chemistry of PCBs, 1986, Butterworth, Stoneham, MA.
6. J.S. Waid, PCBs and the Environment, Vols. 1–3, 1986, CRC Pr., Boca Raton, FL.
7. P. De Voogt and U.A.Th. Brinkman, in R.D. Kimbrough and A.A. Jensen, eds., Production, Properties and Usage of Polychlorinated Biphenyls, 1989, Elsevier Science, Amsterdam, Netherlands.
8. A.A. Jensen and K.F. Jørgensen, Sci. Total Environ., 27 (1983) 231–250.
9. A. De Kok, B. Geerdink, G. De Vries, and U.A.Th. Brinkman, Int. J. Environ. Anal. Chem., 12 (1982) 99–122.
10. G. Remberg, P. Sandra, W. Nying, N. Winker, and A. Nikiforov, Fresenius J. Anal. Chem., 362 (1988) 404–408.
11. D.O. Duelbelbeis, Ph. D. dissertation, 1988, University of Missouri, Columbia, MO.
12. World Health Organization (WHO), Environmental Health Criteria, 140: Polychlorinated Biphenyls and Terphenyls, 2d ed., 1993, WHO, Geneva, Switzerland.
13. R.C. Hale, J. Greaves, K. Gallagher, and G.G. Vadas, Environ. Sci. Technol., 24 (1990) 1727–1731.

14. F.I. Onuska, K.A. Terry, S. Rokushika, and H. Hatano, J. High Resol. Chromatogr., 13 (1990) 317–322.
15. L. Canton and J.O. Grimalt, Chemosphere, 23 (1991) 327–341.
16. B. Scholz, Gewässerschutz, Wasser, Abwasser, 88 (1986) 228–244.
17. C. Strasser and H. Schindlbauer, Erdöl, Erdgas, Kohle, 105(12) (1989) 517–519.
18. L. Renberg, G. Sundström, and L. Reutergårdgh, Chemosphere, 6 (1978) 477–482.
19. M.T. Galceran, F.J. Santos, J. Caixach, and J. Rivera, Chemosphere, 27(7) (1993) 1183–1200.
20. M.T. Galceran, F.J. Santos, J. Caixach, F. Ventura, and J. Rivera, J. Chromatogr., 643 (1993) 399–408.
21. P.G. Wester, J. De Boer, and U.A.Th. Brinkman, Environ. Sci. Technol., 30 (1996) 473–480.
22. R.C. Hale, E. Bush, K. Gallagher, J.L. Gundersen, and R.F. Mothershead, J. Chromatogr., 539 (1991) 149–156.
23. M.A. Fernández, L.M. Hernández, M.J. González, E. Eljarrat, J. Caixach, and J. Rivera, Chemosphere, 36(14) (1998) 2941–2948.
24. S.T. González-Barros, M.E. Alvarez, and J. Simal, M.A. Lage, Chemosphere, 35(6) (1997) 1243–1247.
25. L.H. Wright, R.G. Lewis, L.H. Crist, G.W. Sovocool, and J.M. Simpson, J. Anal. Toxicol., 2 (1978) 76–79.
26. I. Watanabe, T. Yakushiji, and N. Kunita, Bull. Environ. Contam. Toxicol., 25 (1980) 810–815.
27. U. Seidel, E. Schweizer, F. Schweinsberg, R. Wodarz, and A.W. Rettenmeier, Environ. Health Perspect., 104 (1996) 1172–1179.
28. J. Caixach, J. Rivera, M.T. Galceran, and F.J. Santos, J. Chromatogr. A, 675 (1994) 205–211.
29. F.J. Santos, M.T. Galceran, J. Caixach, X. Huguet, and J. Rivera, Rapid Comm. Mass Spectrom., 10 (1996) 1774–1780.
30. T. Cairns, E.G. Siegmund, and J.E. Foberg, Biomed. Mass Spectrom., 8 (1981) 568–574.
31. B. Lau, D. Weber, and P. Andrews, Chemosphere, 32(6) (1996) 1021–1041.
32. M.A. Saleh, in G.W. Ware, ed., Toxaphene: Chemistry, Biochemistry, Toxicity and Environmental Fate, in Reviews of Environmental Contamination and Toxicology, vol. 118 (1991) 1–85.
33. World Health Organization (WHO), Environ. Health Criteria, 45, Workshop on the analytical and environmental chemistry of Toxaphene, 1984, WHO, Burlington, Ontario, Canada.
34. H.J. Geyer, A. Kaune, K.-W. Schramm, G. Rimkus, I. Scheunert, R. Brüggemann, J. Atschuh, C.E. Steinberg, W. Vetter, A. Kettup, and D.C.G. Muir, Chemosphere, 39(4) (1999) 655–663.
35. M.A. Saleh, in J.D. Rosen, ed., Applications of New Mass Spectrometry Techniques in Pesticides Chemistry, 1987, J. Wiley, New York, pp. 34.
36. E.C. Voldner and Y.F. Li, Chemosphere, 27 (1993) 2073–2078.
37. M.A. Saleh, J. Agric. Food Chem, 31 (1983) 748–751.

38. G.A. Pollock and W.W. Kilgore, in F.A. Gunther and J. Davies Gunther, eds., Residue Reviews, vol. 69, 1978, Springer-Verlag, New York, pp. 87–111.
39. W. Vetter, Chemosphere, 26 (1993) 1079–1084.
40. H. Parlar, G. Fingerling, D. Angerhöfer, G. Christ, and M. Coelhan, in R. P. Eganhouse, Molecular Markers in Environmental Geochemistry, ACS Symposium Series 671, 1997, American Chemistry Society, Washington DC, 1997, pp. 346–364.
41. J. Burhenne, D. Hainzl, L. Xu, B. Vieth, L. Alder, and H. Parlar, Fresenius J. Anal. Chem., 346 (1993) 779–785.
42. L. Xu, D. Hainzl, J. Burhenne, and H. Parlar, Chemosphere, 28 (1994) 237–243.
43. V.A. Nikiforov, V.G. Tribulovich, and V.S. Karavan, Organohalogen Compd., 26 (1995) 3.•3–396.
44. V.G. Tribulovich, V.A. Nikiforov, V.S. Karavan, S.A. Miltsov, and S. Bolshakov, Organohalogen Compd., 19 (1994) 97–101.
45. M. Oehme and R. Kallenborn, Chemosphere, 30 (1995) 1739–1750.
46. P. Andrews and W. Vetter, Chemosphere, 31 (1996) 3879–3886.
47. P.G. Wester, H.-J. De Geus, J. De Boer, and U.A.Th. Brinkman, Chemosphere, 35(6) (1997) 1187–1194.
48. D.C.G. Muir and J. De Boer, Chemosphere, 27 (1993) 1827–1834.
49. W. Vetter, G. Scherer, M. Schlabach, B. Luckas, and M. Oehme, Fresenius J. Anal. Chem., 349 (1994) 552–558.
50. P. Andrews, K. Headrich, J.-C. Pilon, B. Lau, and D. Weber, Chemosphere, 31 (1995) 4393–4402.
51. P.G. Wester, H.-J. De Geus, J. De Boer, and U.A.Th. Brinkman, Chemosphere, 35(6) (1997) 2857–2864.
52. D.C.G. Muir and J. De Boer, J. Trends Anal. Chem., 14 (1995) 56–66.
53. J. De Boer and P.G. Wester, Chemosphere, 27 (1993) 1879–1890.
54. M. Shoeib, K. A. Brice, and R.M. Hoff, Chemosphere, 39(5) (1999) 849–871.
55. T.F. Bidleman, R.L. Falconer, and M.D. Wells, Sci. Total Environ., 160/161 (1995) 55–63.
56. R.A. Raport and S.J. Eisenrcich, Environ. Sci. Technol., 22 (1988) 931–944.
57. D.L. Swackhamer and R.A. Hites, Environ. Sci. Technol., 22 (1988) 543–548.
58. D.L. Swackhamer, R.F. Pearson, and S.P. Schottler, Chemosphere, 37 (1998) 2545–2561.
59. J.R. Kucklick and E. Baker, Environ. Sci. Technol., 32 (1998) 1192–1198.
60. W.C. Brumley, C.M. Brownrigg, and A.H. Grange, J. Chromatography, 633 (1993) 177–183.
61. R.F. Pearson, D.L. Swackhamer, S.J. Eisenreich, and D.T. Long, Environ. Sci. Technol., 31 (1997) 3523–3529.
62. D.L. Swackhamer, M.J. Charles, and R.A. Hites, Anal. Chem., 59 (1987) 913–917.
63. L. Alder, H. Beck, S. Khandker, H. Karl, and I. Lehmann, Chemosphere, 34 (1997) 1389–1400.
64. M. Alawi, H. Barlas, D. Hainzl, J. Burhenne, M. Coelhan, and H. Parlar, Fresenius Environ. Bull., 3 (1994) 350–357.
65. J. Zhu and R.J. Norstrom, Chemosphere, 27 (1993) 1923–1936.

66. B.T. Hargrave, D.C.G. Muir, and T.F. Bidleman, Chemosphere, 27 (1993) 1949–1963.
67. R. Baycan-Keller and M. Oehme, J. High Resol. Chromatogr., 21 (1998) 298–302.
68. J. De Boer, H-J. De Geus, and U.A.Th. Brinkman, Environ. Sci. Technol., 31 (1997) 873–879.
69. W. Vetter and V. Schuring, J. Chromatogr. A, 774 (1997) 143–175.
70. L. Alder, R. Palvinskas, and P. Andrews, Organohalogen Compd., 28 (1996) 410–415.
71. U. Klobes, W. Vetter, B. Luckas, and G. Hottinger, Chromatographia, 47(9/10) (1998) 565–569.
72. W. Vetter, U. Klobes, and B. Luckas, Chromatographia, 44 (1997) 65–73.
73. H. Parlar, D. Schulz-Jander, G. Fingerling, G. Koske, D. Angerhöfer, and J. Burhenne, Organohalogen Compd., 35 (1998) 221–224.
74. J. De Boer, Organohalogen Compd., 33 (1997) 7–11.
75. H. Parlar, D. Angerhöfer, M. Ceolhan, and L. Kimmel, Organohalogen Compd., 26 (1995) 357–362.
76. T.R. Schwartz, D.E. Tillitt, K.P. Feltz, and P.H. Peterman, Chemosphere, 26(8) (1993) 1443–1460.
77. D.C.G. Muir, C.A. Ford, N.P. Grift, R.E.A. Stewart, and T.F. Bidleman, Environ. Pollut, 75 (1992) 307–316.
78. L. Alder and B. Vieth, Fresenius J. Anal. Chem., 354 (1996) 81–92.
79. G.A. Stern, D.C.G. Muir, J.B. Westmore, and W.D. Buckannon, Biol. Mass Spectrom., 22 (1993) 19–24.
80. F.J. Santos, M.T. Galceran, J. Caixach, J. Rivera, and X. Huguet, Rapid Commun. Mass Spectrom., 11 (1997) 341–348.
81. K. Oetjen and H. Karl, Chemosphere, 37(1) (1998) 1–11.
82. G.A. Stern, M.D. Loewn, B.M. Miskimmin, D.C.G. Muir, and J.B. Westmore, Environ. Sci. Technol., 30 (1996) 2251–2258.
83. W. Vetter, U. Klobes, B. Krock, B. Lúckas, D. Glotz, and G. Scherer, Environ. Sci. Technol, 31 (1997) 3023–3028.
84. B. Lau, D. Weber, and P. Andrews, Rapid Commun. Mass Spectrom., 8 (1994) 849–853.
85. E.A. Stemmler and R. Hites, Anal. Chem., 57 (1985) 684–691.
86. U.S. Gill, H.M. Schwartz, and B. Wheatley, Intern. J. Environ. Anal. Chem., 60 (1995) 153–161.
87. F. van der Valk and P. Wester, Chemosphere 22(1–2) (1991) 57–66.
88. G.A. Stern, D.C.G. Muir, C.A. Ford, N.P. Grift, E. Dewalley, T.F. Bidleman, and M.D. Walla, Environ. Sci. Technol., 26 (1992) 1838–1840.
89. W. Vetter, C. Natzeck, B. Krock, and G. Heidemann, Chemosphere, 30(9) (1995) 1685–1696.
90. J. Zhu, M.J. Mulvihill, and R.J. Norstrom, J. Chromatogr. A, 669 (1994) 103–117.
91. C.P. Rice, Chemosphere, 27 (1993) 1937–1947.
92. F.I. Onuska, K.A. Terry, A. Seech, and M. Antonic, J. Chromatogr. A, 665 (1994) 125–132.
93. J.P. Boon, M. Helle, M. Dekker, H.M. Sleiderink, H.J. Klamer, B. Govers, P.G. Wester, and J. De Boer, Organohalogen Compd., 28 (1996) 389–394.

94. D.G. Patterson, Jr., W.E. Turner, S. Isaacs, and L.R. Alexander, Chemosphere, 20 (1990) 829–836.
95. T.L. Wade, L. Chambers, P.R. Gardinali, J.L. Sericano, T.J. Jackson, R.J. Tarpley, and R. Suydam, Chemosphere, 34(5–7) (1997) 1351–1357.
96. W.H. Newsome and J.J. Ryan, Chemosphere, 39(3) (1999) 519–526.
97. F.I. Onuska and K.A. Terry, J. Chromatogr., 471 (1989) 161–171.
98. T. Cederberg, Organohalogen Compd., 31 (1997) 64–68.
99. A. Fromberg, T. Cederberg, and G. Hilbert, Organohalogen Compd., 35 (1998) 259–262.
100. L. Barrie, T.F. Dougherty, P. Fellin, N. Grift, D.C.G. Muir, R. Rosenber, G. Stern, and D. Toan, Chemosphere, 27 (1993) 2037–2046.
101. G.A. Stern, D.C.G. Muir, J.B. Westmore, and W.D. Buchannon, Biol. Mass Spectrom., 22 (1993) 19–30.
102. D.B. Donald, G.A. Stern, D.C.G. Muir, B.R. Fowler, B.M. Miskimmin, and R. Bailey, Environ. Sci. Technol., 32 (1998) 1391–1397.
103. L. Kimmel, D. Angerhöfer, U. Gill, M. Coelhan, and H. Parlar, Chemosphere, 37(3) (1998) 549–558.
104. K.L. Busch, G.L. Glish, and S.A. McLuckey, Mass Spectrometry/Mass Spectrometry. Techniques and Applications of Tandem Mass Spectrometry, 1988, VCH Publishers, New York.
105. M.J. Charles, W.C. Green, and G.D. Marbuny, Environ. Sci. Technol., 29 (1995) 1741–1747.
106. L. Zupancic-Kralj, J. Marsel, B. Kralj, and D. Žigon, Analyst, 119 (1994) 1129–1134.
107. H-R. Buser, M. Oehme, W. Vetter, and B. Luckas, Fresenius J. Anal. Chem., 347 (1993) 502–512.
108. F.J. Santos, M.T. Galceran, J. Caixach, X. Huguet, and J. Rivera, Organohalogen Compd., 35 (1998) 225–228.
109. H.M. Chan, J. Zhu, and F. Yeboah, Chemosphere, 36(9) (1998) 2135–3148.

6

On-Line Solid-Phase Extraction– Capillary Gas Chromatography– Mass Spectrometry for Water Analysis

Thomas Hankemeier
TNO Nutrition and Food Research, Zeist, The Netherlands

Udo A. Th. Brinkman
Free University, Amsterdam, The Netherlands

1. INTRODUCTION

For compounds that are amenable to analysis by means of capillary gas chromatography (GC) without prior derivatization, GC should be one's first choice as a separation method because of three main advantages: excellent separation efficiency, high speed of analysis, and a wide range of sensitive, universal—and often selective—detectors. With regard to these aspects, GC is certainly superior to column liquid chromatography (LC), and supercritical and planar chromatography. In most works published in the past decade, it was then immediately added that GC also has one particular weak spot, viz., sample pretreatment/introduction, irrespective of whether samples ending up as organic extracts, or aqueous samples or sample extracts are considered. Fortunately, this is not true today. The purpose of this chapter is to briefly discuss recent developments in this area and to highlight current procedures for on-line sample preparation–GC. Emphasis is on the use of solid-phase extraction (SPE) for sample preparation, and the combination of SPE–GC with mass spectrometry (MS) because of the identification/confirmation potential of such an integrated system. The main application area discussed is (surface, ground, tap, and waste) water analysis—an area, that attracts increasing attention from governments, health authorities, and legislative bodies.

155

In most instances, the analyte enrichment required to enable the detection, identification, and quantification of the microcontaminants of interest—which are often present at concentrations of 10^{-8} to 10^{-11} g/ml—is carried out by means of liquid–liquid extraction (LLE) or solid–liquid sorption, which is generally called solid-phase extraction (SPE). Even today, the often laborious procedures frequently are performed manually and, what is more disadvantageous, use rather large volumes of organic solvent (especially LLE), and almost invariably involve partial or complete evaporation of the sample extract, which has the inherent danger of analyte losses. The main drawback, however, is that out of the final extract of, typically, 0.1 to 1 ml, generally only 1 to 2 µl are injected into the gas chromatograph (for many workers this is so routine a procedure that they frequently omit to state the injection volume in their papers). In other words, in the final step of the sample treatment procedure, some 98 to 99% of all the collected analyte(s) is discarded. One should note here that the above is true irrespective of the type of sample studied, aqueous or nonaqueous, because all of them end up, in a large majority of cases, as organic extracts.

Fortunately, in the past decade the situation with regard to sample introduction in GC has improved dramatically and several techniques—such as on-column, loop-type and programmed temperature vaporizer (PTV) injection—are available today. These allow the routine introduction of sample volumes of up to, say, 100 µl, which is a convenient upper limit for most real-life applications. A detailed introduction of these large-volume injection (LVI) techniques is outside the scope of this review; for the essentials, the reader is referred to Goosens et al. [1] and Chapter 1 of this book. It will be obvious that LVI-based procedures can effect a substantial reduction of the amount of sample required—and, consequently, of the time of analysis. In addition, this will make on-line sample preparation–GC easier to accomplish, and will stimulate the miniaturization of sample preparation procedures. For the rest, one should realize that LVI is the basic step in any SPE-to-GC transfer of analytes, which in turn is the critical step in all SPE–GC procedures. Suitable solvents for LVI–GC include pentane, hexane, methyl and ethyl acetate, and diethyl or methyl *tert*-butyl ether. From among these, methyl and ethyl acetate are preferably used in SPE–GC because of their excellent desorption characteristics for a wide variety of organic compounds. For more detailed information, the reader is referred to Grob et al. [2] and Engewald et al. [3].

If one compares the general situation of sample pretreatment/introduction in LC and GC, it should be clear that, with regard to this aspect, LC is certainly superior. This is especially true if aqueous samples are considered. The use of either disposable or reusable SPE cartridges (also called precolumns) packed with some 50 mg of an alkyl-bonded silica or highly hydrophobic copolymer sorbent, which are combined on-line with a (reversed-phase) LC system equipped with a diode-array UV-vis., a fluorescence or a mass spectrometric detector, enables efficient analyte enrichment from 10- to 100-ml water samples, without analyte breakthrough. This procedure produces improved detectability, in terms of concentration units, of three orders of magnitude, compared with 10- to 100-μl loop injections. The on-line setup ensures the transfer of all of the collected analytes rather than a 1 to 2% aliquot, and facilitates automation. One such system, the SAMOS LC, was developed by our group, and is commercially available from Hewlett Packard (Palo Alto, CA) [4–6].

On the basis of the above considerations, it will be obvious that, for a really powerful GC-based (water) analyzer to be built, an LC-type trace enrichment procedure such as SPE should be combined on-line with GC separation-plus-detection. If increased selectivity is deemed more important than improved detectability, LLE and reversed-phase LC (RPLC) are alternatives. Whatever the final choice, such on-line water analyzers can be readily automated. They are fully integrated analytical systems, combining a minimum of sample handling with transfer of the total mass of the analytes of interest. In other words, the full range of sensitive and selective GC detectors is available and, with quadrupole and ion-trap mass spectrometric detectors among these, it should be possible to use SPE–GC–MS for the simultaneous trace-level (provisional) identification and determination.

2. ON-LINE LIQUID CHROMATOGRAPHY–GAS CHROMATOGRAPHY TECHNIQUES

Over the years, many approaches for combining "aqueous-type" LC and GC have been studied. An overview is presented in Table 1, where two types of goals are distinguished, *heart-cut operation* and *comprehensive analyte isolation*.

Table 1 Experimental Approaches in On-Line "Aqueous-Type" LC–GC

Heart-cut operation	
Direct introduction of water	Micro-LC
	Loop-type interface
	Special retention gaps or stationary phases
No direct introduction of water	Liquid–liquid extraction (LLE)
	Postcolumn trapping column
Comprehensive analyte isolation	
Using a solvent (mixture)	LLE–GC
Using a sorbent	Solvent desorption (SPE–GC)
	Thermal desorption (SPETD–GC)
Using a GC stationary phase	Solvent desorption (OTT–GC)
	Thermal desorption (SPME–GC)

2.1. Heart-Cut Operation

In early years, most attention was devoted to the on-line combination of RPLC with its aqueous–organic eluents and GC. This was, no doubt, partly due to the fact that the interfacing of normal-phase LC, with its nonaqueous eluents, and GC had rapidly become successful. The goal of the various published approaches invariably was to collect the RPLC eluent fraction containing the analytes of interest, trap and/or process them in a convenient way (see below), and transfer them on-line to the GC part of the system.

2.1.1. Direct Introduction of Water

Most studies on the direct introduction of water involve the on-line coupling of micro-LC and GC because the LC peak volumes and, consequently, the amount of water-containing eluent that has to be transferred, will then be small. For example, Cortes et al. [7], who used an on-column interface and a nondeactivated retention gap, introduced up to 20 μl of acetonitrile–water (50:50, v/v) and even pure water without serious distortion of the peaks of components eluting near the fully concurrent solvent evaporation (see Chap. 1) transfer temperature. One application dealt with the determination of the toxic bacteriostat, N-Serve, in corn. In another study, chlorpyrifos was determined in well water after a direct large-volume (20 μl) injection into the GC system [8]. However, the risk remains that more polar analytes will be adsorbed on the inner wall of the nondeactivated retention gap, which will result in bad peak shapes.

Transfer of (partly) aqueous eluents using a loop-type interface does not require wettability of the retention gap: at the high temperatures used for the transfer, water will not destroy the surface of the retention gap. The technique has been used to determine atrazine in a 150-μl heart-cut [9]. Because of the absence of solvent and phase-soaking effects, the method is only suitable for compounds eluting at temperatures more than 100°C above the transfer temperature; that is, it is limited to analytes eluting at very high temperatures. This problem can be partly solved by adding a high-boiling organic cosolvent such as butoxyethanol, which acts as a temporary stationary phase in the retention gap during transfer. Optimization of the procedure has been described [10], but no applications were reported.

Several studies have been published on the use of special stationary phases (using packed GC columns) or special types of deactivated retention gaps [11,12]. Among these, a series of papers by Goosens et al. [13–15] on the use of a carbowax-deactivated retention gap to transfer acetonitrile–water eluents from the LC to the GC part of the system, merit attention. When using an on-column interface and a solvent vapor exit (SVE), up to 200 μl of aqueous–organic eluent could be introduced, provided that the water content of the eluent did not exceed that of the azeotropic mixture (16 vol%). Otherwise, water will be left in the retention gap after evaporation of the azeotropic mixture and will mar the analysis. In order to remove ion-pairing agents (which would be present in the LC eluents of the applications of interest), an anion-exchange micromembrane device was inserted between the LC and GC parts of the system. The applicability of the setup was briefly illustrated for the potential drug eltoprazine and, with MS detection, for the impurity profile study of the drug mebeverine. Despite several unfavorable technical aspects, an identification level of 0.1% with respect to mebeverine was achieved. Van der Hoff et al. [16] reported promising results of 1-μl injections into a few meters of an OV-1701–coated retention gap after exposure to water-saturated ethyl acetate. However, this retention gap, which can be also used at higher temperatures than a carbowax-deactivated retention gap and which is now commercially available, has not been used for RPLC–GC as yet.

Hyötyläinen et al. [17–19] introduced water-containing eluents from RPLC on-line to a GC system by using a vaporizer chamber–precolumn split–gas discharge interface. The water-containing eluent (methanol–water with maximum 20% water) was transferred to a vaporizer chamber at 300°C, and the vapor was removed through a retaining precolumn and an SVE. Much attention was paid to the recovery of volatile analytes. Therefore, the oven temperature was kept close to the temperature at which recondensation of the eluent occurs. However, the loss of volatiles could not be prevented. The system was used for the determination of phthalates in drinking and surface water (detection limit 5 to 10 ng/L, MS detection) and the determination of pesticides in wine (quantification limit 5 to 10 μg/L, flame ionization detection [FID] detection). Blanch et al. [20] transferred up to 2.5 ml of LC fractions to a GC–FID system using a PTV injector with a Tenax TA-packed liner. Chiral lactones in foods were determined by means of RPLC–GC–FID.

2.1.2. No Direct Introduction of Water

With this approach, two routes have been followed, viz., interfacing via an LLE module or via a trapping column.

From a technical point of view, RPLC–LLE–GC is essentially the same as the analysis of a (total) sample by means of LLE–GC, and for some of the technical aspects one is referred to the pertinent section below. An interesting example of the fairly complex instrumentation for, in this instance, RPLC–LLE–GC–MS is shown in Figure 1. Ogorka's group [21,22] used the technique to identify unknown impurities in pharmaceutical products. Phase separation by means of a sandwich-type phase separator eliminates all problems arising from the water and buffer salts present in the LC eluent. The procedure was used because LC–MS is less sensitive than GC–MS and, generally, does not produce mass spectra that can be used for identification. Degradation products obtained under alkaline stress of the drug substance ENA 713 were identified as 3-hydroxy-acetophenone and 3-hydroxystyrene (Fig. 2). For confirmation, simultaneous on-line trimethylation–bromination was included in the procedure by transferring

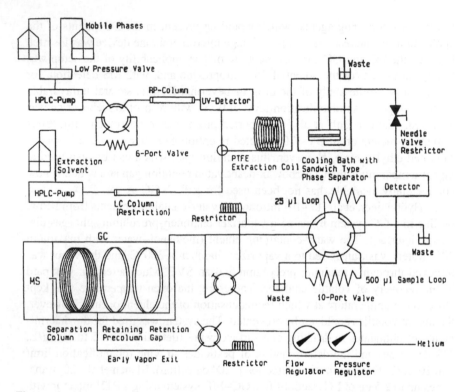

Figure 1 Instrumentation for on-line RPLC–GC–MS using an LLE interface for phase switching. (From Ref. 21.)

the derivatization reagent into the retention gap by means of a second loop via the loop-type interface at the end of the transfer. The same approach was used for the methylation of carboxylic acids with diazomethane, and the extractive benzoylation of 2,6-dimethylaniline during the LLE extraction process [21]. Similar procedures were used by other workers for non-MS applications. The LC-to-GC transfer volumes often are 200 to 500 μl.

In RPLC–trapping column–GC, a short SPE-type column is used for post-separation rather than conventional preseparation purposes [23]. Briefly, the analytes from an up to 2-ml LC heart-cut are retained on the trapping column, the LC eluent is on-line removed by washing with water, and the analytes are desorbed with an organic solvent and transferred to the GC. It is somewhat surprising that this procedure, which tolerates the presence of buffer and other salts in the LC eluent and uses only small volumes of organic solvent (50 to 75 μl), has been studied only once.

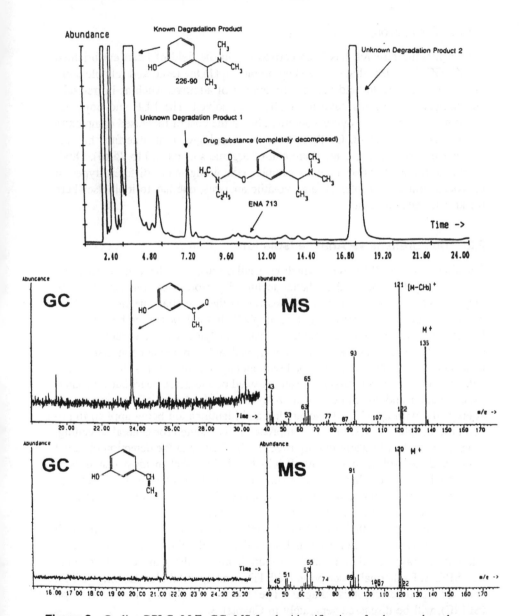

Figure 2 On-line RPLC–LLE–GC–MS for the identification of unknown drug degradation products. (Top) RPLC–UV chromatogram of alkaline-stressed ENA 713. RPLC–LLE–GC–MS obtained after transfer of unknown degradation products 1 (middle) and 2 (bottom). Eluent, aqueous potassium dihydrogen phosphate (0.02 M, pH 5)–acetonitrile–triethylamine (87.5:12.5:0.2 for first 5 min; 82.5:17.5:0.2 for 16 min; 70:30:0.2 for 9 min); flow, 1 ml/min; transfer volume, 500 μl. (From Ref. 21.)

2.1.3.   Conclusion

Although some research has been carried out in the field of heart-cut-oriented
RPLC–GC, it is clear that the two key problems, (1) the rapid and reliable complete removal of water and (2) the elimination of additives such as buffer salts
and ion-pairing reagents have not really been solved. The LLE and trapping-
column interfacing do provide a solution, but at the cost of increased complexity
of the analytical system. Of these two, the trapping-column interface has the
advantage of only using a small amount of organic solvent (50 to 75 μl). Obviously, for the analysis of aqueous samples containing many different types of
microcontaminants of interest, e.g., volatile analytes, one has to look elsewhere
for an adequate solution.

2.2.   Comprehensive Analyte Isolation

The alternatives to RPLC–GC, which we shall discuss now, have attracted attention not only because of their better technical performance, but also because
they provide a more encompassing answer to the essential question, ''Which
microcontaminants are present in this sample?'' In total, or comprehensive, analyte isolation techniques (see Table 1), many analytes with divergent physicochemical characteristics are simultaneously isolated from the aqueous sample by
means of LLE or SPE, then desorbed with an organic solvent or a thermal gradient, and subjected to GC analysis. Thus, instead of a targeted heart-cut-type analysis, isolation and enrichment of a wide variety of (classes of) compounds is
attempted—with the GC–MS separation and identification being applied to unravel the sample composition. This approach is well suited for general screening,
monitoring and early-warning purposes, and is directed at the detection of target
compounds as well as unknowns. It will be obvious that in most instances, a
hydrophobic, i.e., a rather nonselective, sorbent will be selected (if an SPE approach is used), because efficient trapping and enrichment, not preseparation, are
the main purposes of the first step in the procedure.

   The contents of the lower half of Table 1 show that there is one approach
based on solvent extraction (that is briefly discussed below) and there are four
alternatives based on liquid–solid sorption. Three of these are closely related,
i.e., SPE, SPE–thermal description (SPETD), and open tubular trapping (OTT),
and are discussed in this chapter. The fourth, solid-phase microextraction
(SPME), will be discussed separately in Chapter 8.

2.2.1.   Liquid-Liquid Extraction–Gas Chromatography

The on-line incorporation of LLE in an analytical system was reported some two
decades ago. Typically, an aqueous sample plug is injected into an aqueous carrier. The aqueous phase meets an immiscible organic stream in a segmentor, often

a simple T-piece. In the extraction coil, the analytes are distributed between the two phases according to their physicochemical characteristics. Next, the phases are separated in a membrane-type phase separator, which is robust but may cause clogging or carryover problems, or a sandwich-type phase separator, which requires more skill in handling, separation being effected by wettability and/or gravity. Finally, the organic phase is led to the GC system via one of various injection systems (see Chap. 1).

In early continuous-flow LLE–GC studies, pentane was used as extraction solvent for the determination of volatile aromatic hydrocarbons and halocarbons in water [24,25]. More recent applications deal with phenols, N-methylcarbamates, hexachlorocyclohexanes, and organophosphorous pesticides. Occasionally, a derivatization step is included in the LLE–GC procedure to enable the determination of more polar compounds. Phenols, the hydrolysis products of certain methylcarbamates, were derivatized on-line with acetic acid anhydride and pentafluoropropionic anhydride, propionic acid was derivatized by extractive alkylation, and (chlorinated) anilines were acylated with pentafluorobenzoylchloride [26–28]. As regards combination with a spectroscopic detector, i.e., atomic-emission detection (AED), LLE–GC–AED was used for the detection of N-, Cl- and S-containing pesticides and the screening of groundwater samples [29]. Although LLE–GC cannot easily be constructed to enable significant analyte enrichment (because the aqueous/organic volume ratio cannot deviate too much from unity) and is, therefore, not always useful for real trace-level studies, the reader should keep its practicality for monitoring purposes in mind.

An alternative to using a phase separator is to combine micro-LLE and GC at-line by using an autosampler vial. After the addition of an appropriate solvent to the sample and gentle shaking or mixing by repetitive drawing up and subsequent dispensing, a large aliquot of the organic solvent (up to 500 µl) can be injected into the GC system. Venema and Jelink [30,31] added 1 ml of n-pentane to 1 ml of sample in an autosampler vial. After 3 minutes of shaking, the vial was placed in the autosampler tray and 140 µl of the organic extract were injected into the gas chromatograph–mass spectrometer using selected ion monitoring. Hexachlorobenzene and hexachlorobutadiene were detected at the 6 ng/L level.

3. ON-LINE SOLID-PHASE EXTRACTION–GAS CHROMATOGRAPHY

3.1. General Aspects

On-line trace enrichment and clean-up by means of SPE using precolumns or (disposable) SPE cartridges is a popular column-switching technique in LC, and most techniques and much of the hardware now becoming popular for SPE–GC were copied from SPE–LC. A brief introduction is presented here. The SPE cartridges have dimensions of, typically, 10 to 20 mm length × 1 to 4.6 mm inside diameter (ID). In most instances—whether these are LC- or GC-based applications—the cartridges are packed with 10- to 30-μm sorbents such as C18- or C8-bonded silica or a styrene–divinylbenzene copolymer such as PLRP-S. A typical setup of equipment to be used for on-line SPE–GC is depicted in Figure 3. As regards the SPE part, after conditioning of the precolumn in LC-type fashion, i.e., with methanol or another appropriate organic solvent and, next, with water, a sample volume of, often, 5 to 10 ml is loaded at a flow rate of 1 to 4 ml/min. The loading volume is almost invariably distinctly lower than in SPE–LC (typical range, 30 to 100 ml) because of the better detection performance of GC compared with LC detectors. The analytes of interest, and also many other sample constituents, are preconcentrated on the precolumn. After clean-up, which generally is restricted to washing with water and 15 to 30 minutes of drying with nitrogen at ambient temperature, the analytes are desorbed with a small amount, typically 50 to 100 μl, of an organic solvent such as ethyl acetate or methyl acetate and transferred to the GC part of the system for final analysis. Recently, it has even been demonstrated that the transfer volume can be reduced to about 20 μl (see later section on volatile analytes). A brief description of the main steps of such a sample preparation-cum-transfer procedure is presented in Table 2.

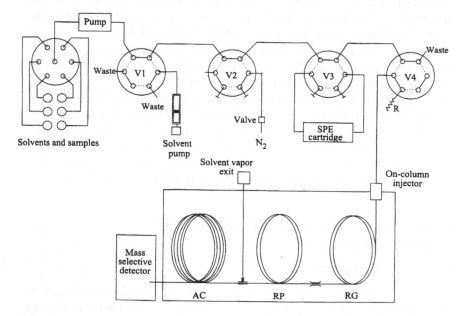

Figure 3   Scheme of on-line SPE–GC–MS system. Abbreviations: AC, analytical column; RG, retention gap; RP, retaining precolumn; R, restriction, V1–4, valves.

Table 2 Typical Time Schedule of Sample Preparation Program of On-Line SPE–GC–MS

Time [min:sec]	Solvent selection valve[a]	Flow[b] [ml/min]	Valves[c]				Auxiliary events[d]			Comment
			V1	V2	V3	V4	1	2	3	
00:00	2	2.5	1	0	0	0	on	off	off	Flush tubing with MeOAc/flush pump with i-PrOH
01:00	1				1					Condition cartridge with MeOAc
02:00			0				off			Condition cartridge with water
03:00	3	5			0					Flush pump/tubing with sample
04:00		2.5			1					Preconcentrate 10 ml of sample
08:00	1	5			0					Clean pump/tubing with water
08:30		2.5			1					Clean-up with water
09:00		0		1				on		Drying cartridge for 20 min
25:00							on			Start of MeOAc pump
29:00								off		Stop drying/depressurize cartridge
29:30			1	0	0					Flush tubing with MeOAc
31:30					1	1			on	Transfer of analytes with 50 μl MeOAc at 120 μl/min
32:25					0	0			off	Clean cartridge with MeOAc
34:30										End of sample preparation

[a] 1, water; 2, i-PrOH; 3, sample.

[b] Flow of solvents/sample pump.

[c] V1–V4: position 0 refers to reference position in Figure 3.

[d] 1, Syringe pump on/off; 2, nitrogen valve on/off; 3, start of SVE controller.

Abbreviations: i-PrOH, isopropanol; MeOAc, methyl acetate (desorption solvent).

Table 3 Selected Applications of On-Line SPE–GC Procedures with Conventional Detection

Analytes	Sample type	Detection	Sample volume (ml)	LOD (ng/l)	Ref.
PCBs, pesticides	River water	ECD	1–12	1	i, ii
Musk ketone and xylene	River water	ECD	10	15	54
Medium polar analytes	Water	FID	1	100	iii, iv
Steroid hormones	Urine	FID	5	100	35
Triazines	Wastewater/orange juice	NPD	10	10	36
Triazines, organo-P pesticides	River water	NPD	2.5–10	1–100	34, 39
Triazines organo-P/S pesticides	River water	FPD	10	1–40	40

i.) E. Noroozian, F.A. Maris, M.W.F. Nielen, R.W. Frei, G.J. de Jong and U.A.Th. Brinkman, J. High Resolut. Chromatogr. Chromatogr. Commun, 10 (1987) 17.

ii.) Th.H.M. Noij, E. Weiss, T. Herps, H. Van Cruchten and J. Rijks, J. High Resolut. Chromatogr. Chromatogr. Commun., 11 (1988) 181.

iii.) J.J. Vreuls, W.J.G.M. Cuppen, E. Dolecka, F.A. Maris, G.J. de Jong and U.A.Th. Brinkman, J. High Resolut. Chromatogr., 12 (1989) 807.

iv.) J.J. Vreuls, W.J.G.M. Cuppen, G.J. de Jong and U.A.Th. Brinkman, J. High Resolut. Chromatogr., 13 (1990) 157.

In early papers on SPE–GC, much attention was devoted to the setup of such an on-line system. Conventional (selective) GC detectors rather than a mass spectrometer were used and essentially technical aspects, such as the design and possible reuse of the SPE cartridges, optimization of the drying step, and the SPE-to-GC transfer, were studied in much detail. To give an impression of the analytical performance then achieved, selected applications are shown in Table 3. The most relevant technical aspects are discussed in the next section.

3.2. Optimization of Solid-Phase Extraction–Gas Chromatography Procedure

3.2.1. Stationary Phase for Trace Enrichment

In SPE–GC, the SPE part of the procedure is in most cases only used to effect trace enrichment and no attempt is made to improve selectivity (cf. below). Consequently, in actual practice only two types of sorbent are used, hydrophobic C18-bonded silicas and highly hydrophobic copolymers. According to abundant literature information, breakthrough volumes of nonpolar and medium-polar analytes, and even many polar analytes, on SPE cartridges packed with these sorbents, and with dimensions of 10 to 20 mm length × 1 to 4.6 mm ID, are much higher than 10 ml. Polystyrene copolymers typically provide 20- to 30-fold more retention than the alkyl-bonded silicas. In other words, it is safe to state that the sorption step of the SPE process will not cause a noticeable loss of analytes. In addition, because of the reliable information already available in the literature, there is no need for the analytical chemist tackling a new problem to start with a rather time-consuming collection of experimentally determined breakthrough data. For a more detailed discussion, the reader is referred to texts such as that by Barceló and Hennion [32].

In recent years, cartridge holders containing one or a few small (diameter 3 to 4.6 mm, thickness 0.5 mm) membrane extraction disks have been recommended as an alternative to conventional precolumns or cartridges for both LC and GC [33,34]. The disks contain approximately 90 wt% of a hydrophobic sorbent held in a polytetrafluoroethylene (PTFE) mesh, and can be loaded at fairly high speed. Drying with nitrogen at room temperature proceeds rapidly. The commercial Empore (3M, St. Paul, MN) extraction disks, which have a diameter of 47 mm, are often recommended for field studies. It will be clear that a single 47-mm ID disk—after having been used for field sampling—can be used for a large number of 4.6-mm-diameter-based SPE–GC analyses!

Abundant experimental evidence shows that analyte desorption from the loaded cartridges or disks (an aspect that is equally important for the final recovery to be obtained as is the sorption step), is easily achieved with less than 100 μl of methyl or ethyl acetate, applied at a flow rate of 50 to 100 μl/min. This is well within the conditions of ordinary LVI procedures and does not present any technical problems.

So far, hydrophobic sorbents have been used in essentially all SPE–GC procedures. One main reason why the additional selectivity provided by modified SPE sorbents, which is frequently studied in SPE–LC, is less popular in GC-based analyses, no doubt is the much higher efficiency of conventional GC compared with LC separations. Still, two studies have been devoted to the on-line combination of immunoaffinity SPE (IASPE) and GC [35,36]. In IASPE, desorption from an antibody-loaded precolumn has to be carried out with, typically, several milliliters of methanol–water (95:5, v/v). Since it is impossible to introduce such a solution directly into the GC part of the system, it is on-line diluted with an excess of HPLC-grade water and the mixture led through a conventional C18-bonded silica precolumn as trapping column, as in RPLC–GC. The gain in breakthrough volume of the analytes, due to increased retention caused by the considerably decreased modifier percentage, easily outweighs the volume increase. Consequently, the analytes are quantitatively trapped on this second precolumn. Desorption from this trapping column and the further procedure are as for conventional SPE–GC. The method was applied to steroid hormones in 5 to 25 ml of urine. The detection limit of 19-β-nortestosterone was about 0.1 μg/L (FID detection). A similar approach, which combined an antibody-loaded first, and a copolymer-packed second precolumn was used for the determination of triazines in river water, wastewater, and orange juice. The detection limits were about 10 ng/L when 10-ml samples were analyzed using nitrogen–phosphorus detector (NPD) detection. It is interesting to add that, although IASPE–GC has not yet been combined with MS detection, setting up such a system is not expected to cause any technical problems. We shall then have an instrumental setup that permits highly selective analyte isolation *and* structure-based identification or, in other words, identification and confirmation, in one run.

3.2.2. Removal of Water by Drying

Problems caused by the presence of water in the retention gap can be overcome by inserting a drying cartridge or drying the SPE cartridge for about 15 to 30 minutes with nitrogen gas (at ambient temperature). Nitrogen drying has the advantage that it is a well-known and simple procedure and is, therefore, generally preferred. Volatile compounds such as chlorobenzene are not lost to a significant extent [37]. Drying of copolymers is distinctly more rapid, and has a somewhat more reliable outcome than that of C18-bonded silicas. The insertion of a drying cartridge containing silica or sodium sulfate between the SPE cartridge and the GC part of a system is an interesting alternative to reduce the drying time. Both silica and sodium sulfate can be reused many times if they are regenerated between runs by (external electrical) heating. Since analyte losses appear to be negligible for a wide variety of compounds, e.g., triazines, alkylbenzenes, chlorobenzenes, and chlorophenols, even at the trace level, the use of a drying cartridge is a viable approach [38–40]. Recently, the design of the drying cartridge was improved to enable higher temperatures during regeneration, and volatile analytes up to tetrachloroethylene could be included (methyl acetate as desorption solvent) [41]. Although molecular sieves and sodium sulfate have a higher drying capacity, silica was found to be the best choice in actual practice.

3.2.3. Solid-Phase Extraction to Gas Chromatography Transfer

When using the SPE cartridges described above, the analytes are generally transferred with 50 to 100 µl of organic solvent (preferably ethyl or methyl acetate) from the cartridge to the GC. This volume is required to desorb the analytes and to prevent memory effects due to adsorption of analytes in the transfer capillary.

If relatively volatile analytes are included in the set of target compounds that has to be determined, on-column interfacing is the preferred technique if the sample extract is not too dirty [42,43]. The main problem is that with highly contaminated samples, such as wastewater, the retention gap easily looses its performance: distorted peak shapes and/or lower analyte responses for, especially, the more polar analytes, can already show up after a few GC runs. To maintain the quality of the analyses, the retention gap should be heated prior to starting the temperature program of the analytical column. This can be done by putting the retention gap in a separate GC oven [44,45] or by wrapping it with heating wire [46]. Recently, the retention gap was placed in a low-weight oven in the GC oven itself and no loss of performance was observed after more than 200 on-line SPE–GC–MS analyses of river water samples [47].

Recent studies recommend the use of a PTV injector in on-line SPE–GC of highly contaminated samples [48]. Exchange of the packed liner of such an injector is straightforward and takes little time. The main drawback is that the separation of volatile analytes from the (desorption) solvent is less satisfactory than with a retention gap [43]. Staniewski et al. used a PTV injector as interface in on-line SPE–GC and 50 µl of ethyl acetate for desorption [49]. Several herbicides were determined in water at about the 1-µg/L level with recoveries of 20 to 90%.

If volatile analytes do not play a prominent role in the samples to be analyzed, a loop-type interface—which is rather easy to use—can be recommended [50,51]. In order to somewhat increase the application range at the volatile end, Noij et al. [50] desorbed the analytes with 500 µl of methyl tert-butyl ether–ethyl acetate (90:10, v/v), and the eluate was injected together with 50 µl of n-decane as cosolvent into the GC. The most volatile analyte included in the study was mevinphos. Another approach to extend the application range of loop-type injections is to use an Autoloop interface (Interchro, Bad Kreuznach, Germany), which essentially consists of an SPE cartridge and two loops for storing organic solvent, which are mounted on a 14-port valve [52–55]. Solvent transfer for desorption to the gas chromatograph is achieved by the carrier gas, which can be diverted via a 6-port valve. The solvent in the first loop effects phase swelling of the retaining precolumn to increase the application range to more volatile analytes. The solvent in the second loop is used to desorb the analytes and transfer them to the retention gap. Because a rather long retention gap and a total transfer volume of about 200 µl are used, a transfer temperature of 90°C can be used, which is lower than that usually applied with ethyl acetate.

As has repeatedly been mentioned, one of the incentives of on-line SPE–GC is that the total analyte-containing fraction is transferred to the GC column. There are, however, a few studies in which this was not done. For example, Ballesteros et al. [56,57] desorbed the analytes from a copolymer SPE cartridge with 100 µl of ethyl acetate and, after homogenization of the extract by means of a mixing coil, injected 5 µl of the eluent via a loop and a capillary in a splitless injector. When 50-ml water samples were analyzed by GC–FID, detection limits of 0.7 to 1 µg/L were achieved for N-methylcarbamates and their phenolic degradation products. The procedure is elegant, but one should keep in mind that using a (small) aliquot of the total sample will, in most instances, cause a considerable loss of detectability expressed in concentration units.

Finally, analyte desorption and, consequently, SPE-to-GC transfer, can also be performed without any organic solvent being used, that is, by thermal desorption (TD). The two options are SPETD, discussed later in this chapter, and SPME, discussed in Chapter 8.

3.2.4. At-Line Operation

Sample enrichment by SPE or LLE and GC analysis can also be integrated into one setup by at-line coupling: the sample extract is transferred from the sample preparation module to the gas chromatograph via, e.g., an autosampler vial using an ASPEC (Gilson, Villiers-le-Bel, France) or a PrepStation (Hewlett-Packard, Palo Alto, CA). The main disadvantage of most of the published procedures is that, after elution of the SPE cartridge and collection of the solvent in a vial, only an aliquot is injected. In some studies, such as those on organochlorines and pyrethroids in surface water [58,59] and benzodiazepines in plasma [60], 100 to 200 µl were injected out of 2 to 5 ml. In another paper, only 1 µl out of 1 ml was injected to determine barbiturates in urine [61]. Not surprisingly, detection limits in the latter PrepStation–GC–MS study were in the sub- to low-mg/L range. Similar sensitivity problems were encountered [62] when serum had to be analyzed and, by other workers, in a PrepStation-based study on the determination of organic acids [63].

In view of the above, it is interesting to look at another study [64,65], which was directed at improving the performance of the PrepStation–GC setup by introducing several modifications, viz., (1) increasing the aqueous sample volume from 1.5 to 50 ml, (2) using 50-µl "at-once" on-column LVI rather than 1-µl injections, and (3) decreasing the desorption volume to 300 µl by reducing the amount of sorbent in the SPE cartridge. The redesigning was markedly successful: an overall 300-fold improvement had been calculated, and a 150- to 300-fold improvement was observed in actual practice. An initial problem was that, as a result of the increased sensitivity, the PrepStation was found to be less inert than expected: several interferences from impurities extracted from the septa and also the commercial cartridges showed up. A cartridge made from stainless steel and polychlorotrifluoroethylene and a 2-needle system were constructed to eliminate these interferences. Several micropollutants were detected in 50 ml of (unfiltered) river water at the 0.2 to 400 ng/L level using full-scan MS acquisition (Fig. 4).

In summary, with carefully designed at-line sample preparation–GC systems, analyte detectability can be made similar to that in on-line SPE–GC set-ups. However, interferences due to contamination and analyte losses will always be more serious.

3.2.5. Medium-polar and Nonpolar Analytes

Most conventional GC and also LVI–GC and SPE–GC procedures have been designed for the determination of the medium-polar compound range, which comprises analytes from, typically, simazine and N,N-dimethylaniline to trichlorobenzene and dibutylphthalate. With such sets of compounds, it is highly unlikely that any sort of technical problems will be encountered. If the range of analytes has to be extended to include really nonpolar compounds, such as organochlorine pesticides, ethion, or bromophos-ethyl, one should add 20 to 30 vol% of methanol to the aqueous sample to prevent adsorption of these analytes to the inner walls of capillaries and valves [66,67]. It will be clear that, in the end, a situation may arise in which too wide a range of analytes has to be determined. One can then of course compromise with regard to the recoveries on either the polar or nonpolar end. However, for a more robust operation, it is recommended to carry out two separate runs, with conditions optimized for the former (purely aqueous sample) and the latter (modifier addition) group, respectively. For example, the addition of 30 vol% of methanol to surface water samples did not interfere with the determination of a set of nonpolar and medium-polar pesticides. However, the most polar analyte in the test set, dimethoate, was largely lost [66].

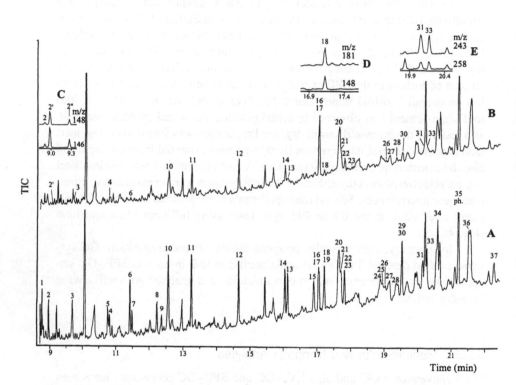

Figure 4 Full-scan PrepStation–GC–MS of 50 ml of Meuse River water (B) without and (A) with spiking with 37 micropollutants at the 0.18 µg/L level; 100 µl out of the 300 µl extract were injected. The inserts show the reconstructed-ion chromatograms of two characteristic masses of 1,3-dichlorobenzene (C), 2-methylthiobenzothiazole (D), and Musk G and T (E). Compounds detected (0.2 to 430 ng/l) in the nonspiked river water: 2, 1,3-dichlorobenzene; 2′, 1,4-dichlorobenzene; 2″, 1,2-dichlorobenzene; 3, acetophenone; 4, decamethyl-cyclopentasiloxane; 6, naphthalene; 10, isoquinoline; 11, 2-methylquinoline); 12, 2,4,7,9-tetramethyl-5-decyne-4,7-diol; 13, dibenzofuran; 14, triisobutyl phosphate; 15, N,N′-diethyl-3-methylbenzamide; 16, 2,2,4-trimethylpentane-1,3-dioldi-isobutyrate; 17, diethyl phthalate; 18, 2-methylthiobenzothiazole; 20, tetraacetylethylene-diamine; 21, tributyl phosphate; 22, ethyl citrate; 23, desethylatrazine; 25, simazine; 26, atrazine; 27, tris(2-chloroethyl)phosphate; 28, N-butylbenzenesulfonamide; 30, tris(2-chloroisopropyl)phosphate; 31, Musk G; 33, Musk T. (From Ref. 65.)

3.2.6. Volatile Analytes

Recently, our understanding of the processes involved in the on-column LVI–
GC [68–71] and on-line SPE–GC [37] analysis of volatile compounds has improved. Due to the pressure drop along the solvent film in the retention gap,
which occurs when an SVE is used, solvent evaporation takes place not only at
the rear end, but also along the whole length, and even at the front of the solvent
film. Loss of volatiles can therefore be severe, especially in SPE–GC, because
these analytes are mainly present in the front part of the desorption solvent. To
improve performance, after conventional sample loading and drying with nitrogen, one should introduce some pure organic solvent, the so-called presolvent,
into the retention gap prior to the actual desorption (using the lower-boiling
methyl rather than ethyl acetate). A solvent film will now have been formed
before the (volatile) analytes arrive as a result of the SPE-to-GC transfer, and
will ensure their retention also during evaporation of the solvent film. The beneficial effect is vividly illustrated in Table 4: with some 30 µl of presolvent, ana-

Table 4 Dependence of Analyte Recoveries of On-Line SPE–GC Transfer on
Amount of Presolvent[a]

	Recoveries (%) for a presolvent volume of			
Compound	0 µl	10 µl	20 µl	30 µl
Volatile				
Monochlorobenzene	5	7	70	**97**
p/m-Xylene	7	8	72	**95**
Styrene	22	34	**86**	**100**
o-Xylene	9	14	**85**	**99**
Methoxybenzene	40	66	**95**	**100**
o-Chlorotoluene	33	54	**94**	**96**
Semi-volatile				
Benzaldehyde	70	**102**	**100**	**103**
1,2-Dichlorobenzene	62	**89**	**95**	**97**
Indene	64	**94**	**96**	**101**
Nitrobenzene	88	**100**	**95**	**100**
Naphthalene	89	**96**	**95**	**99**
High-boiling				
Methylnaphthalene	**95**	**94**	**95**	**96**
Acenaphthene	**99**	**100**	**99**	**101**
Metolachlor	**99**	**100**	**102**	**99**

[a] Pure methyl acetate introduced as presolvent into the GC prior to desorption with 50 µl methyl
  acetate [37].
[b] Recovery values above 80% are shown in bold print.

lytes as volatile as monochlorobenzene are quantitatively recovered even at the 0.5 µg/L level [37]. Recently, it was found that satisfactory recoveries (70 to 90%) can be obtained for volatile analytes such as monochlorobenzene and the xylenes with a mere 20 to 25 µl of methyl acetate even without the use of a presolvent. Preconditions are the use of an 0.53 mm ID retention gap and closure of the SVE during SPE-to-GC analyte transfer [47].

3.2.7. Self-Controlled Setup

When using an on-column interface for the SPE-to-GC transfer, until recently, the injection speed and the timing of the start of the transfer and the SVE closure had to be determined prior to analysis. It has now been demonstrated that the SVE closure can also be performed in an automated fashion [68]. As an example, Figure 5 shows the carrier gas (helium) and solvent vapor flow (FID response) profiles during a 54-µl on-column injection of ethyl acetate into a 0.53-mm ID retention gap. At the start of the injection, the helium flow sharply decreases and the solvent flow sharply increases, while at the end of the evaporation process the helium flow sharply increases and the solvent flow decreases. The sharp in-

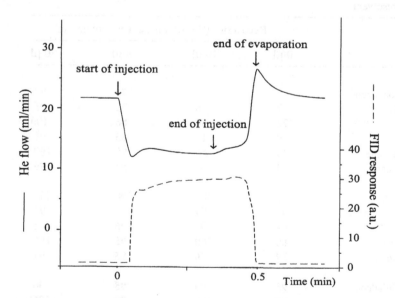

Figure 5   Helium flow rate and solvent peak profile for injections of ethyl acetate into a 0.53-mm ID retention gap. Injection time, 20 sec; injection speed, 160 µl/min; evaporation rate, 130 µl/min; 10 µl were left as solvent film in the retention gap at end of injection. Helium flow is measured by means of a flow meter in the carrier gas tubing, and the solvent flow by means of an FID at the end of the retention gap. (From Ref. 68.)

crease of the carrier gas flow (or, better, of its first derivative) observed when evaporation is complete, can be used to trigger SVE closure. For 30-μl injections of n-alkanes (C_7 to C_{20}) in n-hexane under partially concurrent solvent evaporation (PCSE) conditions, no analytes were lost with the automated procedure when compared with (the conventional) closing of the SVE 0.1 minute prior to completion of the evaporation.

Automation not only makes preoptimization superfluous, but also improves the robustness of the SPE–GC procedure (and any LVI–GC procedure in general): if the evaporation time slightly changes due to, e.g., small changes of the injection speed or injection volume, the SVE will still be closed just in time without undue loss of volatiles or a significant change of the solvent peak width at the detector. The latter aspect is important when working with a mass-selective detector, because the delay time for switching on the filament can now be kept constant. As additional advantages, the repeatability of the retention times of volatile analytes is improved [47] and the capacity of the retention gap is significantly larger when the SVE is closed at the last possible moment [71].

As regards desorption of the SPE, the introduction of the sorption solvent into the retention gap will cause a similar decrease of the helium flow as depicted in Figure 5 for the start of the sample introduction. This allows the introduction of the desired volume of desorption solvent without the need of (re)assessing the proper timing of the start of the introduction when using, e.g., a new type of SPE cartridge. The transfer is stopped after the preprogrammed delay time by switching the transfer valve [47].

The final parameter to be optimized in SPE–GC using an on-column interface is the injection speed. As regards this aspect, recent research [47] has shown that (1) with an optimized desorption-plus-transfer-line-flushing strategy only 20 to 25 μl of methyl acetate are required per run, and (2) the closure of the SVE at the very end of the evaporation process considerably increases the capacity of the retention gap (cf. above). Both improvements allow significant reduction of the amount of solvent evaporated during injection without using too long a retention gap. This implies that the injection speed will, in any case, be higher than the evaporation rate—which makes injection speed optimization superfluous!

There is now no parameter left that has to be optimized when exchanging the retention gap or the SPE cartridge. In other words, one has a truly "self-controlled system."

3.3. Open Tubular Trapping–Gas Chromatography

Another possibility to extract analytes from an aqueous phase is by retaining them in the stationary phase of an open tubular trapping (OTT) column, which essentially is a piece of GC column. After enrichment from 0.2 to 10 ml of water, clean-up with 0.15 to 0.5 ml distilled water and removal of the water by a slow nitrogen flow during 1 minute, the analytes are desorbed with an organic solvent and directly transferred to the GC via a PTV injector as interface [72]. To obtain acceptable breakthrough volumes of 0.5 ml or higher for nonpolar analytes at reasonable flow rates (approximately 0.2 ml/min) and with desorption volumes compatible with LVI (75 to 250 µl), a narrow-bore column of 2 m × 0.32 mm ID with a polysiloxane film of 5 µm thickness was used. With coiled or stitched columns, much higher sample flow rates up to 4 ml/min can be used, because the secondary flow in deformed capillaries enhances radial dispersion [73].

The breakthrough volume and, thus, the sample volume can be significantly increased if the stationary phase is substantially swollen prior to sampling by a solvent, which is generally the desorption solvent [74]. Obviously, this solvent should not dissolve in water, nor should water dissolve in the organic solvent. Open tubular trapping columns swollen with pentane were found to be the most promising option with nonpolar analytes; breakthrough volumes exceeded 10 ml when using a 2-m OTT column. Chloroform was the preferred solvent for swelling if more-polar compounds such as dimethylphenol, dimethylaniline, and atrazine had to be included. Figure 6 shows the OTT–GC–FID analysis of 2.25 ml of Dommel River water spiked with 14 analytes at the 5 µg/L level and using chloroform as swelling agent. Analyte recoveries were satisfactory and detection limits were at the low- and sub-ng/ml level. The OTT–GC–FID technique was also applied to the determination of compounds such as toluene, trichlorobenzene, and lindane in urine and serum. Satisfactory repeatabilities (1.5 to 10%) and quantitative recoveries were obtained, which indicates that no adverse matrix effects occurred.

The main advantage of OTT–GC is the ease of water elimination and the absence of clogging problems. It is somewhat surprising that this rather simple approach has not attracted more attention.

3.4. Solid-Phase Extraction–Thermal Desorption–Gas Chromatography

An alternative to desorption of the analytes trapped on an SPE cartridge with an organic solvent is thermal desorption (SPETD). In the first on-line SPETD setup, 100 to 500 µl of aqueous sample were injected into the packed liner of a PTV [75]. The water was evaporated at a high gas flow rate and a backflush setup inserted between the PTV and the analytical GC column ensured that no water entered the analytical column. Next, the analytes were desorbed and transferred to the analytical column by rapidly increasing the injector temperature. Tenax GR and TA were found to be suitable sorbents [76], that is, they combined sufficient retention power for analytes in the liquid phase (sorption), poor interaction with water (drying), and good thermal stability (desorption). The potential of the method was demonstrated by the analysis of 10 µg/L solutions of n-alkanes and

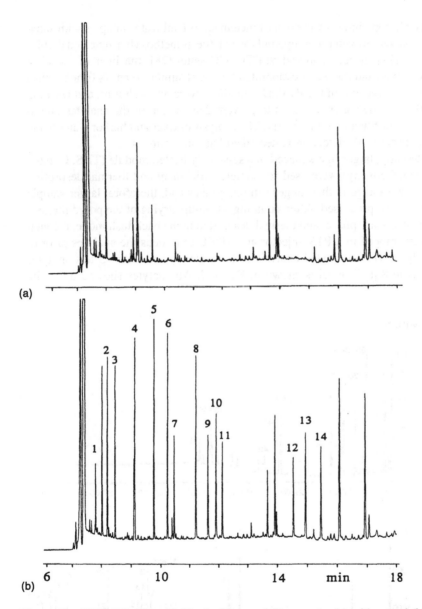

Figure 6 OTT–GC–FID of 2.25 ml of Dommel River water (a) without and (b) with spiking at the 5 µg/L level. Peak assignment: 1, toluene: 2, ethylbenzene; 3, methoxybenzene; 4, p-dichlorobenzene; 5, dimethylphenol; 6, dimethylaniline; 7, p-chloroaniline; 8, indole; 9, dichlorobenzonitrile; 10, trichlorophenol; 11, dinitrobenzene; 12, trifluralin; 13, atrazine; 14, phenanthrene. (From Ref. 74.)

phenols. Micropollutants were determined in up to 1-ml water samples with satisfactory recoveries even for compounds as volatile as methoxybenzene and dichlorobenzene [77]. In an improved SPETD–GC setup [78], the liner of the PTV was redesigned and the water evaporated by a vent similar to an SVE rather than through the purge exit of the PTV. Drying times were much shorter, and washing with HPLC-grade water was used to prevent degradation of the analytes due to remaining constituents of the matrix. The analysis of river and harbor water using ion-trap tandem MS detection is described later in section 4.2.

Recently, larger liners placed in a separately heated module (TDS, Gerstel, Mühlheim, Germany) were used for sample enrichment and thermal desorption [79]. The advantage is that longer sampling tubes and, therefore, larger sample volumes can be processed. After sampling, clean-up, drying of the polydimethylsiloxane stationary phase, and thermal desorption in the backflush mode, the analytes were trapped in a PTV injector at −100°C to refocus the analytes prior to GC analysis. The automated analysis of tap water spiked with 45 test analytes at the 0.4 to 8 µg/L level is shown in Figure 7. All analytes showed up in the

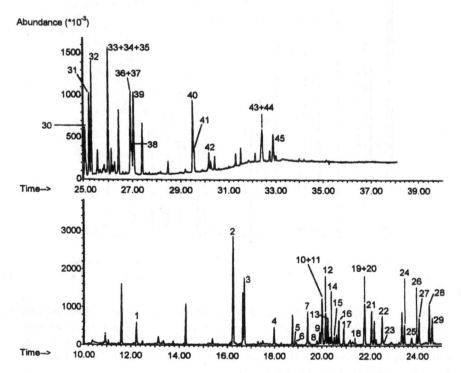

Figure 7   SPETD–GC–MS of 10 ml tap water spiked at the 0.4–8 µg/L level. Stationary phase, 340 mg of 240 to 400 µm particles of polydimethylsiloxane; ID of liner, 4 mm; thermal desorption, 50°C − 1°C/sec − 225°C (5 min). (From Ref. 79.)

GC–MS chromatogram; however, many disturbing peaks caused by the polydimethylsiloxane also showed up. Even though they did not interfere with the pesticides and PAHs of interest, their presence seems to indicate a certain instability of the setup. With 10-ml samples, the procedure gave detection limits of about 10 ng/L using full-scan MS detection.

Finally, analytes can also be enriched on a stationary phase in the PTV insert after the evaporation of the water in the PTV [80–82]. Also in this mode, Tenax appeared to be the best choice [83]. However, sampling took a rather long time due to the low evaporation rate of water (approximately 10 µl/min) [81], and the insert had to be exchanged rather often, because deposition of matrix constituents such as salts in the liner caused the decomposition of chemically less stable compounds [82]. If proper precautions were taken, analyte recoveries were, however, fully satisfactory (80 to 104%), and detection limits of 10 to 20 ng/L were obtained (0.5-ml samples; NPD detection). Actually, the procedure seems to be interesting mainly for compounds with high water solubility.

In one application [84], OTT trapping has been combined with thermal desorption on two different GC columns. Volatile halocarbons were extracted from 8 ml of water by a 95-µm film trap. Next, the analytes were thermally desorbed and refocused by a cold trap in the second GC oven prior to analysis by GC–FID.

4. SOLID-PHASE EXTRACTION–GAS CHROMATOGRAPHY USING SPECTROMETRIC DETECTION

4.1. Solid-Phase Extraction–Gas Chromatography–Mass Spectrometry

Next to trace-level detection, unambiguous confirmation of the presence of target compounds and the provisional identification of unknowns is rapidly gaining importance. Some five years ago, several papers were published that demonstrated that SPE–GC–MS can do just that. In an early study [Ref. i of Table 6], atrazine and simazine were determined by means of SPE–GC–MS, in both the multiple-ion-detection (MID) and full-scan (FS) modes, as well as SPE–GC–NPD, using 1- and 10-ml samples. Quantification of the microcontaminants at levels of approximately 10 to 50 ng/L presented no real problems with relative standard deviations (RSDs) of 3 to 8% (n = 4) and limits of detection as low as 0.5 ng/L (MID), 3 ng/L (FS) and 4 ng/L (NPD) for 10-ml samples. The MS- and NPD-based data showed good agreement (with differences generally being less than 10 ng/L at the 10 to 70 ng/L level) and linear calibration curves were obtained in both instances.

Attention was also devoted to nontarget analysis. A mere 1-ml sample of Rhine River water was spiked with 1 µg/L of each of a mix of 168 microcontaminants, and a 1-minute window selected for further study [Ref. i of Table 6]. By selecting a number of individual masses found in the mass spectra of the apex of each of the six major peaks, it was possible to record rather undisturbed mass traces such as those shown in Figure 8. From these traces, it was apparent that at least nine peaks eluted in the selected time window. The mass spectra of the nine peaks were recorded at their apexes and compared with the NBS library. The data of Table 5 show that the final result was quite satisfactory. Problems started to occur when the peak maxima were merely 0.01 to 0.02 minute apart, as is demonstrated by the pair 1-nitronaphthalene/3-chloro-4-nitrophenol.

In subsequent studies, the development and use of an automated benchtop instrument received much attention. The final system (see Fig. 3) consisted of a Prospekt (Spark Holland, Emmen, Netherlands) automated sample-handling module for trace enrichment, drying of the SPE cartridge, and analyte transfer under PCSE conditions using an SVE and an on-column injector, coupled on-

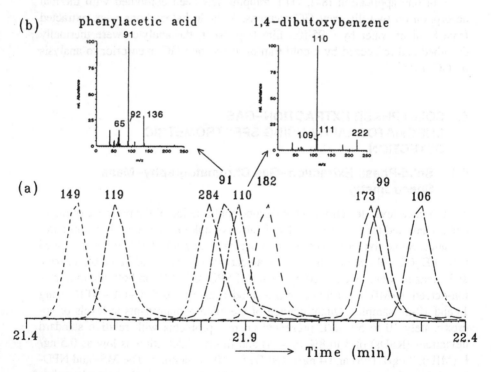

Figure 8   (a) Nine reconstructed ion traces (masses indicated above each peak) of on-line SPE–GC–MS of 1 ml of Rhine River water spiked at the 1 µg/L level with each of 168 micropollutants and (b) mass spectra of phenylacetic acid and 1,4-dibutoxybenzene, which elute at 21.83 and 21.86 min, respectively. (Ref. i of Table 6.)

Table 5 Compounds Identified in the Time Window of 21.4–22.4 Minutes after SPE–GC–MS of 1 ml of River Water Spiked at 1 µg/l

Retention time (min)	Main mass (m/z)	Compound	Library search fit[a]
21.50	149	Diethyl phthalate	0.94
21.58	119	N,N-Diethyl-3-methylbenzamide	0.91
21.80	284	Hexachlorobenzene	0.85
21.83	91	Phenylacetic acid	0.95
21.87	110	1,4-Dibutoxybenzene	0.96
21.93	182	Benzophenone	0.95
22.17	173	1-Nitronaphthalene	0.66
22.18	99	3-Chloro-4-nitrophenol	0.61
22.28	106	3-Aniliniopropionitril	0.95

[a] Library search fit factor on a scale of 1.00.
Source: Ref. i of Table 6.

line to a GC–MS. The total system was completely software-controlled under Microsoft Windows and was used to analyze a variety of water samples. The further development and subsequent upgrading of that setup, with a self-controlled system as the ultimate goal, was discussed previously in section 3.2. The discussions below, therefore, mainly emphasize the detection/identification performance that can be achieved when SPE is on-line combined with GC.

Table 6 summarizes relevant analytical data on the sensitivity of SPE–GC–MS of, chiefly, aqueous environmental samples such as tap water, surface water, and wastewater. The main conclusions that can be drawn from the data are rather promising. Sample volumes of about 10 ml suffice to obtain full-scan MS traces such as are shown in Figure 9. Detection limits were in the 20 to 50 ng/L range or lower for essentially all compounds. As a demonstration of the identification power of the procedure, the traces of the four characteristic ions of peak 11 (benzaldehyde) in the raw, i.e., nonspiked, water are included. Comparison with the 0.5-µg/L spiked trace shows that benzaldehyde was present at a level of approximately 40 ng/L. This system is well suited for the screening of rather volatile as well as high(er)-boiling compounds.

While monitoring studies often aim at detecting "all" microcontaminants present above a threshold level (estimated from, e.g., FID or a total-ion-current trace), there are also situations in which targeted analysis is the main goal. Then, it is beneficial to use (time-scheduled) selective ion monitoring (SIM) and related techniques. Figure 10 shows a relevant example. The detectability of the compounds of interest improved 3- to 10-fold upon going from the total ion chromatogram to post-run ion extraction, and improved a further 10-fold upon going from

Table 6 Selected Applications of On-Line SPE–GC with MS or Tandem MS Detection for River and Tap Water

Analytes	MS detection	Sample volume (ml)	LOD (ng/l)	Ref.
Atrazine, simazine, various micropollutants	SIM	10	0.5	i
	Full-scan		3	
Various micropollutants	Full-scan	10	2–50	37, 41, 52, 67, 87, 88, ii
Pesticides, phenols	SIM	10	1–20	55, 66
Chlorinated pesticides	EI full-scan	100	1–30	89
	NCI full-scan		0.1–3	
Pesticides	MS/MS	10	0.01–2	53
Hetero-atom containing pesticides/micropollutants	AED/full-scan	5–50	1–15	95, 96
	MS[a]			

[a] At-line setup because of necessary repeated injections for AED analysis.

Abbreviations: SIM, selected ion monitoring; NCI, negative chemical ionization; if not stated otherwise, EI ionization was applied.

i.) A.-J. Bulterman, J.J. Vreuls, R.T. Ghijsen and U.A.Th. Brinkman, J. High Resolut. Chromatogr., 16 (1993) 397.

ii.) A.J.H. Louter et al., Intern. J. Environ. Anal. Chem., 56 (1994) 49.

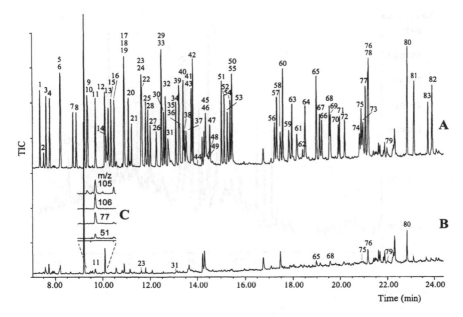

Figure 9 Total ion chromatogram for SPE–GC–MS of 10 ml of Rhine River water (B) nonspiked and (A) spiked at the 0.5 μg/L level with 86 microcontaminants. 50 μl of methyl acetate were used as presolvent. The insert (C) shows the extracted ion chromatograms of four characteristic masses of benzaldehyde (m/z 51, 77, 105 and 106). (From Ref. 41.)

full-scan acquisition to SIM detection (two ions per analyte) [67]. Detection limits of 0.2 to 1.1 ng/L were achieved for 10-ml surface water samples. However, one should always consider that the improved selectivity and detectability are accompanied by a serious loss of information on the general composition of the sample.

Jahr [55] used Autoloop–GC–MS for the trace analysis of phenols in water at the low-ng/L level. The phenols were derivatized by in-sample acetylation with acetic acid anhydride prior to automated SPE–GC–MS. The method was validated with 26 alkyl- , chloro- , and mononitrophenols; these included 4-nonylphenol and 17-ethinylestradiol. Repeatability was good and the sensitivity in the time-scheduled SIM mode was excellent.

On-line dialysis–SPE–GC–MS was developed for the determination of benzodiazepines in plasma [85]. Clean-up was achieved by dialysis of 100-μl samples for 7 minutes using water as the acceptor and trapping the diffused analytes on an SPE column. After drying, the analytes were desorbed with 375 μl of ethyl acetate on-line to the GC–MS via a loop-type interface. Sample clean-up was very efficient and offered the possibility of adding chemical agents that

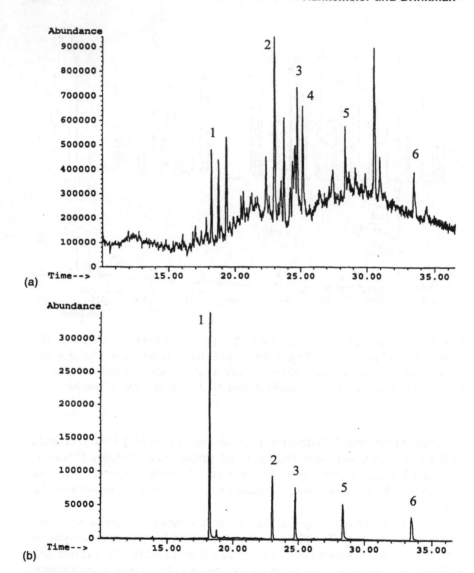

Figure 10 SPE–GC–MS chromatograms of 10 ml of Rhine River water spiked at the 0.1 µg/L level. (a) Full-scan mode (m/z 50–375) and (b) time-scheduled MID. Peak assignment and ions used: 1, mevinphos (m/z 127/192); 2, diazinon (m/z 197/204); 3, fenitrothion (m/z 277/260); 4, fenthion (not determined with MID); 5, triazophos (m/z 161/257); 6, coumaphos (226/362). (From Ref. 67.)

can help to reduce drug–protein binding. The benzodiazepines were determined in plasma at the 1 ng/ml level which is relevant for forensic or pharmacokinetic studies.

4.2. Solid-Phase Extraction–Gas Chromatography–Tandem Mass Spectrometry

In recent years, GC–ion-trap detection (ITD) systems that can perform tandem MS (MS–MS) on a routine basis have become commercially available [86]. Because ITD provides good sensitivity as well as increased selectivity in the MS–MS mode, an on-line SPE–GC–ITD system was optimized for the trace-level determination of polar and apolar pesticides [53]. The Autoloop interface (see section 3.2.3) was operated at an injection temperature of 90°C, which permitted the determination of thermolabile pesticides such as carbofuran and carbaryl. With sample volumes of 10 to 30 ml and a copolymer SPE cartridge, linear calibration curves were obtained for several pesticides over the range of 0.1 to 500 ng/L. Fully satisfactory tandem mass spectra were obtained at levels as low as 0.1 ng/L level in tap and river water. The system was used to analyze water from European and Asian rivers, and the determination of microcontaminants at 8 to 16 ng/L levels did not cause any problems (Fig. 11). Relevant analytical data are presented in Table 7. One conclusion may be that, for this target-compound type of analysis, a sample volume of 1 ml *or less* will be sufficient to comply with governmental directives.

In another application [78], SPETD–GC–ion-trap MS–MS was optimized for alachlor and metolachlor. Appropriate precursor ions with a high m/z value were selected from the electron impact (EI) and positive chemical ionization (PCI) spectra, and the CID voltage optimized so that the highest abundance of a selective product ion was observed to achieve maximum sensitivity. Detection limits of 0.1 µg/L were reported for alachlor and metolachlor for 100-µl samples. As an example, Figure 12 shows the analysis of 100 µl of Rotterdam harbor water, in which metolachlor was suspected to be present. Figure 12a shows the extracted ion SPETD–GC–MS chromatogram of mass m/z 162 and the mass spectrum of the peak at the retention time of metolachlor. The result cannot be called satisfactory. If, however, the analysis was performed in the MS–MS mode (Fig. 12b), the sample background had completely disappeared, and identification of the compound as metolachlor was perfectly straightforward due to the much higher selectivity. Quantification on the basis of the response of m/z 162 gave closely similar results for MS and MS–MS detection, i.e., 1.3 and 1.2 µg/L, respectively. The presence of metolachlor was also confirmed by SPETD–GC–PCI–MS(–MS).

In many monitoring programs, a mixture of compounds must be addressed, a number of which can be analyzed by means of GC–MS, while others require

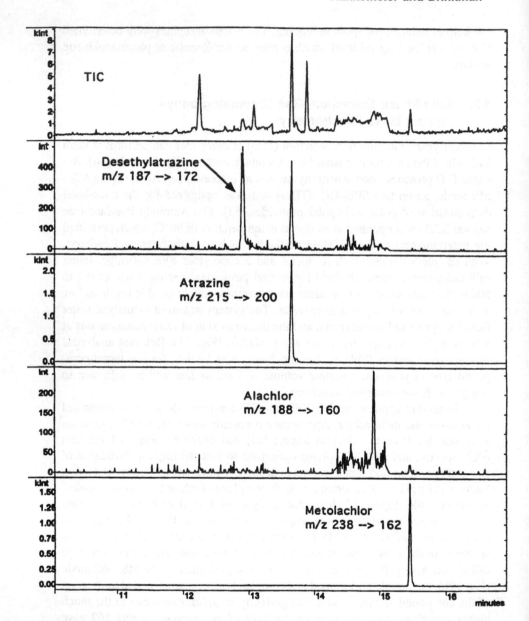

Figure 11 Total ion current and reconstructed ion chromatograms obtained after SPE–GC–MS/MS of 10 ml Rhine River water at m/z 172 (desethylatrazine), 200 (atrazine), 160 (alachlor), and 162 (metolachlor). (From Ref. 53.)

Table 7 SPE–GC–MS-MS of Pesticides in 10 ml Tap Water

Analyte	Linear range (ng/l)	R^2	RSD[b] (%)	LOD (ng/l)
Desethylatrazine[a]	2–200	0.9941	18	0.5
Atrazine	1–200	0.9993	10	0.2
Metolachlor	1–200	0.9997	6	0.04
Trifluralin	0.1–200	0.9993	6	0.01
Carbofuran	0.1–200	0.9981	7	0.1
Parathion-methyl	2–200	0.9991	7	1
Alachlor	0.1–200	0.9996	6	0.05
Fenitrothion	0.1–200	0.9961	7	0.1
Fenthion	0.1–200	0.9969	6	0.1
Parathion-ethyl	5–200	0.9993	6	2
Carbaryl	1–200	0.9979	6	0.1

[a] Less good results for this polar analyte mainly due to integration problems.
[b] RSD determined at 10 ng/l analyte concentration (n = 7).
Source: Ref. 53.

an LC-based approach. If, on the LC side, a particle beam (PB) interface is used, the two techniques can be combined in one setup, sharing the sample-handling unit as well as the MS detector. With this so-called Multianalysis system [87,88], two subsequent runs are performed per sample. First, the analytes from an approximately 10-ml sample trace enrichment are desorbed and sent to the GC–MS; in the next run, a 100- to 200-ml sample (with the larger volume compensating for the lower sensitivity) is preconcentrated, desorbed, and analyzed by LC–diode array–UV detector–MS. In both instances, classical EI spectra are generated that can be searched by using any GC–MS library. In one example, nine test compounds, triazines, anilides, and organophosphorous pesticides, were added to tap water. Detection limits were 0.005 to 0.1 µg/L for SPE–GC–MS (10-ml samples) in the full-scan mode, and 0.5 to 7 µg/L for SPE–LC–PB–MS (100-ml samples) in the full-scan and 0.05 to 1 µg/L in the SIM mode. Obviously, there is an urgent need to improve the performance of the PB interface, but this is a topic outside the scope of the present discussion. When using negative chemical ionization (NCI) MS, with methane as a reagent gas, the detection limits for SPE–GC–MS and SPE–LC–PB–MS could be improved 10- to 30-fold for most of the chlorinated pesticides [89]. The Multianalysis system was used to monitor the pollution at a number of sampling sites along the Nitra River (Slovak Republic), a tributary of the Danube, during a 2-year surveillance program.

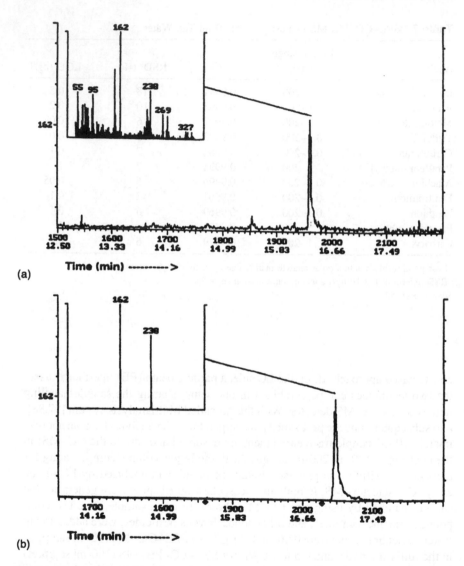

Figure 12 (a) SPETD–GC–MS ion chromatogram (m/z 162) of 100 µl of Rotterdam harbor water. Insert shows the mass spectrum of the prominent peak. (b) SPETD–GC–MS/MS daughter chromatogram (m/z 162; parent mass, m/z 238) of same sample. Insert shows the mass spectrum of the prominent peak. (From Ref. 78.)

4.3. Solid-Phase Extraction–Gas Chromatography–Atomic Emission Detection–Mass Spectrometry

A recent extension of the scope of SPE–GC concerns the use of atomic emission detection. With its distinct element selectivity, the atomic emission detector can provide information that is complementary to that of a mass-selective detector. An additional advantage of AED is that the response per mass unit of an element is more or less independent of the structure of the analyte of interest. This allows the use of the universal calibration concept [90,91], although this statement is

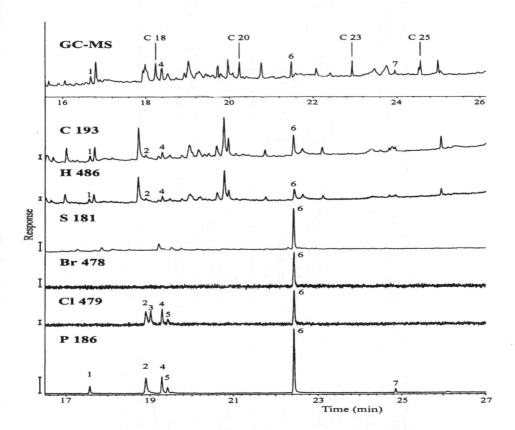

Figure 13 Element-selective SPE–GC–AED chromatograms and full-scan SPE–GC–MS chromatogram of 7 ml of wastewater (C, H, S, P, Cl and Br traces). The sample was spiked with bromophos-ethyl (no. 6) at the 1.86 µg/L level. The sample analyzed by GC–MS was also spiked with some n-alkanes. For peak assignment, see Table 8. The bars indicate the peak height of a (hypothetical) compound containing one of each hetero-atoms at the 0.2 µg/L level. Numbers indicate wavelength in nm. (From Ref. 95.)

Table 8 Identification and Quantification of Hetero-Atom-Containing Microcontaminants by Means of Their Partial Formulas and Universal Calibration (AED) and Corresponding Mass Spectra and Standard Addition (MS) in Wastewater

	SPE-GC-AED		SPE-GC-MS				Concentration (µg/l)				
		Partial	Library search[b]				AED				MS
No.	RI	formula	RI	Identified compound	Q[b]	Formula	S	P	Cl	Br	
1	1662	P	1663	Tributylphosphoric acid	64	$C_{12}H_{27}O_4P$		0.10			0.21
2	1782	$Cl_{2.5}P$	1781	Tris(2-chloroethyl) phosphate	91	$C_6H_{12}Cl_3O_4P$		0.36	0.30		0.54
3	1792	Cl	1778	Hexachlorocyclohexane[a]	32	$C_6H_6Cl_6$			0.12		
4	1818	$Cl_{2.5}P$	1814	Tris(2-chloro-1-methyl-ethyl) phosphate		$C_9H_{18}Cl_3O_4P$		0.25	0.21		
5	1831	$Cl_{2.3}P$	1828	Bis(2-chloro-1-methyl-ethyl) (2-chloropropyl) phosphate		$C_9H_{18}Cl_3O_4P$		0.09	0.06		
6	2148	$Br_{1.1}Cl_{1.9}P_{1.1}S$	2135	Bromophos-ethyl[c]	38	$C_{10}H_{12}BrCl_2O_3PS$	1.1	1.2	1.0	1.2	
7	2427	P	2425	Tris(2-butoxyethyl) phosphate	40	$C_{18}H_{39}O_7P$		0.07			

[a] Isomer not further identified.
[b] NBS library was used for library search and determination of match qualifier (Q) on a scale of 100. Identified compounds were always no. 1 on hit list.
[c] Bromophos-ethyl, spike of 1.0 µg/l corrected for low recovery.
Source: Ref. 95.

not uncontested for, at least, some elements [92]. The atomic emission detector is not as delicate a detector as is sometimes thought: injections of 100 μl of ethyl acetate, and also of other organic solvents such as iso-octane, toluene, and even dichloromethane, do not cause flame-out problems [93]. Consequently, an on-line SPE–GC–AED setup was constructed [94]. With 10-ml water samples, detection limits of 5 to 20 ng/L were obtained for organophosphorous pesticides.

Because of the promising early results, the combination of AED and MS was attempted next. Data obtained by on-line SPE–GC–AED and on-line SPE–GC–MS using a similar setup are ideally suited for the (nontarget) screening of hetero-atom-containing compounds [95]. First, the partial molecular formula for a peak detected at the same retention time in one or more of the AED traces was calculated using the universal calibration concept. Next, the corresponding mass spectrum, i.e., at the same retention index, was obtained from the SPE–GC–MS chromatogram. After (provisional) identification by an MS library search and with the partial formula obtained by GC–AED, the compound was quantified by means of AED-based universal calibration. This concept was successfully used for the detection of S-, Br-, Cl- and/or P-containing compounds present above the 0.02-μg/L level in tap water, and above the 0.2-μg/L level in wastewater. Figure 13 shows that, in the latter case, seven peaks were detected above the threshold concentration(s) for the SPE–GC–AED analysis of 7 ml of the effluent of a municipal wastewater treatment plant. All peaks could be identified on the basis of the combined partial molecular formulas and mass spectral information (Table 8). In a more advanced setup, AED and MS were combined in a single SPE–GC–AED/MS instrument with a split of approximately 1:1 of the GC effluent [96]. Retention times could then be kept the same to within 0.5 second, which made correlation even more straightforward. The system was applied to the analysis of vegetables and the (nontarget) analysis of river water. Several hetero-atom-containing microcontaminants were identified and quantified down to the 20-ng/L level. Ongoing work on the analysis of wastewater confirms the practicality of this approach [97].

5. DISCUSSION AND CONCLUSIONS

This chapter discussed a number of sample preparation techniques that can be coupled on-line to GC. Special attention was devoted to SPE–GC as the preparation-plus-separation procedure, to the analysis of aqueous samples, and to analyte identification, i.e., to the use of MS detection.

One way of subdividing on-line sample preparation–GC techniques is into solvent-free techniques and techniques requiring the introduction of some 10 to 100 μl of an organic solvent into the GC. As regards the latter group, 10 years ago, such amounts of solvent were called "very large." Today, LVI has become a really routine procedure. This, by itself, is a major step forward for essentially all GC-based separation procedures and can be called a distinct breakthrough. As a consequence, LLE–phase separator–GC is an easily accessible technique. However, it does not provide noticeable analyte enrichment and is, therefore, not always well suited for trace-level analysis. Normal-phase LC–GC (not a topic discussed in this paper) has also become markedly successful but, here, the main limitations are the heart-cut nature of the setup and the restricted use of normal-phase LC separations. The counterpart of normal-phase LC–GC, reversed-phase LC–GC, has yet failed to become an analytical technique of interest. Clearly, the problems created by most typical reversed-phase LC eluents keep ruining the LC-to-GC interface performance and, thus, the LVI into the GC. This leaves us with today's favored option within the LVI-based subgroup, SPE–GC.

The history of SPE–GC shows that introducing new technology invariably requires much time and effort. As was remarked earlier [1], setting up an on-line technique is not simply the coupling of two well established "sub"techniques, but will generally require adaptation and subsequent optimization as well as a profound knowledge of the underlying principles. Several recent studies [68–71] clearly demonstrate just that. A better understanding of the LVI–GC and SPE-to-GC transfer processes has allowed optimization of the analytical procedures and, next, a higher sample throughput, wider application range, and improved robustness, and has made optimization essentially superfluous. Thus, SPE–GC has now become a mature technique, making traditional off-line SPE-plus-GC correspondingly less attractive (mainly because of better analyte detectability, improved precision, easier miniaturization and automation, and less solvent consumption). In many cases, SPE–GC should also be preferred over the at-line combination of micro-LLE and LVI–GC for much the same reasons and, in addition, because of its wider application range. One can now fully profit from (1) the wealth of information provided in the LC-oriented literature on SPE to facilitate method development, and (2) the essentially quantitative isolation of analytes that cover a very wide polarity/boiling-point range to perform extensive screening (and, with MS, identification/confirmation) studies at ultratrace levels.

A rather similar type of picture emerges for the solvent-free subgroup. Although supercritical fluid extraction (SFE)–GC (not discussed in this chapter) seemed to be a most promising option a decade ago, especially because interfacing was considered to be straightforward, on-line SFE–GC has not become a viable approach. Operational and design problems, and the unexpectedly complicated nature of supercritical fluid–analyte–matrix interactions are primarily responsible for this outcome. Another addition to the list, SPME, has, on the other hand, become remarkably successful in a few years (see Chap. 8). As a solvent-free technique, SPME can be coupled rather easily with GC, and sorption is rapid, and close to exhaustion for BTEX, substituted benzenes, and other, similar analytes. The main drawback of SPME is that extending the application range to medium-polar or higher-molecular-weight analytes often causes a dramatic increase in the sample preparation time. In addition, the nonequilibrium conditions frequently used, the disturbing role of sample matrix constituents, and the rather high viscosity of the sample solution after the, often necessary, addition of sodium chloride make relatively extensive optimization unavoidable. Still, one can conclude now that SPME–GC is an attractive technique. Actually, because of the preferred application range of SPME indicated above, SPME–GC and SPE–GC may well become complementary rather than competitive techniques. In this context, it is interesting to mention SPETD–GC as an alternative: SPETD combines the solvent-free operation of SPME with the exhaustive analyte isolation of SPE. Further exploration of this "middle of the road" option is, in our opinion, indicated.

As regards the final step of the analytical procedure, until quite recently conventional selective GC detectors were predominantly used. Today, an element-selective detector is a proper choice for screening studies in which the number of negatives far exceeds that of positives. However, with the advent of much less expensive MS detectors with, simultaneously, improved analytical characteristics, and the increasing demand for analyte identification/confirmation on the basis of structural information, GC–MS is becoming the method of choice for a rapidly increasing number of applications. With the target-compound-oriented ion-trap-type MS–MS detection, analyte identification at or below the 1-ng/L level becomes possible for approximately 10-ml aqueous samples.

With SPE–GC–MS rapidly becoming a recognized standard, there is also enhanced interest in alternative combined procedures, notably in the use of atomic emission detection with its multielement detection capability and unusually high selectivity. The recently introduced SPE–GC–AED/MS with its confirmation-plus-identification potential in one run has been remarkably successful for pesticide analysis in a variety of samples. This, in turn, brings another alternative, infrared (IR) detection, to our attention. Several studies show that, with deposition-based detection, e.g., with cryotrapping, trace-level identification is possible at or below the 1 µg/L level with approximately 10-ml aqueous samples [98,99]. This is, by itself, a rewarding result. In view of the wide application range of IR and its frequent complementariness to MS, one would like to see more studies regarding this detection mode being performed in the near future.

Finally, let us take SPE–GC–MS as the model system to attempt quantification of what has been achieved, and to indicate a few trends that start to emerge. In many instances, aqueous samples of about 10 ml are analyzed. Sample preparation and separation-plus-detection each take about 30 minutes so that, with the parallel setup almost invariably used, throughput can be 40 to 45 samples per day. Detection limits in full-scan MS (of special importance if unknowns are expected) are on the order of 5 to 50 ng/L (TIC) or 1 to 10 ng/L (reconstructed ion chromatogram), and with target-orientated MS–MS, 0.1 to 1 ng/L. Since regulatory requirements often are in the range of 0.1 to 1 µg/L, it will be obvious that sample volumes can frequently be reduced 10- to 100-fold. This is a rather challenging conclusion, because it opens the possibility to design completely different ways of sample introduction, and removal of (only a few hundreds of microliters) of water. On the other hand, designing improved, i.e., faster, sample-handling procedures is becoming urgent because of the introduction of fast GC techniques that can effect an approximately 10-fold shorter GC run time: with a 5- to 6-minute GC run-plus-reequilibration time, a similarly reduced sample handling/introduction time span is required. Loading smaller volumes and accelerating the drying step will obviously be especially important. Finally, with such advanced systems, MS detection has to be accelerated correspondingly. Here, time-of-flight MS will probably be the solution to the problem. It is, as yet, too early to give a final verdict. However, it seems fair to state that the rapidity of data acquisition of this MS technique meets all present-day demands [100]; the sensitivity, on the other hand, is still less than that of conventional mass spectrometers.

In summary, we have arrived at the stage where modest sample volumes are sufficient for the automated analysis of a variety of water samples. From among the approaches available, SPE–GC–MS probably has the widest application range, and detection/identification limits are consistently below those set by governmental bodies. At the same time, there are several indications that each of the three steps involved—sample introduction, separation, and detection— will be considerably improved in the near future. Should this indeed be true, then a next challenge will become immediately apparent: how should one interpret, use, and store the avalanche of data produced?

REFERENCES

1. E.C. Goosens, D. de Jong, G.J. de Jong, and U.A.Th. Brinkman, Chromatographia, 47 (1998) 313.
2. K. Grob, in W. Bertsch, W.G. Jennings, and P. Sandra, eds., On-line coupled LC–GC, 1991, Hüthig, Heidelberg.
3. W. Engewald, J. Teske, and J. Efer, J. Chromatogr. A, 842 (1999) 143.
4. J. Slobodnik, E.R. Brouwer, R.B. Geerdink, W.H. Mulder, H. Lingeman, and U.A.Th. Brinkman, Anal. Chim. Acta, 268 (1992) 55.
5. J. Slobodnik, M.G.M Groenewegen, E.R. Brouwer, H. Lingeman, and U.A.Th. Brinkman, J. Chromatogr., 642 (1993) 359.
6. P.J.M. van Hout and U.A.Th. Brinkman, Europ. Water Poll. Control, 3 (1993) 29.
7. H.J. Cortes, C.D. Pfeiffer, G.L. Jewett and B.E. Richter, J. Microcol. Sep., 1 (1989) 28.
8. B.B. Gerhart and H.J. Cortes, J. Chromatogr., 506 (1990) 377.
9. K. Grob Jr. and Z. Li, J. Chromatogr. A, 473 (1989) 423.

10. K. Grob and E. Müller, J. Chromatogr. A, 473 (1989) 411.
11. D. Duquet, C. Dewaele, M. Verzele, and M. McKinley, J. High Resolut. Chromatogr. Chromatogr. Commun., 11 (1988) 824.
12. G. Audunsson, Anal. Chem., 60 (1988) 1340.
13. E.C. Goosens, D. de Jong, G.J. de Jong, and U.A.Th. Brinkman, J. Microcol. Sep., 6 (1994) 207.
14. E.C. Goosens, I.M. Beerthuizen, D. de Jong, G.J. de Jong, and U.A.Th. Brinkman, Chromatographia, 40 (1995) 267.
15. E.C. Goosens, K.H. Stegman, D. de Jong, G.J. de Jong, and U.A.Th. Brinkman, Analyst, 121 (1996) 61.
16. G.R. van der Hoff, P. van Zoonen and K. Gorb, J. High Resolut. Chromatogr., 17 (1994) 37.
17. T. Hyötyläinen, K. Grob, M. Biedermann, and M.-L. Riekkola, J. High Resolut. Chromatogr. 20 (1997) 410.
18. T. Hyötyläinen, K. Grob, and M.-L. Riekkola, J. High Resolut. Chromatogr., 20 (1997) 657.
19. T. Hyötyläinen, K. Janho, and M.-L. Riekkola, J. Chromatogr. A, 813 (1998) 113.
20. G.P. Blanch, M.L. Ruiz del Castillo, and M. Herraiz, J. Chromatogr. Sci., 36 (1998) 589.
21. P. Wessels, J. Ogorka, G. Schwinger, and M. Ulmer, J. High Resolut. Chromatogr., 16 (1993) 708.
22. J. Ogorka, G. Schwinger, G. Bruat, and V. Seidel, J. Chromatogr. A, 626 (1992) 87.
23. J.J. Vreuls, V.P. Goudriaan, G.J. de Jong, and U.A.Th. Brinkman, High Resolut. Chromatogr., 14 (1991) 475.
24. J. Roeraade, J. Chromatogr. A, 330 (1985) 263.
25. E. Fogelqvist, M. Krysell, and L.-G. Danielsson, Anal. Chem., 58 (1986) 1516.
26. E. Ballesteros, M. Gallego, and M. Valcárcel, Anal. Chem., 65 (1993) 1773.
27. E. Ballesteros, M. Gallego, and M. Valcárcel, J. Chromatogr. A, 633 (1993) 169.
28. E.C. Goosens, M.H. Broekman, M.II. Wolsters, R.E. Strijker, D. de Jong, G.J. de Jong, and U.A.Th. Brinkman, J. High Resolut. Chromatogr., 15 (1992) 243.
29. E.C. Goosens, D. de Jong, G.J. de Jong, F.D. Rinkema, and U.A.Th. Brinkman, J. High Resolut. Chromatogr., 18 (1995) 38.
30. A. Venema and J.T. Jelink, in P. Sandra, ed., Proceedings of the 17th International Symposium on Capillary Chromatography, Riva del Garda, September 1994, 1994, Hüthig, Heidelberg, p. 1035.
31. A. Venema and J.T. Jelink, J. High Resolut. Chromatogr., 19 (1996) 234.
32. D. Barceló and M.-C. Hennion, Trace determination of pesticides and their degradation products in water, 1997, Elsevier, Amsterdam.
33. E.R. Brouwer, D.J. van Iperen, I. Liska, H. Lingeman, and U.A.Th. Brinkman, Intern. J. Environ. Anal. Chem., 47 (1992) 257.
34. P.J.M. Kwakman, J.J. Vreuls, U.A.Th. Brinkman, and R.T. Ghijsen, Chromatographia, 34 (1992) 41.
35. A. Farjam, J.J. Vreuls, W.J.G.M. Cuppen, G.J. de Jong, and U.A.Th. Brinkman, Anal. Chem., 63 (1991) 2481.

36. J. Dallüge, Th. Hankemeier, J.J. Vreuls, G. Werner, and U.A.Th. Brinkman, J. Chromatogr. A, 830 (1999) 377.
37. Th. Hankemeier, S.P.J. van Leeuwen, J.J. Vreuls, and U.A.Th. Brinkman, J. Chromatogr. A, 811 (1998) 117.
38. J.J. Vreuls, R.T. Ghijsen, G.J. de Jong, and U.A.Th. Brinkman, J. Chromatogr. A, 625 (1992) 237.
39. Y. Picó, J.J. Vreuls, R.T. Ghijsen, and U.A.Th. Brinkman, Chromatographia, 38 (1994) 461.
40. Y. Picó, A.J.H. Louter, J.J. Vreuls, and U.A.Th. Brinkman, Analyst, 119 (1994) 2025.
41. Th. Hankemeier, A.J.H. Louter, J. Dallüge, J.J. Vreuls, and U.A.Th. Brinkman, J. High Resolut. Chromatogr., 21 (1998) 450.
42. J.J. Vreuls, H.G.J. Mol, J. Jagesar, R. Swen, R.E. Hessels and U.A.Th. Brinkman, in P. Sandra and G. Devos, eds., Proceedings of the 17th International Symposium on Capillary Chromatography, Riva del Garda, September 1994, (1994,) Hüthing, Heidelberg, p. 1181.
43. K Grob and M Biedermann, J. Chromatogr. A, 750 (1996) 11.
44. E.C. Goosens, D. de Jong, J.H.M. van den Berg, G.J. de Jong, and U.A.Th. Brinkman, J. Chromatogr., 552 (1991) 489.
45. G. Hagman and J. Roeraade, J. Microcol. Sep., 5 (1993) 341–346.
46. J.F. Hiller, T. McCabe, and P.L. Morabito, J. High Resolut. Chromatogr., 16 (1993) 5.
47. Th. Hankemeier, S. Ramalho, S.J. Kok, J.J. Vreuls, and U.A.Th. Brinkman, submitted for publication.
48. H.G.J. Mol, M. Althuizen, H.-G. Janssen, C.A. Cramers, and U.A.Th. Brinkman, J. High Resolut. Chromatogr., 19 (1996) 69.
49. J. Staniewski, H.-G. Janssen, C.A. Cramers and J.A. Rijks, J. Microcol. Sep., 4 (1992) 331.
50. Th.H.M. Noij and M.M.E. van der Kooi, J. High. Resolut. Chromatogr., 18 (1995) 535.
51. A.J.H. Louter, P. Jones, D. Jorritsma, J.J. Vreuls, and U.A.Th. Brinkman, J. High Resolut. Chromatogr., 20 (1997) 363.
52. A.J.H. Louter, S. Ramalho, J.J. Vreuls, and U.A.Th. Brinkman, J. Microcol. Sep., 8 (1996) 469.
53. K.K. Verma, A.J.H. Louter, A. Jain, E. Pocurull, J.J. Vreuls, and U.A.Th. Brinkman, Chromatographia, 44 (1997) 372.
54. P. Enoch, A. Putzler, D. Rinne, and J. Schlüter, J. Chromatogr. A, 822 (1998) 75.
55. D. Jahr, Chromatographia, 47 (1998) 49.
56. E. Ballesteros, M. Gallego, and M. Valcárcel, Environ. Sci. Technol., 30 (1996) 2071.
57. M.A. Crespín, E. Ballesteros, M. Gallego, and M. Valcárcel, Chromatographia, 43 (1996) 633.
58. G.R. van der Hoff, S.M. Gort, R.A. Baumann, and P. van Zoonen, J. High Resolut. Chromatogr., 14 (1991) 465.
59. G.R. van der Hoff, F. Pelusio, U.A.Th. Brinkman, R.A. Baumann, and P. van Zoonen, J. Chromatogr. A, 719 (1996) 59.

60. A.J.H. Louter, E. Bosma, J.C.A. Schipperen, J.J. Vreuls, and U.A.Th. Brinkman, J. Chromatogr. B, 689 (1997) 35.

61. A. Namera, M. Yashiki, K. Okada, Y. Iwasaki, M. Ohtani, and T. Kojima, J. Chromatogr. B, 706 (1998) 253.

62. A. Namera, M. Yashiki, Y. Iwasaki, M. Ohtani, and T. Kojima, J. Chromatogr. B, 716 (1998) 171.

63. I.M. Bengtsson, D.C. Lehotay, J. Chromatogr. B, 685 (1996) 1.

64. Th. Hankemeier, P.C. Steketee, J.J. Vreuls, and U.A.Th. Brinkman, J. Chromatogr. A, 750 (1996) 161.

65. Th. Hankemeier, P.C. Steketee, J.J. Vreuls, and U.A.Th. Brinkman, Fresenius J. Anal. Chem., 364 (1999) 106.

66. E. Pocurull, C. Aguilar, F. Borrull, and R.M. Marcé, J. Chromatogr. A, 818 (1998) 85.

67. A.J.H. Louter, C.A. Beekvelt, P. Cid Montanes, J. Slobodnik, J.J. Vreuls, and U.A.Th. Brinkman, J. Chromatogr. A, 725 (1996) 67.

68. Th. Hankemeier, S.J. Kok, J.J. Vreuls, and U.A.Th. Brinkman, J. Chromatogr. A, 811 (1998) 105.

69. E. Boselli, B. Grolimund, K. Grob, G. Lercker, and R. Amadò, J. High Resolut. Chromatogr., 21 (1998) 355.

70. B. Grolimund, E. Boselli, K. Grob, R. Amadò, and G. Lercker, J. High Resolut. Chromatogr., 21 (1998) 378.

71. Th. Hankemeier, S.J. Kok, J.J. Vreuls, and U.A.Th. Brinkman, J. Chromatogr. A, 841 (1999) 75.

72. H.G.J. Mol, J. Staniewski, H.-G. Janssen, C.A. Cramers, R.T. Ghijsen, and U.A.Th. Brinkman, J. Chromatogr. A, 630 (1993) 201.

73. H.G.J. Mol, H.-G. Janssen, C.A. Cramers, and U.A.Th. Brinkman, J. Microcol. Sep., 7 (1995) 247.

74. H.G.J. Mol, H.-G. Janssen, and C.A. Cramers, J. High Resolut. Chromatogr., 16 (1993) 413.

75. J.J. Vreuls, U.A.Th. Brinkman, G.J. de Jong, K. Grob, and A. Artho, J. High Resolut. Chromatogr., 14 (1991) 455.

76. J.J. Vreuls, R.T. Ghijsen, G.J. de Jong, and U.A.Th. Brinkman, J. Microcol. Sep., 5 (1993) 317.

77. H.G.J. Mol, H.-G.M. Janssen, C.A. Cramers, and U.A.Th. Brinkman, J. High Resolut. Chromatogr., 16 (1993) 459.

78. A.J.H. Louter, J. van Doornmalen, J.J. Vreuls, and U.A.Th. Brinkman, J. High Resolut. Chromatogr., 19 (1996) 679.

79. E. Baltussen, F. David, P. Sandra, H.-G. Janssen, and C.A. Cramers, J. Chromatogr. A., 805 (1998) 237.

80. G. Schomburg, E. Bastian, H. Behlau, H. Husmann, and F. Weeke, J. High Resolut. Chromatogr. Chromatogr. Commun., 7 (1984) 4.

81. J. Teske, J. Efer and W. Engewald, Chromatographia, 46 (1997) 580.

82. J. Teske, J. Efer and W. Engewald, Chromatographia, 47 (1998) 35.

83. S. Müller, J. Efer, and W. Engewald, Chromatographia, 38 (1994) 694.

84. S. Blomberg and J. Roeraade, J. High Resolut. Chromatogr., 13 (1990) 509.

85. R. Herraez Hernandez, A.J.H. Louter, N.C. van de Merbel, and U.A.Th. Brinkman, J. Pharm. Biomed. Anal., (1996) 1037.
86. A. Saraf and L. Larsson, J. Mass Spectrom., 31 (1996) 389.
87. J. Slobodnik, A.C. Hogenboom, A.J.H. Louter, and U.A.Th. Brinkman, J. Chromatogr. A, 730 (1996) 353.
88. J. Slobodnik, A.J.H. Louter, J.J. Vreuls, I. Liska, and U.A.Th. Brinkman, J. Chromatogr. A, 768 (1997) 239.
89. A.J.H. Louter, A.C. Hogenboom, J. Slobodnik, J.J. Vreuls, and U.A.Th. Brinkman, Analyst, 122 (1997) 1497.
90. S. Pedersen-Bjergaard, T.N. Asp, and T. Greibrokk, Anal. Chim. Acta, 265 (1992) 87.
91. N.L. Olson, R. Carrell, R. Cummings, R. Rieck, and S. Reimer, J. Assoc. Offic. Anal. Chem. Intern., 78 (1995) 1464.
92. J. Th. Jelink and A. Venema, J. High Resolut. Chromatogr., 13 (1990) 447.
93. E.C. Goosens, D. de Jong, G.J. de Jong, F.D. Rinkema, and U.A.Th. Brinkman, J. High Resolut. Chromatogr., 18 (1995) 38.
94. Th. Hankemeier, A.J.H. Louter, F.D. Rinkema, and U.A.Th. Brinkman, Chromatographia, 40 (1995) 119.
95. Th. Hankemeier, J. Rozenbrand, M. Abhadur, J.J. Vreuls, and U.A.Th. Brinkman, Chromatographia, 48 (1998) 273.
96. H.G.J. Mol, Th. Hankemeier, and U.A.Th. Brinkman, LC–GC Intern., 12 (1999) 108.
97. L.L.P. van Stee, P.E.G. Leonards, R.J.J. Vreuls, and U.A.Th. Brinkman, Analyst, 134 (1999) 1547.
98. Th. Hankemeier, H.T.C. Van der Laan, J.J. Vreuls, M.J. Vredenbregt, T. Visser, and U.A.Th. Brinkman, J. Chromatogr. A, 732 (1996) 75.
99. Th. Hankemeier, E. Hooijschuur, R.J.J. Vreuls, U.A.Th. Brinkman, and T. Visser, J. High Resolut. Chromatogr. 21 (1998) 341.
100. R.J.J. Vreuls, J. Dallüge, and U.A.Th. Brinkman, J. Microcol. Sep 11, 1999, 663.

7

Gas Chromatography–Mass Spectrometry in Occupational and Environmental Health Risk Assessment with Some Applications Related to Environmental and Biological Monitoring of 1-Nitropyrene

P. T. J. Scheepers, R. Anzion, and R. P. Bos
University of Nijmegen, Nijmegen, The Netherlands

1. INTRODUCTION

In the Western world occupational and environmental exposures to toxic substances have had a general tendency to decrease over the last three decades. This has not made life easier or safer because in the same period of time numerous new chemical substances and technologies have been introduced. In pace with these developments, the interest of scientists active in the field of occupational and environmental health has shifted somewhat from the study of direct physical injury and acute intoxications resulting from accidental or frequent high occupational and environmental exposures (to single substances) to the prevention of effects resulting from long-term, sometimes intermittent, exposures to low concentrations of components (of complex mixtures) originating from multiple sources. A second trend is the increasing interest to determine the internal dose as an expression of exposure to a xenobiotic substance, rather than measuring concentrations in ambient air, surface water, soil, or food. These changes of interest have consequences for the technical support of the investigations in terms of the need to detect lower concentrations of hazardous chemical substances from more complex matrices, i.e., body fluids. It is unlikely that the individual contribution of each of these chemical substances in these complex exposures to human health will be understood without the use of component-specific approaches such as offered by mass spectrometry (MS) applications. Modern occupational and environmental health risk assessment therefore relies more and more on the use of sensitive and specific analytical techniques that feature detection of trace levels of toxic compounds from complex matrices combined with positive identification of these compounds. Therefore, more sophisticated approaches such as molecular dosimetry have been developed to detect xenobiotic substances in body fluids or exhaled air. Structural characteristics of xenobiotic substances (or their metabolites) interacting with biomolecules may be used to unravel the mechanisms of toxicity and elucidate the etiology of a forthcoming disease.

In this chapter, the use of gas chromatography–mass spectrometry (GC–MS) in the field of occupational and environmental health is discussed with emphasis on the detection of traces of nitrated polycyclic aromatic hydrocarbons (nitro-PAHs) from complex source emissions (diesel engine exhaust) in the ambient atmosphere and in personal air samples (workplace atmosphere). The uptake, and distribution, covalent binding to proteins, and excretion of urinary metabolites are addressed in terms of specific MS-applications. The analysis of drinking water and foodstuffs for contaminants or residues is not discussed; these topics are presented elsewhere in this book.

2. AMBIENT AIR AND HUMAN BREATH ANALYSIS

Many xenobiotic toxic substances enter the body via inhalation. Therefore, standards and guidelines for protection from health hazards in the workplace and in the general environment are based on regulation of air concentrations of toxic gases, vapors, or particles. Air polluted with these toxic substances may be collected in evacuated containers or canisters or in prepared sample bags for direct analysis from the gas phase of high concentrations of volatile organic compounds [1]. In some cases, the analytes may be captured from the gas phase in the laboratory, but in most cases, the toxic substances are immobilized during the process of (field) sampling: substances in the gas phase or semivolatile analytes are collected on solid sorbents (Amberlite XAD-2 or XAD-4, Tenax, charcoal). This happens after diffusion to the solid phase (passive air sampling), [2,3] or capture by drawing air through a cartridge, filter, or foam prepared with the solid sorbent (active air sampling). Both gas phase and particulate air contaminants may be captured in a liquid, either aqueous or organic solvent, in a midget impinger. For ambient air sampling of diisocyanate aerosols a solution of di-n-butylamine in toluene is used [4]. Alternatively, particle-associated air contaminants (and to a certain extent also semivolatile substances) are collected by drawing air through a membrane filter, causing the air contaminants to impact (particulate matter) or adsorb (semi volatile compounds) onto the surface of the membrane. Usually, these membranes have a large adsorption surface because of their fiber structure, allowing chemical substances to impregnate the fiber surface. Materials commonly used for fiber filters are Teflon, polypropylene, mixed cellulose ester, glass fiber, and quartz. For air sampling of semivolatile compounds such as PAHs, polyurethane foam may be used. Membrane filters used for capturing particulate matter are usually preceded by a system to select a predefined size fraction of the dust, such as inhalable, thoracic, or respirable dust. The solid sorbent may be impregnated with a reagent that converts the compound of interest with the air contaminant to form an inert product, i.e., in the case of aldehydes and aliphatic polyamines [5–7]. Solid-phase microextraction (SPME) has been developed for the determination of volatile organic compounds from ambient air, providing limits of determination down to the sub–ppb and ppt level.

For the GC analysis of (semi)volatile organic compounds, usually flame ionization detection (FID) or electron capture detection (ECD) (for halogenated compounds) offer sufficient sensitivity. Complex mixtures of organics do not always result in baseline separation of the individual contaminants as is the case for organic solvent mixtures, engine fuel vapors, or emissions from combustion processes. Also, the matrix may cause poor peak quality in the chromatographic separation. In these cases, MS can offer the selectivity to compensate for these limitations.

Recent cases, in which MS–MS and negative-ion chemical ionization (NCI) MS proved to be valuable tools in the characterization of exposure of workers to complex mixtures, include the detection of vulcanization products such as aniline and o-toluidine during rubber production [8]; the analysis of muramic acid and 3-hydroxy fatty acids from bacteria in agricultural environments [9]; *Fusarium* toxins in dust from drying and milling of grain [10]; resin acids from soldering flux [11]; 2,4- and 2,6-toluene diisocyanate, 4,4'-dimethylenediphenyl diisocyanate, or 4,4'-methylenedianiline from production or thermal degradation of plastics [12–14]; di(2-ethylhexyl)phthalate as a plasticizer [15]; PAHs in chimney sweeping [16], silicon carbide production [17], and from charcoal grilling [18]; nitro-PAHs from diesel exhaust and kerosene heaters [19,20]; polychlorinated biphenyls (PCB) in urban aerosols [21], and C_2 to C_8 alkenes in indoor smoky air and from car interiors [22]. In some cases, electrospray-MS was used, e.g., for the characterization of allergic or irritant volatile organic air pollutants derived from rape-seed oil [23].

In this chapter, the analysis of nitro-PAHs from combustion sources is used to show some problems and solutions related to the use of GC–MS in the analysis of source emissions and ambient air. The analysis of human breath focuses on traces of xenobiotic volatile organic solvents.

2.1. Nitropyrene in Some Source Emissions

Nitro-substituted polycyclic aromatic hydrocarbons (nitro-PAHs) are formed during the combustion of fossil fuels at high temperatures with a vast supply of combustion air. In this reaction, conversion of nitrite (NO_2) to nitric acid is an important intermediate step. Another source of nitro-PAHs is the photochemical radical–mediated conversions of parent PAHs to nitro-derivatives. Combustion at high temperatures with a vast supply of combustion air may lead to the formation of 1-nitropyrene (1-NP), whereas photochemical conversion of pyrene gives rise to 2- and 4-nitropyrene [24,25].

Recovery of nitro-PAH compounds from the filter surface and particles can be achieved in a one-step extraction with a suitable organic solvent, e.g., acetone or methylene chloride, by sonication, Soxhlet extraction, or supercritical fluid extraction. The organic solvent can be removed by a gentle flow of inert gas, e.g., nitrogen. To speed up evaporation, the extracts can be placed in a thermostatic water bath. Some (semi) volatile compounds may be lost easily at the point of complete dryness, which must be prevented by ceasing evaporation or (more practical) addition of a small volume of a stayer. Any losses during this and other steps can be monitored and corrected for by using a suitable internal standard, added just before commencing extraction. The use of deuterated analogs of the analytes is state-of-the-art. For the analysis of small quantities of nitro-PAHs that are associated with combustion particles, extensive pretreatment may be required. In order to prevent loading of the analytical column with a substantial amount of nonpolar organic compounds such as the parent PAHs, the slightly polar nitro-PAHs can be easily isolated using silica solid-phase cartridges [26]. Nitro-PAHs do not separate nicely on GC. Their volatility can be improved by reduction to their amino analogs and subsequent derivatization with heptafluorobutyric anhydride (Fig. 1). Conditions for GC–MS analysis of nitro-PAHs are given in Table 1.

2.2.2. Road Vehicle Emissions

The 1-NP content was analyzed in particulate matter collected from the exhaust of more than 20 diesel-powered road vehicles during the Dutch emission compliance program for passenger cars (TNO Road-Vehicles Research Institute) in 1996. Particulate matter was collected on polytetrafluoroethylene (PTFE)-coated glass fiber filters according to a European standard procedure (91/542/EC), during a simulated driving cycle on a chassis dynamometer. These fresh exhaust samples contained typically more volatile organic compounds than ambient air samples.

Substance	$[M]^+$	$[M-COC_3F_7]^+$	$[M-CNHCOC_3F_7]^+$
1-Aminopyrene	413	216	189

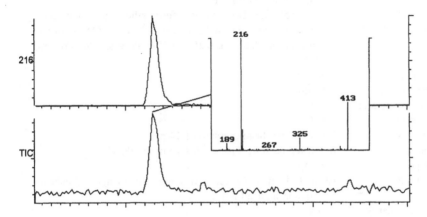

Figure 1 GC–MS–MS analysis of 1-aminopyrene derivatized with HFBA (mass spectrum in insert). For system description and analysis conditions see Table 1. (From Ref. xiii in Table 6.)

The 1-NP content of the particles ranges over two orders of magnitude. Interestingly, Japanese cars produce particles with much higher 1-NP content compared with the exhaust from the cars of European manufacture (Table 2). There is also a tendency for most cars of increasing 1-NP content when comparing the driving conditions from left to right: UDC ± UDC-hot < EUDC, and FTP-cold < FTP-stab < FTP-hot (cf. Table 2). This difference can be explained by the rise in temperature of the engine and the combustion mixture during the preceding driving cycles. In the FTP driving pattern, with the engine preheated during a 9-minute transient cycle (FTP-cold), followed by a 15-minute stabilized cycle (FTP-stab) and 10-minute stationary phase (FTP-hot), the production of oxides of nitrogen (NO_x) is gradually increasing due to an increase in combustion temperature. A higher emission of NO_x is also observed in the extended extra urban driving cycle (EUDC) with the speed going up to 120 km/hr in simulated highway driving. Because of an overall increase of the NO_x production, a higher rate of formation of 1-NP is observed in the last part of the driving cycle (Table 3).

2.2.2. Tobacco Smoke

Active and passive smoking is one of the most important sources of coexposure in inhalation exposure assessment. Therefore, it is important to verify if active

Table 1   Description of System and Conditions Used During the Analysis of Nitro-PAH

GC	Varian 3400 CX
MS	Varian Saturn 4D Ion Trap
Pre-column	5 m × 0.53 mm ID deactivated fused silica "retention gap"
Analytical column	30 m × 0.25 mm ID capillary fused silica DB-5MS column with film thickness of 0.25 μm (J & W Scientific, Folsom, CA)
Injection rate	2.0 μl/sec
Hot needle time	0.1 min
Injector temperature	200°C
Transfer line	290°C
Carrier gas	He at column head pressure of 14 psi
GC programme	Initial temperature 110°C, 10°C/min to 300°C
MS type	ion trap
MS settings	EI mode/EI mode/selective reaction monitoring/ nonresonant collision induced mode
Ionization energy	70 eV
Isolation of m/z	413, 422
Window	m/z 1
Excitation RF	120 amu
Excitation voltage	94 V (noncollision-induced dissociation)
Background mass	m/z 99
Mass range	m/z 180–430
Scan speed	0.6 sec/scan

smoking contributes to 1-NP exposure. So far, 1-NP has not been detected as a constituent of cigarette smoke condensate in MS analysis, presumably because the tobacco smoke mixture is a reductive mixture resulting in the appearance of amino compounds rather than their nitro analogs. It was not observed at 1 ng per cigarette by El-Bayoumy et al. [27], and not at a limit of detection of 10 pg on-column by Williams et al. [28]. The occurrence of 1-NP was verified in tar samples derived from some of the most popular cigarette types in the Netherlands. The precleaning of the tar extracts by silicon dioxide (SiO_2) solid-phase fractionation appeared insufficient for a sensitive analysis. To prevent loading of the analytical column with substantial organic debris, a GC-liner was placed before the analytical column (Alltech Applied Science Group, Emmen, The Netherlands), thus providing scavenging of most of the contamination. In Caballero Plain (13.1 mg tar/ cigarette), Camel Filter (12.8 mg tar/cigarette), Marlboro Filter (12.7 mg tar/cigarette), Marlboro Light (7.9 mg tar/cigarette), and Barclay

Table 2 1-NP Analyzed in Samples of Particulate Matter Collected from Diluted Tailpipe Exhaust of Diesel-Powered Passenger Vehicles Tested on a Chassis Dynamometer

Car	Odometer (km)	1-Nitropyrene (µg/g)		
		UDC[a]	UDC-hot[b]	EUDC[c]
Citroeh BX 1.9 TZD	188513	4.3	7.0	12.3
Citroen BX 1.9 TZD	168484	2.1	7.2	11.9
Peugeot 106	11514	0.3	2.9	15.5
Peugeot 106	27136	L1	5.1	11.1
Peugeot 205	131367	1.4	3.8	8.0
Renault Safrane	53480	4.5	5.1	7.7
Renault Safrane	25541	1.8	4.3	6.5
Opel Astra	5436	1.7	2.3	3.5
Opel Astra	4906	2.6	3.1	4.1
Opel Corsa	18095	2.2	3.1	6.5
Opel Corsa	32444	3.3	4.4	15.6
Mazda 626 CX	30111	22.7	17.6	63.0
Mazda 626 CX	31782	21.2	15.5	67.7
Mazda 626 CX	33344	19.8	31.0	143.4
Chrysler Voyager	96560	6.7	8.1	33.5
		FTP-cold[d]	FTP-Stab[e]	FTP-Hot[f]
Nissan Primera	41902	16.5	6.6	29.5
Nissan Primera	31418	20.3	33.6	41.4
Nissan Primera	11222	20.4[g]	10.4[g]	30.3[g]
Nissan Primera	11222	25.3[g]	15.4[g]	33.3[g]
Nissan Primera	66516	59.4	26.5	90.0
Mercedes C 200 D	16260	1.7	8.2	29.4
Mercedes C 200 D	40473	7.5	12.7	22.0
Mercedes C 200 D	8958	8.7	29.3	33.2

[a] UDC, urban driving cycle.
[b] UDC-hot, UDC hot start.
[c] EUDC, extra urban driving cycle (0–400 sec).
[d] FTP-cold, federal testing procedure—cold transition period (0–505 sec).
[e] FTP-stab, federal testing procedure–stabilized period (505–1372 sec).
[f] FTP-hot, federal test procedure–hot transition period (1372 -1877 sec).
[g] Collection and analysis of particulate matter in duplicate.

Table 3 Determination of the Content of 1-NP (pg/m^3) in Samples of Airborne Particulate Matter According to Two Different Nitro-Reduction Pretreatment Steps and Ion-Trap MS and High-Resolution MS (Both in EI Mode)

Reduction pretreatment System		On-line zinc Ion-trap MS Varian Saturn 1	NaSH Ion-trap MS Varian Saturn 1	NaSH High-resolution MS VG Autospec Q
Mode		EI	EI	EI
Ionisation energy (eV)		70	70	35
Scanning rate (scan/s)		1	1	3
Emission source[a]	Air volume (m^3)			
Outdoor ambient air:				
Background	427	ND[b]	ND	1.7
Lawn mower	215	ND	ND	6.6
Military vehicles	345	ND	ND	12
Passing traffic	420	ND	34	36
River boat	385	93	ND	31
Airport vehicles	260	ND	ND	42
Indoor workplace atmosphere:				
Trucks	340	110	86	80
Forklift trucks	465	150	88	71
Train engines	345	390	160	280
Forklift trucks	200	780	1,600	1,200

[a] Diesel-powered combustion source dominating the air quality.
[b] ND, not detected.

(4.4 mg tar/cigarette), 1-NP and 1-aminopyrene (1-AP) were not detected at a limit of determination of 30 pg per cigarette.

2.2. Ambient Air

For the determination of concentrations of 1-NP in ambient air, a GC–MS based method was developed. At first, an on-line zinc pretreatment was combined with GC–MS. The preparation of zinc columns resulted in a varying reduction performance. To overcome these problems, a simple one-tube reduction (using a 10% solution of sodium hydrosulfide (NaSH) in water for 30 minutes) was introduced, based on previous work by Hisumatsu et al. [29]. In Table 3, the two reduction steps are compared in the analysis of a series of ambient air samples involving emissions from different diesel-powered engines. A comparison was also made of the detection performance of the ion-trap MS and the high-resolution MS, both operated in electron ionization (EI) mode. It is apparent from these results that the outdoor samples contain small traces of 1-NP that require the more sensitive and/or selective high-resolution MS approach. In more recent work, this has been achieved using ion-trap tandem mass sepctrometry (MS–MS) [30].

In ambient air samples analyzed by GC–MS, different peaks of isomers of nitropyrenes and nitrofluoranthenes can be observed (Fig. 2). Some of these products are combustion-derived, while others primarily originate from atmospheric reactions. Some previously reported extremely high concentrations of 1-NP [31] might be partly explained by poor separation and identification of different isomers of 1-NP in high-performance liquid chromatography (HPLC) analysis equipped with fluorescence detection.

2.3. Human Breath

Human breath contains numerous volatile substances derived not only from endogenous metabolism but also from external exposure to vapors and gases or their metabolites. Because of their volatility, these substances are cleared through the respiratory system and may reflect the uptake of xenobiotics or endogenous production and their concentration in the circulation. Instantaneous partitioning of these substances between the blood perfusing the lungs and air ventilating the alveoli supplies accurate data on the systemic bioavailability and dose of the substance or metabolite in the blood circulation. From an analytical perspective, the analysis of contaminants from exhaled air is attractive because it provides the advantage of a relatively simple matrix compared with other biological samples, such as blood. Moreover, its collection is noninvasive, and both collection and laboratory treatment bears no known risk of infection to the laboratory personnel. For some substances, high correlations of alveolar air concentrations with blood and ambient air concentrations were found, such as for hexane [32].

In-the-workplace breath analysis is used for exposure monitoring. Some health authorities have introduced standards (biological exposure indices) for toxic substances in exhaled air [33,34]. Environmental exposures may also be studied by analysis of exhaled air [35,36]. Usually, only the end-exhaled air (end-tidal air) volume is collected for analysis, avoiding the mixing of this volume of air that has been in equilibrium with the bloodstream in the alveoli with so-called dead-space air. The air may be collected in an evacuated stainless steel ('single breath') canister [37] or in a glass pipette or another breath-sampling device [38] with a volume ranging from 50 to 1000 ml. Mixed exhaled air may be collected in a Tedlar bag. The air sample passes a cryogenic trap and is then analyzed on a gas chromatograph. Some investigators flush the air sample from the canister or pipette over a cartridge containing Tenax [39]. For this purpose, a disposable system has been introduced providing safe storage of analytes from human breath on Tenax in a stainless tube that can be sent to the laboratory by the end-user. The analytes are thermally desorbed prior to GC analysis using an automated

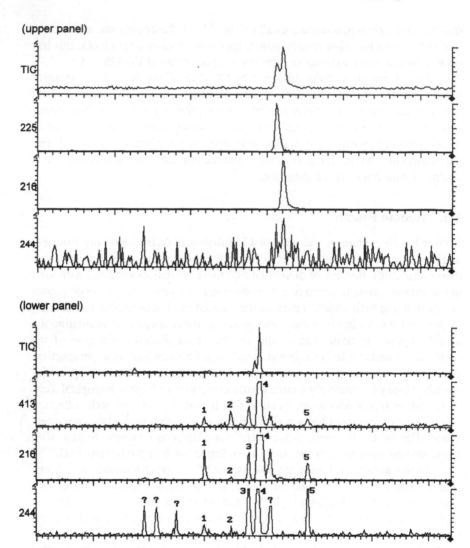

Figure 2 Chromatograms of GC–MS–MS analyses of emissions from a diesel engine (upper panel) and from an ambient air samples (lower panel) under similar conditions. The product ion m/z 216 corresponds to HFB-derivatized 1-AP, whereas the product ion m/z 225 corresponds to the derivatized deuterated internal standard D_9-1-AP (upper panel). The lower panel shows nine different products of which one was identified as 1-NP (4), presumably deriving from combustion sources. Four other peaks were tentatively identified (by cochromatography with a reference standard) as 3-nitrofluoranthene (1), 4-nitropyrene (2), 2-nitrofluoranthene (3), and 2-nitropyene (5), presumably primarily from atmospheric origin. The four remaining peaks (in the product ion trace m/z 244) did not coelute with available reference compounds.

thermal desorber or a microwave device [40]. For the analysis of human breath, a selected-ion flow tube technique was developed [41]. Positive ions are created in a microwave discharge ion source, containing an appropriate gas or gas mixture. Primary ions are selected from the current of ions, e.g., H_3O^+, NO^+, and O_2^+, using a quadrupole mass filter and sampled from the helium (He) carrier gas stream for mass analysis. The ions are then detected by a channeltron/pulse counting system. This results in a swift response of the instrument to reflect near to real-time changes in breath composition down to the 10-ppb level. Measurements of ammonia, isoprene, methanol, and acetone were obtained from active-smoking healthy subjects and from patients suffering from renal failure and diabetes [40].

A specific contribution of the application of MS in the field of breath analysis is based on the suggestion made by Kohlmuller and Kochen [42] regarding the inaccurate or unreliable reports of n-pentane concentrations in exhaled air. n-Pentane is quantitated in human breath to reflect, beside endogenous sources, the influence of exposure to oxidating and redox-cycling compounds from the environment. These compounds produce reactive oxygen species that are involved in lipid peroxidation, resulting in the release of small alkanes in exhaled air. The authors suggest that these previous reports of n-pentane in human breath are, at least in part, the result of erroneous identifications of isoprene that could not be separated from n-pentane on analytical GC columns commonly used at that time.

3. ANALYSIS OF URINARY METABOLITES

Xenobiotic substances that are absorbed into the body may be primarily excreted in feces or urine, depending mostly on the physicochemical properties and molecular mass of the substance or its metabolite. Glutathiocne-S-transferase-conjugated products such as mercapturic acids have been used for the assessment of exposure to alkylating agents [43–45]. Urine may also contain products derived from unstable DNA adducts, e.g., derived from N-7-alkylguanine and N-3-alkyladenine, which can be used as dosimeters for alkylating damage caused by electrophilic carcinogens. Urine represents a biological sample that can be more easily collected and analyzed than feces. The chemical substances may appear in the urine unchanged, but in most cases biotransformation leads to the formation of more polar products and/or conjugates. Often these products are not suitable for direct GC separation although an application has been reported with direct or split injection of urine into the GC at 150 to 250°C [46]. Conjugates usually need to be hydrolyzed, either chemically or enzymatically, e.g., by arylsulfatase and/ or β-glucuronidase. Urinary products may be detected and quantitated without extensive sample workup by HPLC equipped with UV or fluorescence detection.

Alkylpurines derived from DNA adducts may also be detected by HPLC equipped with fluorescence or electrochemical detection, and immunochemical methods. Mass spectrometry techniques are frequently used to obtain structural confirmation of the identity of a urinary product quantitated using GC–nitrosamine-selective detection [47], GC-FID [48,49], or immunoassays [50,51].

The homogeneity of the urine sample needs special attention. Thorough mixing of the sample after thawing from storage temperatures of - 20°C and reheating to the physiological temperature of 37°C before recovery of the analyte may be required because otherwise the analyte may become adsorbed to or enclosed in a protein or salt precipitate [52].

For initial recovery of the analytes from urine, liquid–liquid extraction may be used. Alternatively, ion exchange or immunoaffinity solid-phase or SPME extractions may be used, offering the possibility of enrichment of the analyte. If the solid sorbent is carefully selected and the cartridge is properly activated, loaded, and rinsed, it is possible to eliminate substantial quantities of the undesired products. For polynuclear aromatic compounds, Blue Cotton and Blue Rayon have been used successfully in the extraction of small quantities of planar molecules from large volumes of urine or other aqueous samples. The analytes of interest may be retrieved from the solid phase using a suitable eluent. Often the SPE step is laborious and may lead to substantial sample-to-sample variability, unless suitable internal standards are used. Therefore, in many applications ^{13}C or deuterated isotope analogs of the analyte are added as internal standards [52–55]. Next, the analytes of interest may be derivatized to improve volatility and the chromatographic behavior of the analytes during GC separation. Multiple fluorinated derivatization reagents may be used to improve selective MS detection by creating a mass of the analyte that can be detected at a higher m/z, with less interference from matrix components.

For the detection of the most prominent urinary metabolites of 1NP, GC–MS based methods have been developed [56]. Two different derivatization techniques have been compared (Figs. 3 and 4). The first approach was based on treatment of the metabolites with potassium *tert*-butoxide (KtBO) and iodomethane. 6- and 8-Hydroxy-*N*-acetyl-AP yielded one stable double methylated derivatization product. The methylation of 3-hydroxy-*N*-acetyl-AP was poor, only showing traces of a single methylated product. Probably, this is a result of steric hindrance, since the hydroxyl function is situated adjacent to the acetylamine

Figure 3 Spectra from GC–MS analyses in EI mode of some urinary metabolites of 1-NP following methylation with KtBO/iodomethane. Urine samples were obtained from a Sprague-Dawley rat administered a single intragastric dose of 1 mg of 1-NP in troctanoin per kg of bodyweight (From Ref. 56.)

Derivatization reagent KtBO/KCH₃I	Substance	Parent ion *m/z*	Daughter ion *m/z*
	1-AP	245	230
	x-OH-NP	277	247
	x-OH-AP	275	232
	x-OH-NAAP	303	248

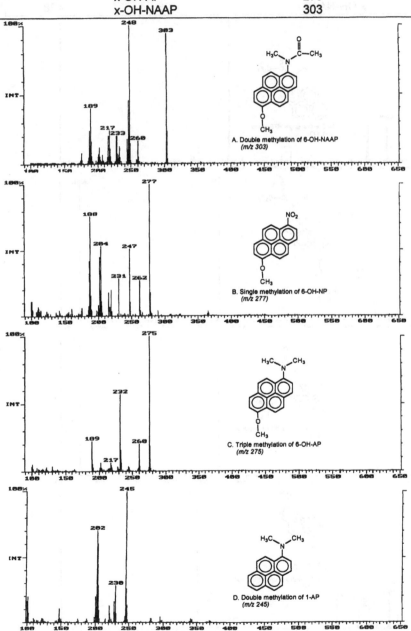

A. Double methylation of 6-OH-NAAP (m/z 303)

B. Single methylation of 6-OH-NP (m/z 277)

C. Triple methylation of 6-OH-AP (m/z 275)

D. Double methylation of 1-AP (m/z 245)

Derivatization reagent	Substance	Parent ion m/z	Daughter ion m/z
HFBI	1-AP	413	216
	x-OH-NP	459	429
	x-OH-AP	626	428
	x-OH-NAAP	626	428

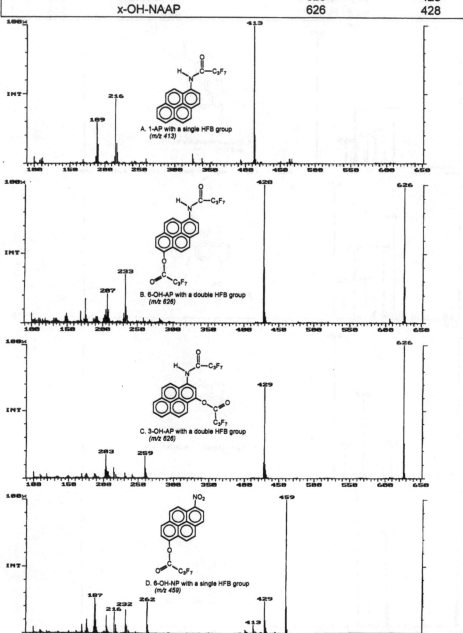

A. 1-AP with a single HFB group (m/z 413)

B. 6-OH-AP with a double HFB group (m/z 626)

C. 3-OH-AP with a double HFB group (m/z 626)

D. 6-OH-NP with a single HFB group (m/z 459)

Substance	[M]$^+$	[M-COC$_3$F$_7$-H]$^+$
6-OH-NAAP	626	428
8-OH-NAAP	626	428
d$_9$-6-OH-NAAP	635	436

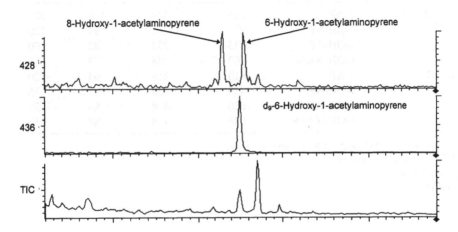

Figure 5 GC–MS–MS chromatogram of double HFB derivatized urinary metabolites derived from 1-nitropyrene from a worker exposed to diesel exhaust from trucks in an expedition department. The spot urine sample was collected on Friday directly after the shift. The exposure to 1-NP on preceding Tuesday was 737 pg/m^3 and on Friday was 201 pg/m^3 (time-weighted average exposure in the breathing zone). The subject is a smoker and was classified as a slow acetylator; the GSTM1 genotype is deficient (From: submitted paper).

function. 3-, 6-, and 8-Hydroxy-AP yielded a major single-methylated product, but also minor double- and triple-methylated products, each showing different fragmentation patterns during MS analysis in EI mode (Fig. 3, top three panels). With respect to the double-methylated product, it remained unclear whether the two methyl groups were both situated at the N function, or at the N and at the O function. Derivatization of 1-AP resulted in one product (not shown), double-derivatized at the N function (Fig. 3, bottom panel D).

Figure 4 Spectra from GC–MS analyses in EI mode of some urinary metabolites of 1-NP following derivatization with HFBI. Urine samples were obtained from a Sprague-Dawley rat administered a single intragastric dose of 1 mg of 1-NP in troctanoin per kg of bodyweight (From Ref. 56.)

Table 4 Optimised Detector Parameters for Sensitive GC–MS–MS Determinations of Urinary Metabolites of 1-NP Following Derivatization Using KtBO/Iodomethane or HFBI

Derivatization reagent	Metabolite	Precursor ion (m/z)	Product ion (m/z)	CID[a] (V)	RF[b] (m/z)
KtBO/ CH₃I	1-AP	245	230	ND[c]	ND
	x-OH-NP	277	247	94	120
	x-OH-AP	275	232	82	100
	x-OH-NAAP	303	248	77	135
HFBI	1-AP	413	216	94	120
	x-OH-NP	459	429	92	135
	x-OH-AP	626	428	88	135
	x-OH-NAAP	626	428	88	135

[a] Excitation amplitude applied for nonresonant CID.
[b] Excitation radiofrequency.
[c] ND, not determined.
Source: Ref. 56.

The second approach was based on the substitution of a heptafluorobutyric group to the amine function of 1-AP, using heptafluorobutyric imidazole (HFBI) at 50°C for 60 minutes. Heptafluorobutyric anhydride (HFBA) has also been applied successfully, but much higher limits of determination were encountered [26]. For some metabolites such as 6- and 8-hydroxy-1-AP, double-derivatized products were retrieved. In N-acetyl-1-AP and hydroxylated N-acetyl-1-APs, the acetyl group was removed and substituted by the derivatization reagent, leading to the formation of the same products as in the derivatization of 1-AP and the hydroxylated 1-APs (Figs. 4 and 5).

Table 4 shows the experimental parameters, optimized for the sensitive detection by GC–MS–MS of the metabolites derivatized with KtBO/iodomethane or HFBI. Optimal conditions were defined as fragmentation to an optimal amount of the product ion without losing the precursor ion completely. Under these optimized conditions, detection by GC–MS–MS is the most sensitive approach for all metabolites of 1-NP, except for the hydroxy-1-NPs that are detected more sensitively after reduction and subsequent fluorescence detection (Table 5). Derivatization with HFBI is the preferred pretreatment for sensitive determination of all of the acetylated and nonacetylated aminometabolites of 1-NP by MS.

Because of the deacetylation and substitution by HFBI, hydroxy-1-NPs and hydroxy-N-acetyl-APs will be detected as hydroxy-APs, whereas 1-NP and N-acetyl-AP will be detected as 1-AP. This conversion of metabolites to a limited set of analytes will facilitate the detection of 1-NP from human subjects with

Table 5 Limits of Determination Based on a Signal-to-Noise Ratio of 3

	Limit of determination (fmol on-column)[b]		
Metabolite	HPLC-fluorescence	Methylation and GC–MS–MS	HFBI and GC–MS–MS
3-OH-NAAP	149.1	50.1	9.2
6-OH-NAAP	151.0	13.5	1.9
8-OH-NAAP	145.2	11.6	1.8
3-OH-AP	—[a]	—[b]	5.1
6-OH-AP	—[a]	—[b]	4.0
8-OH-AP	—[a]	—[b]	3.9
3-OH-NP	16.8	16.8	101.1
6-OH-NP	19.7	48.0	78.2
8-OH-NP	22.3	30.6	66.8
1-AP	11.1	—[c]	2.9
1-NP	11.4	—[d]	—[d]

[a] Instable.
[b] Several products following derivatization, intra-day standard deviation >20%.
[c] Not determined.
[d] Not detected.
Source: Ref. 56.

low exposure to 1-NP as occurring in diesel exhaust exposure. However, information is lost on the relative excretion of specific metabolites. Inorganic or organic metals may be determined using inductively coupled plasma MS [57].

4. ANALYSIS OF DNA AND PROTEIN ADDUCTS

Once systemically available in the body, toxic substances or their reactive intermediates may interact with macromolecules such as deoxyribonucleic acid (DNA), ribonucleic acid (RNA), and proteins. Some of these substances or their electrophilic metabolites are capable of covalent binding to nucleophilic sites in endogenous macromolecules. Others may be noncovalently associated with plasma proteins and be released in target organs. This may lead to tissue injury. In some cases, the binding to DNA causes mutations (different expression of (onco)genes) or chromosomal changes (micronuclei, chromosomal aberrations, sister chromatid exchanges). This may be a first change in a multistep process eventually leading to the conversion of a normal cell into a tumor cell. Most adducts of carcinogens have not been measured in target tissue but in sites where no tumors will occur, such as in lymphocytes. These non–target-site adducts

Table 6 Haemoglobin Adducts Positively Identified by GC–MS in Humans

Agent	Metabolite/substance	Amino acid	Cleavage reagent	Analyte	Detector	Ref.
Acetaldehyde		val[a]				
4-Aminobiphenyl	N-hydroxy-4-aminobiphenyl	cys		4-aminobiphenyl		i
Benzene	Benzene oxide	cys	tfaa/msa	?	NCI	53, ii
Benzo[a]pyrene	benzo[a]pyrenediolepoxide	asp/glu		Benzo[a]pyrene-tetrols	NCI	iii
1,2- and 1,4-Benzoquinone		cys[b]	tfaa/msa	O,O',S-tris-trifluoro acetylhydroquinone	NCI	iv, v
1,3-Butadiene		val[a]				
Ethylene oxide	ethylene oxide	val[a]	pentafluorophynyl thiohydantoin		NCI	vi, vii
Hexahydrophthalic anhydride		lys	acid/Pronase E	pfbb	NCI	viii, ix
Isoprene	2-ethenyl-2-methoxirane 2-(1'-methylethenyl)oxirane	val[a] val[a]	mod. Edman degr.	(95%) (5%)		x
4-(methylnirosamino-1-(3-pyridyl)-butanone	4-Hydroxy-1-(3-pyridyl)-1-butanone		alkaline hydrolysis		ECMS	xi
N'-nitrosonornicotine	4-Hydroxy-1-(3-pyridyl)-1-butanone		alkaline hydrolysis		ECMS	v
N,N-dimethylformamide	N-methyl-carbamoylated	val[a]				xii
1-Nitropyrene	1-nitrosopyrene	cys	alkaline hydrolysis	hfb-1-aminopyrene	MS-MS; NCI	xiii, xiv
2-Nitrofluorene	2-nitrosofluorene	cys	alkaline hydrolysis	hfb-2-aminofluorene	NCI	xiv

3-nitrofluoranthene		cys	alkaline hydrolysis	hfb-3-aminofluoranthene	NCI	xiv
9-nitrophenanthrene		cys	alkaline hydrolysis	hfb-9-aminophenanthrene	NCI	xiv
6-nitrochrysene		cys	alkaline hydrolysis	hfb-6-aminochrysene	NCI	xiv
Propanil	3,4-dichloroaniline		alkaline hydrolysis	3,4-dichloroaniline	NCI	xv
2,4-Toluenediisocyanate		val[a]/lys				xvi
Styrene	styrene-7,8-oxide	cys	Raney nickel	1- and 2-phenylethanol	NCI	xvii

Abbreviations: tfaa = trifluoroacetic anhydride, msa = methanesulfonic acid; hfb = hexafluorobutyric acid; ECMS = electron capture MS, pfbb = pentafluorobenzyl bromide.

[a] N-terminal valine.

[b] Commerically available human Hb.

i. K. Yeowell-O'Connell, N. Rothman, M.T. Smith, R.B. Hayes, G. Li. S. Waidyanatha, M. Dosemeci, L. Zhang, S. Yin, N. Titenko-Holland and S.M. Rappaport, Carcinogenesis, 19 (1998) 1565–1571.

ii. S. Tas, J.P. Buchet, R. Lauwerys, Int. Arch. Occup. Environ. Health, 66 (1994) 343–138.

iii. S. Waidyanatha, K. Yeowell-O'Connell, S.M. Rappaport, Chem. Biol. Interact., 115 (1998) 117–139.

iv. K.P. Braun, J.G. Pavlovich, D.R. Jones, C.M. Peterson, Alcohol Clin. Exp. Res., 21 (1997) 40–43.

v. S. Osterman-Golkar and J.A. Bond, Environ. Health Perspect. 104 (1996) 907–915.

vi. P.B. Farmer, E. Bailey, S.M. Gorf, M. Törnqvist, G. Osterman-Golkar, A. Kautiainen, D.P. Lewis-Enright, Carcinogenesis, 7, 637–640.

vii. E. Bailey, A.G.F. Brooks, C.T. Dollery, P.B. Farmer, B.J. Passingham, MA. Sleightholm, D.W. Yates, Arch. Toxicol. 62, 247–253.

viii. C.H. Lindh and B.A. Jonsson Toxicol. Appl. Pharmacol. 153 (1998) 152–160

ix. C.H. Lindh and B.A. Jonsson, J. Chromatogr. Biomed. Sci. Appl., 710 (1998) 81–90.

x. E. Tareke, B.T. Golding, R.D. Small, M. Törnqvist, Xenobiotica, 28 (1998) 663–672.

xi. S.E. Atawodi, S. Lea, F. Nyberg, A. Mukeria, V. Constatinescu, W. Ahrens, I. Bureske-Hohfeld, C. Fortes, P. Boffetta, M.D. Frisen, Cancer Epidemiol. Biomarkers Prev., 7 (1998) 817–821.

xii. J. Angerer, T. Goen, A. Kramer, H.U. Kafferlein, Arch. Toxicol., 72 (1998) 309–313.

xiii. Y.M. van Bekkum, P.T.J. Scheepers, P.H.H. van den Broek, D.D. Velders, J. Noordhoek, R.P. Bos, J. Chromatogr. B, 701 (1997) 19–26.

xiv. Zwirner-Bayer and H.-G. Neumann, Mutat. Res. 441 (1999) 135–144.

xv. R. Pastorelli, G. Catenacci, M. Guianci, R. Fanelli, E. Valoti, C. Minoia, L. Airoldi, Biomarkers 3 (1998) 227–233.

xvi. D. Schütze, O. Sepai, J. Lewalter, L. Miksche, D. Henschler, G. Sabbioni, Carcinogenesis, 16 (1995) 572–582.

xvii. S. Fustinoni, C. Colosio, A. Colombi, L. Lastrucci, K. Yeowell-O'Conell, S.M. Rappaport, Int. Arch. Occup. Environ. Health 71 (1998) 35–41.

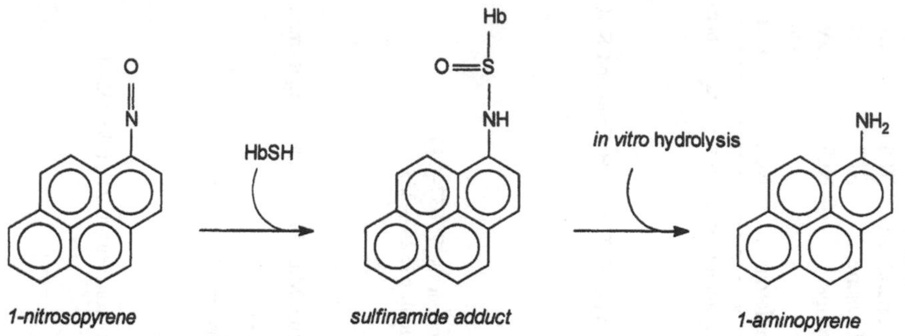

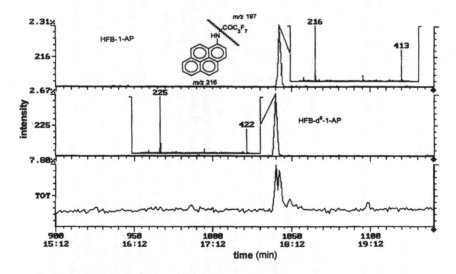

Figure 6 Proposed mechanism of Hb adduct formation by 1-NP, and decomposition of the adduct by acid or basic hydrolysis during pretreatment of the sample prior to GC–MS analysis.

Substance	[M]$^+$	[M-COC$_3$F$_7$]$^+$	[M-CNHCOC$_3$F$_7$]$^+$
HFB-1-AP	413	216	189
HFB-d$_9$-1-AP	422	225	197

Figure 7 GC–MS–MS mass chromatogram of the product ions of heptafluorobutyryl-1-aminopyrene (HFB-1-AP) at m/z 216 (top) and HFB-D$_9$-AP at m/z 225 (middle) with mass spectrum, and total ion chromatogram (bottom) of hydrolyzed pooled Hb of rats exposed to a single dose of 1 mg or 10 mg of 1-nitropyrene in trioctanoin per kg body weight by gavage. Description of system and conditions in Table 2. (From Ref. xiii in Table 6.)

are good measures of the internal dose of the active genotoxic compound. In experimental animal models, target-site DNA adduct levels have been found to correlate with the appearance of tumors for four different classes of carcinogens: N-nitrosoamines, aflatoxines, aromatic amines, and PAHs [58,59]. However, a relationship between the DNA adduct level at a non–target site with the adduct level in target tissue must be established before DNA adduct levels can be used as a basis of establishing risk estimates [60]. Deoxyribunucleic acid adducts may be removed and return the nucleic acids to their original state by cell turnover or by enzymatic repair.

In contrast to DNA adducts, adducted proteins are not known to be subject to enzymatic repair. Their life span may approach the life span of the unadducted protein (in humans: 120 days for hemoglobin and 20 days for albumin). Long-term exposure to low concentrations of an adduct-forming substance may be reflected in an accumulated protein adduct level. Nucleophilic sites such as amine, thiol, and carboxylic functional group are available in different amino acids such as N-terminal valine, cysteine, histidine, aspartic acid, and glutamine, for which adducts have been identified in vivo. For PAHs and aromatic amines, linear dose–response relationships have been established between the dose administered in animal models and hemoglobin (Hb) or albumin adduct levels [61]. On the other hand, a high correlation was observed for some aromatic amines between levels of Hb adducts and DNA adducts in animal models [62]. Bladder cancer patients have been found to a have an elevated 4-aminobiphenyl (4-ABP) adduct level [63], and the 4-ABP–Hb adduct level correlated with the 4-ABP–DNA adduct level in the target tissue (bladder epithelium) [64].

Adducts of DNA, RNA, and proteins such as Hb and albumin have been characterized with LC–MS and GC–MS. Table 6 presents an overview of Hb adducts from humans mostly analyzed by LC–MS or GC–NCI-MS. Usually these determinations help to understand the mechanism of toxicity and may be used for the assessment of uptake and systemic availability of toxic substances in exposure monitoring using biomarkers of exposure. Efforts have been made to use these biomarkers also in risk calculations [65].

The formation of protein and DNA adducts may be studied in vitro by incubation of the macromolecule with a reactive intermediate or with the parent compound in the presence of a metabolizing system, or in vivo in experimental animals. In most studies, radiolabels are used that may be analyzed by liquid scintillation counting but more recently also by a specific MS-based technique (see section 5 of this chapter).

For structure characterization and quantification, adducts to Hb may be cleaved by alkaline or acid hydrolysis (Fig. 6), by enzymatic hydrolysis, e.g., Pronase E, or by Raney nickel. The cleavage products may be separated from the debris by liquid–liquid chromatography, solid-phase extraction, gel filtration, ion-exchange chromatography, or affinity chromatography. In most cases, the

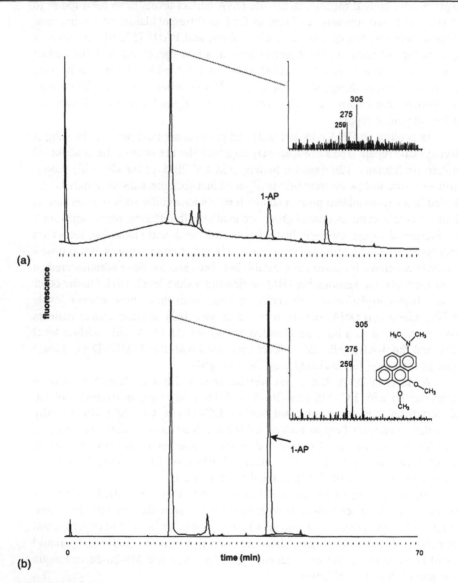

Substance	[M]+	[M-CH3]+	[M-2CH3]+
Me-1-NP-4,5-DHD	305	275	259

Figure 8 Characterization by GC–MS in CI mode (using methane as a reaction gas) of cleavage product from plasma proteins recovered from Sprague-Dawley rats following administration of 1 mg 1-NP per kg of body weight by gavage (a) and of synthesized and subsequently reduced and methylated 1-NP-4,5-dihydrodiol (b). The samples for MS characterization were obtained from HPLC separation monitored by fluorescence detection. The appearance of the peak of 1-AP is indicated. Description of system and conditions in Table 2.

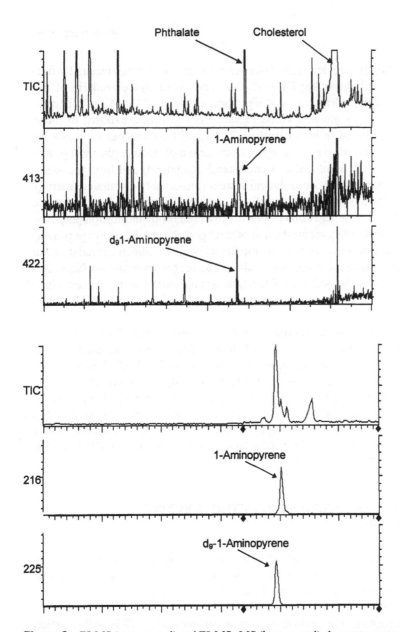

Figure 9 EI-MS (upper panel) and EI-MS–MS (lower panel) chromatograms of a plasma protein hydrolysate from a nonsmoking worker exposed to diesel exhaust in an indoor facility with running truck engines. The blood sample was collected on a Thursday morning. The preceding exposure to 1-NP was 510 pg/m^3 on Monday and 1906 pg/m^3 on Tuesday (time-weighted average exposure during 8 h working period). The subject was classified as having a slow acetylator and slow CYP2A1 phenotype and being GSTM1 deficient. Shown are the total ion chromatogram (top), product ion trace of the HFB derivatized 1-aminopyrene (middle) and of the HFB derivatized internal standard (bottom). In the MS–MS mode, it was possible to quantify the adduct content as 421 fg 1-NP per mg of plasma protein.

cleavage product is then derivatized in a separate step following cleanup by fractionation. Figure 7 shows the GC–MS–MS analysis of hydrolyzed Hb of rats
exposed to 1-NP. In order to assess the performance of the MS–MS system compared with MS in EI mode, a blood sample from a worker exposed to diesel
exhaust was compared in both detectors. The EI-MS chromatogram shows the
appearance of numerous peaks emerging from the sample matrix (the most prominent peak tentatively identified as a cholesterol compound) and from the use of
plastics in the pretreatment or analytical system (presumably a phthalate analog),
whereas the MS–MS chromatogram shows a much cleaner total ion current and
product ion traces with peaks that can be readily used for quantification purposes.

Some investigators report the use of a reagent during the cleavage process
in a combined procedure such as in (modified) N-alkyl Edman degradation for
analysis of adducts to the N-terminal valine. The reagent pentafluorophenyl iso-
thiocyanate cleaves the adducted N-terminal amino acid from the protein chain
as a substituted pentafluorophenyl thiohydantoin, which is subjected to MS analysis [66].

Figure 8 shows a plasma (presumable albumin) adduct of a metabolite of
1-NP, tentatively identified as 1-AP-4,5-dihydrodiol after methylation and the
GC–MS analysis of a product isolated from rats exposed to 1-NP. It is compared
with an identically treated synthetic standard (1-NP-4,5-dihydrodiol).

In the analysis of traces of protein adducts from human blood samples,
MS–MS offers great sensitivity over EI-MS, equaling the sensitivity of NCI-MS
for the analysis of Hb adducts of arylamines (Fig. 9). In the next section, future
developments in the use of MS for adduct analysis in the reconstruction of historical exposure are further described.

5. FUTURE PERSPECTIVES IN RETROSPECTIVE DOSIMETRY

In the past two decades, much effort was invested to detect and identify binding
products of environmental or workplace contaminants with macromolecules such
as DNA, RNA, and blood proteins. In this field of applications, MS-based approaches are competing with immunochemical approaches. For chemical structure characterization and positive identification purposes, MS is still superior,
but in some cases MS detection is lacking the sensitivity of immunoassays or
radiolabeling.

For the study of the formation of adducts to macromolecules in animals
and in humans, accelerator mass spectrometry (AMS) has been introduced in the
field of biomedicine. It is based on the use of a tandem Van de Graaff accelerator
and was developed 20 years ago to measure very small quantities of rare isotopes,
such as in ^{14}C dating studies, e.g., for the Shroud of Turin. This technique is now

used to measure aluminium, calcium, and carbon isotopes down to 10^{-21} to 10^{-18} mole, which is several orders of magnitude more sensitive than liquid scintillation counting. This technique could be useful to study absorption, distribution, metabolism, and excretion of xenobiotics in human volunteers after exposing them to μCi quantities of radio isotope-labeled chemical substances, without using experimental animals [67]. An important limitation of the application of AMS in toxicological studies is the lack of structural information from AMS readings. The toxicokinetics and adduct formation of heterocyclic aromatic amines and aflatoxin B_1 have been studied using AMS. This technique may also be used to develop biomarkers that can be used in retrospective dosimetry of occupational or environmental exposure to radiation [68], e.g., to study dose–response relationships of thyroid cancer in children after the Chernobyl reactor accident by measurement of the stable isotope ^{129}I. It could also play a role to investigate a systematic discrepancy in the neutron dosimetry of neutron irradiation in Hiroshima atomic bomb survivors.

Fast-atom bombardment and electrospray MS and MS–MS have been used (in conjunction with NMR) to characterize synthesized Hb adducts of ethylene oxide and acrylonitrile with the N-terminal tripeptide of the α-chain of human globin for use as internal standards in the MS analysis of Hb adducts of these electrophilic compounds in humans [69]. Micro-LC–electrospray-MS–MS was used to detect Hb adducts and products of unstable DNA adducts (alkylpurines) in the skin and urine of an individual after exposure to the chemical warfare agent sulfur mustard. It is anticipated that electrospray–time-of-flight and matrix-assisted laser desorption/ionization–time-of-flight MS could be powerful tools in the screening for adducts and characterization of these adducts in tryptic digest fragments of proteins containing adducts of xenobiotic origin and comparison to unadducted protein fragments [70]. Because of the physicochemical characteristics of these fragments, LC will be used more instead of GC in conjunction with MS.

Detection of DNA adducts by MS is matching the sensitivity demonstrated in immunochemical assays [71]. Although very sensitive, these immunochemical approaches lack the structural information needed to understand the mechanism of adduct formation and identification of the adducting xenobiotic agent or metabolite. The sensitivity of current MS, however, does not meet the sensitivity of the ^{32}P-postlabeling technique, reaching down to 1 adduct per 10^9 to 10^{11} unadducted DNA nucleotides. Current efforts to improve the sensitivity of MS detection focus on the selective recovery of adducted nucleotides or protein fragments from an overwhelming surplus of unmodified nucleotides or protein fragments in the analytical system. It will be a great challenge to try to match this sensitivity by use of MS, the most promising improvements coming from improved isolation and preconcentration of the analyte and analysis with electrospray MS or electrospray MS–MS hyphenated to capillary-LC, nano-LC, or capillary zone electrophoresis [72–74]. Using these novel approaches, DNA adducts can be analyzed at the nucleoside or even at the nucleotide level These MS-based techniques are expected to provide the structural information that is lacking in the ^{32}P post-labeling approach.

ACKNOWLEDGMENTS

The authors are grateful to Y. M. van Bekkum, M. H. J. Martens, D. D. Velders, and P. H. H. van den Broek for conducting MS analysis presented in this contribution.

REFERENCES

1. J.D. Pleil and M.L. Stroupe, J. Chromatogr. A, 676 (1994) 399– 408.
2. R. Otson and X.L. Cao, J. Chromatogr. A, 802 (1998) 307–314.
3. E. Giller and H.D. Gesser, Environ. Int. 21 (1995) 839–844.
4. M. Chai and J. Pawliszyn, Environ. Sci. Technol., 29 (1995) 693–701.
5. J.-O. Levin, K. Andersson, R. Lindahl, and C.-A. Nilsson, Anal. Chem., 57 (1985) 1032–1035.
6. J.-O. Levin, K. Andersson, I. Fangmark, and C. Hallgren, Appl. Ind. Hyg., 4 (1989) 98–100.
7. H. Tinnerberg, M. Dalene, and G. Skarping, Am. Ind. Hyg. Assoc. J., 58 (1997) 229–235.
8. E.M. Ward, G. Sabbioni, D.G. DeBord, A.W. Teass, K.K. Brown, G.G. Talaska, D.R. Roberts, A.M. Ruder, and R.P. Streicher, J. Natl. Cancer Inst., 88 (1996) 1046–1052.
9. M. Kramer, K. Fox, A. Fox, A. Saraf, and L. Larsson, Am. Ind. Hyg. Assoc. J., 59 1998) 524–531.
10. S. Lappalainen, M Nikulin, S. Berg, P. Parikka, E.L. Hintikka, and A.L. Pasanen, Atmos. Environ. 30 (1996) 3059–3065.
11. P.A. Smith, D.R. Gardner, D.B. Drown, G. Downs, W.W. Jederberg, and K. Still, Am. Ind. Hyg. Assoc. J., 58 (1997) 868–875.
12. M. Dalene, G. Skarping, and P. Lind, Am. Ind. Hyg. Assoc. J., 58 (1997) 587–591.
13. G. Skarping, M. Dalene, and M. Littorin, Int. Arch. Occup. Environ. Health, 67 (1995) 73–77.
14. D. Schutze, O. Sepai, J. Lewalter, L. Mische, D. Henschler, and G. Sabbioni, Carcinogenesis 16 (1995) 573–582.
15. H.A. Dirven, P.H. van den Broek, and F.J. Jongeneelen, Int. Arch. Occup. Environ. Health, 64 (1993) 555–560.
16. U. Knecht, U. Bolm-Audorff, and H.J. Woitowitz, Br. J. Ind. Med. 46 (1989) 479–482.
17. Th. Petry, P. Schmid, and Ch. Schlatter, Ann. Occup. Hyg, 38 (1994) 741–752.
18. A. Dyremark, R. Westerholm, E. Overvik, and J.A. Gustavsson, Atmos. Environ. 29 (1995)1553–1558.

19. P.T.J. Scheepers, M.H.J. Martens, D.D. Velders, P. Fijneman, M. van Kerkhoven, J. Noordhoek, and R.P. Bos, Environ. Mol . Mutagen., 25 (1995) 134–147.

20. J.L. Mumford, J. Lewtas, K. Williams, W.G. Tucker, and and G.W. Traynor, J. Toxicol. Environ. Health, 36 (1992) 151–159

21. R.L. Falconer, T.F. Bidleman, and W.E. Cotham, Environ. Sci. Technol. 29 (1995) 1666–1673.

22. G. Barrefors and G. Petersson, J. Chromatorgr., 643 (1993) 71–76.

23. R.D. Butcher, B.A. Goodman, N. Deighton, and W.H. Smith, Clin. Exp. Allergy, 25 (1995) 985–992.

24. R. Atkinson, J. Arey, B. Zielinska, and A.M. Winer, in M. Cooke, K. Loening, and J. Merritt, eds., Polycyclic aromatic hydrocarbons: measurement, means and metabolism, 1991, Battelle, Columbus, pp. 69–88.

25. B. Zielinska, J. Arey, and R. Atkinson, in P.C. Howard, S.S. Hecht, and F.A. Beland, eds., Nitroarenes, 1986, Plenum, New York, pp. 73–84.

26. P.T.J. Scheepers, D.D. Velders, M.H.J. Martens, J. Noordhoek, and R.P. Bos, J. Chromatogr. 677 (1994) 107–121.

27. K. El-Bayoumy, M. O'Donnell, S.S. Hecht, and D. Hoffmann, Carcinogenesis, 6 (1995) 505–507.

28. R. Williams, C. Sparacino, B. Petersen, J. Bumgarner, R.H. Jungers, and J. Lewtas, Int. J. Environ. Anal. Chem., 26 (1986) 27–49.

29. Y. Hisumatsu, N. Nishimura, K. Tanabe, and H. Matushita, Mutat .Res., 172 (1986) 19–27.

30. Y.M. van Bekkum, P.T.J. Scheepers, P.H.H. van den Broek, R.B.M. Anzion, and R.P. Bos, Cancer Epidemiol, Biomarkers Prev. (in press).

31. IARC Monographs on the evaluation of carcinogenic risks to humans, vol. 46, 1989, IARC, Lyon.

32. F. Brugnone, G. Maranelli, L. Romeo, C. Giuliari, M. Gobbi, F. Malesani, G. Bassi, and C. Alexopoulos, Int. Arch. Occup. Environ. Health, 63 (1991) 157–160.

33. ACGIH, TLVs and BEIs, Threshold Limit Values for Chemical Substances and Physical Agents. Biological Exposure Indices, ACGIH, 1999, Cincinnati, OH, 184 p.

34. Deutsche Forschungsgemeinschaft, List of MAK and BAT Values, 1993, VCH, Weinheim, Germany.

35. L.A. Wallace, E.D. Pellizzari, T.D. Hartwell, V. Davis, L.C. Michael, and R.W. Whitmore, Environ. Res., 50 (1989) 37–55.

36. C.H. Pierce, R.L. Dills, T.A. Lewandeowski, M.S. Morgan, M.A. Wessels, D.D. Shen, and D.A. Kalman, J. Occup. Health, 39, 130-137.

37. J.D. Pleil and A.B. Lindstrom, Clin. Chem., 43 (1997) 723–730.

38. D. Dyne, J. Cocker, and H.K. Wilson, Sci. Total Environ., 20 (1997) 83–89.

39. L.A. Wallace, Toxicol. Ind. Health, 7 (1991) 203–208.

40. K. Riedel, T. Rupper, C. Conze, G. Scherer, and F. Adlkofer, J. Chromatogr. A, 719 (1996) 383–389.

41. D. Smith and P. Spanel, Rapid. Commun. Mass Spectrom., 10 (1996) 1183–1198.

42. D. Kohlmuller, and W. Kochen, Anal. Biochem., 210 (1993) 268–276.

43. B.M. de Rooij, P.J. Boogaard, J.N. Commandeur, N.J. van Sittert, and N.P. Vermeulen, Occup. Environ. Med., 54 (1997) 653–661.

44. D.E. Shuker, V. Prevost, M.D. Friesen, D. Lin, H. Oshima, and H. Bartsch, Environ. Health Perspect., 99 (1993) 33–37.

45. K. Norpoth, and G. Muller, Z. Gesamte Hyg., 35 (1989) 487–490.
46. A.C. Lareo, A. Perico, P. Bavazzano, C. Soave, and L. Perbellini, Int.. Arch. Occup. Environ. Health, 67 (1995) 41–46.
47. W.D. Parsons, S.G. Carmella, S. Akerkar, L.E. Bonilla, and S.S. Hecht, Cancer Epidemiol. Biomarkers Prev. 7 (1998) 257–260.
48. G. Bieniek, Scand. J. Work Environ. Health, 23 (1997) 414–420.
49. A. Bartczak, S.A. Kline, R. Yu, C.P. Weisel, B.D. Goldstein, G. Witz, and W.E. Bechtold, J. Toxicol. Environ. Health, 42 (1994) 245–258.
50. A.D. Lucas, A.D. Jones, M.H. Goodrow, S.G. Saiz, C. Blewett, J.N. Sieber, and B.D. Hammock, Chem. Res. Toxicol., 6 (1993) 107–116.
51. L.O. Henderson, M.K. Powell, W.H. Hannon, B.B. Miller, M.L. Martin, R.L. Hanzlick, D. Vroon, and W.R. Sexson, J. Anal. Toxicol., 17 (1993) 42–47.
52. J.R. Ormand, D.A. McNett, and M.J. Barels, J. Anal. Toxicol., 23 (1999) 35–40.
53. D.B. Barr, and D.L. Ashley, J. Anal. Toxicol. 22 (1998) 96–104.
54. F.F. Hsu, V. Lakshmi, N. Rothman, V.K. Bhatnager, R.B. Hayes, R. Kashyap, D.J. Parikh, S.K. Kashyap, J. Turk, T. Zenser, and B. Davis, Anal. Biochem., 234 (1996) 183–189.
55. R.H. Hill, Jr, D.B. Shealy, S.L. Heal, C.C. Williams, S.L. Bailey, M. Gregg, S.E. Baker, and L.L. Needham., J. Anal. Toxicol., 19 (1995) 323–329.
56. Y.M. van Bekkum, P.H.H. van den Broek, P.T.J. Scheepers, and R.P. Bos, Chem. Res. Toxicol., 11 (1998) 1382–1390.
57. Y.M. Hsue, TY.L. Huang, C.C. Huang, W.L. Wu, H.M. Chen, M.H. Yang, L.C. Lue, and C.J. Chen, J. Toxicol. Environ. Health Part A, 54 (1998) 431–444.
58. M.C. Poirier and F.A. Beland, Chem. Res. Toxicol., 5 (1992) 749-755.
59. F.A. Beland and M.C. Poirier, Environ. Health Perspect. 99 (1993) 5–10.
60. P.B. Farmer, Clin. Chem. 40 (1994) 1438–1443.
61. N.J. Gorelick, D.A. Hutchins, S.R. Tannenbaum, and G.N. Wogan, Carcinogenesis, 10 (1989) 1579–1587.
62. H.-G. Neumann, IARC Sci. Publ., 59 (1984) 115–126.
63. P. Del Santo, G. Moneti, M. Salvadori, C. Saltutti, A. Delle Rose, and P. Dolara, Cancer Lett. 60 (1991) 245–251.
64. G. Talaska, M. Schamer, P. Skipper, S. Tannenbaum, N. Caporaso, L. Unruh, F.F. Kadlubar, H. Bartsch, C. Malaveille, and P. Vineis, Cancer Epidemiol. Biomarkers Prev., 1 (1994) 61–66.
65. M. Törnqvist, D. Segerback, and L. Ehrenberg, in R.C. Garner, P.B. Farmer, G.T. Steel, and A.S. Wright, eds., Human carcinogen exposure. Biomonitoring and risk assessment, 1991, IRC Press, Oxford, pp 141–155.
66. M. Törnqvist, in J. Everse, R. M. Winslow, and K.D. Vandegriff, eds., Methods in Enzymology, Vol 231, Part B (1994), Academic Press, San Diego, CA, pp. 650–657.
67. J. Barker and R.C. Garner (1999) Rapid Commun. Mass Spectrom., 13 (1999) 285–293.
68. T. Straume, L.R. Anspaugh, E.H. Haskell, J.N. Lucas, A.A. Machetti, I.A. Kikhtarev, V.V. Chumak, A.A. Romayukha, V.T. Khrouch, Y.I. Gavrilin, and V.F. Minenko, Stem Cells, 15 (1997) 183–193.
69. R.M. Lawrence, G.M.A. Sweetman, R. Tavares, and P.B. Farmer, Teratogen. Carcinogen. Mutagen. 16 (1996)139–148.

70. P.T.J. Scheepers, Trends Anal. Chem. 16 (1997) V-VI.
71. D.H. Phillips in C.S. Cooper and P.L. Grover, eds., Chemical carcinogenesis and mutagenesis. I. Handbook of experimental pharmacology, 1990, Springer, London, pp. 503–546.
72. K. Vanhoutte, P. Joos, F. Lemière, W. van Dongen, and E.L. Esmans J. Mass Spectrom. Rapid Commun. Mass Spectrom. (1995) S143.
73. K. Vanhoutte, W. van Dongen, I. Hes, F. Lemière, E.L. Esmans, H. van Onckelen, E. van Eeckhout, R.E.J. van Soest, and A.J. Hudson Anal. Chem. 96 (1997) 3161.
74. D.L. Deforce, F.P.K. Ryiers, E.G. van den Eeckhout, F. Lemière, and E.L. Esmans Anal. Chem. 68 (1996) 3575.

8

Application of Solid-Phase Microextraction–Gas Chromatography–Mass Spectrometry in Quantitative Bioanalysis

Paola Manini and Roberta Andreoli
University of Parma, Parma, Italy

1. INTRODUCTION

Solid-phase microextraction (SPME) is a relatively new solvent-free sampling technique. It was developed by Pawliszyn and coworkers at the University of Waterloo (Ontario, Canada) to address the need to facilitate rapid sample preparation for gas chromatography (GC) analysis [1]. The first attempts used optical fibers coated with several types of polymeric films to absorb test analytes contained in aqueous samples [2]. Desorption of the fibers required the opening of the GC injector leading to a pressure loss at the head of the column. Owing to the rapid development of the technique, coated fibers were incorporated into a Hamilton™ microsyringe, giving the first SPME device [3], that allowed sample injection similarly to standard syringe injection. The commercial SPME device now in use is shown in Figure 1. The SPME fiber consists of a fused silica thin solid rod (1 cm long, 0.11 μm OD) coated with an absorbent polymer such as polydimethylsiloxane (PDMS), polyacrylate, or other materials attached to a metal rod. Sampling of organic analytes could be performed from liquid aqueous samples placed in vials sealed with septa by both direct immersion of the fiber and headspace extraction. Organic volatiles could also be sampled in the headspace of solid samples. In the standby position, the fiber is withdrawn into a protective

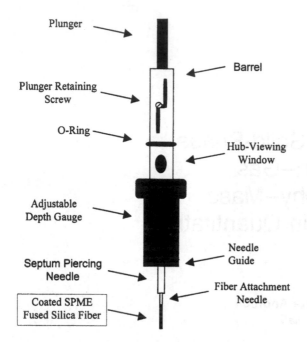

Figure 1   Solid-phase microextraction fiber holder for manual sampling.

sheath, which has also the function of piercing the septum of the sample vial or the GC injector. Analytes are concentrated by adsorption onto the solid phase. Figure 2 shows the adsorption–desorption processes in the case of headspace sampling: after piercing the septum of the sample vial with the fiber housed in the protective steel needle, the fiber is exposed in the headspace of the sample for a certain time at a fixed temperature (see Fig. 2a). Stirring of the sample may help the extraction procedure. The fiber is then pulled into the steel needle before removing the SPME device from the sample vial. The needle is inserted into the GC injector and the fiber immediately exposed for several minutes at high temperature to thermally desorb the extracts from the fiber coating (see Fig. 2b). A special narrow-bore insert (0.75 to 0.8 mm ID) is necessary to obtain a sharp injection band. Desorption in the GC port should be rapid and quantitative to avoid peak splitting effects and carryover, respectively.

Solid-phase microextraction is suitable for rapid extraction and preconcentration of compounds of different polarity and volatility from a variety of matrices. The selectivity of SPME depends on the phase coating of the silica fiber. The volume of the coating determines the capacity and the sensitivity of the method. The choice of the appropriate stationary phase should take into account both volatility and polarity of analytes. Using thick coatings, both extraction and

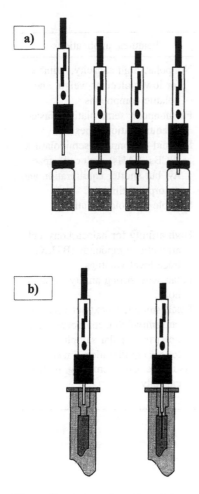

Figure 2 Adsorption (a) and desorption (b) processes in the case of headspace SPME sampling.

desorption become slower and the analysis time longer. Several kinds of stationary phases with different polarity and thickness are now commercially available from Supelco (Bellefonte, PA). Recently, new fibers with combined materials have been proposed for the sampling of a wide range of analytes. Table 1 summarizes the phases available at the present time and their features. Specific experimental coatings have been prepared with special materials, e.g., liquid crystalline films [4], ion exchange coatings [5], and others. Porous layer–coated SPME fibers obtained using silica-bonded C-8 and C-18 particles have been proposed for their

Table 1 Commercially Available SPME Fibers

Stationary phase	Thickness	Features, applications
Polydimethylsiloxane (PDMS)	100 μm[a]	Nonpolar, high capacity, suitable for low-molecular-weight and volatile compounds
	30 μm[a]	For nonpolar semivolatiles, faster in equilibration times
	7μm[b]	For mid-to nonpolar semivolatiles (PCBs, PAHs), reduced capacity, but shorter equilibration and desorption times
Polydimethylsiloxane/ divinylbenzene (PDMS/DVB)	65 μm[d]	Suitable for polar volatiles
Carboxen/PDMS (CAR/PDMS)	75 μm[c]	High affinity for halocarbons and aromatic compounds (BTEX), trace-level volatiles
Carbowax/divinylbenzene (CW/DVB)	65 μm[c]	Polar, with strong afinity for alcohols
Polyacrilate (PA)	85 μm[c]	Polar, suitable to extract very polar semivolatile compounds (phenols) from polar samples, longer equilibration times
Divinylbenzene/carboxen/ polydimethylsiloxane (DVD/ CAR/PDMS)	50–30 μm[d]	Available also 2 cm long, higher capacity

[a] Non-bonded.
[b] Bonded.
[c] Partially cross-linked.
[d] Highly cross-linked.

large capacity and fast analysis times [6]. A GC automated sampling system is commercially available from Varian (Walnut Creek, CA) [7]. Supelco also offers a SPME device for high-performance liquid chromatography (HPLC), but it is still not as widely applied as the one for GC [8].

Versatility, no solvent use, low cost, detection limits in the mid- to low-ppt range, small volumes of samples required for sampling, the possibility of automation, and ruggedness are the main advantages of SPME. The use of SPME in combination with gas chromatography–mass spectrometry (GC–MS) or gas chromatography–tandem mass spectrometry (GC–MS–MS) is a powerful mean for identification of volatile and semivolatile organic compounds, i.e., aliphatic and aromatic hydrocarbons, polychlorinated biphenyls, pesticides, phenols, flavors, and fragrances. Owing to the reproducibility among fibers from different batches, SPME is also recognized as a reliable technique for quantitative analysis. Derivatization before or during extraction can be used to enhance both selectivity and sensitivity of the method [9–10].

The use of SPME has already been applied to the headspace sampling of volatile organic compounds in different matrices of environmental interest, mainly water [10–15], but also soils [16–17], and air [15,18]. More recently, its use has been extended to foods and beverages [19], and biological samples [20–24]. In most cases, validation by comparison of the SPME performances with more established techniques, i.e., purge-and-trap and static headspace, has been conducted and a good correlation was observed [13,25].

2. PRINCIPLES OF SOLID-PHASE MICROEXTRACTION

An excellent and exhaustive description of SPME theory and practice has been given by Pawliszyn [26] in a book published in 1996, whose reading is recommended for a better comprehension of the principles of this technique. Some general concepts useful for method development are reviewed here. Although SPME can also be used for an exhaustive extraction of the analytes, equilibrium extraction is usually preferred in both direct and headspace modes. The direct extraction mode is based on direct immersion of the coated fiber into the sample and direct mass transfer from the matrix to the polymeric phase. In headspace mode, analytes should pass into the vapor phase before the extraction. The choice between direct and headspace extraction mainly depends on the volatility of the analytes and on the kind of matrix to be analyzed. When very dirty samples and/or highly volatile compounds are investigated, headspace SPME should be preferred to direct extraction from the aqueous phase. Fast equilibration times, better reproducibility, no memory effects, and longer lifetime of the fiber are reported in the case of headspace sampling [13–27].

In headspace-SPME a three-phase equilibrium system—the liquid, the headspace, and the fiber coating—is present in the sampling vial. Each equilibrium is governed by a partition coefficient: the coating-headspace partition coefficient (K_{fh}), headspace–sample partition coefficient (K_{hs}) and coating–sample partition coefficient (K_{fs}). The total mass of an analyte after equilibrium is distributed among the three phases:

$$c_0 V_s = c_f V_f + c_h V_h + c_s V_s \tag{1}$$

where c_0 is the initial concentration of the analyte in the sample; V_f, V_h, and V_s are the volumes of the coating, the headspace, and the matrix, respectively, and c_f, c_h, and c_s are the corresponding equilibrium concentrations of the analyte in the three phases.

Rearranging the above equation, the mass (n) of an analyte extracted by the fiber coating can be calculated [28]:

$$n = (K_{fh}K_{hs}V_f c_0 V_s)/(K_{fh}K_{hs}V_f + K_{hs}V_h + V_s) \tag{2}$$

The above relationship is simpler in the case of direct extraction, assuming the vial is fully filled with sample (no headspace):

$$n = (K_{fs}V_f c_0 V_s)/(K_{fs}V_f + V_s) \tag{3}$$

If V_s is very large ($V_s \gg K_{fs}V_f$), the amount extracted is not related to the sample volume:

$$n = K_{fs}V_f c_0 \tag{4}$$

At equilibrium, the amount of an analyte extracted by the fiber coating depends on the its distribution coefficient, K_{fs}, a thermodynamic parameter that is influenced by several factors, i.e., temperature, salting, pH, and the presence of organic solvents. From a kinetic point of view, the speed of the extraction process is governed by diffusion according to Fick's law. Equilibration times depend on the distribution coefficient, the sample temperature, agitation or stirring of the sample, the ionic strength of the solvent, and the thickness of the fiber coating. The roles of all these thermodynamic and kinetic parameters are further discussed and exemplified in section 3.1 below.

3. BLOOD AND URINE ANALYSIS

Most of the SPME work published in the literature is about water analysis. Biological fluids, i.e., blood and urine, are very complex and "dirty" matrices, unsuitable for direct extraction of the analytes, especially in the case of blood. Working with these samples requires particular care in order to avoid fast degradation of the fiber coating and loss in the performance of SPME. To date, only a few studies are reported concerning blood analysis [20–21,29], whereas a number of papers have been published about drugs and their metabolites in urine [21,30–32]. In this chapter, we will describe and comment on a number of SPME–GC–MS methods developed in our laboratory for the quantitative analysis of inorganic and organic compounds in urine and blood. The first example concerns the SPME method development for the simultaneous analysis of inhalation anesthetics, both nitrous oxide (a permanent gas) and the halogenates isoflurane, halothane, and sevoflurane in human urine as biomarkers of occupational exposure in operating room personnel [33]. The second example concerns the analysis of benzene, toluene, ethylbenzene and xylenes (BTEX) in both urine and blood and mainly focuses on calibration and method validation.

3.1 Optimization of Solid-Phase Microextraction Sampling: Analysis of Inhalation Anesthetics in Urine

At equilibrium, analytes are not completely extracted from the matrix. Nevertheless, reliable quantitative analysis is possible by SPME when sampling occurs under reproducible conditions. Optimization of SPME sampling is necessary when a high sensitivity is required, as in the case of halogenated anesthetics. In fact, using traditional static headspace sampling, only nitrous oxide, which is the main component of the anesthetic mixtures used in operating room theaters, can be determined. Urinary concentrations of anesthetics are in fact very low and a high sensitivity is required, especially for the determination of halogenates. Since inhalation anesthetics have never been studied before using SPME, we have optimized the kinetic of extraction and calculated the thermodynamic parameters (distribution coefficients and heats of adsorption), in order to characterize all the equilibria at the various interfaces of sample–headspace–fiber coating. We will use this example to show how to develop a SPME method for a completely new analyte. For the headspace sampling of anesthetics, a 75-μm Carboxen/PDMS-coated fiber was chosen for its very high affinity to halogenated compounds [34]. Samples (10 ml) were collected in 20-ml headspace vials containing 1 g NaCl and 200 μl of H_2SO_4 9N. Analyses were performed by GC–MS using a capillary column with a divinylbenzene porous polymeric stationary phase (RT-QPLOT, 30 m $\times$ 0.25 mm ID).

In principle, optimization of sampling conditions and method validation should be conducted in the real matrix. In the case of urine, sample collection is noninvasive and ethically acceptable. Large amounts of sample are usually available to perform an in-matrix method development.

3.1.1 Extraction Time Profile

Once the appropriate stationary phase and film thickness are chosen, taking into account the volatility and polarity of analytes, the development of a new SPME method starts with the determination of the time needed for the analytes to reach equilibrium between the matrix, the headspace (when present), and the fiber. The extraction time profile can be determined for each analyte by plotting the detector response versus the extraction time. When a further increase of the extraction time does not result in a significant increase in the response, the equilibration time is assumed to be reached. As an example, Figure 3 shows the extraction profile obtained for halogenated anesthetics and for the internal standard (IS) dichloromethane using a 75-μm-Carboxen/PDMS SPME fiber. The steady state is reached after 15 minutes for all the analytes and the IS. The compound chosen as IS should show a similar profile to those of the analytes. Sampling at equilibrium reduces the possibility of errors and ensures reproducible data. Sampling at shorter but precisely measured extraction times is also possible and is recom-

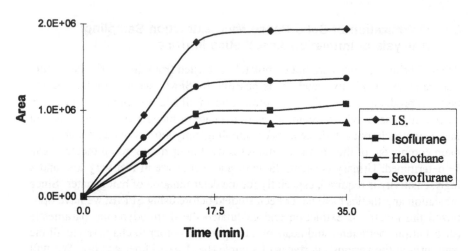

Figure 3 Extraction time profiles of dichloromethane (I.S.), halothane, isoflurane and sevoflurane obtained using a 75-μm-Carboxen/PDMS SPME fiber. Sampling conditions: 25°C, no stirring.

mended when equilibration times are too long. High-molecular-weight compounds with high distribution coefficients, such as polycyclic aromatic hydrocarbons (PAHs), require equilibration times of hours [12].

3.1.2   Effects of Temperature, Agitation, Salting, and pH

An increase in extraction temperature has the double effect of increasing the extraction rate but simultaneously decreasing the distribution constant K_{fs}. In fact, at high temperatures the fiber starts releasing the adsorbed analytes faster than adsorbing. The temperature chosen should be a compromise between these two competitive effects. In general, sampling at room temperature is recommended for highly volatile compounds, for which the main effect is the K_{fs} decrease, while for low volatiles the increase of the extraction rate is the most relevant effect, which is positively influenced by higher temperatures. A modified SPME system, which allows the sampling with an internally cooled fiber from a heated system, has been proposed to enhance the sensitivity [35], but it is not commercially available. Heating the sample can also be useful to help the release of the analytes from a solid matrix, if both the analytes and the matrix are stable at higher temperatures.

   Static sampling is simpler but is limited to volatile analytes and headspace SPME. Agitation of the sample can help the mass transport between the bulk and the fiber, leading to a response increase for low volatile compounds, such as PAHs [28]. Agitation can be obtained by magnetic stirring, intrusive stirring, vortex, fiber movement, flow through, or sonication [10]. Stirring might have the drawback of heating up of the sample. Anesthetics are very volatile compounds, permanent gases or liquids with very low boiling temperatures; for these reasons, we sampled at room temperature without agitation.

Salting can increase the extraction efficiency, especially in the case of polar compounds [14]. Concentrations of salt above 1% are reported to cause a substantial increase in extraction efficiency. Saturation with salt can be used to normalize random salt concentrations in natural matrices [26]. We found that a 10% of salt (w/v) added to urine samples led to a 26 to 35% increase in sensitivity for inhalation anesthetics, whereas a further addition of salt (up to 50%) did not further lower the detection limit. Besides salts, other additives such as nonvolatile acids [35] or water [36] can be used to enhance the extraction of analytes from very complex matrices.

pH control is recommended when dissociable species, acids or bases, are analyzed by SPME [8]. Since only neutral species are extracted, the pH should be two units below or above the pK for acid or basic analytes, respectively. Once the pH of the sample has been adjusted, only headspace sampling is possible [26].

3.1.3 Desorption Temperature and Time

Besides the sampling parameters, the desorption temperature and desorption time also need optimization to avoid sample carryover on the fiber. The desorption temperature must be at least as high as the highest boiling point of the compounds in the mixture. Since desorption becomes faster when the desorption temperature increases, a rapid way to optimize desorption conditions is to set the injector temperature at the maximum allowable temperature (determined by the coating) and then to adjust the desorption time to obtain quantitative desorption in a single run. Carryover can be determined by injecting the fiber again, immediately after the end of the chromatographic run. Compounds with very high molecular weight may lead to carryover problems. Thermal desorption of anesthetics was performed at 240°C for 16 minutes.

3.1.4 Partition Coefficients and Heat of Adsorption

The efficiency of the extraction process and the sensitivity of SPME depend on the fiber coating–sample distribution constant, K_{fs}, a thermodynamic parameter which expresses the affinity of the fiber coating for a target analyte. K_{fs} values for organic compounds may vary widely (3 to 4 orders of magnitude) depending on the coating and the thickness of the SPME fiber. Table 2 compares literature K values obtained with different coatings for BTEX, the most investigated class of organic compounds. For air samples, coating–gas distribution constants, K_{fg} can be estimated using GC retention times on a column with stationary phase

Table 2 Comparison of Distribution Constant K Values Obtained with Different Fiber Coatings and K_{ow} for BTEX

	K PDMS [12]		K Bonded phases [6]		K Carboxen [34]	K_{ow}
	100 μm	7 μm	C-8 70 μm	C-18 70 μm	80 μm	[34]
benzene	199	97	31.8	10.4	8550	135
toluene	758	383	96.3	42.3	10270	537
ethylbenzene	2137	519	n.d.	n.d.	8300	1412
m-xylene	2041[a]	448[a]	227.9	201.7	8600	1412
o-xylene	1819	365	105.6	137.2	9040	1318

[a] m + p-xylene.

identical to the fiber coating or using the linear temperature-programmed retention index system [37] or Kovats retention indices [38]. In the case of aqueous liquid samples, the coating–water distribution constants, K_{fw}, estimated for direct SPME extraction have been correlated with the octanol–water, K_{ow} partition coefficients [39]. The coefficients calculated in pure water or air are modified in the presence of the sample matrix [40].

In headspace mode, K_{fs} determines the absolute SPME response and can be calculated as:

$$K_{fs} = K_{fh}K_{hs} = K_{fg}K_{gs} \tag{5}$$

where K_{hs} is directly related to the Henry's constant of the analyte [1]. K_{fh} has been called K, the calibration factor [41].

In principle, it is not necessary to calculate the partition coefficients, when quantitative analysis is performed with the standard addition method or with isotopically labeled standards. On the other hand, calibration of the sampling device is not necessary when the analyte's K at various temperatures is known [37,42]. In any case, it is always advisable to determine partition coefficients, since knowledge of them may help to better understand all the phenomena occurring at the various interfaces and to predict the effects of a modification of the system.

In the case of halogenated anesthetics, we obtained very high K_{fs} values (110,000 to 190,000). Therefore, we could reach the required sensitivity for our biological monitoring purposes. As expected, both the MS response factor and the affinity of Carboxen/PDMS for nitrous oxide were lower, leading to a 1000-fold lower K_{fs} value (103). Figure 4 shows an SPME–GC–MS chromatogram of a urine sample from an occupationally exposed subject. Despite the low urinary concentrations, the peaks of halothane (1.1 μg/L) and isoflurane (1.3 μg/L) are

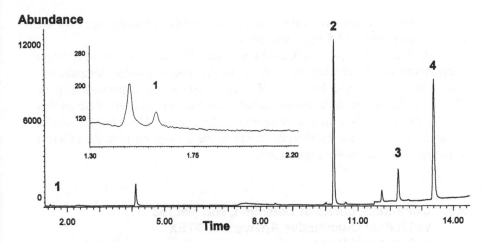

Figure 4 The SPME–GC–MS analysis of a urine sample from a worker exposed to anesthetics. Peak identification: 1, nitrous oxide; 2, I.S.; 3, isoflurane; and 4, halothane.

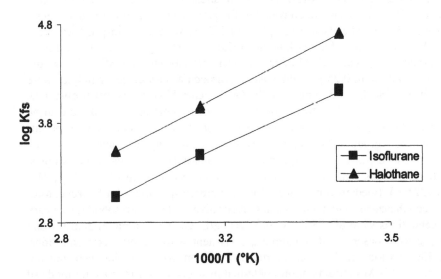

Figure 5 Plot of ln K versus $1000/T$ for isoflurane and halothane.

clearly detectable, whereas in the case of static headspace sampling, only the peak of nitrous oxide (5.5 µg/l) was present.

The heat of absorption, ΔH, can be determined by measuring K_{fs} at different temperatures and by plotting the ln K versus the inverse absolute temperature. The slope of the straight line is $-\Delta H$. Figure 5 shows the experimental plots obtained in the case of isoflurane and halothane, the calculated heats of adsorption being -20.0 and -24.2 kJ/mol, respectively. A negative ΔH, typical of volatile compounds, reflects the exothermic nature of the adsorption process: efficient sampling occurs at lower temperatures.

3.2. Calibration, Limits of Detection, and Method Validation: Quantitative Analysis of BTEX in Blood and Urine

In our laboratory, we have measured the urine and blood concentrations of BTEX by SPME–GC–MS to assess individual exposure to monoaromatic hydrocarbons polluting the urban air [43]. Unlike the use of urine, the use of human blood for method development raises ethical problems, and this strongly limits the number of possible preliminary experiments. Therefore, only few experiments were performed to optimize exposure time, temperature, and stirring rate. On the other hand, relevant data on SPME of BTEX are already available from the literature and it is always advisable to consider all the previous work, although in different matrices [34]. For headspace sampling of BTEX, we used a 75-µm Carboxen/ PDMS SPME fiber, which is known to show a very high affinity for aromatic compounds [34]. The GC–MS analyses were carried out on a HP-5MS column (30 m × 0.25 mm, 0.25-µm film) using hydrogen as carrier gas. Analytes were desorbed at 280°C for 3 minutes in the GC injector. No carryover was observed. The method was applied to the quantitative determination of blood and urine concentrations of BTEX, as biomarkers of exposure to environmental air pollutants, among a group of 24 volunteers cycling for 2 h along city routes. Just before and immediately after the runs, subjects had to provide blood and urine samples. Each sample was analyzed in duplicate. Blood concentrations of 180, 285, 213, and 722 ng/L (median values) were found in prerun specimens for benzene, toluene, ethylbenzene, and total xylenes, respectively. The corresponding urinary concentrations were 89, 280, 73, and 220 ng/L. Changes in blood benzene and toluene were observed after exercise, consistent with airborne concentrations. Significant variations between prerun and postrun urine samples were not observed. A GC–MS chromatogram obtained in selective-ion monitoring mode of a blood sample from a nonsmoker subject obtained at the equilibrium conditions is shown in Figure 6.

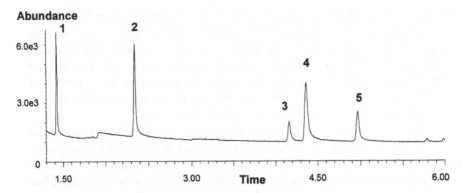

Figure 6 A GC–MS chromatogram in selective ion monitoring mode of a blood sample from a non-smoker subject. SPME fiber: 75-μm-Carboxen/PDMS. Extraction conditions: 37°C, 30 minutes, 500 rpm. Peak identification: 1, benzene + I.S.; 2, toluene; 3, ethylbenzene; and 4, m- + p-xylene; 5, o-xylene.

3.2.1. Calibration

Since blood and urine have a natural content of BTEX, the standard addition procedure was chosen to assess linearity of the SPME–GC–MS method. External calibration and the use of literature K values are only possible with simple and clean matrices, i.e., air or clean water [37,42]. In the case of more complex samples, such as soils or biological fluids, the use of isotopically labeled internal standards or the standard addition is recommended [1,12]. Blood (with heparin) and urine were spiked with BTEX mixtures in order to obtain concentration increases in the 0 to 50 μg/L range. We prepared a calibration standard set in order to spike the calibration samples (2 ml) with the same small volume of methanolic standard (20 μl). The addition of methanol or other organic solvents to aqueous matrices should always be lower than 1%. Moreover, methanol itself contains trace amounts of BTEX. We used $^{13}C_6$-benzene as internal standard to improve the reproducibility of the data. Although not separated from benzene, $^{13}C_6$-benzene has unique ions for quantitation. Since isotopically labeled compounds have chemical and physical properties very similar to those of the unlabeled analogs, they are considered as the best SPME–GC–MS internal standards. Valuable quantitative results at very low concentration levels were obtained using particular care during sample preparation and storage. Samples for calibration and authentic samples were prepared and frozen in a similar way. In the case of blood, we could obtain linear response curves only when the samples were frozen 2 hr after the spiking, indicating the relevant role of the equilibration time of the standard spiked into the matrix. Without equilibration of the analytes in the matrix, no reliable calibration could be obtained. This behavior is peculiar to blood and was not observed in the case of urine, probably due to the different lipophilic components of the two matrices. In general, it is always advisable that the spiked analytes equilibrate with the natural content already present in the matrix and with the matrix itself. Moreover, when extracting conditions are optimized, it should be kept in mind that the interaction between the matrix and native analytes should be stronger than in the case of spiked analytes [44].

With the standard addition method, experimental data fit a linear model $y = a + bx$, where x is the concentration increases and y the chromatographic peak areas to IS area ratio. The correct standard addition procedure requires the standard addition to every sample, to overcome the matrix effect, i.e. the difference in the composition of human fluids due to the interindividual biological variance. As a drawback, standard addition leads to a high number of samples to be processed. In the case of urine, there is also the possibility of minimizing the matrix effects using a pool of urine for calibration or minimizing differences in salt composition by adding salt up to saturation. By contrast, blood samples from different subjects cannot be mixed because they are incompatible with each other. Moreover, in the case of blood, the requirement of a larger volume of sample to perform standard additions is not always acceptable.

According to our study design, a large number of urine and blood samples to be analyzed in a time limited by analyte stability was expected ($n = 192$). For these reasons, we decided to evaluate preliminarily the variability due to matrix effects. Using blood and urine from different subjects ($n = 5$), we repeated the entire standard addition calibration procedure several times. What we found was only a slight matrix effect, comparable with the precision of the method: different blood (or urine) samples used for calibration gave rise to equations with different intercept values (a, the native content), but with similar (CV <10%) slope values (b). We decided to use the equations obtained with standard addition to a limited number of samples as external calibrations to calculate blood and urine concentration of BTEX in all the samples, after y-axis shift ($y = bX$, where $X = x + a/b$). On the other hand, our purpose was to measure BTEX at the ppt level and also to appreciate differences between prerun and postrun blood and urine samples from the same subject. This approach cannot be generalized without the evaluation of the matrix effect case by case. The affinity of the matrix for the analytes strongly varies depending on the chemical and physical properties of both the matrix and the analytes. In some cases, the variability is too large to be acceptable [17]. In any case, when the number of biological samples is limited and a sufficient amount of sample is available, it is always advisable to spike each sample.

Figure 7 shows the calibration graphs obtained in the case of blood BTEX. Linearity was established for concentrations up to 50 µg/L. A very wide linear

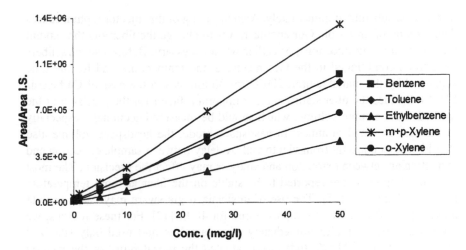

Figure 7 Calibration graphs obtained in the case of blood BTEX.

dynamic range is recognized for SPME–GC using flame ionisation or quadrupole MS detection. The use of ion trap MS can limit the linearity of SPME [45].

3.2.2 Limits of Detection

Since measurable concentrations of BTEX are present in the blank matrix, detection limits were determined using the corresponding deuterated compounds. Isotopically labeled compounds are also useful to improve specificity and to ensure precision in peak assignment. Limits of detection were 5 ng/L for benzene and toluene, and 10 ng/L for ethylbenzene and xylenes. Detection limits of 10 to 20 ng/L were also obtained for halogenated anaesthetics. The sensitivity of SPME depends on K_{fs} values and is not enhanced by larger sample volumes, especially for compounds with $K_{fs} < 500$ [12].

3.2.3 Precision

Precision of the method is controlled by several factors, first of all, the status of the SPME fiber. The mean fiber lifetime is 100 runs when desorbing into an injector heated to 220°C or immersing into water saturated with salt and at pH 2. Direct sampling from dirty matrices may cause faster degradation of the fiber, due to the adsorption of high-molecular-weight species such as proteins, salt crystals, or humic materials. If the process is reversible, the fiber can be soaked with a proper solution [8]. When a fiber used for quantitative analysis starts to degrade, it should be substituted immediately. Also the aging of the injector septum causes deterioration of precision. Our approach was to change the fiber and the septum after 100 sampling-injections, or earlier when necessary. Before use, new fibers have to be conditioned in the GC injector at the temperature and for the time suggested by the manufacturer, 280°C for 30 minutes in the case of Carboxen/ PDMS. Besides the fiber status, another important factor for the precision of the method is the sample volume, which should be measured accurately, especially in the case of small volumes. In headspace mode, the headspace volume also should be kept constant from vial to vial. A delay between sample collection and extraction or between extraction and analysis may affect the precision. The most volatile compounds are reported to be stable on the fiber at room temperature for about 2 minutes [15]. The use of unsilanized glassware may cause analyte losses up to 70% in water samples stored for 48 hr [12]. For these reasons, we always analyzed samples immediately after sampling and used only silanized glass vials. In our BTEX study, we evaluated the repeatability of the method from six consecutive extraction-injections of the same blood or urine sample nonspiked. Repeatability was in the range of 6.5–9.2% (expressed as % RSD) for aromatic compounds. In the case of very volatile analytes, permanent gases or compounds with very low boiling temperatures, such as anaesthetics, the use of gaseous standard solutions instead of liquid solutions was found to improve the reproducibility in standard preparation. We also used a gaseous IS, dichloromethane, to avoid interference arising from other solvents and peaks overlapping with the peak of nitrous oxide. The calculated intraday and interday precisions for all the anesthetics were 3.0 to 7.2% and 6.5 to 12.9%, respectively. Our findings are in agreement with data reported by other authors [14,22,27].

3.2.4 Validation

Validation of a SPME method for target analytes should be performed using standard reference materials with similar matrix, when available. Another possible and frequently used way is validation of a SPME method against well-accepted extraction techniques, such as purge-and-trap [13,25,46] or static headspace [46]. Several interlaboratory studies demonstrated that SPME is a reliable technique for the quantitative analysis of volatile organic compounds [46] and pesticides in water samples [47–48]. We have validated our SPME–GC–MS method for the determination of nitrous oxide in urine by means of the comparison with static headspace [33].

4. CONCLUSIONS

SPME is now a mature sampling technique, extremely powerful for both identification and quantitation of a wide range of analytes in matrices of environmental, biomedical, and forensic interest. Since its introduction in 1993, the number of users is continuously increasing and a considerable number of applications (more than 300) have been developed and published in international journals. In our experience, SPME is the first choice for the sampling of organic volatile compounds in biological fluids, especially when high sensitivity is required.

REFERENCES

1. Z. Zhang, M.J. Yang, and J. Pawliszyn, Anal. Chem., 66 (1994), 844A–853A.
2. J. Pawliszyn and S. Liu, Anal. Chem., 59 (1987) 1475–1478.
3. R. Berlardi and J. Pawliszyn, Water Pollut. Res. J. Can., 24 (1989) 179–191.
4. C.L. Arthur and J. Pawliszyn, Anal. Chem., 62 (1990) 2145–2148.
5. J.L. Liao, C.M. Zeng, S. Hjerten, and J. Pawliszyn, J. Microcol. Sep., 8 (1996), 1–4.
6. Y. Liu, Y. Shen, and M.L. Lee, Anal. Chem., 69 (1997) 190–195.
7. C.L. Arthur, I.M. Killam, K.D. Buchholz, J. Pawliszyn, and J.R. Berg, Anal. Chem., 64 (1992) 1960–1966.
8. H.L. Lord and J. Pawliszyn, LC-GC Internat, Dec. (1998) 776–785.
9. L. Pan and J. Pawliszyn, Anal. Chem., 69 (1997) 196–205.
10. M.-L. Bao, F. Pantani, O. Griffini, D. Burrini, D. Santianni, and K. Barbieri, J. Chromatogr. A, 89 (1998) 75–87.
11. S.P. Thomas, R.S. Ranjan, G.R.B. Webster, and L.P. Sarna, Environ. Sci. Technol., 30 (1996) 1521–1526.
12. J.J. Langenfeld, S.B. Hawthorne, and D.J. Miller, Anal. Chem. 68 (1996) 144–155.
13. T. Nilsson, F. Pelusio, L. Montanarella, B. Larsen, S. Fachetti, and J.O. Madsen, J. High Resolut. Chromatogr., 18 (1995), 617–624.
14. S.S. Johansen and J. Pawliszyn, J. High Resolut. Chromatogr., 19 (1996), 627–632.
15. M. Chai, C.L. Arthur, J. Pawliszyn, R.P. Belardi, and K.F. Pratt, Analyst, 118 (1993) 1501–1505.
16. K.J. Hangeman, L. Mazeas, C.B. Grabanski, D.J. Miller, and S.B. Hawthorne, Anal. Chem., 68 (1996) 3892–3898.
17. A. Fromberg, T. Nilsson, B.R. Larsen, L. Montanarella, S. Facchetti, and J.O. Madsen, J. Chromatogr. A, 746 (1996) 71–81.
18. M. Chai and J. Pawliszyn, Environ. Sci. Technol., 29 (1995) 693–701.
19. T. Peppard and X. Yang, J. Agric. Food Chem., 42 (1994) 1925–1930.
20. P. Okeyo, S.M. Rentz, and N.H. Snow, J.High Resol. Chromatogr., 20 (1997), 171–173.
21. M. Nishikawa, H. Seno, A. Ishii, O. Suzuki, T. Kumazawa, K. Watanabe, and H. Hattor, J. Chomatogr. Sci., 35 (1997), 275–279.
22. S. Fustinoni, R. Giampiccolo, S. Pulvirenti, M. Buratti, and A. Colombi, J. Chromatogr. B, 723 (1999) 105–115.
23. C. Grote and J. Pawliszyn, Anal. Chem., 69 (1997), 587–596.
24. C. Malosse, P. Ramirez-Lucas, D. Rochat, and J.-P. Morin, J. High Resol. Chromatogr., 18 (1995), 669–670.

25. B. MacGillivray, P. Fowlie, C. Sagara, and J. Pawliszyn, J. Chromatogr. Sci., 32 (1994) 317–322.
26. J. Pawliszyn, Solid Phase Microextraction, 1997, Wiley-VCH, New York, pp. 43–140.
27. Z. Zhang and J. Pawliszyn, Anal. Chem., 65 (1993) 1843–1852.
28. Z. Zhang and J. Pawliszyn, J. High Resolut. Chromatogr., 19 (1996) 155–160.
29. T. Kumazawa, H. Seno, X. Lee, A. Ishii, O. Suzuki, and K. Sato, Chromatographia, 43 (1996) 393–397.
30. M. Yashiki, T. Kojima, T. Miyazaki, N. Nagasawa, Y. Iwasaki, and K. Hara, Forensic Sci. Int., 76 (1995) 169–177.
31. T. Kumazawa, K. Watanabe, K. Sato, H. Seno, A. Ishii and O.Suzuki, Japan. J. Forensic Toxicol., 13 (1995) 207–210.
32. A. Namera, M. Yashiki, T. Kojima, and N. Fukunaga, Japan. J. Forensic Toxicol., 16 (1998) 1–15.
33. D. Poli, E. Bergamaschi, P. Manini, R. Andreoli, and A. Mutti, J. Chromatogr. B, 723/1 (1999) 115–125.
34. P. Popp and A. Paschke, Chromatographia, 46 (1998) 419–424.
35. Z. Zhang and J. Pawliszyn, Anal. Chem., 67 (1995) 34–43.
36. K. Buchholz and J. Pawliszyn, Anal. Chem., 66 (1994) 160–167.
37. P.A. Martos, A. Saraullo, and J. Pawliszyn, Anal. Chem., 69 (1997) 402–408.
38. B. Schaefer, P. Hennig, and W. Engewald, J. High Resol. Chromatogr., 20 (1997) 217–221.
39. J.R. Dean, W.R. Tomlinson, V. Makovskaya, R. Cumming, M. Hetheridge, and M. Comber, Anal. Chem., 68 (1996) 130–133.
40. J. Poerschmann, Z. Zhang, F.-D. Kopinke, and J. Pawliszyn, Anal. Chem., 69 (1997) 597–600.
41. R.J. Bartelt, Anal. Chem., 69 (1997) 364–372.
42. P.A. Martos and J. Pawliszyn, Anal. Chem., 69 (1997) 206–215.
43. R. Andreoli, P. Manini, E. Bergamaschi, A. Brustolin, and A. Mutti, Chromatographia, 50 (1999) 167–172.
44. J. Pawliszyn, Trends in Anal. Chem., 14 (1995) 113–122.
45. T. Gorecki and J. Pawliszyn, Anal. Chem., 68 (1996) 3008–3014.
46. T. Nilsson, R. Ferrari, and S. Facchetti, Anal. Chim. Acta, 356 (1997), 113–123.
47. R. Ferrari, T. Nilsson, R. Arena, P. Arlati, G. Bartolucci, R. Basla, F. Cioni, G. Del Carlo, P. Dellavedova, E. Fattore, M. Fungi, C. Grote, M. Guidotti, S. Morgillo, L. Muller, and M. Volante, J. Chromatogr. A, 795 (1998) 371–376.
48. T. Gorecki and J. Pawliszyn, Analyst., 121 (1996) 1381–1386.

9

Gas Chromatography–Mass Spectrometry of Drugs in Biological Fluids After Automated Sample Pretreatment

M. Valcárcel, M. Gallego, and S. Cárdenas
University of Córdoba, Córdoba, Spain

1. INTRODUCTION

The primary goal of today's analytical chemistry is gaining access to as much accurate, reliable chemical information of the highest possible quality by expending increasingly less material, time, and human resources, taking the lowest hazards, and incurring the smallest expenses [1]. Thus, one of the most important research and development targets in analytical sciences is the incorporation of the three general trends in science and technology today (viz., automation, miniaturization, and simplification) into the three stages of the analytical process. Promising specific analytical approaches such as rapid response systems, off-line, on-line, in-line, and in-field measurements, sensors, screening systems, etc., are being developed to this end [2].

As can be seen in Figure 1, preliminary operations (sampling, sample preservation and treatment, subjection to separation techniques) involve intensive human participation and are time consuming (70 to 90% of the time devoted to the overall analytical process is expended on these operations) and the source of major random and systematic errors—which have a decisive influence on the quality of the results. In addition, because of their variability, instrument manufacturers only undertake the automation/enhancement of preliminary operations in those instances where a wide market (e.g., clinical samples) is anticipated.

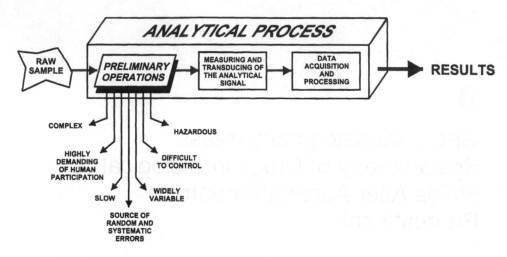

Figure 1   Features of the first stage of the analytical process.

Some "automatic analyzers" only reduce human participation in the simplest of these operations [3]. Each analytical problem calls for a specific experimental setup to be used in the preliminary operations in order to fill the gap between the raw sample and the analytical instrument used for measurement. It is clear that the length of these operations depends on the sample (e.g., the complexity of its matrix) and on the type of instrument used. Figure 2 depicts the more usual links between a real sample and a gas chromatograph. In addition to direct insertion, which is scarcely used, there are three different options depending on whether or not an automated treatment module or autosampler is used, namely: off-line, on-line, and mixed off/on-line. The off-line alternative is commonly used in gas chromatography (GC), even though the mixed option is steadily gaining ground in recent years. Preliminary operations are carried out manually and only cleanup steps are automated. The on-line alternative is used to a lesser extent and occasionally involves a manual step.

Drug determinations in biological fluids can be undertaken for several reasons, such as pharmacokinetics and bioavailability studies, drug overdose, or screening for drugs of abuse. A significant part of the absorbed drug is bound to protein (mainly albumin and α_1-glucoprotein) in the plasma compartment [4]; since the free drug is the metabolically active form, direct measurement of this concentration is usually required. Drug concentrations in plasma are reduced by drug metabolism in the liver (biotransformation) through a variety of organic reactions. Functional groupings in drugs include mainly amino, hydroxyl, and carboxyl groups, which are subjected to oxidation, reduction, hydrolysis, hydroxylation, dealkylation, or conjugation with glucuronic acid or an acid salt

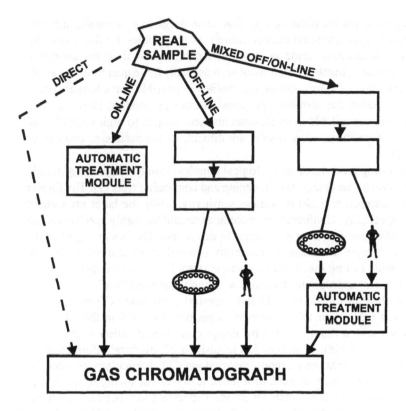

Figure 2 Choices of real sample treatments prior insertion into the gas chromatograph.

(sulfate, phosphate) to form more polar compounds that can be readily excreted by the kidneys [4].

There are many factors to be considered in selecting a specimen for analysis; the choice should be based on availability of the sample and drug concerned. Blood and urine are the most common sample choices in clinical analysis cases and in drug tests on athletes. Stomach contents, liver, bile, brain, and kidney samples are also often used. Blood is the most useful sample for both identifying drugs and quantitation. However, plasma and serum samples are preferred to whole blood because they contain fewer interfering substances; as a result, reported therapeutic concentrations are most often given for either plasma or serum [5]. On the other hand, analyses of these samples are more difficult, expensive, and complex; also, the amount of specimen available is usually smaller. The greatest advantages of using urine are that it affords drug concentrations about 100 times higher than blood and that it is subject to few interferences. On the other hand, urine has the disadvantage that some drugs (e.g., cannabinoids and benzodiazepines) are excreted almost entirely as metabolites by this route [5]. Despite these limitations, urine is the preferred biological fluid for substance abuse screening as collection is noninvasive, reasonably large quantities are available, and detection of drugs and/or metabolites is possible for a longer period than in blood. Saliva has also been proposed in some procedures. However, it is difficult to obtain, available in small amounts, and, subject to high variability in the duration of positives; also, results are difficult or impossible to confirm or quantitate [6].

Drug testing procedures for biological samples (mainly urine) are generally divided into two broad categories: screening and confirming tests [7]. The former is normally designed to yield maximum sensitivity, while the latter are usually focused on specificity. Confirmatory procedures should be highly specific for the drug detected in the screening procedure. In this regard, GC has emerged as the best choice. Its main advantage is versatility (several drugs and metabolites in the same sample can be measured in a single step). Mass spectrometry (MS) is no doubt the most appropriate detector for unequivocal identification. It is used in both the electron ionization (EI) and chemical ionization (CI) modes. The most common practice in this context is to perform GC–MS in the EI mode, in conjunction with a library search for unequivocal identification of drugs and metabolites (if available), as well as for acquisition of structural information on the molecule. As regards the CI mode, positive-ion CI is particularly suitable for these analytes; however, it is used to a much lesser extent owing to the strong influence of operational parameters on the mass spectrum and, hence, to its interpretation being hindered by the absence of libraries and databases. For confirmatory purposes, full-scan acquisition should be used; if higher sensitivity is needed, however, selective ion monitoring (SIM) is more appropriate. The equipment used is described in detail elsewhere in this volume (see Chap. 1).

2. STEPS INVOLVED IN THE PRETREATMENT OF BIOLOGICAL FLUIDS

Although capillary GC–MS has extremely high separation efficiency and provides highly sensitive and selective detection, biological fluids are among the most troublesome extracts in terms of contamination because of the large number of unknown compounds they may content. These hydrophobic, high-boiling interfering matrix compounds may well accumulate on the top of the chromatographic column or be adsorbed on the walls of connecting tubes. Therefore, the assay of drugs in biological fluids by GC–MS poses many difficult analytical problems and frequently involves lengthy sample preparation procedures, which are usually the most complex and difficult to automate. Thus, the main objectives in preparing biological fluids for GC–MS analysis are isolating the analytes from the sample matrix and dissolving them in a suitable solvent, as well removing as many interferences as possible. Preconcentration of the analytes is also important and often requires derivatization to aid separation and/or detection. The steps usually involved in the pretreatment of biological fluids for the determination of drugs by GC–MS are summarized below.

1.   Hydrolysis of conjugate drugs. As stated above, drugs are biotransformed during their metabolic pathway. Most are excreted in the urine as glucuronides. As these derivatives are not appropriate for GC, the urine has to be hydrolyzed prior to analysis in order to obtain the highest possible concentration of the parent drug. This hydrolysis is usually carried out enzymatically with glucuronidase [8–11].

2.   Extraction. Liquid–liquid extraction has frequently been employed for the extraction of analytes from biological fluids. Nowadays, however, the disadvantages inherent in this technique (viz. laborious, time-consuming manipulations, the need to use large volumes of sample and organic solvents, and the formation of emulsions) are all well known. In addition, determinations of drugs in biological fluids should be addressed on the understanding that pH restrictions exist (some drugs decompose at the optimal operational pH). Solid-phase extraction (SPE) has rapidly established itself as an alternative to the pretreatment of biological fluids for drug analysis, and disposable extraction columns in a broad range of column sizes and sorbents have become available. The SPE technique provides many advantages, including high recoveries of analytes, effective concentration, highly purified extracts, the ability to simultaneously extract analytes of widely variable polarity, easy automation, absence of emulsions, and reduced organic solvent consumption (mainly that of chlorinated solvents in order to reduce environmental pollution [12]). Two important developments have been introduced in this direction, viz., disks or membranes for SPE [8,13] and solid-phase microextraction (SPME); the latter, however, is suitable for dedicated, selective extractions [14]. In the last few years, dialysis has also been successfully applied to the determination of analytes in biological fluids; the semipermeable membrane used allows the removal of macromolecular sample constituents [15,16].

3.   Preconcentration and cleanup. Drugs typically occur in the low nanogram per milliliter (pharmacokinetic studies) or microgram per milliliter (therapeutic drug monitoring) concentration range. Because the injected sample volume is typically 1 to 2 µl, some concentration of the sample is usually required in order to improve sensitivity. Preconcentration is also advisable when the analytes have been diluted during extraction (e.g., dialysis and off-line SPE). In drug analysis, sample clean-up is required to separate the analytes of interest from potential interferences. In SPE, this step involves the use of a sequence of pure or mixed solvents to selectively elute part of the extracts retained together with the analytes

of interest, on which the solvent will have no effect. Then, retained compounds are eluted from the column by using an appropriate solvent or mixture.

4.   Solvent changeover. This step usually involves evaporation to dryness under a nitrogen stream and reconstitution of the residue in a few microliters of an organic solvent compatible with the chromatographic system (increasing sensitivity through the volume ratio). Occasionally, the selectivity can be improved if the compounds accompanying the analytes in the final extracts are less soluble in the organic solvent used.

5.   Derivatization. Drug derivatization is usually employed to increase the volatility, improve the thermal stability, or tag the compound with a particular sensitive moiety for mass spectral detection. Silylation, and also methylation to a lesser extent, are the most common derivatization procedures for drug determinations by GC–MS. The formation of different derivatives allows the identification of the series of metabolites of a drug (e.g., for cannabinoids) [17]. On-line derivatization in GC–MS has, however, received little attention. Amphetamine and methamphetamine have been derivatized on-line to their trifluoroacetate derivatives in a GC–MS system by introducing of N-methylbis(trifluoro)acetamide vapors into the GC injector and column inlet [18]. On-line derivatization of non-steroidal anti-inflammatory drugs (NSAIDs) and abuse drugs was accomplished by using flow injection configurations [19–21].

3.   AUTOMATED CLEANUP SYSTEMS FOR BIOLOGICAL FLUIDS

The literature on the use of automated SPE in drug analysis is confusing; thus, papers with titles such as ''automated extraction,'' ''automated system,'' ''automated sample cleanup,'' ''automatic sample preparation,'' etc., can be found, all of which refer to the use of SPE based on disposable cartridges, the other steps involved in the sample preparation procedure being carried out manually. In these extraction systems, the biological fluid can be loaded manually or by means of an autosampler arm onto a conditioned sorbent unit. The advantages of these off-line systems are well known. Thus, several samples can be processed in parallel with high operational flexibility. However, these sample preparation systems are difficult to automate as they require the use of an expensive workstation [22]; in addition, they are time consuming and prone to solute losses. These modules have been extensively used for the determination of drugs in biological fluids; however, they are not automated systems for sample preparation (the subject matter of this chapter), so only a few examples are discussed here.

Human urine containing various model compounds (drugs and metabolites) was extracted by SPE using mixed-phase Bond-Elut Certified LCR columns, C_{18} reverse-phase extraction disks, Plus C_{18} AR/MP$_3$ Multi-Model microcolumns, and also by SPME, using polydimethylsiloxane or polyacrylate fibers. SPR Plus Multi-Model microcolumns was found to be the most suitable method for the broad-spectrum screening, and SPME the most selective extraction method [13].

Typical examples of drug determinations in such real samples are presented in Table 1. Only automated cleanup was done in all instances and several additional sample preparation steps were required to minimize problems caused by matrix effects in GC–MS. Thus, enzymatic hydrolysis for urine samples or centrifugation for blood specimens was necessary before sample cleanup. Various SPE approaches have been adapted for increased sorption efficiency and decreased interferences; columns were conditioned with methanol, distilled water, and buffer solutions of weakly acid pH and, after sorption, were rinsed with water, dilute acid (normally acetic acid) and methanol prior to elution [9,10,23–26]. Of course, methanol was also present as eluent [24] or combined with acid [23] or ammonia (NH_3) aliquots [10] for amphetamines or designer drugs, respectively. The most frequently used eluent for drugs is the dichloromethane (CH_2Cl_2)–isopropanol–NH_3 mixture in variable ratios but always containing greater amounts of dichloromethane [25–27]. Finally, the volatility of some drugs is increased by derivatization with N,O-bis(trimethylsilyl)trifluoroacetamide (BSTFA) to form trimethylsilyl (TMS) derivatives [9,27]; this step involves evaporation to dryness and incubation at approximately 70°C for several hours. The most frequently studied analytes in this context are those potentially involved in fatal cases of drug overdose.

4. FULLY AUTOMATED PRETREATMENT OF BIOLOGICAL FLUIDS

Advances in instrumentation have led to much faster and more selective chromatographic separations with automated sample injection and computer-assisted data analysis, thereby increasing the speed and precision of analyses. Thus, artificial intelligence is being used increasingly in the clinical laboratory, both in the form of stand-alone expert systems for clinical decision making and as knowledge-based systems embedded within laboratory instrumentation [28]. However, sample preparation remains the limiting factor for completely automated analysis. Automation via robotization essentially links a series of off-line approaches for sample preparation. The current market for robotics is dominated by applications for pharmaceutical and biotechnology companies, particularly in the fast-growing fields of drug discovery, using combinatorial chemistry techniques, and high-throughput screening of candidate drugs, such as those derived from natural compounds [29]. Workstations and laboratory robotics are expanding vigorously for liquid handling and sample preparation. Workstations operate best when there are one to three functions being automated; more operations or less defined ana-

Table 1 Selected Determinations of Drugs in Biological Fluids by GC–MS, Using Automated Cleanup Systems

Analyte/sample	Eluent	Comments	Ref.
β-agonists/urine	Ethyl acetate (3% ammonia)	Enzymatic hydrolysis of conjugate metabolites. SPE (Clean Screen DAU cartridge), followed by manual derivatization to TMS. Screening of 13 β-agonists with EI and confirmation by PCI over the range 0.5–10 ng/ml, with RSD of approximately 1.6%.	9
Amphetamines/urine	Methanol (0.1% H$_2$SO$_4$)	SPE (automatic urine injection onto a Clean Screen ODS cartridge) followed by evaporation of the eluate to dryness. Manual derivatization to trifluoroacetate derivatives in CH$_2$Cl$_2$. Calibration graphs (EI) are linear from 0.1 to 10 ng/ml; RSD = 6%.	23
Zatebradine/plasma	Methanol	After centrifugation, 1 ml of the supernatant is automatically cleaned up using the Zymark Benchmate (with Bond-Elut column). The analyte extract in 2 ml methanol is evaporated to dryness and redissolved (preconcentration ratio 2 ml/30 μl). Calibration graphs (PCI) are linear from 0.2 to 30 ng/mL; RSD = approximately 5%.	24
Fentanyl and its analogs/urine, serum, plasma and blood	CH$_2$Cl$_2$/isopropanol/NH$_3$ (39:10:1)	After centrifugation, the supernatant is introduced into an SPE (Clean Screen extraction cartridge). The eluate is evaporated under N$_2$. Recoveries are ≥92%.	25
Designer drugs and ecstacy/urine	Methanol (3% ammonia)	Enzymatic hydrolysis with β-glucuronidase, followed by SPE (mixed-bed of C$_8$ and cation exchange). The extract of drugs in ammoniacal methanol is evaporated and the residue heated with acetic anhydride, reevaporated and dissolved in methanol. Significant masses are tabulated for five compounds.	10
Opiates/serum and blood	CH$_2$Cl$_2$/isopropanol/NH$_3$ (40:20:1)	After centrifugation, the supernatant (in buffer of pH 8.5) is cleaned up by SPE. Four mixed-phase columns were compared. For morphine and metabolites, recoveries are 6 to 124% for 0.05 to 0.5 μg/ml analyte. Large peaks due to plasticizers are observed in the chromatograms.	26
Cocaine and its metabolites/meconium and urine	CH$_2$Cl$_2$/isopropanol/NH$_3$ (40:10:1)	After centrifugation, the supernatant (in buffer of pH 4) is loaded onto SPE columns. After evaporation, the residue in acetonitrile is manually derivatized to TMS. Calibration ranges for cocaine and its metabolites are linear from 1.6 to 1000 ng/ml.	27
Benzodiazepines/urine	Not stated	After enzymatic hydrolysis, samples are loaded onto a Toxi-Lab Spec column containing a 15-mg disk. Analytes are on-disk derivatized to alkylsilyl derivatives.	8

lytical steps need the sophistication of a robotic arm system. Typical workstations—manufactured by Gilson (Villiers-le-Bel, France) (ASTED XL and ASPEC XL) and Zymark (Hopkinton, MA) (Rapid Trace)—and robotic arms (also from Zymark) operate for sample cleanup using dialysis and SPE, among other operations (with pipetting modules, automatic column conditioning, multiple fraction collection, and 100-sample capacity). These systems have been used conventionally for drug analysis using liquid chromatography (LC) and therefore include direct high-performance liquid chromatography (HPLC) injection; few applications have been described for GC–MS, however. Other sources of these systems and companies, such as the annual *Lab-Guide*, should be consulted for a more comprehensive information.

Workstations and robotic systems are very expensive, so inexpensive alternatives such as flow configurations have been developed for automated sample preparation. The earliest flow systems for sample preparation were used for GC determination (with flame ionization detector [FID] or electron capture detector [ECD] detection) of organic compounds, which requires no special extraction or derivatization, in environmental matrices [30–34]. Automated GC–MS systems for the determination of volatiles in water or air [35–38] are the most commonly reported. Detailed descriptions of these systems can be found elsewhere in this book. Few continuous flow systems (CFSs) for the automated pretreatment of biological fluids in combination with GC–MS have been developed to date. The intrinsically discrete nature of the GC–MS sample introduction mechanism makes on-line coupling to continuous flow systems theoretically incompatible for reasons such as the different types of fluids used (liquid and gas) and the fact that the chromatographic column affords volumes of only 1 to 2 μl of cleaned-up extract. Therefore, the organic extracts from CFSs have traditionally been collected in glass vials and aliquots for manual transfer to the GC–MS instrument (off-line approach); only in a few cases is an appropriate interface used to link the CFS to the GC–MS instrument (on-line approach). These are the topics dealt with below.

4.1. Fully Automated Sample Pretreatment Systems Coupled Off-Line to Gas Chromatography–Mass Spectrometry

Dialysis is a barrier separation technique, in addition to osmosis and ultrafiltration, among others. Mass transfer takes place through a semipermeable membrane located between two liquid (donor and acceptor) phases that need not be immiscible. Separations are kinetically controlled and based on the occurrence of concentration gradients between the two liquid phases, isolated by a selective membrane that discriminates solutes according to pore size and/or electric charge [39]. The applications of continuous dialysis as a purification step for biological fluids prior to GC–MS analysis has received little attention despite its significance to routine

clinical laboratories. In addition, this technique has limited capability to imple-
ment preconcentration (normally it is employed for sample dilution and interferent removal), so continuous dialysis systems for drug analysis often include
trace enrichment of the dialysate. Only off-line approaches including autosam-
plers with sample preparation functions for liquid samples before injection into
the GC have been used; thus, the automated sequential trace enrichment of dialy-
sates (ASTED) instruments from Gilson have been employed for dialysis of biological fluids. The earliest attempt in this context was made by Krogh et al. [15]
in the determination of opiates; the ASTED system used is schematically depicted
in Figure 3. The dialysis unit was of the sandwich type and furnished with a
microporous cuprophane membrane. The sample (blood) was held static in the
donor channel of the dialyzer while 5 ml of 10^{-4} M ammonia was delivered in
pulses of 650 μl; ammonia diffused through the membrane and into the sample,

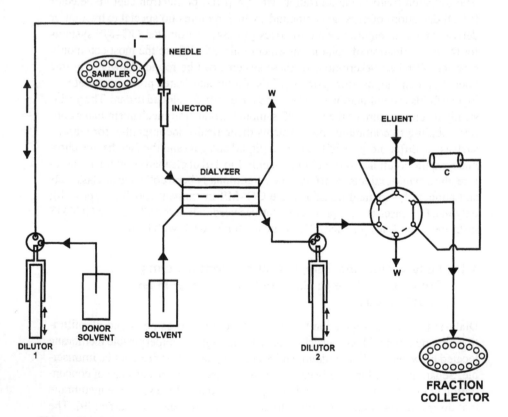

Figure 3 Automated ASTED system for the determination of opiates in human blood.
C = polystyrene–divinylbenzene column; W = waste. See text for details.

which resulted in reduction of the protein-opiate binding without protein precipitation. Drug recoveries ranged from 66 to 80% for 9 minutes of dialysis. The dialysate in ammonia was preconcentrated on a SPE unit including a polystyrene–divinylbenzene column (1 cm × 2 mm); acetonitrile was used as eluent, and the first 100 μl of eluate was discarded and the next 700 μl was collected in glass vials. The eluates were evaporated and then derivatized. Approximately 100 samples can thus be automatically prepared in 24 hr. The analytical features of this approach are shown in Table 2. A similar dialysis–SPE unit was used by Herráez-Hernández et al. [16] for the determination of drugs in plasma, using some benzodiazepines as model compounds. In contrast to the above-described method, no further manual steps were required after elution as no derivatization of the benzodiazepines was necessary. The ASTED system included a similar dialysis unit with the same membrane [15], but the donor and acceptor channel volumes of the dialysis cell were lower (100 and 170 μl, respectively) and both liquid (donor and acceptor) phases were water. For preconcentration, a PLRP-5 column was used and the analytes were eluted with 275 μl of ethyl acetate. On-line experiments were done with FID and nitrogen–phosphorus detector (NPD) detection (linear range 0.1 to 20 μg/ml benzodiazepines; relative standard deviation (RSD) 4 to 16%. With MS, 100-μl volumes of eluate were manually injected; this was only used for confirmation, so no analytical features were given.

Two commercially available automated devices for the preparation of biological fluids based on SPE have been reported. One uses an HP-PrepStation (Hewlett-Packard, Palo Alto, CA) for the extraction and derivatization of narcotics in serum samples after enzymatic hydrolysis [11]; C_{18} cartridges were used for SPE. The other is a fully automated procedure applied to the confirmation by GC–MS of positives of cocaine and benzoylecgonine in previously screened urines, using a laboratory robotic system [22].

Continuous flow systems for the automated pretreatment of biological fluids for GC–MS of drugs have seemingly been developed by the authors' group only. A CFS relying on SPE constitutes the most simple, robust, cheap, and fruitful approach in this context. Sample, reagents, and eluents are introduced into a continuous module furnished with a sorbent column located in the loop of an injection valve. The CFS operation comprises three steps: sorption (sample introduction), elution, and derivatization—the last two can occur simultaneously, however. The final extract obtained can be introduced into the GC–MS (Fig. 4).

The simplest version of such CFS was that developed for the rapid SPE–derivatization, with volume-based sampling (1 ml of urine), of various abuse drugs (barbiturates, opiates, and cocaine, among others) in real human urine [19]. The sorbent material was polymeric Amberlite XAD-2 and there were no restrictions on the retention pH as all the drugs assayed exhibited maximum adsorption within the range 6 to 9, which is within the normal urine pH range. The derivatising reagent for silylation, BSTFA, was added to the eluent (trichloromethane

Table 2 Applications of Automated Sample Pretreatment Systems to Drug Analyses by GC–MS (EI Mode)

Drug	Matrix	System	Separation principle/derivatizing reagent	Linear range (ng/ml)	RSD (%)	Ref.
Opiates	Plasma and whole blood	ASTED	Dialysis and SPE/BSTFA[a]	0.2–5.0 nmol/ml	1.3–7.7	15
Benzodiazepines	Plasma	ASTED	Dialysis and SPE	Not stated	Not stated	16
Narcotics	Serum	HP-PrepStation	SPE/derivatization	2–250	0.5–2.0	11
Cocaine and benzoylecgonine	Urine	Zymate Robot	SPE/derivatization	5–100	1.1–6.0	22
Abuse drugs	Urine	CFS	SPE/BSTFA	1–2000	2.1–6.8	19
Antiinflammatory drugs	Urine and plasma[b] (horse)	CFS	SPE/CH_3I-K_2CO_3	50–3000	3.4–6.0	20
Cocaine and its metabolites	Urine	CFS	SPE/BSA	5–3000	3.1–5.8	21
Benzodiazepines	Urine[c]	CFS	SPE	5–2000	4.5–6.5	42
Abuse drugs	Urine and serum	CFS	SPE	5–2500	2.9–4.6	43
Caffeine	Urine	ASPEC	SPE	Not stated	Not stated	44

[a] Manual derivatization with N,O-bis(trimethylsilyl)trifluoroacetamide (BSTFA).
[b] Plasma sample required manual liquid–liquid extraction before SPE.
[c] Urine was hydrolyzed with 3 M HCl (100°C, 1 hr).
Abbreviations: ASTED, automated sequential trace enrichment of dialysate; CFS, continuous flow system; ASPEC, automated sample preparation with extraction columns; SPE, solid phase extraction; BSA, bis(trimethylsilyl)acetamide.

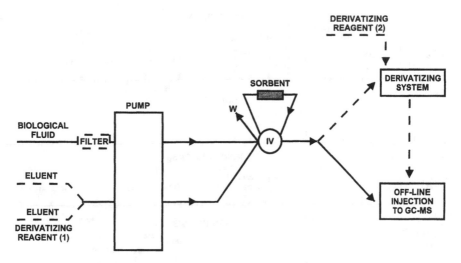

Figure 4 Configurations of an SPE-based CFS for the automated pretreatment of biological fluids. IV = injection valve; W = waste. See text for details.

(CHCl$_3$):acetone, 1:1, v/v) in a 1:1 ratio; the derivatization was carried out without halting the flow and at room temperature. One potential disadvantage of the method may be its unsuitability for on-line hydrolysis of the glucuronide conjugates of the drugs, as the enzymatic reaction requires a long time (over 60 minutes). However, the use of nonhydrolyzed urine allows one to monitor relevant metabolites (e.g., 6-*O*-monoacetyl morphine (6-MAM) from heroin, which would otherwise be decomposed [40] and whose presence is decisive for stating heroin use). Thus, as can be seen in Figure 5, codeine, morphine, and 6-MAM were detected in a heroin-user urine sample following automated sample pretreatment in the system of Figure 4.

An automated system for the SPE of 17 NSAIDs and on-line methylation prior to their off-line introduction into a GC–MS has also been proposed [20]. The ensuing method involves derivatization with methyl iodide (CH$_3$I) in the presence of solid potassium carbonate (K$_2$CO$_3$) as catalyst (official method) and was satisfactorily applied to the determination of these drugs in racehorse urine and plasma. Amberlite XAD-2 was found to be the most convenient sorbent and acetonitrile the best eluent. No pH restrictions regarding NSAIDs were observed as they were completely retained within the range 5.5 to 10; however, elution required conditioning with dilute acetic acid. The major problem was the water traces remaining in the CFS, which hindered development of the derivatization reaction through dissolution of K$_2$CO$_3$. Acetic anhydride (20%, v/v) was added to the eluent to act both as drying agent and conditioner for the resin at an appro-

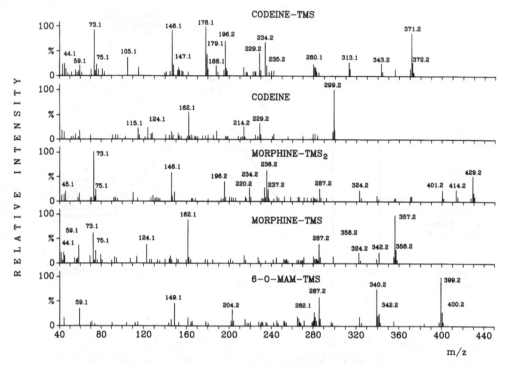

Figure 5 EI mass spectra for opiates detected in a urine sample from a heroin addict. See text for details.

priate pH. As in the previous system, CH_3I (the derivatising reagent) was added to the eluent (25% v/v). The main technical disadvantage of this method results from compaction of the K_2CO_3 column (located after the sorbent column) but was overcome by packing the catalyst with sodium aluminosilicate pellets placed at intervals in the column. Under these conditions, the derivatization reaction was completed within 5 minutes without heating. The analytical features of the method are summarized in Table 2. The rapid one-step extraction–derivatization of the sample and the avoidance of the cleanup and evaporation steps significantly reduce the analysis time and solvent consumption. It is worth noting that the analytes [19,20] were derivatized without the need to use the drastic temperature and time conditions recommended in the literature (60 to 100°C for 0.5 to 3 hr).

Solid-phase extraction, elution, evaporation, and formation of TMS-derivatives is the most frequently used pretreatment method for the determination of cocaine and its metabolites by GC–MS. A CFS for the automated pretreatment of urine samples for the determination of cocaine, ecgonine, ecgonine methyl ester, and benzoylecgonine was developed [21]. In this case, simultaneous elution and derivatization of the ecgonine and benzoylecgonine was impossible under the optimal conditions established, i.e., retention on RP-C_{18} and elution with 2% ammonium hydroxide (NH_4OH) in isopropanol:$CHCl_3$ (25:75, v/v) spiked with the reagent bis(trimethylsilyl)acetamide (BSA). Therefore, both steps must be carried out separately (see Fig. 4) and the eluent removed by evaporation before the derivatizing reaction. Then, an on-line evaporation step must be implemented in the proposed manifold before the automated addition into the vial of BSA for redissolution and subsequent derivatization at 70°C for 15 minutes. The analytical features of the method are shown in Table 2.

Finally, a CFS for confirming the presence of benzodiazepines (through benzophenone formation) in previously screened positive samples was reported. As hydrolysis of the urine samples was required in the screening method [41], hydrolyzed samples were also used for the GC–MS method [42]. This option resulted in several advantages, such as increased thermal stability of the benzo-phenones relative to the parent compounds, and a higher sensitivity, probably due to cleavage of the glucuronide bond, which allowed the use of sample volumes as low as 0.5 ml of urine. The hydrolysis was carried out in a discrete mode, as it required drastic conditions (100°C, 1 hr, 3 M HCl). The CFS developed for this purpose introduced two additional steps: on-line filtration of the urine samples through a home-made cotton column located at the inlet of the sample aspiration channel (see Fig. 4) and adjusting the pH of the analytes prior retention on an XAD-2 column, as sorption was favoured by an alkaline medium. Thus, the urine, in 3 M HCl, was merged with a 4 M NaOH stream to provide a pH greater than 12 after the mixing point prior to retention. Trichloromethane was used as eluent because of its increased selectivity. The analytical features of this method are summarized in Table 2.

4.2. Fully Automated Pretreatment Systems Coupled On-Line to Gas Chromatography–Mass Spectrometry

The chief restrictions arising from the combination of a CFS and a gas chromato-graph are due to the different aggregation states of the carrier and the mobile phase, and also to the pressure difference. The interface must be designed to cause minimal reversible changes in the gas chromatograph in such a way that it can be used in the conventional injection mode, using an injected volume of only a few microliters and adapting the split ratio to it.

The simplest way of directly coupling a home-made CFS to a GC–MS instrument is by using a high-pressure injection valve [32,43]. A schematic dia-gram of the interface is shown in Figure 6. The outgoing organic stream from

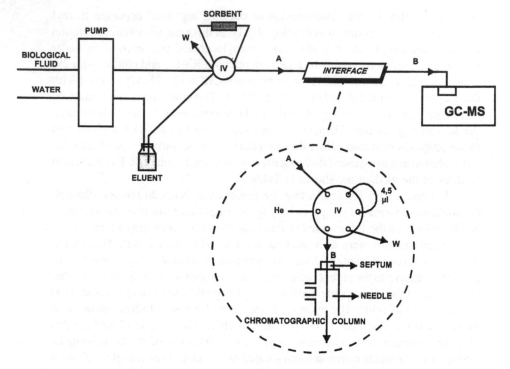

Figure 6 Scheme of a CFS for the preconcentration/determination of abuse drugs in human serum and urine. IV = injection valve; W = waste. See text for details.

the CFS fills the 4.5 μl loop of the injection valve (inner volume 2.5 μl), constructed from polytetrafluoroethylene (PTFE). A single helium input through the injection valve at a flow rate of 0.7 ml/min allows the organic plug to be transferred to the gas chromatograph as the valve is switched, which shuts the carrier inlet to the instrument. A stainless steel tube (70 mm × 0.3 mm ID) furnished with an injection needle was fitted to the carrier outlet and the needle was inserted into the septum of the injection port [43]. The interface unit was kept at room temperature and no instrumental modifications were required. Various drugs of toxicological interest (viz., methamphetamine, alphaprodine, lidocaine, methadone, cocaine, and codeine) were determined in human urine and serum by using this on-line configuration. Dilute samples (equivalent to 1 ml of urine and serum) were continuously aspirated into a CFS similar to those previously described; drugs were adsorbed onto a C_{18} column packed with approximately 40 mg of sorbent and then eluted with $CHCl_3$; under these conditions, the column can be reused for at least three months. The analytical features of the method are summarized in Table 2. When using this on-line configuration, the split ratio of the instrument must be increased at least fourfold in order to avoid saturation of the stationary phase.

Finally, an interface similar to that previously described was developed for the direct coupling of a commercially available automated sample preparation with extraction columns (ASPEC) unit and large-volume PTV–GC–MS [44]. The interface between the ASPEC system and the programmed temperature vaporization (PTV) injector was made from the standard Rheodyne injection valve of the ASPEC unit; however, the eluate was transferred in a discrete manner. Minor instrumental modifications were needed for the coupling; an injected volume of 100 µl was chosen and two gas chromatographs (on-line connected) were used. Although the proposed system was not applied to drug analyses—only to determine caffeine spiked to human urine—it will probably be used for the determination of drugs. A 4-ml sample was passed through a 200-mg C_{18} cartridge previously conditioned with methanol and water. After retention, the cartridge was cleaned up with n-hexane and caffeine was finally eluted with 4 ml of $CHCl_3$. The cartridge cannot be reused. No analytical features were given other than the reproducibility of the large injected volume (100 µl), which was found to be 6.9%

Although the examples clearly show the potential of the method for handling sample volumes in the ml region, one must admit that on-line coupled SPE–GC–MS has not yet become such a routine application as on-line SPE–LC, mainly because of the incompatibility of the chromatographic system with aqueous phases (the final injected volume must contain absolutely no traces of water if deactivation and deterioration of the chromatographic column are to be avoided). This problem can be especially serious when using polar organic solvents (methanol, acetonitrile, etc.) as eluents for the retained compounds.

5. CONCLUSIONS

The high potential of GC–MS for identifying and quantifying drugs and their metabolites in biological fluids depends critically on preliminary analytical operations, which are the source of major errors, both systematic and random, that strongly affect the quality of the results upon which decisions must be based. Matrix problems in biological fluid analysis are the most serious ''bottleneck'' for the corresponding analytical processes. Any improvement in sample preparation also improves the quality of the analytical information derived. This chapter demonstrated the potential of automated systems coupled to GC–MS for such a purpose and the state of the art and prospects in this respect.

REFERENCES

1. M. Valcárcel, Fresenius J. Anal. Chem., 343 (1992) 814–816.
2. M. Valcárcel, M.D. Luque de Castro, and M.T. Tena, Anal. Proc., 30 (1993) 276–279.
3. W.J. Hurst and J.W. Mortimer, Laboratory Robotics, 1987, VCH, New York.
4. Remington Farmacia, 17th ed., vol. 1, 1987, Editorial Médica Panamericana, Buenos Aires, pp. 980–1061.
5. A.C. Moffat (ed.), Clarke's Isolation and Identification of Drugs, 2d ed., 1986, Pharmaceutical Press, London, pp. 111–113.
6. D.G. Deutsch (ed.), Analytical Aspects of Drug Testing, 1989, Wiley, New York, pp. 287–289.
7. M. Valcárcel, S. Cárdenas, and M. Gallego, Trends Anal. Chem., 18 (1999) 685–694.
8. J.W. King and L. J. King, J. Anal. Toxicol. 20 (1996) 262–265.
9. M. Montrade, B. Le Bizec, F. Monteau, B. Siliart, and F. Andre, Anal. Chim. Acta, 275 (1993) 253–268.
10. P. Gerhards and J. Szigan, LaborPraxis, 20 (1996) 212–214.
11. K. Harzer, LaborPraxis, 20 (1996) 44, 47, 48.
12. V. Pichon, M. Bouzies, C. Miege, and M.C. Hennion, Trends Anal. Chem., 18 (1999) 219–235.
13. F. Degel, Clin. Biochem., 29 (1996) 529–540.
14. S.W. Myung, H.K. Min, S.K. Kim, M.S. Kim, J.B. Cho, and T.J. Kim, J. Chromatogr. B, 716 (1998) 359–365.
15. M. Krogh, A.S. Christophersen, and K.E. Rasmussen, J. Chromatogr. B, 621 (1993) 41–48.
16. R. Herráez-Hernández, A.J.H. Louter, N.C. van de Merbel, and U.A.Th. Brinkman, J. Pharm. Biomed. App., 14 (1996) 1077–1087.
17. D.J. Harvey, in: Mass Spectrometry, Clinical and Biomedical Applications, vol. 1, 1992, Plenum Press, New York, pp. 216–246.
18. H. Tsuchihashi, K. Nakajima, M. Nishikawa, K. Shioni, and S. Takahashi, J. Chromatogr. 467 (1989) 227–235.
19. S. Cárdenas, M. Gallego, and M. Valcárcel, Rapid Comm. Mass Spectrom., 11 (1997) 298–306.
20. S. Cárdenas, M. Gallego, M. Valcárcel, R. Ventura, and J. Segura, Anal. Chem., 68 (1996) 118–123.
21. S. Cárdenas, M. Gallego, and M. Valcárcel, Rapid Comm. Mass Spectrom., 10 (1996) 631–636.
22. R.W. Taylor and S.D. Lee, J. Anal. Toxicol., 15 (1991) 276–278.
23. M. Kagati, H. Nishioka, K. Nakajima, M. Nishikawa, H. Tsuchihashi, M. Takino, and K. Yamagushi, Jpn. J. Toxicol. Environ. Health, 41 (1995) 148–154.
24. J. Schmid, A. Buecheler, and B. Mueller, J. Chromatogr. B, 658 (1994) 93–101.
25. F. Caputo, D. Fox, G. Long, C. Moore, S. Mowes, and G. Purnell, LaborPraxis, 19 (1995) 62–64, 67.
26. M.J. Bogusz, R.D. Maier, K.H. Schiwy-Bochat, and U. Kohls, J. Chromatogr. B, 689 (1996) 177–188.

27. J. Oyler, W.D. Darwin, K.L. Preston, P. Suess, and E.J. Cone, J. Anal. Toxicol., 20 (1996) 453–462.
28. J.F. Place, A. Truchaud, K. Ozawa, H. Pardue, and P. Schnipelsky, J. Autom. Chem., 17 (1995) 1–15.
29. A. Newman, Anal. Chem., 69 (1997) 255A–259A.
30. J. Roeraade, J. Chromatogr., 330 (1985) 263–274.
31. E. Fogelqvist, M. Krysell, and L. G. Danielsson, Anal. Chem., 58 (1986) 1516–1520.
32. E. Ballesteros, M. Gallego, and M. Valcárcel, Anal. Chem., 62 (1990) 1587–1591.
33. M. Valcárcel, E. Ballesteros, and M. Gallego, Trends Anal. Chem., 13 (1994) 68–78.
34. J.J. Vreuls, A.J. Bulterman, R.T. Ghijsen, and U.A.Th. Brinkman, Analyst, 117 (1992) 1701–1705.
35. A.K. Vickers and L.M. Wright, J. Autom. Chem., 15 (1993) 133–139.
36. A.C. Lewis, R.E. Robinson, K.D. Bartle, and M.J. Pilling, Environ. Sci. Technol., 29 (1995) 1977–1981.
37. P.G. Simmonds, S. O'Doherty, G. Nickless, G.A. Sturrock, R. Swaby, P. Knight, J. Ricketts, G. Woffendin, and R. Smith, Anal. Chem., 67 (1995) 717–723.
38. M.J. Yang and J. Pawliszyn, Trends Anal. Chem., 15 (1996) 273–278.
39. J.F. Van Staden and A. Van Rensburg, Analyst, 115 (1990) 1049–1054.
40. P. Wernly and W. Thormann, Anal. Chem., 63 (1991) 2878–2882.
41. D. Gambart, S. Cárdenas, M. Gallego, and M. Valcárcel, Anal. Chim. Acta, 366 (1997) 93–102.
42. S. Cárdenas, M. Gallego, and M. Valcárcel, J. Chromatogr. A, 823 (1998) 389–399.
43. S. Cárdenas, M. Gallego, and M. Valcárcel, Rapid Comm. Mass Spectrom., 11 (1997) 973–980.
44. S. Öllers, M. Van Lieshout, H. Janssen, and C.A. Cramers, LC-GC Int., 10 (1997) 435–439.

10
Gas Chromatography–Mass Spectrometry Analysis of Anesthetics and Metabolites Using Multidimensional Detection Strategies

M. Désage and J. Guitton
Université Claude Bernard, Lyon, France

1. INTRODUCTION

1.1. Properties and Clinical Uses of Anesthetics

Anesthetics may be grouped into different families, and here we indicate briefly their pharmacological properties and clinical uses. Propofol is an alkyl phenol administered intravenously. This drug is an induction agent and it is particularly useful for day case anesthesia. It can also be used for the maintenance of anesthesia by continuous infusion. Midazolam (MDZ) is a benzodiazepine that is mainly used for the induction of anesthesia. This drug is also employed for sedation during endoscopy and in the intensive care unit. The inhalational anesthetics are a group of gases and vapors allowing a reversible abolition of the perception of external stimuli. This review focuses only on the halogenated class of volatile anesthetics, which are mainly used in clinical practice. Ketamine is an aryl cyclo-hexylamine used as a racemic mixture administered either intravenously or intramuscularly. Clinical uses of ketamine include anesthesia in radiotherapy treatment and other investigations. Fentanyl is a leader in the opioid anesthetics family, which includes alfentanil (ALF), sufentanil (SUF), and remifentanil. These drugs exhibit several differences in their pharmacokinetic and pharmacological properties, which give them specific clinical uses in the anesthetic and analgesic fields. Local anesthetics (articaine, lidocaine, and bupivacaine) produce a reversible depression of nervous conduction with loss of sensory and motor function. Gamma-hydroxybutyrate (GHB) is an endogenous constituent of mammalian brain. This drug is used as an anesthetic adjuvant in surgery and obstetrics.

1.2. Interest in Anesthetics Analysis

As interest in the clinical applications of these compounds has increased, the necessity to detect them has also increased. At the same time, it is important to study the biotransformations of these compounds in order to understand their metabolism, as well as to quantify them. Moreover, modern anesthetic techniques involve combinations of several drugs that may produce interactions. The knowledge of blood concentrations and metabolic pathways of the different anesthetics provides for safety in daily practice.

In view of this, the use of coupled analytical techniques affords the efficacy of a high-resolution chromatographic system and the high-power identification of spectroscopy. Nevertheless, the capabilities of these techniques may sometimes be insufficient. In these cases, some analysts have implemented other complementary techniques or have developed new analytical strategies to combine the information provided by each approach.

Numerous areas of investigation are discussed in this chapter with the focus on different strategies that can be brought into play thanks to gas chromatography–mass spectrometry (GC–MS). First, an overview of investigations is provided where GC–MS is considered as the technique of choice for direct identification of metabolites or for quantification of the parent drug and/or its metabolites. In this section, we also discuss derivatization steps, which are part of several strategies. Secondly, strategies involving the use of labeled compounds are reviewed that allow numerous investigations in quantification and metabolic areas. The last part focuses on strategies that involve MS and another detection mode coupled with GC, such as atomic emission detection (AED) and Fourier-transform infrared (FTIR) spectroscopy. Each of these discussions presents selected examples of anesthetic drugs to illustrate the multidimensional strategy analysis.

2. COUPLING TECHNIQUES BASED ON SPECTROSCOPIC DETECTION

Molecular analysis methods coupled to analytical separation techniques exhibit a definitive advantage over destructive methods, as they permit the identification of peaks in a chromatogram. Therefore, they take a prominent place in the studies concerning metabolic investigations. Where classical detection, such as flame ionization detection (FID) or nitrogen–phosphorus detection (NPD), can only provide information on retention time, molecular detection methods that provide spectral data are potentially more efficient for identification of unknown peaks such as metabolites or degradation products.

3. GAS CHROMATOGRAPHY–MASS SPECTROMETRY

Mass spectrometry coupled to capillary GC constitutes the ideal method, in classical electron ionization (EI) mode, for the analysis of compounds able to be gas chromatographed, i.e., sufficiently volatile and nonthermally degradable compounds. In recent years, increasing development of new modes of atmospheric pressure ionization processes has created greater opportunities for the coupling of high-performance liquid chromatography (HPLC) to MS. Nevertheless, this very promising technique is limited by the fact that the ionization process results in quasi–chemical ionization (CI) spectra, from which the structural information is poor compared with that obtained by the standard EI fragmentation process. Mass spectrometry takes on particular importance in metabolism studies as reflected by the large amount of literature published in the last decades. Gas chromatography–mass spectrometry is an extremely useful tool in EI mode, which provides valuable structural information, or in CI mode, which allows one to obtain complement data on the pseudomolecular ion. Indeed, the value of MS lies in its capacity to identify compounds arising from parent drugs able to undergo a wide variety of enzymatic transformations leading to, for example, hydroxylation, N-dealkylation, reduction, or the introduction of a novel group. The GC–MS technique can be implemented in two ways: full-scan mode or selective ion monitoring (SIM) mode.

3.1. Full-Scan Mode

The classical full-scan mode permits the identification and detection of parent molecules and metabolites by examination of the spectrum of each peak in the total ion chromatogram (TIC) with very good sensitivity in current instrumentation. The examples given bellow illustrate the basic technique of detection and identification by GC–MS.

3.2. Basic Use of Full-Scan Mode

An example of the use of GC–MS for local anesthetic detection is discussed by Fox et al. [1]. It concerns articaine used in dentistry. The investigation demonstrates the power of this detection technique by proving the presence of the intact molecule in the urine 9 hr after the administration.

The most common biotransformations of volatile anesthetics involve oxidation (either cleavage or dehalogenation) except for halothane, which undergoes reduction as well as oxidation. Nevertheless, there is great variability in the extent of transformation of the drugs. Enflurane biotransformation was investigated in human urine using GC–MS [2]. In this study, the presence of difluoromethoxy-2,2-difluoroacetic acid was demonstrated as its spectrum exhibited fragment ions at m/z 51, 67, 95, and 111. This metabolite was previously described as the result of the oxidative dechlorination at the β-C atom.

Propofol provides a simple example of metabolite detection and identification. This drug and its metabolites are mainly eliminated in vivo in a conjugated form, i.e., as glucuronate or sulfate, requiring a deconjugation reaction prior to GC–MS analysis. The metabolic pathway is simple, since only three phase-I metabolites were described in different species. Identification of 2,6-diisopropyl-1,4-quinol (hydroxylation at the *para* position on the phenyl ring) was carried out with GC–MS in EI mode, leading to a mass spectrum with a molecular ion at m/z 194 and a major fragment at m/z 179 due to the loss of a methyl group [3]. Recently two other hydroxymetabolites [(2-ω-propanol)-6-isopropylphenol, and (2-ω-propanol)-6-isopropyl-1,4-quinol] were identified in humans by using full-scan mode [3a].

Many years ago, the identification of ketamine metabolites was performed in biological fluids from rats using GC–MS [4]. In this study, Kochhar [4] described two metabolites: norketamine (NK) and 5,6-dehydronorketamine (DHNK). No derivatization step was performed and GC was carried out on a glass column packed with 1% Carbowax 20 M. In EI mode, the main fragments were observed at m/z 180, 166, and 153 for ketamine, NK, and DHNK, respectively. Using CI with methane as reagent gas, the main ions were observed at m/z 238 [M + H]$^+$, m/z 207 [(M + H) − NH$_3$]$^+$, and m/z 205 [(M + H) − H$_2$O]$^+$ for ketamine, NK, and DHNK, respectively.

Metabolic pathways of fentanyl (F$_T$I) have been studied in vivo in several mediums, such as blood, saliva, and urine, and in vitro using isolated hepatocytes and microsomes from different species. Fentanyl is metabolized according to three reactions: oxidative *N*-dealkylation, aliphatic hydroxylation, and/or aromatic hydroxylation, leading to 10 metabolites. An eleventh phase-I metabolite was described resulting from hydrolysis of the amide bond. Structures of F$_T$I metabolites described in several studies are shown in Figure 1. Throughout this chapter we use F$_T$I for numerous examples because the metabolism of this drug has been extensively studied with techniques based on several multidimensional detection strategies. First, GC–MS was employed to attempt to identify fentanyl metabolites in rat urine [5]. In this study, the authors concluded that fentanyl was metabolized mainly by a hydrolytic pathway leading to F$_T$III as main metabolite, as confirmed by the mass spectra.

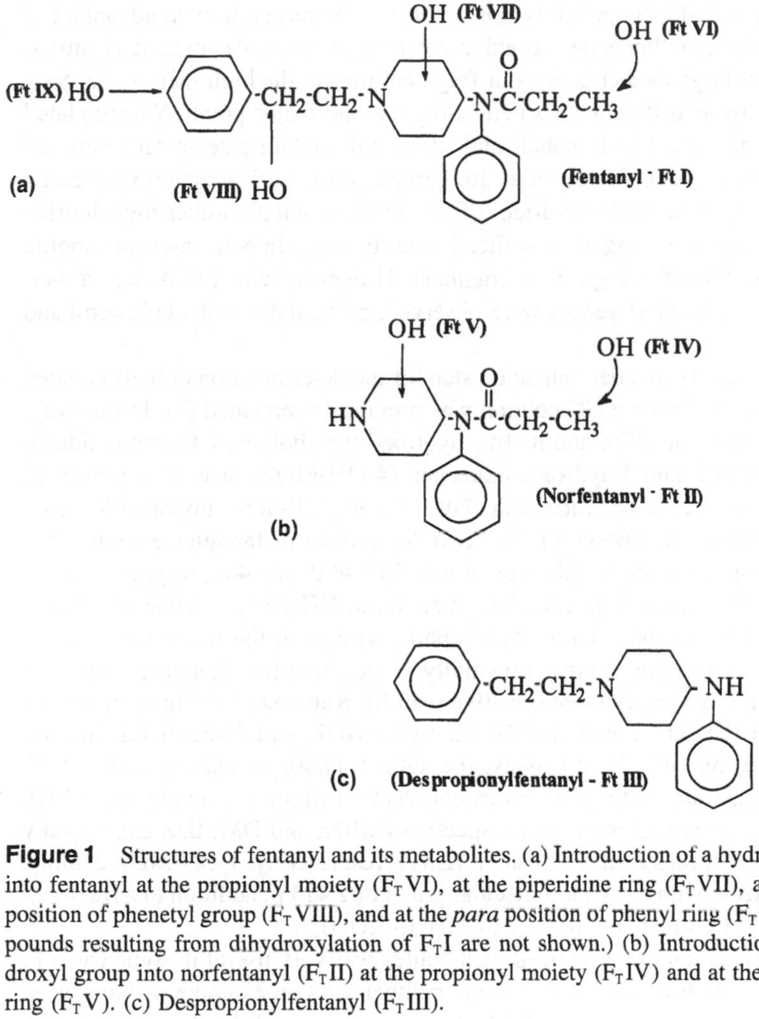

Figure 1 Structures of fentanyl and its metabolites. (a) Introduction of a hydroxyl group into fentanyl at the propionyl moiety (F_T VI), at the piperidine ring (F_T VII), at the alpha position of phenetyl group (F_T VIII), and at the *para* position of phenyl ring (F_T IX). (Compounds resulting from dihydroxylation of F_TI are not shown.) (b) Introduction of a hydroxyl group into norfentanyl (F_T II) at the propionyl moiety (F_T IV) and at the piperidine ring (F_T V). (c) Despropionylfentanyl (F_T III).

At last, surface ionization coupled to GC was reported by a Japanese research group (see Chap. 2) and applied to lidocaine.

3.3. Derivatization

In many cases, the detection of metabolites requires a derivatization step since these compounds are usually too polar. This process may be considered as a tedious step before analysis. However, it can also be considered as an optimization strategy for MS. Indeed, this constraint may be turned into an advantage if a derivatization reaction is performed to create high-molecular-mass derivatives, which exhibit large molecular ions or fragment ions in the high-mass scale. Several studies from different anesthetic drugs support this point. N-dealkylated metabolites of fentanyl, alfentanil, and sufentanil contain a secondary nitrogen that can be derivatized. Valaer et al. [6] compared the derivatization with either pentafluoropropionic acid anhydride (PFPA) or with pentafluorobenzoyl chloride (PFB-Cl) for the screening of these metabolites in urine. In both cases, metabolite derivatives exhibited a high-mass fragment. However, with PFPA derivatives, lower background interferences were observed, particularly with cholesterol and steroids.

The necessity of a derivatization step for the determination of hydroxylated metabolites of MDZ on a GC column was previously indicated [7]. In this way, the determination of MDZ and its two hydroxyl metabolites, 1-hydroxymidazolam (1-OH-MDZ) and 4-hydroxymidazolam (4-OH-MDZ), was also performed with GC–MS after a derivatization with N-methyl-N-(tert-butyldimethylsilyl) trifluoroacetamide (MTBSTFA). The two derivatized metabolites exhibited the same three fragments in the EI mode at m/z 398, 400, and 440, respectively [8].

Furthermore, some specific derivatization of NH group, such as fluoroacylation permit the introduction of electrophilic groups in the molecule ($COCF_3$, COC_2F_5), thus yielding a great sensitivity in negative-ion chemical ionization (NCI), which is generally more sensitive than EI. Rubio et al. [9] have described a method for the determination of MDZ, 1-OH-MDZ, and desmethylmidazolam (DMMDZ) using NCI. In this assay, the authors chose to analyze 1-OH-MDZ after derivatization with N-O-bis(trimethylsilyl)trifluoroacetamide (BSTFA), since MDZ carries a Cl atom. Mass spectra of MDZ and DMMDZ exhibit only pseudomolecular ions at m/z 325 and m/z 311, respectively. 1-OH-MDZ exhibits a mass spectrum with a pseudomolecular ion at m/z 413 in addition to a fragment ion at m/z 323 corresponding to $[M-((CH_3)_3-SiOH)]^+$.

The formation of derivatized metabolites was very useful in some cases to avoid misinformation about a metabolic pathway. As an example, Adams et al. [16] used pentafluoropropionyl (PFP) derivatives to avoid the thermal degradation of hydroxylated metabolites during GC–MS. This degradation applied to DHNK indicated that it is an artifact of the analytical process rather than a major metabolite as previously described [4,10]. Another example is provided by the study of Van Rooy et al. [11], who detected and identified $F_T II$ and $F_T III$ using GC–MS in the plasma of patients treated with fentanyl. Unlike Maruyama and Hosoya [5], Van Rooy et al. [11] introduced a derivative-forming step using acetic anhydride as reagent. Acetyl derivatives of the metabolites happened to be less volatile and no significant losses of the compound during evaporation were observed.

The type of derivatization chosen, however, induces a shift in retention time of the peaks obtained by this process, the intensity of the shift being dependent on the characteristics of the derivatizing group, allowing adjustment of the chromatographic resolution. Some other examples below illustrate the use of the derivatization strategy for the analysis of some anesthetics.

Meuldermans et al. [12] presented a complete study on the metabolic pathways of SUF and ALF from the excreta of rats and dogs. Among other techniques, they used GC–MS in EI and PCI modes for the characterization of major metabolites after derivatization by silylation or acylation. The same laboratory [13] has identified the two major metabolites of ALF in human urine. Lavrijsen et al. [14] have studied the biotransformation of SUF in vitro in different species using GC–MS in EI mode, but also with desorption CI and Townsend discharge ionization. Several metabolites were identified unambiguously on the basis of their MS data, in comparison with the mass spectra of authentic reference compounds.

In some cases, GC–MS offers the possibility of identification by analogy with related structures. Stereoselective metabolism of enflurane to difluoromethoxy-2,2-difluoroacetic acid (DFMDA) was studied in vitro using GC–MS [15]. The DFMDA was determined after derivatization step using ethanolamine. However, authentic DFMDA was not available. Therefore, the authors used mass spectral data of a related ethanolamide derivative structure, i.e., 2-(N-2-chloro-2,2-difluoro-acetamido)ethanol, to predict the mass spectrum of the derivatized DFMDA. In this study, the authors concluded that (R)-enflurane metabolism was more important than the (S)-enflurane metabolism.

Although derivatization was not necessary for analyzing the metabolites of propofol, silylation was achieved. In this case, shifts in retention time and in the m/z of the molecular ion of the major peak were observed. Two prominent ions at m/z 338 and m/z 323 were attributed to (di-TMS) and (M-15) of the 1,4-quinol derivative, respectively. Another unshifted peak was observed and identified as 2,6-diisopropyl-1,4-quinone (1,4-quinone) either in vivo or in vitro. Its mass spectrum exhibited a molecular ion at m/z 192 and a major fragment ion at m/z 149 due to the loss of a propyl group. It seems that 1,4-quinone was not a metabolite of propofol but corresponds to a chemical conversion of 1,4-quinol [3].

3.4. Multiple Strategy: Derivatization and High-Resolution Mass Spectrometry

An example of multistrategy analysis is provided by Adams et al. [16] who performed a thorough study on the biotransformation of ketamine in rat liver microsomes. In this investigation, eight metabolites (products of alicyclic ring hydroxylation of ketamine) were described using low-resolution (m/Δm = 600) mass spectra and high-resolution (m/Δm = 10,000) mass spectra of derivatized metabolites. High-resolution mass spectral data of ketamine, NK, and DHNK, used as pure compounds, allowed the investigators to establish the fragmentation pathways of ketamine metabolites under EI, and finally to obtain accurate knowledge on the sites of the compounds involved in the metabolic reactions. Ketamine metabolites were analyzed in this study as their pentafluoropropionyl (PFP) derivatives, while trifluoroacetyl derivatives were used for quantitative analysis.

3.5. Selective Ion Monitoring

Selective ion monitoring (SIM) detection in GC–MS allows the detection of already known compounds at a lower concentration than the full-scan technique does. It combines enhanced sensitivity due to the use of a smaller number of measured ions, and enhanced specificity due to the choice of more specific high-mass ions.

3.6. Classical Use of Selective Ion Monitoring

A study of a local anesthetic by Tahraoui et al. [17] compared assays of bupivacaine by HPLC and GC–MS, both using pentacaine as internal standard. The GC–MS was performed with SIM at m/z 140 and 154, for the molecule measured and the internal standard, respectively. These ions represent nearly the complete ionic current in the mass spectra of these molecules, thus assuring high sensitivity. In this case, the simplicity of the mass spectrum of the bupivacaine and the internal standard does not permit any modulation of the specificity of the detection, since only one ion is available. Moreover, this ion appears at a low m/z value, which does not provide a high specificity.

An assay for ketamine was described by Feng et al. [18]. The ion-trap detector, used by these authors, confines ions in the ionization chamber before expelling them, which induces ion–molecule reactions. The spectra obtained in this way are often different from conventional EI spectra. For the ketamine, for instance, which should show a molecular ion at m/z 237, an unexpected [M + 1]$^+$ ion at m/z 238 is observed. This artifact appears to result from CI, but in the present case it permits a specific detection of ketamine. The linearity ranges from 25 to 250 ng/ml and the yield of recovery was very good and in an acceptable concentration range.

Two assays for the quantification of propofol have been described. The first one was developed by Stetson et al. [19] using a silylation reaction and monitoring the [M–15]$^+$ ion for quantification, and the molecular ion for the confirmation of each of the silylated derivatives. A liquid–liquid extraction was performed and thymol was used as internal standard. The method was applied in the range 1 to 3000 ng/ml. The second assay was performed in human whole blood after a simple extraction. The propofol and the internal standard thymol were quantified by means of the [M–15]$^+$ ion from the nonderivatized compound. A linear response was obtained in the range of 10 to 10,000 ng/ml [3].

3.7. Indirect Measurement

As previously observed in relation to propofol, the observation of 1,4-quinone in the samples revealed a chemical transformation of 1,4-quinol. Therefore, for the assay of the hydroxy metabolite, the conversion of quinol into quinone, which can be achieved in basic medium, was applied to measure the metabolite, thereby avoiding all supplementary derivatization reactions. The metabolite is measured by monitoring the m/z-149 $[M-43]^+$ ion of quinone and the m/z-135 $[M-15]^+$ ion of thymol used as the internal standard. This fast and appreciable method proved to be linear between 50 and 2500 ng/ml. It enables the estimation of propofol hydroxylation at the phenyl ring during in vitro experiments, using rat microsomes of animals pretreated by different inducers [20].

The acidic GHB, having a hydroxyl group in the gamma position, is susceptible to internal cyclization in acidic medium, resulting in butyrolactone. Ferrara et al. [21], after verification of the absence of this lactone in the plasma of patients treated with GHB, took advantage of this reaction to transform the acid into the lactone in order to avoid the necessity of derivatization. The formed lactone was measured by GC–MS in SIM mode using valerolactone as the internal standard. The separation was performed on a nonpolar column and with split injection. The method is applicable to concentrations in the range of 2 to 200 μg/ml in plasma, and 2 to 150 μg/ml in urine.

3.8. Enhanced Specificity

Podkowik and Masur [22] used two ions each for midazolam and for clinazolam used as the internal standard. One of the two ions was used for quantification, while the second was used as a qualifying ion in order to increase the specificity of the detection. This method is linear between 10 and 500 ng/ml when a volume of only 40 μl of plasma is used. The lower limit of quantification can be reduced to 0.25 ng/ml when a 500-μl sample is used.

3.9. Multiple Internal Standards

In another midazolam assay, Martens and Banditt [8] combined the advantages of a high-mass derivative and the choice of an internal standard as closely related to the target molecule as possible. The authors applied two distinct internal standards, i.e., underivatized medazepam to quantify midazolam, and temazepam, which carries a hydroxyl group amenable to derivatization. The second internal standard was used for quantification of the two hydroxy metabolites, optimizing the conditions of quantification. This technique permits the detection of the minor metabolite 4-OH-MDZ in the range of 0.1 to 5 ng/ml, of 1-OH-MDZ from 1 to 50 ng/ml, and of midazolam itself from 2.5 to 125 ng/ml.

3.10. Use of Labeled Molecules

MS detection also permits the use of stable isotope labeling, a technique that can be used in several ways. In the *tracer technique*, metabolites are detected by the presence of isotope clusters after the in vivo absorption or in vitro incubation of an equimolar mixture of unlabeled and labeled molecules. The labeling can be done with one or more atoms of a heavy isotope, generally ^2H, ^{13}C, ^{15}N or ^{18}O, avoiding the use of radioactive tracers. Under these conditions, the mass spectrum of the parent drug exhibits ions of both the unlabeled and the labeled molecules, producing doublets in the isotope cluster pattern, the intensities of which correspond to the isotopic enrichment in the labeled material. Consequently, all metabolites retaining the labeled atoms will also exhibit spectra containing the isotope cluster, proving unambiguously their metabolic origin. In this way, the analyst can focus only on the spectra showing this particularity. This so-called "isotope cluster," "ion doublet," or "twin-ion" technique greatly facilitates the identification of metabolites.

Goromaru et al. [23] used the stable isotope tracer technique in their study of the metabolism of fentanyl. An equimolar mixture of $F_T I$ and fentanyl labeled with deuterium on the aniline ring (Ft I-d5) was orally administered to rats at a dose of 20 mg/kg. Urinary extracts were derivatized with BSTFA and subsequently analyzed by GC–MS. Five components were found to exhibit the isotope clusters in their mass spectra and were consequently recognized as metabolites of $F_T I$. The mass spectrum of the main peak showed cluster ions at m/z 304: 309 $[M]^+$, m/z 289:294 $[M-15]^+$, m/z 275:280 $[M-C_2H_5]^+$, m/z 247:252 $[M-COC_2H_5]^+$, m/z 231:236, and a single ion at m/z 155. The authors identified this compound as trimethylsilylated $F_T II$, corresponding to N-dealkylation of $F_T I$ (see Fig. 1). Two other peaks exhibited ion clusters in their mass spectra with doublet m/z 377:382 $[M]^+$. This value is 88 Da higher than that of $F_T II$, indicating the introduction of an additional trimethylsilyl ether group (O-TMS) in the $F_T II$ structure. The study of the fragmentation allows identification of the position of hydroxylation. From the ion at m/z 117 $[C_2H_4OTMS]^+$, it could be concluded that the substitution occurs on the propionyl moiety ($F_T IV$). The absence of the m/z-117 ion would indicate that the hydroxylation occurs at the piperidine ring ($F_T V$). Moreover, the presence of isotope clusters separated by 5 Da indicates that the aniline group is not metabolized. Similarly, two other metabolites were identified corresponding to the hydroxylation of $F_T I$ at the propionyl moiety ($F_T VI$) and at the piperidine ring ($F_T VII$). Due to these results, Goromaru et al. [24] succeeded in the identification of those metabolites in humans. Another study from isolated hepatocytes of rat and guinea pig allowed the identification of several other metabolites of fentanyl ($F_T VIII$, $F_T IX$, $F_T X$, $F_T XI$), also with the use of the ion cluster technique [25].

Using D_5-ALF labeled at the phenyl ring, Meuldermans et al. [12] concluded that the hydroxylation reaction occurs at this part of the molecule since the mass spectrum of the metabolite (N-(x-hydroxyphenyl) propanamide) exhibited a shift of 4 Da instead of 5 Da, indicating the substitution of a deuterium atom by a hydroxyl substituent.

3.11. Quantification of Compounds

Quantification of analytes can be performed by the internal standard method, using an isotopically labeled internal standard added to the sample in a precise amount at the early stage of the analytical treatment. By selectively monitoring specific ions of the molecule to be quantified and the corresponding shifted ions of the labeled homologue as reference signal, the precision of the analytical results of GC–MS can be improved. Moreover, random errors during different steps of the sample preparation are minimized. This so-called "isotopic dilution" technique has been widely used in numerous fields. Several examples of this are reported below.

A set of local anesthetics was measured by GC–MS in SIM mode using solid-phase microextraction (SPME) [26] for sample pretreatment. Decadeuterated lidocaine was synthesized and used as the internal standard. This technique permitted the simultaneous quantification of lidocaine as well as its homologues mepivacaine, bupivacaine, prilocaine, and dibucaine in a concentration range of 0.1 to 20 μg/g of tissue, except for dibucaine. This method, which combines a very soft extraction and specific detection in SIM, resulted in very clean profiles. It was thus applicable to the blood samples without previous protein precipitation.

The enantiomers of ketamine exhibit differences in pharmacological activities and in metabolism. Kharasch and Labroo [27] studied the in vitro metabolism by liver microsomes of the isolated isomers of racemate. In a separate experiment, they determined the concentrations of the main metabolite NK. This secondary amino compound was treated with PFPA for derivatization. The hexadeuterated NK was used as the internal standard.

In the early 1980s, in the continuing elucidation of fentanyl metabolism, Goromaru et al. [24] developed a method using fentanyl-D_5 and norfentanil-D_5 as internal standards. Mautz and Labroo [28,29] developed an assay for alfentanil, its major metabolite noralfentanil, and secondary metabolites using deuterated internal standards and specific derivatization. More recently, Valaer et al. [6] developed a screening method for the major N-dealkylated metabolite of these two parent drugs and of sufentanil using synthesized deuterated analogs of the metabolite of each molecule, transformed into their PFPA derivative.

The quantification of remifentanil was studied in SIM using a deuterated internal standard, first by Lessard et al. [30] by employing the deuterated analog of the main desmethyl derivative, and secondly by Grosse et al. [31]. The latter authors used an isotopically purer internal standard of the remifentanil and *high-resolution MS*, which eliminated interference with the analyte. This strategy improved the detection limit to 0.1 ng/ml.

Leung and Baillie [10] studied the *stereoselective metabolism* of NK in rat liver microsomes using racemic NK and its individual enantiomers (R)-NK and (S)-NK labeled with deuterium on aromatic ring ((S)-NK-D_2). Ketamine was used as internal standard and NK and its metabolites were analyzed after PFP derivatization as mono- or bis-PFP derivatives, respectively. They have assessed the production of 4- and 6-hydroxy-norketamine (4-OH-NK, 6-OH-NK) by monitoring the two fragment ions at m/z 304 and 306, which correspond to unlabeled and dideuterated NK, respectively. Norketamine was monitored by two fragment ions at m/z 306 and 308 for the unlabeled (R)-enantiomer and the (S)-labeled enantiomer, respectively. Norketamine was shown to exhibit stereoselective metabolism since 6-OH-NK was preferentially obtained from (S)-NK, and 4-OH-NK was preferentially obtained from (R)-NK.

An isotopic dilution method has been described for the determination of the bioavailability of the ropivacaine and its [$^2H_3$] counterpart in two different preparations in healthy volunteers [32]. Selective ion monitoring of the characteristic ions of the natural molecule and its labeled counterpart was performed. A [$^2H_7$] internal standard was used. Unfortunately, the characteristic fragment of ropivacaine in the EI spectrum does not contain the group carrying the deuterium atoms. Therefore, the distinction between the natural and the labeled molecules could not be made. Consequently, the authors used CI with NH_3 as reagent gas, which provided [M + H]$^+$ ions. This permitted the simultaneous measurement of the natural molecule at [M + 1], the [$^2H_3$]-counterpart at [M + 4], and the [$^2H_7$] internal standard at [M + 8]. The authors investigated the possible cross-contribution between the natural molecule and the labeled analogs, and established a maximum admissible concentration for the internal standard. In addition, the optimum ion-source temperature was determined in order to avoid deuterium exchange. The linearity of the method was established for two concentration ranges, i.e., 30 to 300 nM and 300 to 4000 nM. This technique allowed demonstration of the absence of an isotopic effect of the [$^2H_3$] molecule, which is a prerequisite to all biodisposition studies using stable isotopes. This method was later improved for its application in urine and tissue [33].

In order to improve results in terms of detection limits, some workers added not only a homologous internal standard but also a large amount of another labeled molecule in order to minimize absorption effects to glassware, etc. This practice is known as the *carrier effect*. Recently, an assay for the determination of fentanyl at the sub-ng level was described [34]. The use of sufentanil as internal standard and a large amount of fentanyl-D_5 for the carrier effect was described.

This strategy resulted in improved detection limits, in the range of RIA performance.

4. GAS CHROMATOGRAPHY–ATOMIC EMISSION DETECTION

In atomic emission spectroscopy, the energy source produces the atomization of the sample molecules, resulting in the generation of high-energy atoms. Those atoms spontaneously transmit light by returning to a lower, more stable energy level. The GC–AED technique is performed by passing the effluent from a GC column directly into a microwave-induced plasma. At the same time, the emitted light is monitored at specific wavelengths. Because each kind of atom in the molecule produces light of a unique wavelength, the AED achieves an elemental detection. In addition, the light intensity is proportional to the number of atoms concerned. Therefore, one can easily determine the presence of characteristic atoms in a chromatographic peak, e.g., N, F, Si, S, P, or Cl.

The GC–AED technique should be particularly helpful for the localization of metabolites when the parent drug contains one of the above-mentioned atoms, or to detect molecules, which are derivatized with appropriate reagents containing one of these characteristic atoms. To our knowledge, no results on the qualitative analysis of anesthetics have been reported using GC–AED.

The GC–MS technique is based on molecular response, and depending on the selected ions, it will give different responses to different molecules. Therefore, the determination of the response factor of each target compound is an absolute necessity in quantitative work. With GC–AED on the other hand, the quantification of molecules with similar structures may even be done when the standard molecules are not available [35,36]. The procedure is to compare the responses of a common atom in the molecules to be quantified with that in a reference molecule, for example the parent drug.

In this way, we have applied GC–AED to the quantification of propofol, 1,4-quinone, and 1,4-quinol from an incubation mixture. The responses of the C atom, measured at 193 nm, for these three molecules were related to that of thymol, which was used as the internal standard. At the same time, these compounds were monitored by GC–MS in SIM mode [37]. Assuming that 1,4-quinol and 1,4-quinone are the only compounds formed during incubation, the sum of the signals from residual propofol and those of the appearing compounds always showed a better correlation with the 100% initial signal at t = 0 of the incubation, when measured by GC–AED than by GC–MS. As such, this detection technique presents interesting possibilities for the quantification of new molecules, which are not commercially available, e.g., metabolites. Assuming an equivalent extraction efficiency between the parent drug and its metabolites, GC–AED shows great potential in metabolism studies. In this way, GC–MS, which is able to determine molecular composition, and GC–AED, which can achieve an objective quantitative response, appear to be complementary techniques in a multiple strategy analysis concept.

5. GAS CHROMATOGRAPHY–FOURIER-TRANSFORM INFRARED DETECTION

The preceding examples with fentanyl have shown that GC–MS coupled with the ion cluster technique can be extremely useful for the recognition of metabolites in complex matrices. Nevertheless, the ion cluster technique requires the availability of the labeled parent drug. Infrared (IR) detection also provides three-dimensional information and can be considered as an alternative to MS detection. The use of Fourier-transform infrared (FTIR) detection allows short-time analysis and great sensitivity compared with dispersive IR instrumentation. Consequently, FTIR has become an important detection mode for GC. It can provide the identification of isomeric compounds and information on the functional groups present. Although GC–FTIR is less sensitive than GC–MS, the information obtained from both techniques is complementary.

The GC–FTIR technique was used in conjunction with GC–MS to identify several fentanyl metabolites after derivatization by BSTFA [38]. Whereas the ion cluster technique clearly indicated a peak to be due to a metabolite, examination of the mass spectrum did not always permit conclusions on the filiation of the compound. In this study, FTIR was employed in two ways. First, since different compounds, selected on the basis of their mass spectra, showed the same main characteristic bands in FTIR as fentanyl, it was suggested that these compounds came from this parent drug. Secondly, $F_T VIII$, $F_T IX$, and $F_T X$ exhibit the same EI spectrum as fentanyl, despite the fact that a hydroxyl group is present in the structure [25]. In this respect, the FTIR spectra of these metabolites were particularly useful for determining the most likely position of the hydroxyl group in the molecular structure of $F_T VIII$ and $F_T IX$. The position of the hydroxyl group in $F_T IX$ was suggested by the band at 925 cm^{-1}, assigned to the stretching vibration of the Si—O aromatic group. Moreover, the FTIR spectrum provided information on the probable position of substitution, as indicated by the band at 1510 cm^{-1} observed for v_{cc} when the aromatic ring is *para*-substituted with O-TMS, and by the shift of the O-TMS stretching band from 1100 cm^{-1} to 1250 cm^{-1} when the *para*-substitution occurs. The FTIR spectrum of $F_T VIII$ did not exhibit characteristic absorption bands due to aromatic substitution. It was qualitatively similar to the spectra of $F_T VI$ and $F_T VII$, suggesting a hydroxylation in the alpha position of the phenethyl group.

6. CONCLUSIONS

Gas chromatography–mass spectrometry permits the combination of the high resolution of capillary chromatography and the important molecular information obtained by the mass spectrum. The development of benchtop instruments and the greater capabilities of data systems have widely increased interest in this analytical tool. The GC–MS technique has been largely used in the field of anesthetic drug analysis, for metabolism studies, as well as for quantitative studies on the parent drug and/or its metabolites. Different strategies, e.g., CI, high-resolution MS, specific derivatization, and the use of the ion cluster technique, have been used in order to improve the information obtained. Pharmacokinetic studies on more and more potent drugs, which are given at lower doses, demand the quantification of smaller quantities. The implementation of isotopic dilution methods has significantly improved the SIM technique for the quantification of some classes of compounds. Unfortunately, this use is limited by the availability of suitably labeled standards. Despite the efficiency and the ease-of-use of GC–MS, some questions concerning the structural elucidation of metabolites require more advanced techniques. In quantification, the highly compound-dependent behavior in the ionization process limits the use of GC–MS as a universal quantification technique. No reliable quantification can be achieved if the target compound is not available as standard, which is often the case for metabolites. In these cases, GC–FTIR and GC–AED may be considered as complementary alternatives. The GC–FTIR technique may act as a complementary source of structural information for the unambiguous determination of metabolites, while GC–AED acts as a universal elemental detector that is able to produce quantitative results for any metabolite with a known molecular formula, e.g., obtained from GC–MS.

REFERENCES

1. S.M. Fox, S.C. Chan, and D.H. Van-Middlesworth, J. Chromatogr., 494 (1989), 331–334.
2. M.S. Miller and A.J. Gandolfi, Life Sci., 27 (1980) 1465–1468.
3. J. Guitton, M. Désage, A. Lepape, C.S. Degoute, M. Manchon, and J.-L. Brazier, J. Chromatogr., 669 (1995) 358–365.
3a. P. Favetta, C. Dufresne, M. Désage, O. Païssé, J.P. Perdrix, R. Boulieu, and J. Guitton, Rapid Comm. Mass Spectrom. (2000) in press.
4. M.M. Kochhar, Clin. Toxicol., 11 (1977) 265–275.
5. Y. Maruyama and E. Hosoya, Keio J. Med. 18 (1969) 59–70.
6. A.K. Valaer, T. Huber, S.V. Andurkar, C.R. Clark, and J. DeRuiter, J. Chromatogr. Sci., 35 (1997) 461–466.
7. J. Vasiliades and T. Sahawneh, J. Chromatogr., 228 (1982) 195–203.

8.   J. Martens and P. Banditt, J. Chromatogr. B, 692 (1997) 95–100.

9.   F. Rubio, B.J. Miwa, and W.A. Garland, J. Chromatogr., 233 (1982) 157–165.

10. L.Y. Leung and T.A. Baillie, in E.F. Domino and J.-M. Kamenka (eds.), Sigma and Phencyclidine-Like Compounds as Molecular Probes in Biology, 1988, NPP Books, Ann Arbor, MI, pp. 607–617.

11. H.H. Van Rooy, N.P.E. Vermeulen, and J.G. Bovill, J Chromatogr., 223 (1981) 85–93.

12. W. Meuldermans, J. Hendrickx, W. Lauwers, R. Hurkmans, E. Swysen, J. Thijssen, Ph. Timmerman, R. Woestenborghs, and J. Heykants, Drug Metabol. Disp., 15 (1987) 905–913.

13. W. Meuldermans, A.Van Peer, J. Hendrickx, R. Woestenborghs, W. Lauwers, J. Heykants, G. Vanden Bussche, H. Van Craeyvelt, and P. Van Der Aa, Anesthesiology, 69 (1988) 527–534.

14. K. Lavrijsen, J. Van Houdt, D. Van Dick, J. Hendrickx, W. Lauwers, R. Hurkmans, M. Bockx, C. Janssen, W. Meuldermans, and J. Heykants, Drug Metabol. Disp., 18 (1990) 704–710.

15. K.J. Garton, P. Yuen, J. Meinwald, K.E. Thummel, and E.D. Kharasch, Drug Metab. Dispos., 23 (1995) 1426–1430.

16. J.D. Adams, T.A. Baillie, A.J. Trevor, and N. Castagnoli, Biomed. Mass Spectrom., 8 (1981) 527–538.

17. A.Tahraoui, D.G. Watson, G.G. Skellern, S.A. Hudson, P. Petrie, and K. Faccenda, J. Pharm. Biomed. Anal., 15 (1996) 251–257.

18. N. Feng, F.X. Vollenweider, E.I. Minder, K. Rentsch, T. Grampp, and D.J. Vonderschmitt, Ther. Drug Monit., 17 (1995) 95–100.

19. P.L. Stetson, E.F. Domino, and J.R. Sneyd, , J. Chromatogr., 620 (1993) 260–267.

20. J. Guitton, T. Buronfosse, M. Sanchez, and M. Désage, Anal. Lett., 30 (1997) 1369–1378.

21. S.D. Ferrara, L. Tedeschi, G. Frison, F. Castagna, L. Gallimberti, R. Giorgetti, G.L. Gessa, and P. Palatini, J. Pharm. Biomed. Anal., 11 (1993) 483–487.

22. B.I. Podkowik and S. Masur, J. Chromatogr. B, 681 (1996) 405–411.

23. T. Goromaru, H. Matsuura, T. Furuta, S. Baba, N. Yoshimura, T. Miyawaki, and T. Sameshima, Drug Metab. Dispos., 10 (1982) 542–546.

24. T. Goromaru, H. Matsuura, N. Yoshimura, T. Miyawaki, T. Sameshima, J. Miyao, T. Furuta, and S. Baba, Anesthesiology, 61 (1984) 73–77.

25. T. Goromaru, M. Katashima, H. Matsuura, and N. Yoshimura, Chem. Pharm. Bull., 33 (1985) 3922–3928.

26. T. Watanabe, A. Namera, M. Yashiki, Y. Iwasaki, and T. Kojima, J. Chromatogr. B, 709 (1998) 225–232.

27. E.D. Kharasch and R. Labroo, Anesthesiology, 77 (1992) 1201–1207.

28. R. Labroo and E.D. Kharasch, J. Chromatogr., 660 (1994) 85–94.

29. D.S. Mautz, R. Labroo, and E.D. Kharasch, J. Chromatogr., 658 (1994) 149–153.

30. D. Lessard, B. Comeau, A. Charlebois, L. Letarte, and I.M. Davis, J. Pharm. Biomed. Anal., 12 (1994) 659–665.

31. C.M. Grosse, I.M. Davis, R.F. Arrendale, J. Jersey, and J. Amin, J. Pharm. Biomed. Anal., 12 (1994) 195–203.

32. B.M. Emanuelsson, C. Norsten-Hoog, R. Sandberg, and J. Sjovall, Eur. J. Pharm.-Sci., 5 (1997) 171–177.

33. M. Engman, P. Neidenström, C. Norsten-Höög, S-J. Wiklund, U. Bondesson, and T. Arvidsson, J. Chromatogr. B, 709 (1998) 57–67.

34. B. Fryirs, A. Woodhouse, J.L. Huang, M. Dawson, and L.E. Mather, J. Chromatogr. B, 688 (1997) 79–85.

35. T.J. Barden, M.Y. Croft, E.J. Murby, and R.J. Wells, J. Chromatogr. A, 785 (1997) 251–261.

36. S.V. Kala, E.D. Lykissa, and R.M. Lebovitz, Anal. Chem. 69 (1997) 1267–1272.

37. W. Elbast, J. Guitton, M. Désage, D. Deruaz, M. Manchon, and J.L. Brazier, J. Chromatogr. B, 686 (1996) 97–102.

38. J. Guitton, M. Désage, S. Alamercery, L. Dutruch, S. Dautraix, J.P. Perdrix, and J.L. Brazier, J. Chromatogr. B, 59 (1997) 59–70.

11

Gas Chromatography–Mass Spectrometry in Clinical Stable Isotope Studies: Possibilities and Limitations

F. Stellaard
University Hospital Groningen, Groningen, The Netherlands

1. DEFINITIONS

The scope of this chapter is based on the combination of gas chromatography (GC) with mass spectrometry (MS). In a classical sense, GC is coupled to a scanning mass spectrometer containing a magnet, quadrupole, or ion trap as the analyzer. During the last decade, important developments have been made combining GC with isotope-ratio mass spectrometry (IRMS). First, this has been established directly by capillary tubing for the ^{13}C isotope analysis in breath gas (GC–IRMS, continuous flow IRMS). Thereafter, instruments have been developed enabling on-line ^{13}C and ^{15}N analysis in organic compounds by a capillary combustion(/reduction) interface (GC–C–IRMS). Recently, instrumentation has become commercially available for the purpose of on-line ^2H and ^{18}O analysis in individual organic compounds by a pyrolysis interface (GC–P–IRMS). The terms GC–C–IRMS and GC–P–IRMS are categorized as GC–reaction–IRMS. The GC–MS, GC–IRMS and GC–reaction–IRMS technologies and applications are covered in this chapter in relation to clinical isotope studies. References to relevant publications in these areas are presented, although it has not been the objective to cover all available literature dealing with the various topics.

2. INTRODUCTION

In the clinical chemistry departments of hospitals, mass spectrometry is not a common analytical tool due to the complexity and costs of instrumentation. Specialized laboratories associated with university hospitals or commercial institutes offer mass spectrometry facilities. On the basis of diagnostics, MS is used for qualitative screening of metabolites in body fluids or the quantification of metabolites. In most cases, measurement of metabolite concentrations is sufficient for the diagnosis of an impaired metabolic function. However, a metabolite concentration in, for instance, plasma is determined by the balance of efflux and influx of the metabolite of interest to and from tissues. Therefore, the concentration is a static reflection of the metabolic process studied. More detailed studies are necessary to obtain a dynamic view of the kinetics of the metabolite in the plasma compartment as a reflection of the metabolic function of interest. For this purpose, isotope studies are undertaken in which precursor or product metabolites are administered tagged with an isotopic label. In principle, a choice can be made between radioactive isotopes and stable isotopes. However, in humans, isotope studies with radioactive tracers are not ethically accepted, particularly in children, pregnant women, and more generally in women of child-bearing age. Stable isotopes are preferred and accepted as being harmless when applied in tracer quantities. Another major advantage is that in vivo stable isotope studies can be repeated as often as necessary and within short time intervals. Thus, the subjects can serve as their own control. This reduces the degree of variation in the experimental outcome to a large extent. Detection of stable isotopes is based on the difference in atomic mass and, in biological fluids, this is almost exclusively performed by mass spectrometry.

3. STABLE ISOTOPES

Elements occur in nature in more than one isotopic form (Table 1). The isotopes differ in the number of neutrons in the nucleus, while having the same number of protons and electrons. Therefore, they differ in atomic mass. The isotopic forms are not always similar in chemical and physical behavior. A relatively large difference in atomic mass, such as in the case of the hydrogen isotopes ^{1}H and ^{2}H (deuterium), introduces a large difference in behavior, whereas the chemical and physical behavior is similar when the difference in atomic mass is relatively small, as is the case with the carbon isotopes ^{12}C and ^{13}C. However, when organic compounds are labeled with ^{2}H at one or only a few hydrogen positions, negligible differences are observed in the physiological handling of the labeled and unlabeled material. Interestingly, differences in gas chromatographic behavior are also observed between deuterated and natural metabolites. The shift in retention time has to be taken into consideration when isotope ratio measurements

Table 1. Isotopic Composition of Selected Elements of
Interest for Clinical Studies

Element		Atomic number	Mass number	Natural abundance (%)
Hydrogen	H	1	1	99.985
		1	2	0.015
		1	3*	tr
Carbon	C	6	11*	tr
		6	12	98.89
		6	13	1.11
		6	14*	tr
Nitrogen	N	7	13	tr
		7	14	99.635
		7	15	0.365
Oxygen	O	8	15	tr
		8	16	99.76
		8	17	0.04
		8	18	0.20

tr = trace; * = radioactive isotope.

are performed using GC–MS or GC–reaction–IRMS instrumentation. Again, the
larger the number of ^2H atoms in the molecule, the larger the isotope effect. In
some instances, the chromatographic isotope effect is even used purposely by
applying a heavily deuterated compound as internal standard for quantification
by GC only. This is exemplified by the GC analysis of pipecolic acid in biological
fluids using $^2H_{11}$ pipecolic acid as internal standard [1]. Normally, the GC separation is incomplete, and MS separation of the isotopes within the GC peak is
performed.

4. ENRICHED SUBSTRATES

Stable isotope–labeled compounds can be obtained by chemical modification of
the molecule. Many labeled products can be purchased from commercial suppli-
ers and ordered from the catalog or on special request. Organic compounds of
interest for clinical research are labeled with ^2H, ^{13}C, ^{15}N, and/or ^{18}O at one or
more positions. Multilabeled substrates containing two or three different labels
are also available. The choice of the type of label and the position(s) of labeling
may be critical for the success of the application for two reasons. First, metabolism of the substrate may lead to loss of label, when the label is in the wrong
position. Secondly, unfavorable MS fragmentation may be a cause of loss of

label. High-mass fragment ions and molecular ions are preferred for isotope analysis. For this reason, chemical ionization GC–MS in the positive or negative mode is often used for isotope ratio measurements in order to restore high-mass ions carrying the stable isotope label. On the other hand, in some situations, it may be of particular interest to determine isotope enrichment at specific positions in the molecule, for instance, to elucidate the origin of a metabolite. In this case, suitable fragmentation is necessary to localize the specific part of the molecule, and substrates labeled at specific positions are required.

A special category of labeled products is obtained by biological processing. Algae or plants can be grown under conditions in which nutrients are enriched with a stable isotope. So far, ^{13}C-enriched products have been obtained by the use of $^{13}CO_2$-enriched atmospheric air. The enrichment can be near 100% as obtained with algae, whereby all carbon atoms in the molecule exclusively consist of ^{13}C [2,3]. As examples, saturated and unsaturated long-chain fatty acids, such as palmitic acid (C16:0) and linoleic acid (C18:2), can be purchased labeled with ^{13}C in all 16 or 18 carbon positions. Such a product can be used more generally than the specifically labeled compound, since the problem of loss of label due to metabolic events or MS fragmentation is absent. Also, minimal amounts of these highly enriched products are sufficient for tracer studies in humans. Biologically ^{13}C-enriched plant products described in the literature are rice starch and wheat starch [4–6]. Recently, animal products (eggs and milk protein) have also been obtained by applying ^{13}C or ^{15}N amino acid infusions [7–10]. Truly naturally enriched compounds can be obtained as commercial food products or synthetic products derived from so-called C4 plants. The C4 plants differ from other plants (C3) in the photosynthetic pathway of incorporating carbon dioxide from the air [11]. The deviation in this pathway results in a different degree of natural isotope fractionation leading to a different ^{13}C content. Although this difference is very small (approximately 1.10% ^{13}C for C4 plants vs. 1.08% ^{13}C for C3 plants), it can be measured easily by specific IRMS. Products derived from C4 plants can be used directly as labeled substrate for human in vivo studies or as starting material for the production of other naturally enriched products. In case of in-vivo isotope studies, the ^{13}C abundance will vary between 1.08 and 1.095% ^{13}C depending on the rate of isotope incorporation, the pool size in which the label is diluted, and the turnover rate of the pool. Well-known C4 plants are corn and sugar cane. Pure corn starch, corn glucose, and cane sugar are accepted substrates [12–15]. Feeding corn to cows leads to enriched milk components lactose, protein, and fat [16]. Naturally ^{13}C-enriched lactose has been used in diagnostic studies [17–19]. Furthermore, this ^{13}C-lactose has been converted by an isomerization reaction to ^{13}C-lactulose in order to study bacterial fermentation of carbohydrate to CO_2 [20]. However, it should be emphasized that these naturally enriched substrates can only be used in geographical regions where there is little consumption of products derived from C4 plants.

5. GAS CHROMATOGRAPHY–MASS SPECTROMETRY TECHNIQUES

5.1. Instrument Configurations

For isotope abundance measurements in organic compounds, different techniques can be used (Table 2). In principle, two MS techniques are involved coupled to different peripherals. In the first technique, the classical scanning MS technique can be used in which the analyzer system allows different ions to be transmitted in sequence onto a fixed detector (electron multiplier or photo multiplier). In this mode, selective ion monitoring (SIM) mode is normally used for isotope analysis. A suitable fragment (M) is chosen usually on the basis of ion intensity, and the isotope ions (M + 1, M + 2, . . .) are selected for the measurements. Peaks are registered in the selected mass chromatograms and the area ratios are determined. A calibration curve translates the area ratio into a molar ratio labeled/ unlabeled metabolite. In general, magnetic sector, quadrupole, and ion-trap instruments can be applied for clinical applications. Measurements are performed in EI as well as in positive and negative CI modes.

In the second technique, IRMS is used. The IRMS technique is always based on magnetic sector instrumentation (Nier type). The mass range is limited to about m/z 90 making the technique only applicable to simple, small, gaseous compounds such as CO_2 (for ^{13}C or ^{18}O measurements), N_2 (for ^{15}N measurements), H_2 (for ^2H measurements), and SO_2 (for ^{34}S). The geometry of the instrument is static. This means that the magnet is fixed and allows for continuous registration of isotope ions on fixed detector plates (Faraday cups). In this way, isotope ratios are registered continuously with high precision, which permits determination of variations in natural abundance. For this reason, the technique is used frequently in geological sciences and in the field of food adulteration detection. The high sensitivity for isotope abundance measurements is obtained by the static geometry and by the choice to determine isotope ratios as the absolute

Table 2 Combinations of GC–MS Systems

Type MS	Analyzer	Interface	Abbreviation
Scanning ms	Magnetic sector	Direct	GC–MS
	Quadrupole	Direct	GC–MS
	Ion trap	Direct	GC–MS
Static ms	Magnetic sector	Direct	GC–IRMS (continuous flow IRMS)
		Combustion/ reduction	GC–C–IRMS
		Pyrolysis	GC–P–IRMS

difference (δ value in ‰) between the sample and a calibrated reference gas. In order to use IRMS for isotope abundance measurements in organic molecules, the molecules have to be converted to the gases mentioned above. Originally, this was done off-line by hand using complex and time-consuming techniques. In general, the compound of interest had to be isolated from the biological matrix in a pure form. Thereafter, the compound was converted to CO_2 for ^{13}C or ^{18}O measurements, to N_2 for ^{15}N measurements, or to H_2 for ^2H measurements. Nowadays, this can be done on-line by interfacing IRMS with GC via a capillary reaction system (GC–reaction–IRMS), i.e, by GC–C–IRMS or GC–P–IRMS. For ^{13}C analysis, compounds are converted post–GC-column to CO_2 by oxidation in a capillary cupric oxide/platinum (CuO/Pt) reactor at about 800°C. Although this technique was described already in 1978 [21], the commercial instrumentation became available in 1988. ^{15}N analysis requires oxidation to nitrogen oxides (NO_x) in the CuO/Pt reactor and subsequent reduction of NO_x to N_2. Also, H_2 and CO can be formed from organic compounds by on-line pyrolysis permitting analysis of ^2H and ^{18}O abundances after GC separation. In this process, organic compounds are heated at temperatures above 1400°C under exclusion of O_2. The gas pulses originating from the combustion or pyrolysis reactors are directly transferred to the IRMS through a capillary and an open split interface. Water vapor formed during oxidation is removed by nafion tubing or cryogenic trapping.

5.2. Analytical Considerations for Isotope Abundance Measurements (Table 3)

5.2.1. Gas Chromatography–Mass Spectrometry

To appreciate the analytical demands encountered in clinical isotope studies, one needs to be aware of the isotope distribution in organic compounds. As shown in Table 1, different elements have different isotope abundances. Thus, the isotope distribution in molecular ions or fragment ions is determined by the elemental

Table 3. Characteristics of Various On-Line GC–MS Systems for Isotope Ratio Measurements

	GC–MS	GC–IRMS	GC–C–IRMS	GC–P–IRMS
Quantification limit (mol% isotope enrichment)	0.5%	0.01%	0.01%	0.01%
Material required (ng)	1	1000	10	100
Type of compounds	Organic	CO_2	Organic	Organic
Type of isotopes	$^{13}C, ^2H, ^{15}N, ^{18}O$	$^{13}C, ^{18}O$	$^{13}C, ^{15}N$	$^{18}O, ^2H$

composition. The major contributor of heavier isotopes is carbon with 1.1% ^{13}C. For GC–MS, this means that the M + 1 contribution of a given ion, M, as a percentage of the intensity of M will be close to the number obtained by multiplying the number of carbon atoms by 1.1%. A fragment containing 20 carbon atoms will exhibit a M + 1/M ratio of approximately 23%. The M + 2, M + 3, M + 4, etc., contributions are proportionally lower, creating a typical isotope cluster. One has to realize that derivatization of the molecule introduces additional C, H, O, N, and S atoms and possibly also other atoms such as Si that may have very abundant isotope contributions. Thus, derivatization may lead to an increase in the baseline M + X/M ratio. When a labeled substrate is administered containing one ^{13}C atom or one ^{2}H atom, the M + 1 abundance will increase. If the labeled substrate contains two ^{13}C atoms or two ^{2}H atoms, the M + 2 abundance will increase, etc. At any time, the enrichment has to be measured on top of the natural isotope abundance. The more labeled atoms are present in the substrate, the larger the mass difference and the lower the natural background. On one hand, this low background is advantageous to measure low enrichments. On the other hand, it demands good sensitivity and linear range to accurately determine M + X/M isotope ratios down to <1%. The choice of the type and combination of labelled atoms (^{13}C, ^{2}H, ^{15}N, etc.) and the number and positions of labeled atoms is determined by the research question and the commercial availability rather than analytical considerations. As a consequence, the GC–MS technique has to be adapted to the GC behavior and MS behavior of the derivatized metabolite, the position and type of labeling, as well as the metabolite concentration and the size and matrix of the biological sample. The labeling aspects as well as the aspects related to sample size and metabolite concentration are determinants in the choice of derivative and the mode of ionization. In general, the GC–MS applicability for stable isotope work is characterized by the facts that, on one hand, the technique is sensitive and selective in terms of required mass, and, on the other hand, the technique is limited in the accuracy of the isotope ratio measurement. Usually molar enrichments <0.5% cannot be determined with GC–MS with sufficient accuracy.

5.2.2. Gas Chromatography–Reaction–Isotope-Ratio Mass Spectrometry

In a number of cases, this minimal stable isotope enrichment necessary for GC–MS measurements cannot be reached. This may occur, for instance, when the isotope-labeled substrate is diluted in a large metabolic pool, as is the case with in vivo protein labeling applying labeled amino acids. Infusion of high enough amounts of labeled amino acid may not be possible due to physiological reasons, high costs, or unavailability of highly enriched substrates. In these occasions, GC–reaction–IRMS can be used. So far, two types of instruments are available:

GC–C–IRMS or GC–P–IRMS can be used for ^{13}C and ^{15}N, and for ^{2}H and ^{18}O, respectively. The first generation GC–C–IRMS, introduced around 1988, enabled combustion allowing ^{13}C analysis. The second generation GC–C–IRMS instrumentation (1994) added a capillary reduction oven behind the combustion oven, allowing on-line ^{15}N/^{14}N isotope ratio measurements. Deuterated (^{2}H) substrates are very common in stable isotope studies applying GC–MS. However, a third generation IRMS instrument (1998) had to be developed for on-line ^{2}H/^{1}H ratio measurements, due to the problem of interference of He[+] ions (m/z 4) on the Faraday cup detecting ([2]H[1]H)[+] ions (m/z 3). Extended ion optics and/or mass filtering is necessary to fully separate the He[+] ions from the ([2]H[1]H)[+] ions. Furthermore, the on-line conversion of organic compounds to H_2 had to be established. These adaptations were accomplished only recently [22–29]. Both H_2 and CO production are obtained by a pyrolysis reaction. Conversion rates of >99% are now being obtained. Through the formed H_2, ^{2}H/^{1}H isotope ratios and through CO the ^{18}O/^{16}O isotope ratios are measured. Software adaptations had to be made to correct for chromatographic isotope fractionation, in particular for ^{1}H and ^{2}H, and for the effect of H_3^+ ions produced in the ion source. The general advantage of GC–C–IRMS and GC–P–IRMS over GC–MS is to be found in the extreme sensitivity in measuring differences in isotope abundance. Molar isotope enrichments down to 0.01% can be measured. However, the disadvantage of both techniques compared with GC–MS is the insensitivity with regards to the required amount of metabolite. GC–MS analysis of isotope abundance can usually be performed applying a low-ng / high-pg range of material in EI and positive CI mode and even pg amounts in electron-capture negative CI mode [30–32]. The GC–C–IRMS technique requires at least 10 ng of organic material for ^{13}C measurements and even more for ^{15}N and [2]H (100 ng). The relative low sensitivity for ^{15}N is mainly due to the fact that organic compounds generally contain fewer N than C atoms. Furthermore, the degree of ionization of N_2 and H_2 is lower than that of CO_2. For CO_2, approximately 1 ion is detected per 1000 molecules while only approximately 1 H_2^+ ion is produced per 10,000 H_2 molecules. Real sensitivity is dependent on many factors related to the application, e.g., molecular weight of the (derivatised) molecule, the elemental composition, the chromatographic behavior, and interference of related matrix compounds, as well as MS behavior i.e., ionization efficiency of the gas molecules.

6. POSSIBILITIES AND LIMITATIONS

6.1. Gas Chromatography–Mass Spectrometry

In GC–MS in the SIM mode, isotope ratios are determined as ratios of peak areas obtained for the metabolite in the selected mass chromatograms. Normally, the intensity ratio between a high isotope ion (M + X) and the basic fragment ion M is requested. The quality of the area ratio is determined by the quality of the peaks and mainly by the quality of the smallest peak, i.e., the peak obtained for the highest isotope ion (M + X). The quality of a mass chromatographic peak is set by the intensity, signal-to-noise ratio, and the number of data points collected over the GC peak. The intensity of the peak and the signal-to-noise ratio are affected by the dwell time used for monitoring the isotope ion. Longer dwell times are beneficial as long as sufficient data points, normally at least 20, are collected over the GC peak [33]. The optimal conditions are dependent on the gas chromatographic behavior of the metabolite of interest. Rapid gas chromatographic elution creating very narrow GC peaks requires a fast scanning instrument to ensure sufficient data points to guarantee adequate GC peak description. The requirements for isotope ratio measurements are twofold, high accuracy and high precision, i.e., the degree of agreement with the theoretical isotope abundance value. This theoretical value can be calculated from the elemental composition and the isotopic compositions of the elements involved. Optimization of the analytical parameters as mentioned above will lead to maximization of the accuracy of the measurement. Obviously, the accuracy will decrease with increasing isotope mass (M + X), since the ratio M + X/M declines in this order. Also, low mass fragments have lower M + X/M ratios than high-mass fragments, mainly caused by a smaller number of carbon atoms. Thus, high-mass fragments are preferred for this reason but also for the reason of higher selectivity due to lower chemical interference caused by biological matrix, solvent and reagent effects, and column bleed. Therefore, chemical ionization may be beneficial for isotope enrichment measurements.

Interestingly, high accuracy does not necessarily means high precision. This is clearly demonstrated by our own work studying the isotope clusters of bile acid methyl ester trimethylsilyl (TMS) derivatives applying electron impact ionization GC–MS. Cholic acid as a methyl TMS derivative shows a mass spectrum with the three most abundant fragment ions m/z 253, 368, and 458. The isotope cluster analysis for the ions m/z 368 and 458 gives the results shown in Table 4. Isotope measurements at m/z 368 result in values that are clearly precise but wrong, whereas the values for m/z 458 are all precise and reasonably accurate. Measurements performed with increased mass resolution, other derivatives, and various GC columns do not improve the agreement with the calculated values obtained for the fragment ion m/z 368. This effect is probably caused by internal source events as described by Fagerquist and Schwarz [34,35] for the isotope cluster analysis of palmitic acid. Their data indicate that gas-phase chemistry can be a dominant cause of inaccuracy and imprecision in isotope ratio analysis. From this, it may be concluded that it is crucial to select the most suitable derivative, ionization mode, and fragment ion to establish accurate and correct isotope abundance values. Also, the linear range of the detector system is important. Isotope ratio measurements are carried out within a defined intensity range. The lower

Table 4. Isotope Clusters of the Methyl TMS Derivative of Cholic Acid Measured with EI-GC–MS at Mass Fragments m/z 368 and 458

	Isotope Ratio (%), mean ± SD		
	M1/M0	M2/M0	M3/M0
m/z 368	42.13 ± 0.21	8.50 ± 0.10	2.44 ± 0.03
Theoretical	28.20	4.23	0.45
m/z 458	38.17 ± 0.07	10.90 ± 0.05	2.53 ± 0.04
Theoretical	37.52	10.71	2.09

and upper limits of this range need to be determined or corrections are to be made [36,37]. The data system–controlled conversion of the recorded data into area ratios is another determinant in the accuracy of the final results. Definition of peak area and noise level must be consistent and unaffected by sample-to-sample variation. Individual attempts have been made, as by Bluck and Coward [38], who wrote their own algorithm to interpret recorded data obtained by a commercial instrument in order to improve accuracy of the GC–MS isotope ratio measurements. The minimal isotope enrichment that can be detected on top of the natural abundance value is dependent on the accuracy of the measurement and the level of the natural abundance value. As a rule of thumb, the minimal enriched value that can be detected accurately is 3 times the standard deviation of the measurement. Thus, a standard deviation of 0.3% relates to a minimal quantification level of 0.9%. This difference may be measured on top of a 1% baseline value (M + 3) leading to a 90% increase or on top of a 30% baseline value (M + 1) creating a 3% increase. Improvement may be obtained when the relative isotope abundance of isotope ion M + X is not measured against M but against M + 1 or M + 2 [39]. This decreases the required linear range. However, it needs to be realized that in the background the signal intensity for M is then overloaded and a constant M + 1/M or M + 2/M ratio is assumed.

6.2. Gas Chromatography–Reaction–Isotope-Ratio Mass Spectrometry

The variation in isotope abundance measurement in GC–C–IRMS and GC–P–IRMS is much less than in GC–MS. Therefore, molar enrichments of 0.01% can still be quantified. For this reason, IRMS techniques can replace MS techniques at low enrichment levels or can be added to the MS technique in order to extend an isotopic decay curve to enable multicompartment analysis. However, the latter application may be unnecessary since linearity of GC–C–IRMS is usually excellent. Linearity has been shown from 0.01 to 25% molar enrichment for valine

[40,41]. However, GC–reaction–IRMS techniques also have limitations. In contrast to GC–MS, GC–reaction–IRMS techniques lack MS mass selectivity. For instance, GC–C–IRMS detects an overall CO_2 response due to general oxidation of both background carbon (mainly column bleeding) and carbon from eluting organic compounds. Therefore, compounds of interest are detected on top of background signal and have to be baseline separated from each other and from other eluting compounds. Deconvolution of overlapping peaks is not yet possible with commercial systems. Individual attempts appear to be successful [42,43] but have not yet led to commercial implementation. The design of the reaction interface and the flow-dependent requirements for the dimensions of the GC columns may have effects on the GC peak shape and peak width. Much more attention should be paid to chromatography when GC–reaction–IRMS is used than in the case of GC–MS. Also, derivatives have to be selected more carefully, for instance, Si-containing reagents and derivatives may release SiO_2 in the combustion oven, which leads to diminished combustion performance. Completely combustible carbon derivatives such as alkyl derivatives are preferred. Derivatization also introduces atoms similar to those present in the metabolite itself. For instance, ^{13}C enrichment in the metabolite will therefore be diluted with natural abundance ^{13}C in the derivatizing agent. The ^{13}C abundance can be drastically shifted due to derivatization when the derivatizing agent is manufactured from fossil-derived starting material. The ^{13}C abundance of fossil material is very low, i.e., around 1.06%. In the example of penta-acetylated glucose, the measured enrichment in the derivative is only approximately 30% of the real enrichment in the glucose molecule.

The GC–MS methodology and sensitivity are independent of the label used in the experiment. Only the number of labeled atoms is a determinant in the choice of the isotope ions. In contrast, GC–reaction–IRMS requires different interfacing techniques for the different isotopic labels ^{13}C, ^{15}N, ^{18}O, and ^2H. Also, the sensitivity for the various isotopic labels is different. This is true in terms of required mass as well as for the minimal quantifiable enrichment. One also needs to realize that GC–P–IRMS on nitrogen-containing organic compounds requires modification of the methodology since CO and N_2, which are formed during pyrolysis, create the same molecular mass ion m/z 28. An on-line post-reaction GC column separation of the gases must then be performed.

7. APPLICATIONS

7.1. General Comments

Clinical stable isotope studies cover a wide range of applications. Most are related to studies on nutrients and related endogenous compounds (Table 5). The fields of proteins, fats, lipids, and carbohydrates are widely covered. Both GC–MS and

Table 5. Fields of Application for In Vivo Clinical Stable Isotope Studies

Digestion	Carbohydrates, triglycerides, proteins
Absorption	Monosaccharides, fatty acids, amino acids, cholesterol
Metabolism	Glucose, fatty acids, amino acids
Biosynthesis	Fatty acids, glucose, cholesterol, bile acids, proteins, lipoproteins, urea
Pool size	Bile acids
Turnover	Bile acids, cholesterol
Organ functions	
Liver	Cytosolic, microsomal, mitochondrial enzyme activities
Intestine	Digestive enzyme activities
	Transit time
	Bacterial overgrowth
Stomach	*Helicobacter pylori* detection
	Gastric emptying

GC–C–IRMS are being used, while GC–P–IRMS is now being explored. The applications generally deal with (1) digestion (carbohydrates, fats, proteins) and absorption (sugars, fatty acids, cholesterol) from the intestinal tract and (2) metabolism (glucose, fatty acids, amino acids, cholesterol). In addition, organ functions are studied in order to establish the diagnosis of dysfunction (liver, stomach, intestine). Special studies are performed related to synthetic functions mainly in the liver dealing with glucose (glycolysis, glycogenesis, gluconeogenesis), fatty acids (lipogenesis), and lipids such as cholesterol (cholesterogenesis) and bile acids. The choice between using GC–MS or GC–C–IRMS is determined by a number of factors. If isotope clusters are to be analyzed, GC–C–IRMS cannot be used; GC–MS is then the only alternative. This is the case when Mass Isotopomer Distribution Analysis (MIDA) is used. MIDA is increasingly being applied for the measurement of relative or absolute synthesis rates of endogenous polymers, such as glucose, fatty acids, cholesterol, and bile acids. Thus, by infusion with a ^{13}C-labeled precursor (glycerol, acetate), the incorporation of label into the polymer can be monitored in time. The strong point of the MIDA technique is that the relative enrichments of the different isotope ions (M + 1, M + 2, M + 3) enable calculation of the precursor enrichment. This is an advantage because in most cases the precursor pool is located intracellularly and cannot be sampled.

When general isotope enrichment must be determined, the pool dilution and kinetics are parameters. If the degree of pool dilution of the isotope is small and the kinetics of label disappearance is reasonably slow, GC–MS can be used. Kinetics related to measurements in a large pool requires more sensitive techniques unless an excessive amount of label is introduced. In this situation, GC–reaction–IRMS is the preferred method.

Of clinical interest also is the measurement of energy expenditure and body composition. Among other techniques, $2H_2^{18}O$, and ^2H_2O are used for this purpose and isotope enrichments are measured in body water. So far, the technique required sample preparation steps converting water into H_2 and exchanging oxygen between water and CO_2. Possibly GC–P–IRMS will soon replace these procedures and offer rapid on-line measurement of ^2H and ^{18}O directly in body water.

A particular application consists of breath tests. In this case, ^{13}C substrates are administered orally or intravenously. The ^{13}C label is positioned at the C atom(s) that enter the oxidation process. As such, the ^{13}C label appears in breath CO_2. The rate of appearance of label in breath reflects the speed of the slowest metabolic step in the total process of digestion, absorption, intermediate metabolism, and oxidation. The $^{13}CO_2$ in breath is easily determined by GC–IRMS, also called continuous flow IRMS.

7.2. Carbohydrates

Digestion of dietary carbohydrates in the small intestine is measured by oral application of a ^{13}C labeled carbohydrate, i.e., a disaccharide (lactose, maltose, sucrose) or a complex carbohydrate (starch). Although synthetically labeled monosaccharides and disaccharides are commercially available, naturally labeled products are also used for all carbohydrates in Europe where the natural ^{13}C background in humans is low. In some instances, more highly enriched biologically obtained starch substrates have been used (rice, wheat). The response can be monitored by measurement of the ^{13}C enrichment of CO_2 in breath [12,18,19,44]. It should be realized that many factors, such as energy metabolism and colonic bacterial metabolism, influence the outcome of the test, leading to false negative test results in a number of cases. To overcome these problems, the response can be measured as the increase in ^{13}C abundance in blood glucose. This technique is under development as a diagnostic test for lactose maldigestion [45]. The ^{13}C-glucose response in blood as a measure for starch digestion has been documented by Normand et al. [46,47] and been implemented as a test for the exogenous glycemic index for starchy foods, which expresses the bioavailability of glucose [48]. So far, measurement of fructose absorption has been studied only semi quantitatively by measuring the $^{13}CO_2$ response in breath after intake of ^{13}C-fructose by mouth [49]. Glucose absorption has been studied quantitatively by simultaneous administration of ^{13}C-glucose (oral) and ^2H-glucose (IV) and measurement of the ratio of the two enrichments in blood [50,51]. Glucose kinetics is especially important in patients with dysfunctions in the regulation of glucose metabolism (diabetes, glycogen storage disease). Determination of glucose production and turnover is possible by infusing ^{13}C-glucose or ^2H-glucose and measuring the dilution of label [52–54]. Gluconeogenesis can be measured using the MIDA method as described above infusing ^{13}C-glycerol [55,56] or ^{13}C-lactate

[56,57]. Glucose oxidation is measured as the $^{13}CO_2$ response in breath during ^{13}C-glucose infusion [58].

7.3. Fats

Triglycerides form a major contribution to the diet and are also formed endogenously to remove fatty acids from the liver. For the in vivo processing, dietary triglycerides are digested mainly by pancreatic lipase, thereby releasing fatty acids. These are transported over the intestinal mucosa as chylomicrons and transported by the lymph to the systemic circulation. Fatty acids are released by lipoprotein lipase and taken up by tissues for storage or metabolism. A major part is oxidized. The different steps in processing can be studied with stable isotopes. The intestinal digestion is monitored using the artificial ^{13}C–mixed triglyceride (^{13}C-MTG) breath test [59]. At the sn2-position it contains ^{13}C-octanoic acid. After cleavage by pancreatic lipase, the released ^{13}C-octanoate is rapidly absorbed and oxidized. Thus, the $^{13}CO_2$ response is considered to reflect pancreatic lipase activity. The absorption of fatty acids is studied by use of uniformly ^{13}C-labeled fatty acids, such as $^{13}C_{16}$-palmitic acid or $^{13}C_{18}$-linoleic acid, and monitoring of the $^{13}CO_2$ response in breath or the ^{13}C-fatty acid response in blood [60–62]. To study oxidation of fatty acids specifically, ^{13}C-fatty acids are administered by intravenous infusion and the $^{13}CO_2$ response in breath is measured [63–67]. Metabolism by chain shortening, chain elongation, and desaturation can be monitored by administration of U-^{13}C fatty acids and measurement of ^{13}C enrichment of the particular metabolic products in blood [68–71]. Chain elongation is of particular interest for the metabolism of the essential unsaturated fatty acids linoleic acid (C18:2ω6) and α-linolenic acid (C18:3ω3) to arachidonic acid (C20: 4ω6) and docosahexanoic acid (C22:6ω3) [72–74]. Physiologically important is the endogenous synthesis (lipogenesis) of the fatty acids palmitic acid and stearic acid as polymers of acetate units. For the understanding of the regulation of fat metabolism quantification of lipogenesis is performed using the MIDA technique [75]. ^{13}C-acetate is infused and the rate of incorporation of ^{13}C label into the plasma fatty acids is determined.

7.4. Proteins and Lipoproteins

Proteins are major nutrients of both animal and plant origin. They are mainly digested by pancreatic trypsin. The final end products, amino acids, are absorbed and incorporated into endogenous proteins, metabolized to glucose, or oxidized directly. Synthesis and breakdown of endogenous protein should be in a positive balance. Measurement of intestinal protein digestion has not been possible for a long time because of a lack of labeled substrate with high enough enrichment. Recently, Evenepoel et al. [7,8] infused chickens with ^{13}C-leucine, therewith labeling egg white protein to a ^{13}C enrichment (APE) of about 1%. So far, absorption studies with ^{13}C–egg white protein have been monitored by the $^{13}CO_2$ response in breath using GC–IRMS. Applying the same principle to cows, Casseron et al. [9,10] labeled milk protein with ^{15}N-ammonium sulfate or ^{13}C-leucine. The ^{15}N enrichment (APE) of casein reached 0.25%. Much older than the protein digestion studies are the studies on protein metabolism. In this case, ^{13}C- or ^{15}N-labeled amino acids (leucine, glycine, valine, phenylalanine) are infused intravenously and the rate of incorporation of label in specific endogenous proteins is determined. For this purpose, the protein of interest is isolated from the biological matrix (plasma, muscle) in a pure form and hydrolyzed into the amino acids. The amino acids are then derivatized. The enrichment measurement is performed by GC–MS or GC–C–IRMS [76–79]. Oxidation of amino acids is quantified by measurement of the breath $^{13}CO_2$ response during a constant infusion of ^{13}C-labeled amino acid [80,81]. Excretion of nitrogen from the body occurs mainly via urea production, the catabolic end product from amino acid degradation. Measurement of urea production can be performed with stable isotope studies involving ^{13}C- or ^{15}N-urea [82–85]. Kloppenburg et al. [85] elegantly apply enzymatic hydrolysis of urea in plasma, converting ^{13}C-urea into $^{13}CO_2$, which can be sampled directly from the gas phase of the test tube and measured with continuous flow GC–IRMS. No extensive extraction and derivatization procedures are required.

Lipoproteins are proteins associated with transport of lipids. As an example, apolipoprotein B100 (apo B100) is synthesized in the liver and used to transport very low density lipoprotein (VLDL) particles containing triglycerides and cholesterol from the hepatocyte into the systemic circulation. The rate of VLDL secretion is quantified by labeling the apo B100 moiety via intravenous infusion of ^{13}C-labeled amino acid, such as leucine or valine, and measurement of ^{13}C enrichment in the leucine or valine present in the VLDL apo B100. GC–MS and GC–C–IRMS can be applied [40,86–90]. Because of the difficult separation of leucine and isoleucine, valine is chosen as the preferred substrate in conjunction with GC–C–IRMS [91,92].

7.5. Lipids

Usually, lipids are defined as general lipophilic compounds, including phospholipids, cholesterol, other neutral sterols, and bile acids. In particular, techniques enabling the investigation of interrelationships between cholesterol and bile acid metabolism are of interest. For a good understanding, it is important to know that cholesterol is synthesized mainly in the liver as a polymeric product of acetate units. In the hepatocyte, newly synthesized cholesterol merges with cholesterol entering the liver after absorption from the small intestine. Part of hepatic cholesterol is transformed to the bile acids cholic acid (CA) and chenodeoxycholic acid (CDCA), while the rest is secreted into the systemic circulation and bile. Bile acids are also secreted into bile and reabsorbed from the small intestine. Hepatic cholesterol and bile acids regulate cholesterol synthesis and catabolism, i.e., bile acid synthesis. Therewith, cholesterol levels in the body are controlled. Thus, important parameters in cholesterol homeostasis are cholesterol absorption, cholesterol synthesis, and bile acid absorption and synthesis. For research purposes, these parameters need quantification in order to study the effects of dietary or drug treatment.

Stable isotope methodology to determine cholesterol absorption has been described by Bosner et al. [93,94] applying administration of $^2H_6$-cholesterol orally and $^{13}C_5$-cholesterol intravenously and measurement of the $^2H_6/^{13}C_5$ isotope ratio by GC–MS. Another approach was chosen by Lütjohann et al. [95,96] who combined the oral administration of $^2H_6$-cholesterol with that of $^2H_4$-sitostanol and measured the ratio $^2H_6$-cholesterol/$^2H_4$-sitostanol in feces collected after administration.

Two techniques are in use determining the in vivo rate of cholesterol synthesis. One is performed by oral administration of ^2H_2O and measurement of the rate of incorporation of ^2H in plasma cholesterol [97–99]. Since ^2H enrichment in cholesterol is very low using acceptable levels of ^2H_2O, IRMS is necessary. So far, these measurements have been performed after off-line preparation of samples. The near future will certainly show application of on-line GC–P–IRMS for this purpose. When the MIDA technique is used applying infusion of ^{13}C-acetate, GC–MS is used to measure the time-dependent isotope cluster profile in plasma cholesterol [100–102]. The rate of ^{13}C incorporation is then a measure of cholesterol synthesis. Since bile acids are formed by modification and shortening of the cholesterol molecule, MIDA can also be applied to bile acids [103]. In this way, the origin of cholesterol precursor pools (newly synthesized vs. circulating cholesterol) used for bile acid synthesis may be identified. The actual steady state bile acid synthesis rate can be measured by the bile acid isotope dilution technique. ^{13}C- or ^2H-labeled bile acids are administered orally and the decay of isotope enrichment is measured in the bile acid pool. Isotope enrichment can be determined in 2 to 4-ml blood samples [104,105] applying methyl TMS derivatives and EI or 100-μl blood samples applying pentafluorobenzyl TMS derivatives and negative CI [30,106,107].

7.6. Organ Functions

Organ functions are normally characterized by blood concentrations of metabolites or enzyme activities. In vivo stable isotope tests are considered as dynamic and (semi)quantitative tests. Dynamic reflection of organ functions can therefore be obtained by application of stable isotopes. A stable isotope–labeled precursor for a specific function is administered and the rate or degree of metabolism is determined by measuring the isotope enrichment of the metabolite. In order to reduce the invasive character of the test and simplify the analysis, ^{13}C breath tests are applied. The CO_2 is separated from other breath gases by GC and the CO_2 pulse is analyzed for ^{13}C abundance with IRMS. The clinical areas of interest below are being approached using [13]CO_2 breath tests.

7.6.1. Liver

Various liver functions are measured using different types of labeled substrates. In Table 6, substrates are linked to specific liver functions, and literature references describing applications or evaluations of the tests are provided.

7.6.2. Intestine

Intestinal processes studied with [13]CO_2 breath tests are carbohydrate (lactose and starch) digestion (lactase and α-amylase activity), lipid digestion (lipase activity), protein digestion (trypsin activity), sugar absorption, fatty acid absorption, small intestinal bacterial overgrowth, oro-cecal transit time. The substrates used are indicated in Table 6.

Table 6. Organ Function Testing with [13]CO_2 Breath Tests

Organ	Function	[13]C substrate	References
Liver	Cytosol	Galactose	108
	Microsomes	Aminopyrin, methacetin, phenacetin, caffeine, erythromycin, etc.	109–111
	Mitochondria	α-Ketoisocaproic acid	112
Intestine	Digestion:		
	Carbohydrate	Lactose, starch	12,17–19,44
	Fat	Triglycerides	59,60,113,114
	Protein	Egg or milk protein	7,115
	Absorption		
	Carbohydrate	Fructose	49
	Fatty acids	Various fatty acids	60,113
	Bacterial overgrowth	Glycocholate, xylose	116–118
	Oro-cecal transit time	Lactose-ureide	119,120
Stomach	*Helicobacter pylori*	Urea	121,122
	Gastric emptying		
	Solids:	Octanoate	123,124
	Liquids:	Glycine, acetate, sodium bicarbonate	125–128

7.6.3. Stomach

Two types of breath tests are devoted to the stomach. First, the ^{13}C-urea breath test is used frequently in order to detect the presence of the *Helicobacter pylori* bacteria in the stomach. These bacteria are characterized by high urease activity. Presence of the bacteria is associated with the appearance of ^{13}CO$_2$ in breath after oral intake of ^{13}C-urea. Gastric emptying of solid meals is measured applying ^{13}C-octanoate, whereas ^{13}C-acetate or ^{13}C-glycine are generally used to measure the gastric emptying of liquid meals. Disturbances in emptying kinetics are detected and the effect of drugs is monitored (see Table 6).

7.7. Diagnostic Value

Breath tests applying GC–IRMS are often described as attractive alternatives for the diagnosis of organ malfunctions. Particularly in children, breath tests are recommended. So far, application in general clinical practice is limited to the ^{13}C-urea breath test for the detection of *Helicobacter pylori* infection in the stomach, considered to be a major cause of the development of gastric and duodenal ulcers. This test has been in use since 1987 and is now recognized as the gold standard. Other tests as mentioned in Table 6 do not show similar excellent diagnostic power and compete with many traditional tests of similar or better quality. In general, only a large degree of malfunction can be detected with certainty.

The GC–C–IRMS analyses of isotope enrichments in individual metabolites in biological fluids have not yet been used for the diagnosis of organ malfunctions. This is mainly due to the fact that the technique is relatively new and complex. Also, extensive sample preparation is required. It is expected that in a number of situations the measurement of ^{13}C enrichment in a metabolite in blood may have superior diagnostic power compared with the corresponding ^{13}C breath test. Thus, in the future, GC–C–IRMS may develop into a diagnostic tool, when blood sample collection can be reduced to a minimum. The GC–P–IRMS technique is still in an experimental phase. The GC–MS technique in combination with stable isotopes is regularly being used for diagnostic purposes, but only for the purpose of accurate quantification applying stable isotopes as internal standards. In vivo use of stable isotopes has not yet resulted in diagnostic tests.

REFERENCES

1. T. Zee, F. Stellaard and C. Jakobs, J. Chromatogr., 574 (1992), 335–339.
2. H.K. Berthold, L.J. Wykes, F. Jahoor, P.D. Klein and P.J. Reeds, Proc. Nutr. Soc. 53, (1994), 345–354.
3. H.K Berthold., D. Hachey, P.J. Reeds, and P.D. Klein, FASEB J. 4, (1990) A806 (abst.).

4. C.H. Lifschitz, B. Torun, F. Chew, T.W. Boutton, C. Garza, and P.D. Klein, J. Pediatr., 118 (1991) 526–530.
5. C. Rambal, C. Pachiaudi, S. Normand, J.P. Riou, P. Louisot, and A. Martin, Carbohydr. Res., 236 (1992) 29–37.
6. C. Rambal, C. Pachiaudi, S. Normand, J.P. Riou, P. Louisot, and A. Martin, Br. J. Nutr., 73 (1995) 443–454.
7. P. Evenepoel, M. Hiele, A. Luypaerts, B. Geypens, J. Buyse, E. Decuypere, P. Rutgeerts, and Y. Ghoos, J. Nutr., 127 (1997) 327–331.
8. P. Evenepoel, D. Claus, B. Geypens, B. Maes, M. Hiele, P. Rutgeerts, and Y. Ghoos, Aliment. Pharmacol. Ther. 12 (1998) 1011–1019.
9. J. Rubert-Aleman, G. Rychen, F. Casseron, F. Laurent, and G.J. Martin, J. Dairy. Res., 66 (1999) 283–288.
10. F. Casseron, G. Rychen, X. Rubert-Aleman, G.J. Martin, and F. Laurent, J. Dairy. Res., 64 (1997) 367–376.
11. S.v. Caemmerer, Plant, Cell Environ., 15 (1992) 1063–1072.
12. M. Hiele, Y. Ghoos, P. Rutgeerts, and G. Vantrappen, Biomed. Environ. Mass Spectrom., 16 (1988) 133–135.
13. G.P. Leese, J. Thompson, C.M. Scrimgeour, and M.J. Rennie, Eur. J. Appl. Physiol., 72 (1996) 349–356.
14. W.H. Saris, B.H. Goodpaster, A.E. Jeukendrup, F. Brouns, D. Halliday, and A.J. Wagenmakers, J. Appl. Physiol., 75 (1993) 2168–2172.
15. J. Decombaz, D. Sartori, M.J. Arnaud, A.L. Thelin, P. Schurch, and H. Howald, Int. J. Sports Med., 6 (1985) 282–286.
16. T.W. Boutton, H.F. Tyrrell, B.W. Patterson, G.A. Varga, and P.D. Klein, J. Anim. Sci. 66 (1988) 2636–2645.
17. R.D. Murray, T.W. Boutton, P.D. Klein, M. Gilbert, C.L. Paule, and W.C. MacLean, Jr., Am. J. Clin. Nutr. 51 (1990) 59–66.
18. M. Hiele, Y. Ghoos, P. Rutgeerts, G. Vantrappen, H. Carchon, and E. Eggermont, J. Lab. Clin. Med., 112 (1988) 193–200.
19. H.A. Koetse, F. Stellaard, C.M. Bijleveld, H. Elzinga, R. Boverhof, R. van der Meer, R.J. Vonk, and P.J. Sauer, Scand. J. Gastroenterol., 34 (1999) 35–40.
20. H.A. Koetse, R.J. Vonk, S. Pasterkamp, J. Pal, S. de Bruin, and F. Stellaard, Scand. J. Gastroenterol., 35 (2000) 607–611.
21. D.E. Matthews and J.M. Hayes, Analy. Chem. 50 (1978) 1455–1473.
22. D.H. Chace and F.P. Abramson, Biomed. Environ. Mass Spectrom., 19 (1990) 117–122.
23. D.H. Chace and F.P. Abramson, Anal. Chem., 61 (1989) 2724–2730.
24. S.J. Prosser and C.M. Scrimgeour, Analyt. Chem., 67 (1995) 1992–1997.
25. I.S. Begley and C.M. Scrimgeour, Rapid Commun. Mass Spectrom., 10 (1996) 969–973.
26. I.S. Begley and C.M. Scrimgeour, Analyt. Chem., 69 (1997) 1530–1535.
27. J. Koziet, J. Mass Spectrom., 32 (1997) 103–108.
28. H.J. Tobias and J.T. Brenna, Anal. Chem., 68 (1996) 3002–3007.
29. H.J. Tobias and J.T. Brenna, Anal. Chem., 69 (1997) 3148–3152.
30. F. Stellaard, S.A. Langelaar, R.M. Kok, and C. Jakobs, J. Lipid Res., 30 (1989) 1647–1652.

31. F. Stellaard, H.J. ten Brink, R.M. Kok, L. van den Heuvel, and C. Jakobs, Clin. Chim. Acta, 192 (1990) 133–144.
32. H.J. ten Brink, F. Stellaard, C.M. van den Heuvel, R.M. Kok, D.S. Schor, R.J. Wanders, and C. Jakobs, J. Lipid Res., 33 (1992) 41–47.
33. F. Stellaard and G. Paumgartner, Biomed. Mass Spectrom., 12 (1985) 560–564.
34. C.K. Fagerquist, R.A. Neese, and M.K. Hellerstein, J. Am. Soc. Mass Spectrom., 10 (1999) 430–439.
35. C.K. Fagerquist and J.M. Schwarz, J. Mass Spectrom., 33 (1999) 144–153.
36. B.W. Patterson, G. Zhao and S. Klein, Metabolism, 47 (1998) 706–712.
37. B.W. Patterson and R.R. Wolfe, Biol. Mass Spectrom., 22 (1993) 481–486.
38. J.C. Bluck and W.A. Coward, J. Mass Spectrom., 32 (1997) 1212–1218.
39. A.G. Calder, S.E. Anderson, I. Grant, M.A. McNurlan, and P.J. Garlick, Rapid. Commun. Mass Spectrom., 6 (1992) 421–424.
40. M.G. de Sain-van der Velden, T.J. Rabelink, M.M. Gadellaa, H. Elzinga, D.J. Reijngoud, F. Kuipers, and F. Stellaard, Anal. Biochem., 265 (1998) 308–312.
41. W.W. Wong, D.L. Hachey, S. Zhang, and L.L. Clarke, Rapid Comm. Mass Spectrom., 9 (1995) 1007–1011.
42. K.J. Goodman and J.T. Brenna, Anal. Chem., 66 (1994) 1294–1301.
43. K.J. Goodman and J.T. Brenna, J. Chromatogr. A, 689 (1995) 63–68.
44. M. Hiele, Y. Ghoos, P. Rutgeerts, G. Vantrappen, and K. de Buyser, Gut, 31 (1990) 175–178.
45. R.J. Vonk, Y. Lin, H.A Koetse., C. Huang, C. Zeng, H. Elzinga, J.-M. Antoine, and Stellaard, F. Eur. J. Clin. Invest., 30 (2000) 140–146.
46. S. Normand, C. Pachiaudi, Y. Khalfallah, R. Guilluy, R. Mornex, and J.P. Riou, Am. J. Clin. Nutr., 55 (1992) 430–435.
47. S. Tissot, S. Normand, R. Guilluy, C. Pachiaudi, M. Beylot, M. Laville, R. Cohen, R. Mornex, and J.P. Riou, Diabetologia, 33 (1990) 449–456.
48. R.J. Vonk et al., Digestion of so-called resistant starch sources in the human small intestine. Am. J. Clin. Nutr. (2000) Aug 72 (2) 432–438.
49. J.H. Hoekstra, J.H. van den Aker, C.M. Kneepkens, F. Stellaard, B. Geypens, and Y.F. Ghoos, J. Lab. Clin. Med., 127 (1996) 303–309.
50. E. Ferrannini, O. Bjorkman, G.A. Reichard, Jr., A. Pilo, M. Olsson, J. Wahren, and R.A. DeFronzo, Diabetes, 34 (1985) 580–588.
51. Y.T. Kruszynska, A. Meyer-Alber, F. Darakhshan, P.D. Home, and N. McIntyre, J. Clin. Invest., 91 (1993) 1057–1066.
52. S.C. Kalhan, S.M. Savin, and P.A. Adam, J. Lab. Clin. Med., 89 (1977) 285–294.
53. S.C. Kalhan, D.M. Bier, S.M. Savin, and P.A. Adam, J. Clin. Endocrinol. Metab., 50 (1980) 456–460.
54. K.Y. Tserng and S.C. Kalhan, Am. J. Physiol., 245 (1983) E476–82.
55. H.K. Berthold, F. Jahoor, P.D. Klein, and P.J. Reeds, J. Nutr., 125 (1995) 2516–2527.
56. R.A. Neese, J.M. Schwarz, D. Faix, S. Turner, A. Letscher, D. Vu. and M.K. Hellerstein, J. Biol. Chem., 270 (1995) 14452–14466.
57. J. Katz, P. Wals, and W.N. Lee, J. Biol. Chem., 268 (1993) 25509–25521.
58. H.N. Lafeber, E.J. Sulkers, T.E. Chapman, and P.J. Sauer, Pediatr. Res., 28 (1990) 153–157.

59. G.R. Vantrappen, P.J. Rutgeerts, Y.F. Ghoos, and M.I. Hiele, Gastroenterology, 96 (1989) 1126–1134.
60. J.B. Watkins, P.D. Klein, D.A. Schoeller, B.S. Kirschner, R. Park, and J.A. Perman, Gastroenterology, 82 (1982) 911–917.
61. D.M. Minich, M. Kalivianakis, R. Havinga, H. van Goor, F. Stellaard, R.J. Vonk, F. Kuipers, and H.J. Verkade, Biochim. Biophys. Acta, 1438 (1999) 111–119.
62. M. Kalivianakis, D.M. Minich, C.M. Bijleveld, W.M. van Aalderen, F. Stellaard, M. Laseur, R.J. Vonk, and H.J. Verkade, Am. J. Clin. Nutr., 69 (1999) 127–134.
63. R.R. Wolfe, J.E. Evans, C.J. Mullany, and J.F. Burke, Biomed. Mass Spectrom., 7 (1980) 168–171.
64. B.D. Kossak, E. Schmidt-Sommerfeld, D.A. Schoeller, P. Rinaldo, D. Penn, and J.H. Tonsgard, Neurology, 43 (1993) 2362–2368.
65. D.E. MacDougall, P.J. Jones, J. Vogt, P.T. Phang, and D.D. Kitts, Eur. J. Clin. Invest., 26 (1996) 755–762.
66. M.A. Nada, C. Vianey-Saban, C.R. Roe, J.H. Ding, M. Mathieu, R.S. Wappner, M.G. Bialer, J.A. McGlynn, and G. Mandon, Prenat. Diagn., 16 (1996) 117–124.
67. C. Maffeis, F. Armellini, L. Tato, and Y. Schutz, J. Clin. Endocrinol. Metab., 84 (1999) 654–658.
68. H. Demmelmair, T. Sauerwald, B. Koletzko, and T. Richter, Eur. J. Pediatr., 156(Suppl. 1) (1997) S70–4.
69. H. Demmelmair, M. Baumheuer, B. Koletzko, K. Dokoupil, and G. Kratl, J. Lipid Res., 39 (1998) 1389–1396.
70. P. Szitanyi, B. Koletzko, A. Mydlilova, and H. Demmelmair, Pediatr. Res., 45 (1999) 669–673.
71. V.P. Carnielli, E.J. Sulkers, C. Moretti, J.L. Wattimena, J.B. van Goudoever, H.J. Degenhart, F. Zacchello, and P.J. Sauer, Metabolism, 43 (1994) 1287–1292.
72. V.P. Carnielli, D.J. Wattimena, I.H. Luijendijk, A. Boerlage, H.J. Degenhart, and P.J. Sauer, Pediatr. Res., 40 (1996) 169–174.
73. B. Koletzko, T. Decsi, and H. Demmelmair, Lipids, 31 (1996) 79–83.
74. H. Demmelmair, U. von Schenck, E. Behrendt, T. Sauerwald, and B. Koletzko, J. Pediatr. Gastroenterol. Nutr., 21 (1995) 31–36.
75. M.K. Hellerstein, M. Christiansen, S. Kaempfer, C. Kletke, K. Wu, J.S. Reid, K. Mulligan, N.S. Hellerstein, and C.H. Shackleton, J. Clin. Invest., 87 (1991) 1841–1852.
76. P.J. Garlick and E. Cersosimo, Baillieres. Clin. Endocrinol. Metab., 11 (1997) 629–644.
77. K.E. Yarasheski, K. Smith, M.J. Rennie, and D.M. Bier, Biol. Mass Spectrom., 21 (1992) 486–490.
78. M.J. Rennie, R.H. Edwards, D. Halliday, D.E. Matthews, S.L. Wolman, and D.J. Millward, Clin. Sci., 63 (1982) 519–523.
79. P.J. Garlick, J. Wernerman, M.A. McNurlan, P. Essen, G.E. Lobley, E. Milne, G.A. Calder, and E. Vinnars, Clin. Sci., 77 (1989) 329–336.
80. K.J. Motil, A.R. Opekun, C.M. Montandon, H.K. Berthold, T.A. Davis, P.D. Klein, and P.J. Reeds, J. Nutr., 124 (1994) 41–51.
81. S.C. Denne, E.M. Rossi, and S.C. Kalhan, Pediatr. Res., 30 (1991) 23–27.

82. B. Kaplan, Z. Wang, O. Siddhom, T.K. Henthorn, and S.K. Mujais, Artif. Organs, 23 (1999) 44–50.

83. E.J. Freyse and S. Knospe, Isotopes. Environ. Health Stud., 34 (1998) 107–118.

84. D.E. Matthews and R.S. Downey, Am. J. Physiol., 246 (1984) E519–27.

85. W.D. Kloppenburg, B.G. Wolthers, F. Stellaard, H. Elzinga, T. Tepper, P.E. de Jong, and R.M. Huisman, Clin. Sci., (Colch.), 93 (1997) 73–80.

86. M.G. de Sain-van der Velden, G.A. Kaysen, H.A. Barrett, F. Stellaard, M.M. Gadellaa, H.A. Voorbij, D.J. Reijngoud, and T.J. Rabelink, Kidney Int., 53 (1998) 994–1001.

87. D. Halliday, S. Venkatesan, and P. Pacy, Am. J. Clin. Nutr., 57 (1993) 726S-730S; discussion 730.

88. P.J. Reeds, D.L. Hachey, B.W. Patterson, K.J. Motil, and P.D. Klein, J. Nutr., 122 (1992) 457–466.

89. K.G. Parhofer, P. Hugh, R. Barrett, D.M. Bier, and G. Schonfeld, J. Lipid Res., 32 (1991) 1311–1323.

90. B.W. Patterson, D.L. Hachey, G.L. Cook, J.M. Amann, and P.D. Klein, J. Lipid Res., 32 (1991) 1063–1072.

91. D.J. Reijngoud, G. Hellstern, H. Elzinga, M.G. de Sain-van der Velden, A. Okken, and F. Stellaard, J. Mass Spectrom., 33 (1998) 621–626.

92. W. Kulik, J.A. Meesterburrie, C. Jakobs and K. de Meer, J. Chromatogr. B., 710 (1998) 37–47.

93. M.S. Bosner, R.E. Ostlund, Jr., O. Osofisan, J. Grosklos, C. Fritschle, and L.G. Lange, J. Lipid Res., 34 (1993) 1047–1053.

94. M.S. Bosner, L.G. Lange, W.F. Stenson and R.E. Ostlund, Jr., J. Lipid Res., 40 (1999) 302–308.

95. D. Lütjohann, C.O. Meese, J.R.3.Crouse, and K. von Bergmann, J. Lipid Res., 34 (1993) 1039–1046.

96. D. Lütjohann, I. Bjorkhem, U.F. Beil, and K. von Bergmann, J. Lipid Res., 36 (1995) 1763–1773.

97. W.W. Wong, D.L. Hachey, A. Feste, J. Leggitt, L.L. Clarke, W.G. Pond, and P.D. Klein, J. Lipid Res., 32 (1991) 1049–1056.

98. P.J. Jones, Can. J. Physiol. Pharmacol., 68 (1990) 955–959.

99. P.J. Jones and D.A. Schoeller, J. Lipid Res., 31 (1990) 667–673.

100. R.A. Neese, D. Faix, C. Kletke, K. Wu, A.C. Wang, C.H. Shackleton, and M.K. Hellerstein, Am. J. Physiol., 264 (1993) E136–47.

101. R.H. Bandsma, F. Stellaard, R.J. Vonk, G.T. Nagel, R.A. Neese, M.K. Hellerstein, and F. Kuipers, Biochem. J., 329 (1998) 699–703.

102. W.N. Lee, S. Bassilian, H.O. Ajie, D.A. Schoeller, J. Edmond, E.A. Bergner, and L.O. Byerley, Am. J. Physiol., 266 (1994) E699–708.

103. R.H. Bandsma, F. Kuipers, R.J. Vonk, R. Boverhof, P.J. Sauer, G.T. Nagel, H. Elzinga, R. Neese, M. Hellerstein, and F. Stellaard. Biochim. Biophys. Acta, 483 (2000) 343–351.

104. F. Stellaard, M. Sackmann, T. Sauerbruch, and G. Paumgartner, J. Lipid Res., 25 (1984) 1313–1319.

105. F. Stellaard, M. Sackmann, F. Berr, and G. Paumgartner, Biomed. Environ. Mass Spectrom., 14 (1987) 609–611.

106. K.M. Gibson, F. Stellaard, G.F. Hoffmann, D. Rating, M. Hrebicek, and C. Jakobs, Clin. Chim. Acta, 217 (1993) 217–220.

107. T. Miyara, N. Shindo, M. Tohma, and K. Murayama, Biomed. Chromatogr., 4 (1990) 56–60.

108. F. Mion, M. Rousseau, J.Y. Scoazec, F. Berger, and Y. Minaire, Eur. J. Clin. Invest., 29 (1999) 624–629.

109. C.S. Irving, D.A. Schoeller, K.I. Nakamura, A.L. Baker, and P.D. Klein, J. Lab. Clin. Med., 100 (1982) 356–373.

110. F. Mion, P.E. Queneau, M. Rousseau, J.L. Brazier, P. Paliard, and Y. Minaire, Hepatogastroenterology, 42 (1995) 931–938.

111. M.J. Arnaud, A. Thelin-Doerner, E. Ravussin and K.J. Acheson, Biomed. Mass Spectrom., 7 (1980) 521–524.

112. A. Witschi, S. Mossi, B. Meyer, E. Junker, and B.H. Lauterburg, Alcohol Clin. Exp. Res., 18 (1994) 951–955.

113. P.J. Jones, P.B. Pencharz, and M.T. Clandinin, J. Lab. Clin. Med., 105 (1985) 647–652.

114. M. Kalivianakis, H.J. Verkade, F. Stellaard, M. van der Were, H. Elzinga, and R.J. Vonk, Eur. J. Clin. Invest., 27 (1997) 434–442.

115. P. Evenepoel, B. Geypens, A. Luypaerts, M. Hiele, Y. Ghoos, and P. Rutgeerts, J. Nutr., 128 (1998) 1716–1722.

116. C.E. King and P.P. Toskes, Crit. Rev. Clin. Lab. Sci., 21 (1984) 269–281.

117. S.F. Dellert, M.J. Nowicki, M.K. Farrell, J. Delente, and J.E. Heubi, J. Pediatr. Gastroenterol. Nutr., 25 (1997) 153–158.

118. G.W. Hepner, A.F. Hofmann, J.R. Malagelada, P.A. Szczepanik, and P.D. Klein, Gastroenterology 66 (1974) 556–564.

119. W.E. Heine, H.K. Berthold, and P.D. Klein, Am. J. Gastroenterol., 90 (1995) 93–98.

120. K.D. Wutzke, W.E. Heine, C. Plath, P. Leitzmann, M. Radke, C. Mohr, I. Richter, H.U. Gulzow, and D. Hobusch, Eur. J. Clin. Nutr., 51 (1997) 11–19.

121. D.Y. Graham, P.D. Klein, D.J. Evans, Jr., D.G. Evans, L.C. Alpert, A.R. Opekun, and T.W. Boutton, Lancet, 1 (1987) 1174–1177.

122. P.D. Klein, H.M. Malaty, R.F. Martin, K.S. Graham, R.M. Genta, and D.Y. Graham, Am. J. Gastroenterol., 91 (1996) 690–694.

123. Y.F. Ghoos, B.D. Maes, B.J. Geypens, G. Mys, M.I. Hiele, P.J. Rutgeerts, and G. Vantrappen, Gastroenterology, 104 (1993) 1640–1647.

124. G. Veereman-Wauters, Y. Ghoos, S. van der Schoor, B. Maes, N. Hebbalkar, H. Devlieger, and E. Eggermont, J. Pediatr. Gastroenterol. Nutr., 23 (1996) 111–117.

125. D.J. Bjorkman, J.G. Moore, P.D. Klein, and D.Y. Graham, Am. J. Gastroenterol., 86 (1991) 821–823.

126. B.D. Maes, Y.F. Ghoos, B.J. Geypens, M.I. Hiele, and P.J. Rutgeerts, Aliment. Pharmacol. Ther., 9 (1995) 11–18.

127. S. Mossi, B. Meyer-Wyss, C. Beglinger, W. Schwizer, M. Fried, A. Ajami, and R. Brignoli, Dig. Dis. Sci., 39 (1994) 107S–109S.

128. B. Braden, S. Adams, L.P. Duan, K.H. Orth, F.D. Maul, B. Lembcke, G. Hor, and W.F. Caspary, Gastroenterology, 108 (1995) 1048–1055.

12

Clinical Steroid Analysis by Gas Chromatography–Mass Spectrometry

Stefan A. Wudy
Children's Hospital of the University of Giessen, Giessen, Germany

Janos Homoki, and Walter M. Teller*
University of Ulm, Ulm/Donau, Germany

1. INTRODUCTION

1.1. Clinical Steroid Analysis and Gas Chromatography–Mass Spectrometry

In the field of steroid analysis, both gas chromatography (GC) and mass spectrometry (MS) are scientific tools with long-standing "tradition." In the steroid laboratory of the Children's Hospital of Ulm (Germany), GC has now been used for nearly 25 years and GC–MS was introduced more than a decade ago. Both techniques are successfully applied to delineate steroid-related disorders in children and adults. This chapter is written from the viewpoint of pediatric endocrinologists and therefore concentrates on the current methods used in the authors' laboratory and the clinical applications of steroid analysis by GC–MS. It is not our intention to review the many milestones of instrumental progress and method development. The reader is referred to further comprehensive reviews [1,2,3]. Interestingly, there has always been a close relationship between clinical endocrinology, particularly pediatric endocrinology, and laboratories spearheading techniques of GC–MS for steroid analysis. As many of these laboratories are located adjacent to pediatric hospitals, application of these techniques to pediatric endo-

---

* This chapter is dedicated to the memory of Professor Dr. Walter M. Teller (1928–1999), a pioneer of pediatric endocrinology.

309

crine problems seems to have proven advantageous, and it is our goal to explain the underlying reasons in the following pages.

The potential of GC–MS in determining a multitude of steroid metabolites simultaneously in a single "steroid profile" is still unsurpassed. In particular, GC bears the greatest potential for separating steroids, and MS allows for the highest specificity in determining steroid metabolites. The benchtop GC–MS systems currently available are easy to operate, can be tuned automatically, and present the most robust developments in the MS instrumental field, currently and in the near future. Furthermore, reasonable prices and computerized data management make benchtop instruments suited for routine clinical use.

In this chapter, urinary steroid profiling is discussed first, followed by a description of the state of the art concerning clinical profiling of steroid hormones in plasma by stable isotope dilution (ID) GC–MS. The reason is merely a historical one: the art of urinary steroid profiling matured earlier and has found more widespread use. The early attempts to measure plasma steroids by GC–MS were rapidly surpassed by the introduction of immunoassay techniques with which MS could not compete, especially with respect to analytical run time and cost. However, it was not until recently that it was realized that the lack of specificity of immunoassays might initiate a renaissance for clinical MS techniques in steroid analysis.

1.2. Biosynthesis and Catabolism of Steroids

Steroid hormones are derived from the cyclopentanophenanthrene ring, termed gonane, the 17 carbons of which are saturated by hydrogen [4]. The addition of

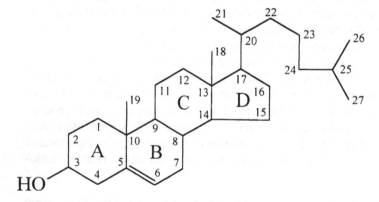

Figure 1 Structure of cholesterol and numbering of carbon atoms of the steroid skeleton.

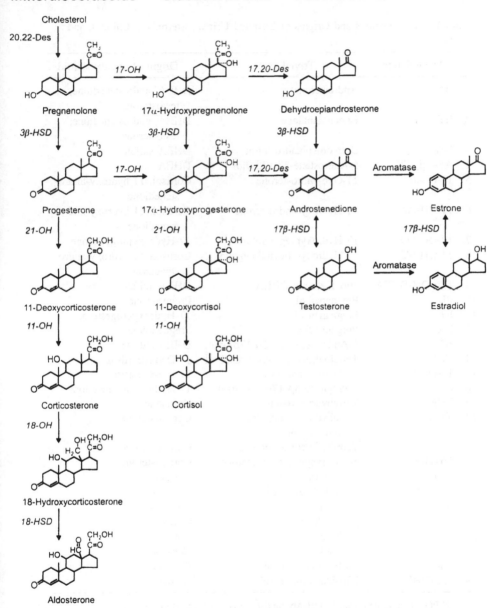

Mineralocorticoids   Glucocorticoids   Sex Hormones

Figure 2 Biosynthesis of steroid hormones. Abbreviations of names of enzymes: 20, 22-Des, 20,22-desmolase; 17,20-Des, 17,20-desmolase; 3β-HSD, 3β-hydroxysteroid-dehydrogenase; 17β-HSD, 17β-hydroxysteroid-dehydrogenase; 18-HSD, 18-hydroxysteroid-dehydrogenase; 17α-OH, 17α-hydroxylase; 21-OH, 21-hydroxylase; 11β-OH, 11β-hydroxylase; 18-OH, 18-hydroxylase.

Table 1 Abbreviations and Origins of Excreted Urinary Steroids in Children and Adults

	Abbreviation	Trivial name	Origin of urinary steroid
1	AN	Androsterone	DHEA, androstenedione, testosterone
2	ET	Etiocholanalone	DHEA, androstenedione, testosterone
3	DHEA	Dehydroepiandrosterone	DHEA-sulfate
4	A5–3β,17β	5-Androstene-3β,17β-diol	DHEA
5	11-O-An	11-Oxo-androsterone	Cortisol, 11-hydroxy-androstenedione
6	11-OH-AN	11-Hydroxy-androsterone	Cortisol, 11-hydroxy-androstenedione
7	17-OH-PO[a]	17-Hydroxypregnanolone	17-Hydroxyprogesterone
8	11-OH-ET	11-Hydroxy-etiocholanolone	Cortisol, 11-hydroxy-androstenedione
9	16α-OH-DHEA	16α-Hydroxy-DHEA	DHEA-sulfate
10	PD	Pregnanediol	Progesterone
11	PT	Pregnanetriol	17-Hydroxyprogesterone
12	P⁵D	Pregnenediol	Pregnenolone
13	A⁵T-16α	5-Androstene-3β,16α,17β-triol	DHEA-sulfate
14	THS	Tetrahydro-11-deoxycortisol	11-Deoxycortisol
15	11-O-PT	11-Oxo-pregnanetriol	21-Deoxycortisol
16	P⁵T	5-Pregnene-3β-17α,20α-triol	17-Hydroxypregnenolone
17	THE	Tetrahydrocortisone	Cortisone
18	THA	Tetrahydro-11-dehydro-corticosterone	Corticosterone
19	THB	Tetrahydrocorticosterone	Corticosterone
20	5α-THB	5α-Tetrahydrocorticosterone	Corticosterone
21	THF	Tetrahydrocortisol	Cortisol
22	5α-THF	5α-Tetrahydrocortisol	Cortisol
23	α-CL	α-Cortolone	Cortisone
24	β CL[b]	β-Cortolone	Cortisone
25	α-C	α-Cortol	Cortisol
26	F	Cortisol	Cortisol
27	6β-OH-F	6β-Hydroxycortisol	Cortisol
28	20α-DHF	20α-Dihydrocortisol	Cortisol

[a] 17-OH-PO not separated from 11-OH-AN (under the GC conditions given).
[b] β-CL not completely separated from β-C (under the GC conditions given).

an 18th carbon atom results in the estrane ring (C_{18}), the addition of a 19th carbon results in the androstane ring (C_{19}) and the addition of a side chain produces the pregnane (C_{20}–C_{21}), cholane (C_{20}–C_{24}), or cholestane (C_{20}–C_{27}) skeleton (Fig. 1).

The adrenal cortex and the gonads (ovaries and testes) present the main steroid-producing endocrine glands. A scheme of human steroid biosynthesis is depicted in Figure 2. The polycyclic carbon ring of steroid hormones is not degraded during metabolism. Following diverse chemical modifications, it is eliminated for the most part in urine, with a minor quantity passing into the feces. Catabolism consists of a series of reductions and hydroxylations, and the steroid is finally conjugated either with a glucuronic or sulfuric acid. The hydrosoluble compounds thus formed are eliminated. Biological activity is irreversibly lost after the first or one of the first chemical transformations (mostly reductions) of the hormonal molecule [5,6]. Table 1 summarizes important steroid hormones and their principal urinary metabolites. The complexity of steroid metabolism requires specialist steroid biochemical knowledge and expertise to correctly interpret steroid profiles.

2. PROFILING URINARY STEROIDS BY GAS CHROMATOGRAPHY–MASS SPECTROMETRY

2.1. Characteristics of Urinary Steroid Profiling by Gas Chromatography–Mass Spectrometry

Urinary steroid analysis by GC with flame ionization detection (FID) or GC–MS using repetitive scanning mode are nonselective methods for the quantitative determination of excreted steroid metabolites of adrenocortical and gonadal origin. The result of this multicomponent chromatographic analysis is the urinary steroid profile [2,7]. The major metabolites found in urine of healthy individuals are metabolites of dehydroepiandrosterone sulfate, progesterone, corticosterone, and cortisol, conjugated with glucuronic or sulfuric acid (see Table 1).

During early infancy (first 3 mo of life) 3β-hydroxy-5-ene steroid sulfates and free or glucuronidated metabolites of cortisol predominate in the urinary steroid profile (Table 2) [8,9]. In infants up to 3 mo of age, separation of steroid glucuronides and free steroids from steroid monosulfates by LH-20 chromatography is necessary prior to quantitative steroid profiling [3].

The measurement of steroid excretion rates in a 24-hr urine sample (quantitative urinary steroid profile) represents the integrated output of adrenocortical and gonadal steroid production. The identification of steroids using GC–MS is more specific than relying solely upon the retention time of a peak recorded during GC analysis with FID. Quantitative urinary steroid profiling permits detection of the different types of congenital adrenal hyperplasia (CAH), adrenal or gonadal tumors, Cushing's syndrome, and states of adrenal insufficiency or sup-

Table 2 Abbreviations and Names of Urinary Steroids Excreted by Newborns and Young Infants

Abbreviation	Name
3β-Hydroxy-5-ene steroid sulfates	
16α-OH-DHEA	16α-Hydroxy-DHEA
A^5T-16β	5-Androstene-3β,16α,17β-triol
16β-OH-DHEA	16β-Hydroxy-DHEA
15β,16α-OH-DHEA	15β,16α-Dihydroxy-DHEA
16-O-A^5D	16-Keto-5-androstene-3β,17β-diol
A^5T-16α	5-Androstene-3β,16α,17β-triol
16α,18-OH-DHEA	16α,18-Dihydroxy-DHEA
15β,17α-OH-P^5-olone	5-Pregnene-3β,15β,17α-triol-20-one
16α-OH-P^5-olone	5-Pregnene-3β,16α-diol-20-one
A^5-tetrols	5-Androstene-3β,15β,16α,17β-tetrol
	5-Androstene-3β,16α,17β,18-tetrol
P^5-tetrol-15β	5-pregnene-3β,15β,17α,20α-tetrol
21-OH-P^5-olone	5-pregnene-3β,21-diol-20-one
P^5-3β,20α,21-triol	5-pregnene-3β,20α,21-triol
P^5-3β,16α,20,21-tetrol	5-pregnene-3β,16α,20,21-tetrol
Free and glucuronidated metabolites of cortisol	
THE	Tetrahydrocortisone
6α-OH-THE	6α-Hydroxy-THE
1β-OH-THE	1β-Hydroxy-THE
β-CL	β-Cortolone
6α-OH-α-CL	6α-Hydroxy-α-cortolone
6α-OH-β-CL	6α-Hydroxy-β-CL
1β-OH-β-CL	1β-Hydroxy-β-CL
1β-OH-α-CL	1β-Hydroxy-α-CL

pression. Gas chromatography–mass spectrometry urinary steroid profiling from spot urine samples allows us to diagnose inborn errors of steroid biosynthesis by identifying characteristic steroid metabolites and by calculating ratios between precursor metabolites and product metabolites [2].

Quantitative urinary steroid profiling by GC–MS using selective ion monitoring (SIM) provides much higher sensitivity, but is a more selective approach. It enables one to further determine aldosterone and 18-hydroxylated cortisol metabolites in hypertension research [10], and in different steroid enzyme deficiencies [3].

2.2. Method

In this section, we give a description of the current procedure used for urinary steroid profiling in our laboratory. Twenty milliliters of a 24-hr urinary specimen or a spot urine sample are extracted using a Sep-pak C_{18} cartridge [11]. The extract is dried, reconstituted in 4 ml of 0.1 M acetate buffer (pH 5.0) and hydrolyzed for 48 hr at 37°C with 12 mg of Type I powdered Helix pomatia enzyme (SIGMA). Then, the resulting free steroids are reextracted using a Sep-pak C_{18} cartridge, dried and taken up in 2 ml of methanol. An internal standard mixture (5α-androstane-3α,17α-diol, stigmasterol, and cholesteryl butyrate, 2.5 µg of each) is added to 200 to 500 µl of this solution. The sample is dried and 3 drops of 2% methoxyamine hydrochloride in pyridine are added. The mixture is derivatized at 60°C for 2 hr. Then, the pyridine is blown off and 7 drops of trimethylsilylimidazole are added. Derivatization proceeds overnight at 100°C. The derivatized extract is purified by gel chromatography on a Lipidex 5000 minicolumn. A Pasteur pipette, plugged with glass wool, is filled up to two-thirds with Lipidex 5000 suspended in cyclohexane. After washing the column with 3 ml of cyclohexane, the sample is taken up in 1 ml of cyclohexane and put on the column. The eluate is collected, and the column is further washed with an additional 3 ml of solvent. The cyclohexane eluate is then concentrated under a stream of nitrogen to a volume of about 200 µl of which 1 to 2 µl are analyzed by GC or GC–MS.

2.2.1. Steroid Conjugate Fractionation

In neonates up to 3 mo of age, Sephadex LH-20 chromatography is used before hydrolysis and derivatization. Four grams of Sephadex LH-20 in a glass column are allowed to swell in chloroform–methanol (1:1 v/v; saturated with sodium chloride). Then, the slurry is filled into a glass column. After Sep-pak C_{18} extraction, the dried sample is taken up in 2 ml of the solvent system and applied on the column. The free and glucuronidated conjugates are eluted with 30 ml of the solvent system and the steroid sulfates are eluted with a further 50 ml of methanol. Both fractions are then dried, reconstituted in 4 ml of 0.1 M acetate buffer, hydrolyzed and derivatized as described above.

2.2.2. Gas Chromatography

Gas chromatography is performed on a Carlo Erba 6000 Vega 2 gas chromatograph with autosolid injector. Hydrogen is used as carrier gas. The gas chromatograph houses a chemically bonded dimethylpolysiloxan fused–silica column (Macherey-Nagel; Optima 1; 25 m × 0.2 mm ID, film 0.1 µm). The injection temperature is 60°C, which is held for 3 minutes. Then, a rapid increase in temperature (30°C/min) up to 190°C follows. Thereafter, the steroids of interest are eluted at a rate of 2.1°C/min up to 290°C. For detection, FID is used.

2.2.3. Gas Chromatography–Mass Spectrometry

Two GC–MS systems are available in the authors' laboratory. The GC–MS system I consists of a DANI 6500 gas chromatograph equipped with an autosolid injector (Carlo Erba) [12]. The gas chromatograph is directly interfaced with a Hewlett Packard 5970 mass selective detector (MSD). The GC–MS system II consists of a Hewlett Packard 5890 series gas chromatograph equipped with a Hewlett Packard 6890 autosampler (splitless/split autoliquid injection) interfaced to a Hewlett Packard 5972 A MSD. Helium is used as carrier gas. For GC column and temperature program, see the GC method above.

2.2.4. Quantification

Quantification is achieved by relating the peak areas of individual components to the areas of the internal standard peaks, i.e., androstanediol (AD) and stigmasterol (SS). The absolute amounts are calculated by multiplication with dilution factors and expressed in µg/day.

We have recently evaluated our method for urinary steroid profiling using repetitive scanning analysis on our GC–MS system II. Calibration plots showed excellent linearity ($r = 0.989$ to 0.998) in the range between a ratio of $1:10$ and $10:1$ of analyte and internal standard. Sensitivity was highest for pregnanetriol (PT, 0.25 ng) and lowest for tetrahydro-11-deoxycortisol (THS, 5 ng). When the same samples were worked up and analyzed repetitively ($n = 5$), coefficients of variation for the individual steroids ranged between 3.8 and 6.0%. After spiking the samples with analytes up to concentrations 2- to 10-fold greater than in unspiked specimens of healthy individuals, the relative errors showed a median of 3% (range -14% to $+18\%$).

2.3. Urinary Steroid Excretion Pattern in Healthy Individuals During Childhood and Adolescence

In their first days of life, neonates excrete large amounts of placental estrogen and progesterone metabolites. Later, the excretion of these compounds decreases rapidly. The excretion of 3β-hydroxy-5-ene steroids, produced by the fetal zone of the adrenal glands, decreases from birth to the fourth month of life from 10.3 to 2.8 mg/day. The excretion of the metabolites of cortisol, which is a product of the adult zone of the adrenal glands, increases from 0.9 to 2.1 mg/day (Table 3). The fraction of free and glucuronidated steroids from a healthy newborn contains predominantly cortisol metabolites. It is characteristic for this period of time that cortisol metabolites show 1β-, 6α-, and 6β-hydroxyl groups, except tetrahydrocortisone (THE) (see Fig. 3a).

Table 3 Urinary Excretion Rates of Steroids in Early Infancy (μg/day)

Age	Σ 3β-Hydroxy-5-ene steroids, median (range)	Σ Cortisol metabolites, median (range)
1–12 days	2820 (521–10276)	343 (149–897)
4–8 weeks	2318 (810–3791)	1042 (410–1394)
3–4 months	382 (65–2766)	1015 (848–2116)

The urinary steroid excretion changes fundamentally between 4 and 7 mo of age. The 3β-hydroxy-5-ene steroids disappear and "adult type" metabolites of cortisol (5α-THF, THF) become the predominant urinary steroids. At age 6 to 8 yr, the urinary steroid profile is changing again. Increasing amounts of adrenal androgens are excreted in girls and in boys, too. Later, at age 14 to 16 yr, the urinary steroid profile resembles the adult pattern in both sexes [10,13]. The ratio of adrenal androgens to cortisol metabolites is normally <0.8 [14].

The excretion of cortisol metabolites increases progressively with age and body size. When the excretion rates of cortisol metabolites are corrected for body surface, it can be seen that cortisol metabolite excretion rates are constant from childhood to adulthood (Fig. 4a). The excretion rates for the major adrenal androgen metabolites increase sharply during childhood, a phenomenon termed adrenarche (Fig. 4b). The major difference in urinary steroid profiles between males and females is the changing higher excretion of pregnanediol (PD) and PT according to the menstrual cycle in females [15]. In pregnant women, the excretion of PD and estriol predominates [16].

2.4. Clinical Indications for Urinary Steroid Profiling

A wide range of clinical symptoms, such as ambiguous genitalia, sodium-loosing states, precocious pseudopuberty, hirsutism, virilization, arterial hypertension, hypokalemia, trunkal obesity, primary amenorrhea, and hypoglycemia, are the most important clinical indications for the use of urinary steroid profiling for diagnostic purposes (Table 4) [17]. Urinary steroid profiling is also useful to monitor treatment and compliance in CAH due to enzyme deficiencies and after surgery of adrenal or gonadal tumors. The advantages of the method are nonselectivity and independence of circadian rhythm of most steroid hormones secreted. Another advantage, which bears special importance for pediatric endocrinology, is that the procedure is noninvasive, because the analytical sample can easily be obtained.

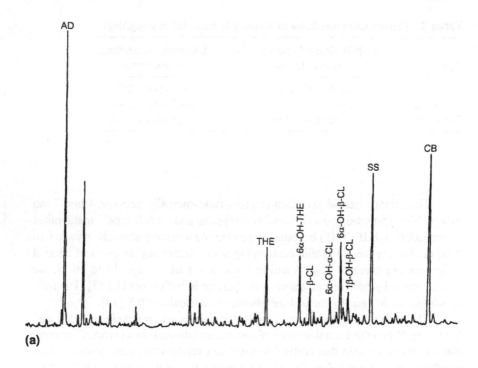

(a)

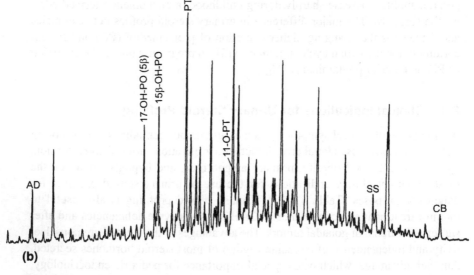

(b)

Figure 3 (a) Steroid profile (FID) of a healthy female neonate at age 2 wk (free and glucuronide fraction). AD, SS, CB: internal standards. For abbreviations see Table 2. (b) Urinary steroid profile (FID) in a 3-wk-old male neonate with 21-hydroxylase deficiency. 15β-OH-PO: 5β-pregnane-3α,15β,17α-triol-20-one.

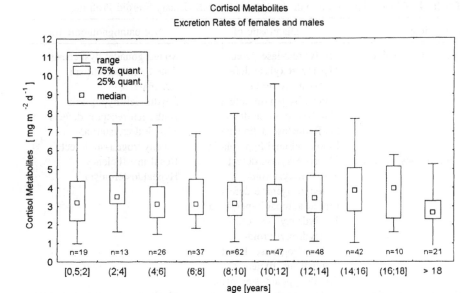

(a)

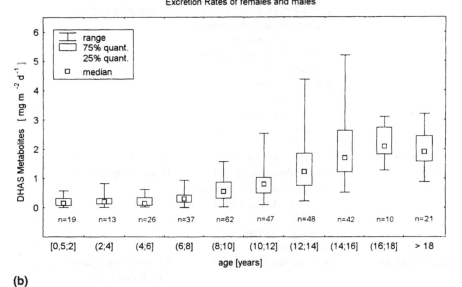

(b)

Figure 4 Cortisol (a) and androgen (b) metabolite excretion rates (corrected for body size) from childhood to adulthood. Since no significant sex differences have been observed, data have been combined.

Table 4   Clinical Indications For the Use of GC-MS Urinary Steroid Profiling

Indications	Diagnostic of	Not pathognomonic for

Ambiguous genitalia	21-Hydroxylase defect
	11β-Hydroxylase defect
	3β-Hydroxysteroid-
	  dehydrogenase defect
	17α-Hydroxylase defect
	5α-Reductase deficiency
	Lipoid adrenal hyperplasia
		Mixed gonadal dysgenesis
		True hermaphroditism
		XX male
		Leydig cell hypoplasia
		Androgen receptor defects
		17β-Hydroxysteroid-
		  dehydrogenase defect

Sodium-loosing states	21-Hydroxylase defect
	3β-hydroxysteroid-
	  dehydrogenase defect
	Lipoid adrenal hyperplasia
	18-Hydroxylase defect
	18-Hydroxysteroid-
	  dehydrogenase defect
	Adrenal insufficiency
	Pseudohypoaldosteronism
		Renal insufficiency
		Hypoaldosteronism

Precocious pseudo-
  puberty
	12-Hydroxylase defect (sim-
	  ple virilizing, late onset)
	11β-Hydroxylase defect
	Adrenal tumors
	Gonadal tumors
		McCune-Albright's syndrome
		Testotoxicosis
		β-hCG Excess

Hirsutism, virilization	21-Hydroxylase defect (sim-
	  ple virilizing, late onset)
	11β-Hydroxylase defect
	3β-Hydroxysteroid dehydro-
	  genase defect
	Adrenal tumors
	Gonadal tumors
	Polycystic ovaries
	Cortisone reductase defi-
	  ciency
		Idiopathic hirsutism

Hypertension, hypo-
  kalemia
	11β-Hydroxylase defect
	17α-Hydroxylase defect
	Apparent mineralocorticoid
	  excess (AME) Type I
	  and II
	Cushing's disease
	Adrenal tumors
		Renal hypertension
		Phaeochromocytoma

Hypoglycaemia	Adrenal insufficiency	Islet-cell hyperplasia Nesidioblastosis

2.4.1. Gas Chromatography–Mass Spectrometry Urinary
        Steroid Profiling in Patients with Congenital Enzyme
        Defects of Steroid Biosynthesis

2.4.1.1. 21-Hydroxylase Deficiency. The most common cause of ambiguous genitalia in the female newborn (46,XX) is CAH due to 21-hydroxylase
deficiency. Decreased cortisol synthesis induces excess adrenocorticotropic hormone (ACTH) secretion and overproduction of 17α-hydroxyprogesterone and
21-deoxycortisol. The low aldosterone production causes hyperkalemia and hy-
ponatremia (salt-loosing CAH). This salt-loosing state presents a major symptom
in the affected male newborn, as well as hyperpigmentation of genitalia. Patients
with simple virilizing and late-onset CAH have accelerated growth rate and bone
age with precocious pseudopuberty in males and virilization in females. The urinary steroid profile in newborns with salt-loosing CAH is dominated by 17-
hydroxpregnanolones, pregnanetriol, 11-oxo-pregnanetriol, and 15β-hydroxy-
pregnanolone (see Fig. 3b). In simple virilizing and late-onset CAH, additional
detectable or low amounts of cortisol metabolites are excreted [18,19].

2.4.1.2. 11β-Hydroxylase Deficiency. The clinical signs for 11β-hy-
droxylase deficiency are clitoromegaly and ambiguous genitalia in genetically
female newborns, and hyperpigmentation of the genital area in male newborns.
Postnatally, in both sexes rapid somatic growth, advanced bone age, progressive
clitoral or penile enlargement, premature pubarche, and hypertension occur. The
urinary steroid profile of newborns is characterized by excretion of 6-hydroxy-
tetrahydro-11-deoxycortisol (6α-OH-THS) [20]. In older infants and adults, excretion of metabolites of 11-deoxycortisol is increased, while excretion of cortisol
metabolites is low or absent [21].

2.4.1.3. 3β-Hydroxysteroid Dehydrogenase Deficiency. The clinical
spectrum of 3β-hydroxysteroid dehydrogenase (3β-HSD) deficiency at birth includes both salt-loosing and non–salt-loosing forms, independent of the extent
of genital ambiguity [22,23]. The urinary steroid profile in the salt-loosing form
has a characteristic fingerprint: dehydroepiandrosterone (DHEA), 16α-hydroxy-
DHEA, PT, and 17α-hydroxypregnenetriol are the major excreted steroids because of the virtually absent or very low excretion of cortisol metabolites. In
late-onset 3β-HSD deficiency, i.e., during adolescence or adulthood, varying degrees of hypogonadism occur in males [24], and hirsutism, irregular menses, and
polycystic ovaries occur in females. Patients with simultaneous elevation of post-
ACTH serum 17α-hydroxypregnenolone to 17α-hydroxprogesterone ratio and
elevated basal urinary 5-ene steroid excretion have mild 3β-HSD deficiency [25].

2.4.1.4. 17α-Hydroxylase/17,20-Lyase Deficiency. Male pseudoherma-
phroditism or ambiguous genitalia in 46,XY-individuals, absent pubertal development in 46,XX-individuals with hypokalemia, and development of hyperten-

sion are the signs for 17α-hydroxlase deficiency. Deficiency of 17α-hydroxylase leads to reduced production of both gonadal sex steroids and cortisol, with accompanying overproduction of corticosterone and 11-deoxycorticosterone. The urinary steroid profile is dominated by metabolites of corticosterone and its precursors and lacking in cortisol metabolites [26,27].

2.4.1.5. 18-Hydroxylase Deficiency. The 18-hydroxylase deficiency can present as infection-triggered, life-threatening salt-loosing state with hyperkalemia in newborns and young infants. It is characterized by aldosterone biosynthetic defects both in 18-hydroxlase deficiency (CMO I) and in 18-hydroxysteroid dehydrogenase deficiency (CMO II) [28]. The urinary steroid profile in 18-hydroxylase deficiency is characterized by increased excretion of free corticosterone and metabolites of corticosterone, while 18-hydroxylated corticosterone metabolites are absent or very low. The excretion of cortisol metabolites is normal (Fig. 5).

2.4.1.6. 18-Hydroxysteroid Dehydrogenase Deficiency. In CMO II, the urinary steroid profile shows, in addition to high amounts of corticosterone metabolites, 18-hydroxylated corticosterone metabolites (18-OH-THA, 18-OH-THB) (Fig. 6) [29].
    Pseudohypoaldosteronism, another condition associated with severe salt-wasting, can be detected by profiling urinary steroids [30].

2.4.1.7. Adrenal Insufficiency. Low excretion of glucocorticoid metabolites and adrenal androgen metabolites characterize the steroid profile. However, the different causes (Addison's disease, lipoid adrenal hyperplasia, and congenital adrenal hypoplasia) cannot be differentiated using urinary steroid profiling.

2.4.1.8. 11β-Hydroxysteroid Dehydrogenase Deficiency. Failure to thrive, polyuria, polydipsia, hypertension, hypokalemia, and nephrocalcinosis are the symptoms for apparent mineralocorticoid excess due to 11β-hydroxysteroid dehydrogenase deficiency (cortisol oxidase deficiency) and/or a steroid ring A reductase defect [31]. In the urinary steroid profile, the excretion of THE is much too low compared with the high THF, 5α-THF and free cortisol excretion [32].

2.4.1.9. Cortisone Reductase Deficiency. In hirsutism and virilization in females, cortisone reductase deficiency has been described. Patients with this disorder convert all their cortisol into cortisone. This gives rise to an apparent cortisol deficiency. Adrenocorticotropic hormone increases and stimulates the adrenal steroid synthesis. The urinary steroid profile is characterized by a very high excretion of THE, cortolones, and adrenal androgens, and low excretion of THF and 5α-THF [33].

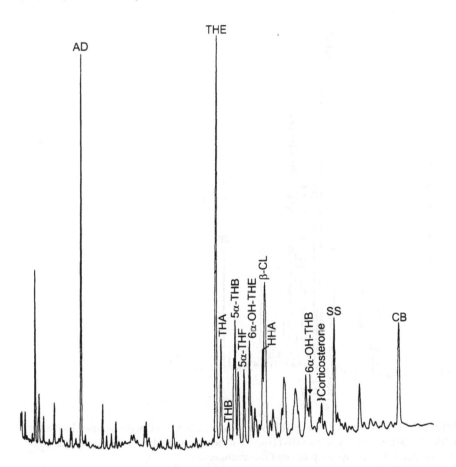

Figure 5 Urinary steroid profile (FID) in 18-hydroxylase deficiency (CMO I). Male, age 7 wk.

2.4.1.10. 5α-Reductase Deficiency.   In genetically male infants, ambiguous genitalia at birth and pubertal virilization are the clinical symptoms in inherited 5α-reductase deficiency. In urine, extremely low excretion of 5α-THF in young infants, and additional low excretion of androsterone and 11β-hydroxyandrosterone in older children indicate 5α-reductase deficiency [34].

2.4.2. Further Pathological Conditions

2.4.2.1. Polycystic Ovary Syndrome.   Polycystic ovary (PCO) syndrome is characterized by menstrual irregularity and hirsutism. It is a common cause of anovulatory infertility. The urinary steroid profile is dominated by adrenal

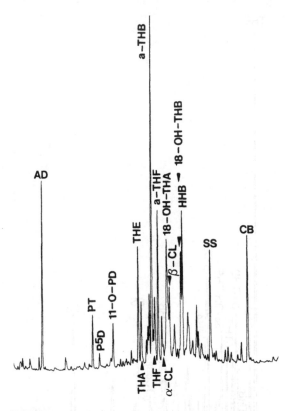

Figure 6 Urinary steroid profile (FID) in 18-hydroxysteroid-dehydrogenase deficiency (CMO II). Male, age 5 yrs. 18-OH-THA: 17-hydroxy-tetrahydro-11-dehydrocorticosterone. 18-OH-THB: 18-hydroxy-tetrahydrocorticosterone.

androgens and high excretion of androsterone and 5α-THF. In urinary steroid metabolites, the 5α/5β ratio is significantly increased (Fig. 7) [35].

2.4.2.2. Cushing's Syndrome. Prolonged hypercortisolism leads to Cushing's syndrome. The major symptoms are growth failure, obesity, predominantly of the trunk and the neck, and a full moonlike face. Hyperglycemia, purple striae, hirsutism, osteoporosis, muscular weakness, and hypogonadism are the additional signs. The etiology of Cushing's syndrome is frequently iatrogenic (long-term treatment with synthetic corticoids). In these cases, the urinary steroid profile shows absence or low excretion of adrenal steroids and excretion of metabolites of the prescribed corticoid drug. In the remaining cases, pituitary adenoma, adrenal adenoma and carcinoma, primary adrenocortical nodular dysplasia, McCune–Albright's syndrome, and ectopic ACTH- or corticotropin-releasing

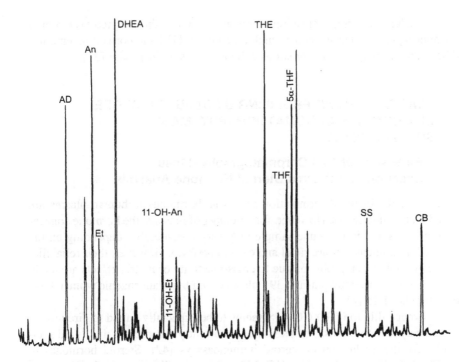

Figure 7 Chromatogram (FID) of urinary steroids in polycystic ovary syndrome Female, age 14 yrs.

factor (CRF)-secreting tumors can all cause Cushing's syndrome. Independent of the pathogenetic cause, the urinary steroid profile is characterized by increased excretion of THF and free cortisol, 6β-hydroxycortisol and 20α-dihydrocortisol [14].

2.4.2.3. Adrenocortical Tumors. Virilization, hypercortisolism, feminization, abdominal pain, palpable abdominal tumor, or combinations of these features are clinical symptoms of adrenocortical tumors. Both types of adrenocortical tumors (carcinoma and adenoma) can produce a wide variety of steroid hormones. This is a consequence of multiple enzyme deficiencies in tumor tissues. The tumor cells are capable of synthesizing large amounts of steroid hormone precursors independent of ACTH stimulation. Excessively high amounts of DHEA and other 3β-hydroxy-5-ene steroids characterize the urinary steroid profile in children with adrenocortical carcinoma, but similar profiles can also be produced by adrenal adenomas. Elevated 11β-hydroxy-androsterone excretion alone or combined with high excretion of cortisol metabolites or 3β-hydroxy-5-ene steroids are characteristic of the urinary steroid profile for adrenocortical adenomas [36,37].

In adults suffering from adrenocortical carcinoma, the elevated excretion of 3β-hydroxy-5-ene steroids and/or high amounts of THS and cortisol metabolites (THF) are the major characteristics of the urinary steroid profile [38].

3. PLASMA STEROID PROFILING BY STABLE ISOTOPE-DILUTION GAS CHROMATOGRAPHY–MASS SPECTROMETRY

3.1. Relevance of Gas Chromatography–Mass Spectrometry Plasma Steroid Hormone Analysis

Most methods for the determination of steroid hormones in human plasma are based on immunoassays. However, for a number of reasons, the hormone concentrations measured in the same sample may vary considerably depending on the kit used. Immunoassays are rapid and easy to perform. However, their reliability is questionable when problems due to cross-reactivity or matrix effects are likely to arise, e.g., in neonatal plasma [39], plasma from hyperandrogenic women [40], or follicular fluid [41].

Isotope dilution (ID) GC–MS seems to be especially suited to circumvent these analytical problems. The technique is currently regarded as the ''gold standard'' for the evaluation of steroid immunoassays [42]. Steroid hormones in plasma have been analyzed by GC–MS since its earliest days [43]. However, many of the steroids quantified have not been those required for clinical evaluation. Furthermore, the technique has still not been utilized in a routine clinical setting. In contrast to immunoassays, where only a single steroid can be measured at a time, GC–MS offers the advantageous capability of determining a whole spectrum of steroid hormones simultaneously in a profile. Meanwhile, vast technical improvements have rendered the development of reliable, labor- and cost-effective benchtop GC–MS instruments possible. The introduction of SIM allowed the determination of steroids at the nanogram and sub-nanogram levels. Therefore, SIM GC–MS can compete with radioimmunoassays [44].

In our attempt to evaluate disorders of androgen metabolism—some of the most common endocrinopathies—a specific, accurate, and sensitive method for the determination of plasma androgens was needed. We will describe our ID-GC–MS method, developed for profiling key steroids of androgen metabolism, by which it is up to now possible to simultaneously determine seven different steroid hormones in a single sample: testosterone (4-androstene-17β-ol-3-one), 4-androstenedione (4-androstene-3,17-dione), 17α-hydroxyprogesterone (4-pregnene-17α-ol-3,20-dione), 17α-hydroxypregnenolone (5-pregnen-3β,17α-diol-20-one), dehydroepiandrosterone (5-androstene-3β-ol-17-one), androstanediole (5α-androstane-3α,17β-diol), and 5α-dihydrotestosterone (5α-androstane-17β-ol-3-one).

3.2. Stable Isotope–Labeled Internal Standards

Stable isotope–labeled internal standards are almost ideal internal standards, because they have the advantage of showing practically the same chemical and chromatographic properties as the corresponding analytes. Furthermore, they allow procedural losses to be disregarded and they can easily be distinguished from the unlabeled compounds in MS by monitoring different ions.

In the field of steroid biochemistry, stable isotope–labeled steroid hormones have increasingly found application in metabolism studies and particularly in the quantification of steroid hormones in body fluids and tissues. Unfortunately, many stable isotope–labeled steroid hormones are currently not commercially available. Therefore, they have to be synthesized by the user. Two stable isotopes, ^{13}C or deuterium (^{2}H), can be used for the synthesis of labeled compounds. ^{13}C-Labeled steroids have major advantages for metabolism studies in vivo because of the stability of the label and the avoidance of isotope effects. However, the synthesis of these compounds is rather complex, laborious, and expensive. Furthermore, their use as internal standards seems to be limited, because the isotope enrichment achieved in the synthesis products rarely exceeds 90%. Deuterium, however, is relatively inexpensive and available at high isotopic purity in a larger number of chemical reagents. Compared with ^{13}C-labeling techniques, the incorporation of deuterium in the steroid molecule is generally less difficult. Since the late 1970s, a variety of synthetic pathways have been reported that lead to highly enriched deuterium-labeled steroid analogs. Synthetic routes leading to deuterium-labeled steroid hormones have been summarized in a special review [45].

We have recently described a first synthetic scheme leading to deuterium-labeled 17α-hydroxypregnenolone for use as internal standard [46]. In brief, the 17α-hydroxy-group of 17α-hydroxyprogesterone, which served as cheap and readily available starting material, was protected by tetrahydropyranylation. Introduction of deuterium by base-catalysed enolization and protection of the 3-hydroxygroup by acetate formation yielded [2,2,4,21,21,21-^{2}H$_6$]3,17α-dihydroxypregna-3,5-diene-20-one 3-acetate 17-tetrahydropyranyl ether. After reductive deuteration of the 3-ene and removal of the protecting groups, [2,2,4,4,21,21,21-^{2}H$_7$]17α-hydroxypregnenolone was obtained. Nuclear magnetic resonance (NMR) and MS studies confirmed the positions of the labels.

3.3. Method

The main steps of our ID-GC–MS method [47] for plasma steroid hormone profiling are: addition of the internal standard to the plasma sample, incubation, extraction, purification, derivatization, GC–MS analysis, and subsequent quantification.

3.3.1. Sample Workup Procedure

Plasma (0.1 to 1.0 ml) is incubated with 50 μl of an internal standard cocktail, i.e., a methanolic solution containing the deuterated steroid analogs in amounts similar to the expected plasma values. The samples are allowed to equilibrate. Then, the sample is loaded onto an Extrelut™ column (Merck, Darmstadt, Germany) and the steroids of interest are eluted with 3 × 5 ml of ethyl acetate. The combined organic extracts are evaporated to dryness under a gentle stream of nitrogen. The dried extract is purified by gel chromatography on Sephadex LH-20 minicolumns (50 mm × 5 mm ID) using cyclohexane/ethanol (9:1 v/v) as mobile phase. After an initial waste of 1 ml, the next 4 ml are collected, taken to dryness, and derivatized.

3.3.2. Derivatization Procedure

100 μl of acetonitrile and 20 μl of heptafluorobutyric anhydride (HFBA) are added to the dry plasma extract and the mixture is left for 1 hr at ambient temperature. Thereafter, the excess of reagent is removed under a stream of nitrogen, and the residue is dissolved in 50 μl of isooctane. A 2-μl portion of the solution is analyzed by GC–MS.

3.3.3. Gas Chromatography–Mass Spectrometry

In the authors' laboratory, steroid analyses can be performed on either of two Hewlett Packard benchtop GC–MS instruments (see section 2.2). The following temperature program is used for plasma steroid profiling: the initial column temperature is set at 60°C and after 6 minutes the temperature increases at a rate of 30°C/min up to 230°C. This is maintained for 2 minutes. Then, at a rate of 3°C/min, the column temperature is raised to 290°C. Quantification is performed using the peak area ratios between the ion pairs of the analytes and their corresponding labeled analogs. Portions of typical SIM chromatograms are shown in Figure 8.

The judicious choice of an appropriate derivatization is of primary concern not only for GC analysis, but also because it plays an essential role concerning sensitivity as well as specificity of MS. Therefore, monitoring of ions that have high abundance is desirable. For our analytes, derivative formation with the acylation reagent HFBA proved best for simultaneously derivatizing all C_{19} and C_{21} steroid analytes. The HFB esters are stable and give rise to a major mass increase, thus diminishing interfering background noise. They also have the great advantage over trimethylsilyl derivatives that there is less contribution from endogenous isotopes, e.g., from silicone.

When derivatized under the conditions described above, 4-ene-3-one steroids yield almost exclusively the 3,5-dien-3-HFB-ester. Testosterone forms the 3,5-dien-3,17β-di-HFB-ester. The mass spectrum of this compound reveals a molecular ion at m/z 680.2 of high relative intensity. 4-Androstenedione gives a thermally stable 3,5-dien-3-mono-HFB-ester. Its spectrum shows a very intense molecular ion peak at m/z 482.1. We have for the first time described the successful application of perfluoroacylation for 17α-hydroxyprogesterone, a C_{21} steroid. Its spectrum is shown in Figure 9. In this case, we monitor m/z 465.1, which is the base peak of the spectrum. It corresponds to the steroid nucleus (3,5-dien-3-HFB-ester) after loss of the side chain and the 17α-hydroxy group [48].

The 3β-hydroxy-5-ene steroids 17α-hydroxypregnenolone and dehydroepiandrosterone form the 3-mono-HFB derivatives. In case of dehydroepiandrosterone, we chose m/z 270.1 $[M-214]^+$ in SIM. This is the base peak of the spectrum and results from loss of the 3-hydroxy group. Regarding 17α-hydroxypregnenolone, we monitored m/z 467.1 $[M-43]^+$, again the base peak of the spectrum. It represents the steroid nucleus after loss of the 17α-hydroxy group and the side chain.

Derivatization of 5α-androstane-3α,17β-diol with HFBA yielded the 3α,17β-di HFB-ester. For SIM, the base peak of the spectrum $[M-214]^+$ at m/z 470 was chosen. 5α-Dihydrotestosterone gave the 17β-mono-HFB derivative. For SIM, we monitored m/z 414 $[M-72]^+$, the base peak of the spectrum.

With respect to the purity of our samples, the extraction procedure using Extrelut columns and ethylacetate appeared to be most satisfactory. A purification step using a LH-20 minicolumn proved to be beneficial in two ways. It significantly decreased the amount of plasma lipids that otherwise would interfere with MS analysis. In addition, we found that polar, conjugated steroids were retained under the conditions chosen. This is particularly important for dehydroepiandrosterone sulfate, which is present in much higher amounts than its unconjugated form. It is thus prevented from contaminating the sample.

Specificity was assessed by visually evaluating the chromatographic peak shape and by ensuring that likely steroid contaminants would be chromatographically resolved. It can be seen from Figure 8 that the peaks are essentially clear.

For a GC–MS assay using benchtop instruments, the method has excellent sensitivity. Sensitivity is lowest for testosterone with a signal-to-noise ratio (S/N) of 2.4 for 10 pg, and highest for 5α-androstane-3α,17β-diol with a S/N of 17 for 10 pg. For the steroids studied, the intra- and interassay coefficients of variation are between 2.7 and 7.6%, respectively. Accuracy was determined by spiking plasma with known amounts of steroids. The agreement between the values found and the amounts added was excellent with relative errors less than 7.5%. Standard curves were prepared by analysis of a selection of reference analyte/internal standard mixtures of known concentrations. All standard plots were linear.

The requirement for small amounts of plasma, the rapid, convenient workup, and the application of the benchtop GC–MS prove that our method is suited for routine clinical use in adults and children. The method does not require

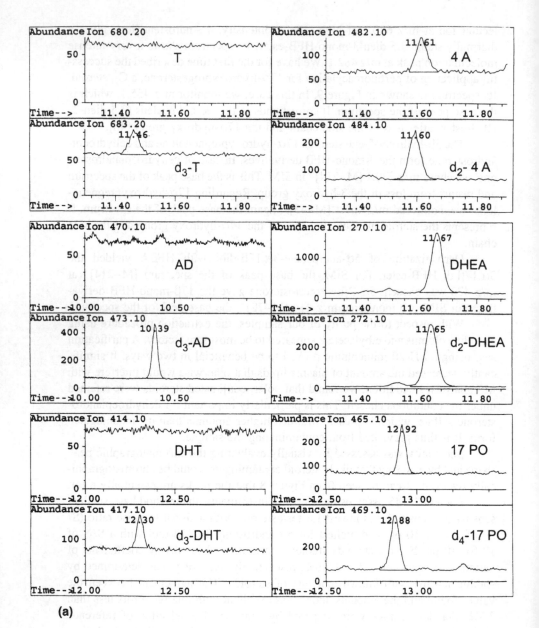

(a)

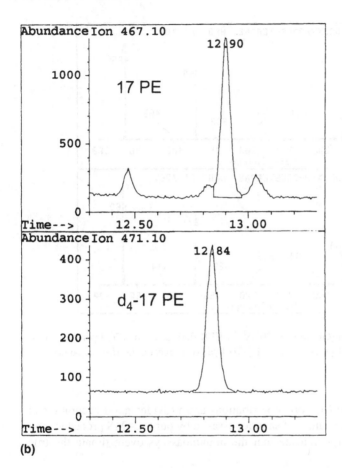

(b)

Figure 8 Selected ion monitoring of seven steroid hormones profiled in the plasma sample (0.1 ml of plasma analyzed) of a female neonate: (a) testosterone (T, m/z 680.2) not detected, [16,16,17-^{2}H$_3$]testosterone (d$_3$-T, m/z 683.2), 4-androstenedione (4A, m/z 482.1), [7,7-^{2}H$_2$]4-androstenedione (d$_2$-4A, m/z 484.1), 17α-hydroxyprogesterone (17PO, m/z 465.1), [11,11,12,12-^{2}H$_4$]17α-hydroxyprogesterone (d$_4$-PO, m/z 469.1), 5α-androstane-3α,17β-diol (AD; m/z 470.1) not detected, [16,16,17-^{2}H$_3$]5α-androstane-3α, 17β-diol (d$_3$-AD, m/z 473.1), 5α-dihydrotestosterone (DHT, m/z 414.1) not detected, [16,16,17-^{2}H$_3$]5α-dihydrotestosterone (d$_3$-DHT, m/z 417.1), dehydroepiandrosterone (DHEA, m/z 270.1), [7,7-^{2}H$_2$]dehydroepiandrosterone (d$_2$-DHEA, m/z 272.1). (b) Further SIM chromatograms of 17α-hydroxypregnenolone (17PE, m/z 467.1) and its corresponding internal standard [2,2,4,4,21,21,21-^{2}H$_7$]17α-hydroxypregnenolone (d$_4$-17PE, m/z 471.1).

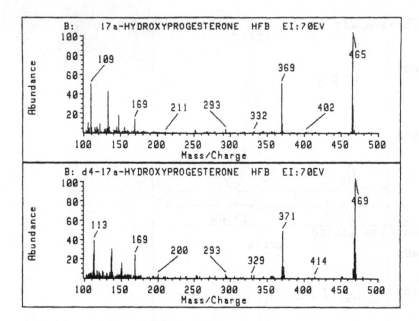

Figure 9 Mass spectra (GC-MS, EI 70 eV) of HFB-derivatives of 17α-hydroxyproges-terone (upper panel) and [11,11,12,12-^{2}H$_4$]17α-hydroxyprogesterone (lower panel).

complex corrections for isotope contributions and provides good accuracy and precision. Comparative values of samples assayed by our GC–MS procedure and by direct immunoassays indicate that the immunoassays overestimate the true concentration.

3.4. Clinical Applications

The many clinical indications of steroid hormone analysis by ID-GC–MS are at best demonstrated by explaining the diagnostic significance of the respective steroid hormones (see Fig. 2).

3.4.1. 17α-Hydroxyprogesterone

17α-Hydroxyprogesterone is the most valuable parameter for diagnosis and monitoring of 21-hydroxylase deficiency, the most frequent enzyme defect in steroid hormone biosynthesis leading to hyperandrogenism. We were the first to publish concentrations of plasma 17α-hydroxyprogesterone in neonates (Fig. 10), children, and adolescents [49,50] determined by GC–MS. The reliability of 17α-hydroxyprogesterone determinations by radioimmunoassays in the neonatal pe-

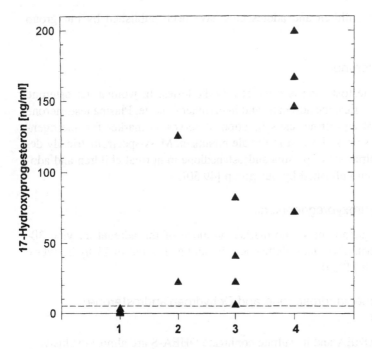

Figure 10 17α-Hydroxyprogesterone (ng/ml) in plasma of neonates: (1) healthy neonates (age 2 to 20 days; $n = 26$; median 1.5 ng/ml; range 0.0 to 5.6 ng/ml), (2) stressed (after convulsions) neonates without 21-hydroxylase deficiency, (3) neonates with simple virilizing 21-hydroxylase deficiency (3 to 13 days old), and (4) neonates with salt-loosing 21-hydroxylase deficiency (5 to 7 days old).

riod and early infancy is continually debated [51]. Cross-reacting steroids from the fetal zone of the adrenals may yield falsely elevated 17α-hydroxyprogesterone concentrations which, in turn, may cause misdiagnosis. Umbilical cord plasma contains high amounts of steroids from the fetal placental unit. We could determine concentrations of 17α-hydroxyprogesterone in umbilical cord plasma by ID-GC–MS and show that they are much lower than previously thought [49].

3.4.2. 4-Androstenedione

4-Androstenedione is the most important precursor hormone of androgens in females. Plasma levels are often increased in hirsute women. Next to 17α-hydroxyprogesterone, it is an important diagnostic parameter of 21-hydroxylase deficiency in plasma and amniotic fluid [52]. A decreased ratio between testosterone and androstenedione is indicative of 17β-hydroxysteroid dehydrogenase deficiency. The first mass-spectrometrically determined concentrations of plasma androstenedione in children and adolescents have been published by our group [49,50].

3.4.3. Testosterone

In men, 80% of testosterone is produced by the testes. In women, testosterone is produced by ovaries, the adrenals, and in peripheral tissue. Plasma testosterone is an indicator of endocrine testis function. It serves as marker for androgen-producing tumors or is elevated in female hirsutism. Mass-spectrometrically determined concentrations of plasma androstenedione in normal children and adolescents have been published by our group [49,50].

3.4.4. 17α-Hydroxypregnenolone

17α-Hydroxypregnenolone is the marker hormone of the adrenal enzyme 3β-hydroxysteroid dehydrogenase deficiency. It can be elevated in 21-hydroxylase deficiency as well [49,50].

3.4.5. Dehydroepiandrosterone and Dehydroepiandrosterone Sulfate

The C_{19} steroid DHEA and its sulfate conjugate DHEA-S are almost exclusively of adrenal origin. In contrast to DHEA, the plasma concentration of DHEA-S is about 500-fold higher and does not reveal a circadian rhythm. Considered the leading marker of adrenal androgen secretion, plasma DHEA-S is an important parameter for the evaluation of adrenal androgen production. Both metabolites are indicative of 3β-hydroxysteroid dehydrogenase deficiency. Also, DHEA-S serves as marker for adrenal androgen-producing tumors. It can further be elevated in hirsutism. Furthermore, it has been suggested to be an indicator of ACTH secretion, and a tumor marker in women with breast cancer. We have published mass-spectrometrically determined concentrations of plasma androstenedione in normal children and adolescents [49,50].

3.4.6. 5α-Dihydrotestosterone

5α-Dihydrotestosterone is the most potent human androgen. Its production is decreased in 5α-reductase deficiency. This condition is diagnosed by an increased ratio between testosterone and 5α-dihydrotestosterone. Our group has published mass-spectrometrically determined concentrations of plasma androstenedione in normal children and adolescents [49,50].

3.4.7. Androstanediol and Androstanediol Glucuronide

Androstandediol and androstanediol glucuronide are end metabolites of 5α-dihydrotestosterone. We have analyzed the developmental patterns of androstanediol and androstanediol glucuronide in children and adults. We have reported the first mass-spectrometrically determined normal values [50,55]. Both parameters have been suggested as markers of androgenicity [53].

3.5. Profiling Amniotic Fluid Steroids by Isotope-Dilution Gas Chromatography–Mass Spectrometry

The use of ID-GC–MS is not restricted to plasma only. Recently, we demonstrated application of this technique to the determination of steroid hormones in amniotic fluid [52]. We have profiled 17α-hydroxyprogesterone, androstenedione, testosterone, dehydroepiandrosterone, androstanediole, and 5α-dihydrotestosterone in amniotic fluid of midgestation in normal fetuses and fetuses at risk for 21-hydroxylase deficiency. Thus, the first reference values for amniotic fluid concentrations of the above-mentioned hormones could be established. Furthermore, our results showed that 17α-hydroxyprogesterone and androstenedione were the diagnostically most valuable steroids in the prenatal hormonal diagnosis of 21-hydroxylase deficiency. Prenatal hormonal diagnosis of 21-hydroxylase deficiency presented a highly accurate diagnostic procedure.

3.6. Determination of Conjugated Steroids by Isotope-Dilution Gas Chromatography–Mass Spectrometry

We are able to determine two steroid conjugates, 5α-androstane-3α,17β-diol glucuronide (ADG) and DHEA-S by ID-GC–MS. Cleavage of the conjugate group is required prior to GC [2]. The released free steroids are analyzed according to the GC–MS conditions described above.

Determination of ADG [55] is accomplished by including an enzymatic hydrolysis step after the addition of internal standard, $[16,16,17-^2H_3]5\alpha$-androstane-3α,17β-diol, in the method described above (see Section 3.3). We use β-glucuronidase (type B3, SIGMA), a preparation without sulfatase activity, according to the conditions given by Horton [54].

For the analysis of DHEA-S [56], a dideuterated internal standard, $[7,7,-^2H_2]$-DHEA, was prepared by sulfation of dideuterated DHEA. Prior to HFBA derivatization, the sample (10 μl of plasma) was subjected to a short-time methanolysis.

4. CLINICAL STEROID ANALYSIS BY GAS CHROMATOGRAPHY–MASS SPECTROMETRY: SIGNIFICANCE AND PERSPECTIVE

Hormonal analysis is indispensable to monitoring endocrine diseases. In the diagnosis of enzyme defects of steroid biosynthesis, hormonal testing and molecular biology have been shown to constitute important complementary techniques. Therefore, the development and improvement of reliable quantitative analytical methods for hormone determination presents a decisive field of endocrinology. In the field of steroid analysis, analytical techniques based on MS offer highest specificity. Therefore, the major advantage of GC–MS steroid profiling is the high degree of proof for every steroid analyzed.

Urinary steroid profiling using GC–MS is a nonselective multicomponent analysis of very high diagnostic potential. It allows definitive delineation of practically all disorders of steroid metabolism. It is best suited as a confirmatory technique after positive screening values (elevated 17α-hydroxyprogesterone) in screening neonates for 21-hydroxylase deficiency. Furthermore, the technique is noninvasive and rapid. The constellation of urinary steroid metabolites allows the diagnosis of most steroid-related disorders from spot urine samples, whereas determination of excretion rates of steroid metabolites requires 24-hr urine samples. In newborns and patients with steroid-secreting tumors, unusual steroids are produced for which specific serum assays are not available, but their metabolites can be monitored by urinary profiling.

The potential of ID-GC–MS in clinical quantitative plasma steroid analysis has not yet been exploited to its full extent. Drawbacks concerning the use of ID-GC–MS in a clinical setting have primarily been the huge cost of the sophisticated instrumentation, complex sample workup procedures, and the unavailability of most stable isotope–labeled internal standards. However, during the last decade, vast technical improvements have rendered the development of reliable labor- and cost-effective GC–MS instruments possible. With respect to appropriate internal standards, several suitable pathways leading to nonradioactive internal standards have been published. Our method for profiling seven steroids by ID-GC–MS has been found to be clinically applicable. Regarding the possible number of samples to be analyzed, GC–MS cannot compete with direct immunoassays. The latter techniques require only minimal sample preparation and allow analysis of numerous samples in batch assays in contrast to only serial assays possible with GC–MS. Plasma steroid analysis by GC–MS will not replace immunoassays in general. Besides its role as a reference methodology, we suggest application of ID-GC–MS in a clinical setting, whenever problems from matrix effects or cross-reactivity are likely to arise, and/or when suspicious results need to be rechecked. The GC–MS technique should provide the chance of a complementary analytical technology with highest specificity.

To keep costs within a reasonable limit, we suggest establishing priorities for urinary and plasma steroid analysis by GC–MS. Thus, steroid analyses are at best carried out in a small number of specialized supra-regional laboratories (reference centers) equipped with the analytical instrumentation and where support of specialist biochemists or clinicians is available.

Over the last decade, steroid determination by immunoassays has faced considerable loss of reliability due to preference of direct immunoassays. Most assays currently available cannot be used in analytically critical periods such as the neonatal period. For steroids rarely requested, hardly any assays are offered any more. This development has been favored by uncritical efforts of saving money in the biomedical disciplines. In this context, it is important to point out that clinical steroid analysis by GC–MS fulfills all criteria of a good clinical assay. It still has to be realized that quality is not equivalent to luxury.

ACKNOWLEDGMENTS

Stefan A. Wudy and Janos Homoki acknowledge the research grants that have been awarded to them from the Deutsche Forschungsgemeinschaft (DFG). The authors gratefully acknowledge the help of their Ph.D. students Michaela Hartmann, Claudia Solleder, and Ulrich Wachter. The skilled technical assistance of Heide Pinzer and Edith Ambach, the graphical support of Frank Wörsinger and the expert secretarial work of Heidrun Richter are gratefully acknowledged.

REFERENCES

1. C.H.L. Shackleton and W. Chai, Endocr. Rev., 6 (1985) 441–486.
2. C.H.L. Shackleton, J. Chromatogr., 379 (1986) 91–156.
3. C.H.L. Shackleton, J. Merdinck, and A.M. Lawson, in C.N. McEwen and B.S. Larsen (eds.), Mass Spectrometry of Biological Materials, 1990, Marcel Dekker, New York, pp. 297–378.
4. A.E. Kellie, in H.L.J. Makin (ed.), Biochemistry of Steroid Hormones, 2d ed., 1984, Blackwell Scientific, London, pp. 1–19.
5. D.B. Gower, in H.L.J. Makin (ed.), Biochemistry of Steroid Hormones, 2d ed., 1984, Blackwell Scientific, London, pp. 117–206.
6. D.B. Gower and J.W. Honour, in H.L.J. Makin (ed.), Biochemistry of Steroid Hormones, 2d ed., 1984, Blackwell Scientific, London, pp. 349–408.
7. J.W. Honour, Ann. Clin. Biochem., 34 (1997) 32–44.
8. C.H.L. Shackleton, Clin. Chim. Acta, 76 (1976) 287–298.
9. J.W. Honour, J. Kent, and C.H.L. Shackleton, Clin. Chim. Acta, 129 (1983) 229–232.
10. C.H.L. Shackleton, J. Steroid. Biochem. Molec. Biol., 45 (1993) 127–140.
11. C.H.L. Shackleton and J.O. Whitney, Clin. Chim. Acta, 107 (1980) 231–243.
12. C.H.L. Shackleton and J.W. Honour, Clin. Chim. Acta, 69 (1976) 267–274.
13. C.W. Weykamp, T.J. Penders, N.A. Schmidt, A.J. Borburgh, J.F. van de Calseyde, and B.J. Wolthers, Clin. Chem., 35 (1989) 2281–2284.
14. J. Homoki, R. Holl, and W.M. Teller, Klin. Wochenschr., 65 (1987) 719–726.
15. A. Ros and I.F. Sommerville, J. Obstet. Gynaecol. Br. Comm. 78 (1971) 1096–1107.

16. E. Hähnel, S.P. Wilkinson, and R. Hähnel, Clin. Chim. Acta, 151 (1985) 259–271.
17. J.W. Honour and C.G.D. Brook, Ann. Clin. Biochem., 34 (1997) 45–54.
18. J. Homoki, J. Sólyom, U. Wachter, and W.M. Teller, Eur. J. Pediatr., 151 (1992) 24–28.
19. J. Homoki, J. Sólyom, and W.M. Teller, Eur. J. Pediatr., 147 (1988) 257–262.
20. J.W. Honour, J.M. Anderson, and C.H.L. Shackleton, Acta Endocrinol., 103 (1983) 101–109.
21. S.A. Wudy, J. Homoki, U.A. Wachter, and W.M. Teller, Dtsch. Med. Wschr. 122 (1997) 3–10.
22. A.M. Bongiovanni and A. Clark, J. Clin. Invest., 41 (1962) 2086–2092.
23. R.L. Rosenfield, B.H. Rich, J.I. Wolfsdorf, F. Cassarola, S. Parks, A.M. Bongiovanni, C.H. Wu, and C.H.L. Shackleton, J. Clin. Endocrinol. Metab., 51 (1980) 345–353.
24. G. Schneider, M. Genel, A.M. Bongiovanni, A.S. Goldmann, and R.L. Rosenfield, J. Clin. Invest., 55 (1975) 681–690.
25. J. Sólyom, Z. Halasz, E. Hosszú, E. Glaz, R. Vihko, M. Orava, J. Homoki, S.A. Wudy, and W.M. Teller, Horm. Res., 44 (1995) 133–141.
26. M. D'Armiento, G. Reda, C. Kater, C.H.L. Shackleton, and E.G. Biglieri, J. Clin. Endocrinol. Metab., 56 (1983) 697–701.
27. C.E. Fardella, D.W. Hum, J. Homoki, and W.L. Miller, J. Clin. Endorinol. Metab., 79 (1994) 160–164.
28. S. Ulick, J. Clin. Endocrinol. Metab., 43 (1976) 92–96.
29. B.P. Hauffa, J. Sólyom, E. Glaz, C.H.L. Shackleton, G. Wambach, P. Vecsei, H. Stolecke, and J. Homoki, Eur. J. Pediatr., 150 (1991) 149–153.
30. J.W. Honour, M.J. Dillon, and C.H.L. Shackleton, J. Clin. Endocrinol. Metab., 54 (1982) 325–331.
31. F. Mantero, M. Palermo, M.D. Petrelli, R. Tedde, P.M. Stewart, and C.H.L. Shackleton, Steroids, 61 (1996) 193–196.
32. J. Müller-Berghaus, J. Homoki, D.U. Michalk, and U. Querfeld, Acta Paediatr., 85 (1996) 111–113.
33. G. Phillipou and B.A. Higgins, J. Steroid Biochem., 22 (1985) 435–436.
34. J. Imperato-McGinley, T. Gautier, M. Pichardo, and C.H.L. Shackleton, J. Clin. Endocrinol. Metab., 63 (1986) 1313–1318.
35. P.M. Stewart, C.H.L. Shackleton, G.H. Beastall, and C.R.W. Edwards, Lancet, 335 (1990) 431–433.
36. J.W. Honour, D.A. Price, N.F. Taylor, H.B. Marsden, and D.B. Grant, Eur. J. Pediatr., 142 (1984) 165–169.
37. E.M. Malunowicz, M. Ginalska-Malinowska, T.E. Romer, A. Ruszczynska-Wolska, and M. Dura, Horm. Res., 44 (1995) 182–188.
38. S. Gröndal, B. Eriksson, L. Hagenäs, S. Werner, and T. Curstedt, Acta Endocrinol. (Copenh), 122 (1990) 656–663.
39. T. Wong, C.H.L. Shackleton, T.R. Covey, and G. Ellis, Clin. Chem., 38 (1992) 1830–1837.
40. F.I. Chasalow, S.I. Blethen, D. Duckett, S. Zeitlin, and J. Greenfield, Steroids, 54 (1989) 373–383.

41. P. Silberzahn, L. Dehennin, I. Zwain, and A. Reiffsteck, Endocrinology, 117 (1985) 2176–2181.
42. J.G. Middle, in H.L.J. Makin, D.B. Gower, and D.N. Kirk (eds.), Steroid Analysis, 1995, Blackie Academic & Professional, London, pp. 647–696.
43. J. Sjövall, Steroids, 7 (1966) 447–453.
44. R. Knuppen, O. Haupt, W. Schramm, and H.O. Hoppen, J. Steroid. Biochem., 11 (1979) 153–161.
45. S.A. Wudy, Steroids, 55 (1990) 463–471.
46. C. Solleder, T. Schauber, J. Homoki, and S.A. Wudy, J. Labelled Cpd. Radiopharm., 41 (1998) 557–565.
47. S.A. Wudy, U.A. Wachter, J. Homoki, W.M. Teller, and C.H.L. Shackleton, Steroids, 57 (1992) 319–324.
48. D. Stöckl, L.M. Thienpont, V.I. DeBrabandere, and A.P. DeLeenheer, J. Am. Soc. Mass Spectrom., 6 (1995) 264–276.
49. S.A. Wudy, U.A. Wachter, J. Homoki, and W.M. Teller, Ped. Res., 38 (1995) 76–80.
50. S. A. Wudy, M. Hartmann, C. Solleder, U.A. Wachter, J. Homoki, in J.R. Heys and D.G. Melillo (eds.), Synthesis and Applications of Isotopically Labeled Compounds 1997, 1998, John Wiley & Sons, Chichester, pp. 575–579.
51. P. Bodlaender, Ann. Clin. Biochem., 28 (1991) 423–425.
52. S.A. Wudy, H.G. Dörr, C. Solleder, and J. Homoki, J. Clin. Endocrinol. Metab., 84 (1999) 2724–2728.
53. S.A. Wudy, U.A. Wachter, J. Homoki, and W.M. Teller, Eur. J. Endocrinol., 134 (1996) 87–92.
54. R. Horton, D. Endres, and M. Galmarini, J. Endocrinol. Metab., 59 (1984) 1057.
55. S.A. Wudy, U.A. Wachter, J. Homoki, and W.M. Teller, Eur. J. Endocrinol., 134 (1996) 87–92.
56. S.A. Wudy, U.A. Wachter, J. Homoki, and W.M. Teller, Horm. Res., 39 (1993) 235–240.

13

Gas Chromatography–Mass Spectrometry for Selective Screening for Inborn Errors of Metabolism

Jörn Oliver Sass
Leopold Franzens University Innsbruck, Innsbruck, Austria

Adrian C. Sewell
Johann Wolfgang Goethe University, Frankfurt, Germany

1. INTRODUCTION

Inborn errors of metabolism represent a group of diseases, which are individually rare, but collectively numerous [1]. In order to detect treatable inborn diseases as early as possible, during the last decades most European and North American countries have established neonatal laboratory screening tests, which search only for a small number of inborn errors of metabolism [2]. However, for the great majority of metabolic diseases, biochemical investigations are only performed in selected cases, after clinical symptoms or family history have indicated that a metabolic disease might be present (selective screening). Selective screening may involve the analysis of amino acids in plasma and urine, organic acids, purines, pyrimidines, oligosaccharides, and mucopolysaccharides in urine; as well as free carnitine, acylcarnitines, and very long-chain fatty acids in plasma. Among the positive results of selective screening, detection of organic acidurias accounts for a major portion [3]. More than 50 phenotypically different diseases are known to present with organic aciduria at least during metabolic decompensation [4]. A variety of clinical features are known. Patients may present with clinical symptoms including vomiting, muscular hypotonia, failure to thrive, or a sepsis-like picture. If an organic aciduria is suspected on clinical grounds, organic acid analysis is a vital part of the investigations and represents the main application of gas chromatography–mass spectrometry (GC–MS) in selective screening.

Organic acids, as understood in this chapter, are intermediates in the degradative metabolic pathways of amino acids, fats, and carbohydrates. Organic acids contain one or more carboxylic or acidic phenolic groups, but do not contain any primary amino groups which might react with ninhydrin. Therefore, although their transamination products (oxoacids) are addressed, amino acids per se are not considered here. Very long-chain fatty acids and inorganic metabolites are likewise not considered. Organic acids may also result from exogenous sources, such as diet, drugs, or bacterial contamination. In addition to the naturally occurring urinary organic acids which may be increased in various disease states, there are many abnormal acids, the presence of which is indicative of a particular disease (e.g., propionic acidemia) (Fig. 1), together with glycine [5] and glucuronide conjugates [6].

Since most organic acids undergo efficient renal excretion and urine is often obtained more easily from children than other body fluids, urine represents the preferred sample for organic acid analysis. However, in special cases or if no other sample should be available, plasma, serum, cerebrospinal fluid, or vitreous

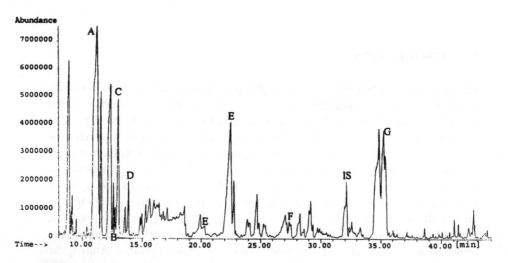

Figure 1 Urinary organic acid chromatogram of a patient with propionic acidemia. Peak A is 3-hydroxypropionic acid, B is 2-methyl-3-hydroxybutyric acid, C is 3-hydroxy-isovaleric acid, D is 3-hydroxy-*n*-valeric acid, peaks E represent propionylglycine, peak F is tiglylglycine, peaks G are methylcitric acid. IS = internal standard (tricarballylic acid).

humor (the last obtained post mortem) can also be analyzed. Urinary concentrations of organic acids are usually adjusted for the creatinine content of a sample.

2. SAMPLE PREPARATION

2.1. Extraction of Organic Acids

Organic acids have frequently been extracted from biological fluids by use of organic solvents. This is still the most commonly used means of extraction because of its relative speed and simplicity. Solvent extraction is usually performed with ethyl acetate or diethyl ether, aiming at extraction of the entire group of compounds. While the former solvent extracts predominantly hydrophilic acids, the latter improves the extraction of hydrophobic acids. However, extraction of the acids into organic solvents is not quantitative for several metabolites [7]. The extraction of urinary organic acids has also been accomplished by the use of weak anion exchange resin (diethylaminoethyl–Sephadex) [8], which has been reported to give higher recoveries for the more polar acids and to yield higher reproducibility [9,10]. Extraction on disposable columns filled with a trimethyl-aminopropyl phase (strong anion-exchange material) has been advocated by Verhaeghe et al. [11], while Mardens et al. [12] have expressed a more critical view. Use of solid-phase extraction on normal-phase silica columns from which organic acids can be eluted by a mixture of chloroform and t-amyl alcohol has been reported to be both rapid and reliable [13]. In Figure 2, two chromatograms of the same urine sample are presented, obtained either after solvent extraction with ethyl acetate (Fig. 4A) or with a silica column according to the procedure of Bachmann et al. [13] (Fig. 4B). Although MS has become indispensable for urinary organic acid analysis, there are nevertheless problems with artifacts due to either medication or dietary manipulation. Organic acid excretion patterns in urine of patients receiving valproic acid as an anticonvulsant are exceedingly complex and may well mask other important metabolites. Infants who receive milk formulas supplemented with medium-chain triglycerides exhibit a characteristic urinary organic acid pattern, which may be confused with that of patients with a mitochondrial fatty acid β-oxidation defect [14]. Even with rather sophisticated extraction methods aiming at efficient extraction of a wide range of compounds of clinical importance, there may be difficulties in identifying some components either because they occur in very small amounts or because they may not be completely extracted prior to derivatization. An example of the former is 4-hydroxybutyric acid excreted in succinic acid semialdehyde dehydrogenase deficiency [15]. As a trimethylsilyl (TMS) derivative, this compound elutes very close to the urea peak and because some patients excrete very small amounts, may well be missed. Patients with Canavan's disease excrete N-acetylaspartic

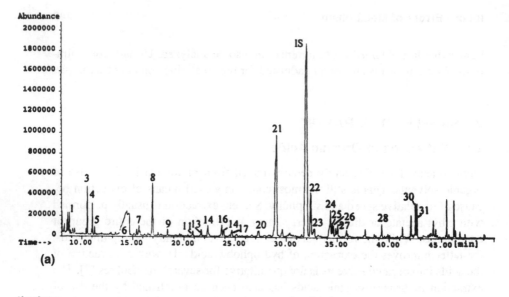

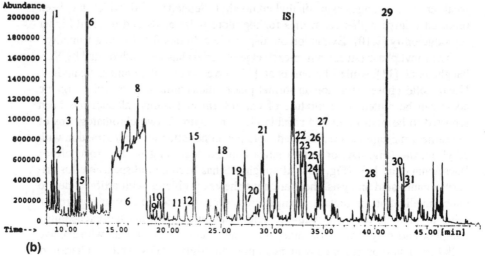

Figure 2 Normal organic acid chromatograms of the same urine sample, obtained (a) after solvent extraction with ethyl acetate or (b) after solid-phase extraction with a silica column according to the procedure of Bachmann et al. [13]. Peaks: 1, lactic acid; 2, glycolic acid; 3, oxalic acid; 4, pyruvic acid; 5,3-hydroxybutyric acid (*n* + iso); 6, urea; 7, phosphoric acid; 8, succinic acid; 9, fumaric acid; 10, 2,3-dihydroxybutyric acid; 11, glutaric acid; 12, 2,4-dihydroxybutyric acid; 13, 3-methylglutaric acid; 14, 3-methylgluta-conic acid; 15, 3,4-dihydroxybutyric acid; 16, mevalonic acid; 17, adipic acid; 18, pyroglutamic acid; 19, 2,3,4-trihydroxy butyric acid; 20, 2-hydroxyglutaric acid; 21, 2-oxoglutaric acid; 22, *cis*-aconitic acid; 23, homovanillic acid; 24, hippuric acid; 25, citric acid; 26, isocitric acid; 27, 3-hydroxyphenylhydracryllic acid; 28, 3-hydroxyhippuric acid; 29, uric acid; 30, 4-hydroxyhippuric acid; 31, stearic acid; IS = internal standard (tricarballylic acid).

acid [16], and extraction of urine from such patients with ethyl acetate results in only approximately 30% being extracted. Another example is that of succinyl acetone, a key marker for tyrosinemia type 1 [17]. This substance may disappear in alkaline urine samples or, when present in small amounts, may be missed. Attempts to perform the time-consuming sample preparation for determination of urinary organic acids with an automated robotic workstation have so far not been successful [18].

2.2. Derivatization of Organic Acid Extracts

As organic acids are polar, thermally unstable, and have low volatility, it is necessary for GC analysis to convert them into nonpolar, volatile, and thermally stable derivatives. Such derivatization is usually achieved by esterification. The most common method is trimethylsilylation with N,O-bis(trimethylsilyl)trifluoroacetamide (BSTFA) containing 1% trimethylchlorosilane. However, methylation with diazomethane or methanolic hydrogen chloride/boron trifluoride in methanol is also possible [19,20]. The derivatization to other esters has also been reported, but is rather uncommon in routine analysis [21]. Oxoacids should be stabilized by formation of an oxime prior to the esterification described above. Otherwise, formation of multiple/nonstable derivatives would greatly impair the determination of this class of compounds. The oximes can then react with BSTFA to yield TMS-oxime TMS esters.

Not only hydroxylamine hydrochloride, but also substituted derivatives such as methyl- or ethyl-hydroxylamine hydrochloride can be used. O-(2,3,4,5,6-pentafluorobenzyl)hydroxylamine hydrochloride has been used to convert oxoacids, aldehydes, and ketones into the corresponding oximes and has been considered advantageous because the subsequently formed TMS esters are eluted later, in a less crowded region of the chromatogram [22].

The use of at least one suitable internal standard is essential in organic acid analysis, regardless of whether quantitative analysis of certain acids is performed or whether semiquantitative analysis is done to assess the metabolite profile. The internal standard(s) should be added at the very start of the sample preparation. Compounds used for this purpose include tricarballylic acid [13,22], hexadecanedioic acid [22], pentadecanoic acid [23], heptadecanoic acid [24], isopropylmalonic acid [20], and malonic acid [25]. The latter, however, would interfere with the detection of malonic aciduria in malonyl-CoA decarboxylase deficiency. Only oxoacids such as 2-oxocaproic acid allow the checking of the analysis with regard to this group of compounds. To overcome analytical problems mentioned above, for both succinyl acetone [26] and N-acetylaspartic acid [27], improved analysis using stable isotope dilution is available. However, in view of their limited availability, the rather high price, and the wide spectrum of compounds determined in organic acid analysis, stable isotope dilution is not routinely used in selective screening for organic acidurias, but is of major importance in their prenatal diagnosis by analysis of metabolites in amniotic fluid. The TMS derivatives of organic acids are very stable if stored in the refrigerator and protected from moisture and air. Volatile organic acids, such as unsubstituted short-chain aliphatic acids, are usually lost during sample workup.

In the course of selective screening for organic acid disorders, the presence of abnormal metabolites usually suffices to make a tentative diagnosis; however, for monitoring dietary therapy, quantitation is necessary. For some diseases this is not possible because the reference compounds are not commercially available. In contrast, compounds such as methylmalonic acid, glutaric acid, and others can be quantitated with construction of calibration curves and then subjected to quality control. In Europe, external quality assurance is provided for a limited range of organic acids by the ERNDIM foundation [28]; however, the interlaboratory correlation for some organic acids is extremely disappointing.

3. GAS CHROMATOGRAPHY–MASS SPECTROMETRY INSTRUMENTATION

The first reported use of MS in the study of organic acid metabolism was by Klenk and Kahlke in 1963 [29] to identify phytanic acid in serum of a patient with Refsum's disease. However, it was only when, in 1966 [30], GC–MS was used to identify 3-hydroxyisovaleric acid in urine of a patient with the hitherto unknown disease isovaleric aciduria that the use of this analytical technique opened up the field for further study. The ability to analyze a wide range of organic acids in biological fluids using GC [31,32] led to the concept of "metabolic profiling" and screening for metabolic diseases. Early work utilized GC–MS instruments without on-line computerized data systems, with only selected peaks in the profile being scanned with the MS by manual activation. The real value of metabolic profiling was only realized with the introduction of on-line computers to control and maintain the MS in a repetitive scanning mode in which spectra are recorded continuously over a defined mass range at a preselected scan rate with the large volume of data created also being stored by the same computer system [33,34]. Such methods were pioneered by Jellum et al. [35,36], Mamer et al. [37], and Chalmers et al. [38,39]. These analyzes were carried out using packed GC columns yielding some 50 to 60 peaks from a single 50-minute chromatogram with a scan rate of 1 spectrum per 5 sec across the mass range required for most organic acids (m/z 50 to 500) (i.e., 5 sec^{-1}). This resulted in approximately 500 to 600 spectra, often of mixed components, which required not only the use of automatic peak identification, background subtraction, and library search routines, but also considerable skill and experience of the analyst in sorting out the mixed spectra. Today, storage of complete data allows retrospective processing of spectra and the use of extracted ion profiles to identify individual components obscured by overlapping peaks.

3.1. Gas Chromatograph Columns

Optimization of chromatographic performance is dependent upon a number of factors, most notably column selection. The characteristics of bonded fused-silica capillary columns have been well established, providing a range of stationary phases designed for polar or nonpolar analytes. These flexible polyamide-coated fused silica columns offer advantages in GC–MS by allowing direct introduction of the column into the ion source and obviating the need for molecular separators. For the analysis of TMS derivatives of extracted organic acids, columns coated with the dimethylsiloxane-type phase, such as DB-1, DB-5, or DB-5MS, are most suitable. In considering the length, diameter, and film thickness of the column it is necessary to account for sample capacity, speed of analysis, required separating power, ease of setting up, and requirements of the MS. Column length is usually of minor importance—a 30-m column is suitable for organic acid analysis. Column diameter depends upon the MS requirements. Widebore (0.32 mm ID) columns are normally used, however, megabore (0.53 mm ID) columns offer the advantage that they provide considerable sample capacity using high carrier gas flow [40]. Their drawback is that the separation power is limited and they require extra connections to the MS. Narrowbore (0.22 mm ID) columns are suitable for split/splitless injection where the concentration of components is low to avoid overloading the column. Film thickness can have a significant effect on resolution, with thick films increasing resolution, retention, and sample capacity. However, for routine analyzes, a standard film thickness of 0.25 to 0.5 µm is adequate. Most GC–MS systems for organic acid analysis utilize helium as carrier gas operating at a flow rate of 1 to 2 ml/min.

3.2. Injection and Temperature Programming

For multicomponent analysis where a wide range of components are to be separated and identified, temperature programming is the norm. For capillary columns, sample injection is usually split/splitless with splits ranging from 1:20 to 1:60. Injection volume is typically 1 µl at an injection temperature of approximately 200 to 250°C. For high sample throughput, an autosampler is indispensable. Temperature programs are controlled by a computer, also used for data collection and processing, and vary according to individual requirements. Sweetman et al. [40] used a starting temperature of 70°C with a ramp of 4°/min to 290°C for the separation of pentafluorobenzyloxime-TMS esters of oxoacids, aldehydes and ketones [22,40]. A similar program (65 to 260°C at 4°/min) was used by Lehnert [20] for methyl esters and by one of the authors (70°C to 250°C at 2°/min) for TMS-oximes [41]. The mass transfer line is normally approximately 280°C and a solvent delay is used to prevent unnecessary contamination of the ion source.

3.3. Mass Selective Detection

The use of capillary GC columns for organic acid analysis has in turn introduced new MS requirements with much faster scan rates and cycle times being necessary to cope with the fast eluting components, a peak eluting in some 5 to 20 sec and scan rates of 0.5 to 1 sec^{-1} ideally being required. Although analysis times are shorter on capillary columns, with these scan rates some 1800 to 3600 spectra will be obtained in a single analysis. The introduction of benchtop MS instruments has revolutionized the field of metabolic profiling by allowing the analysis of urinary organic acids to become the cornerstone of selective screening for inborn errors of metabolism. The Mass Selective Detector (MSD, Hewlett Packard) was conceived as a stand-alone capillary or packed column GC detector requiring no cooling water, compressed air, or facilities other than a main power supply. It is a quadrupole instrument with either EI or CI modes and is controlled exclusively by a computer-operated system with application programs for tuning, data acquisition, data editing, and reports. Important features of the system are its reduced demand for specialized technical skill for operation and that instrumental conditions, once logged, can be used repeatedly and reproducibly. Quadrupole instruments (HP MSD and Fisons MD 800) are moderate-cost instruments and their success stems from the ease with which ion masses can be manipulated by the computer. As a result they are particularly suited to selective ion monitoring. A typical acquisition setup for organic acid analysis in scan mode with a MSD is as follows: impact voltage 70 eV, electron multiplier voltage 2200, scan range m/z 50 to 550 at 2 scans/1.3 sec. The MS is tuned using heptacosafluorotributyl-amine as calibrant in most cases.

The ion-trap detector (ITD) (Finnigan-MAT 800) was designed as a GC detector for use with capillary columns to give highly specific MS data for detection, identification, and quantitation. It is constructed from a ring electrode and two end cap electrodes. Ions, formed and trapped in the central cavity following electron impact, become unstable as the voltage on the ring electrode is increased and exit axially through the end electrodes to be detected by a conventional electron multiplier.

4. DATA PROCESSING AND INTERPRETATION

The repetitive scanning used by benchtop MS instruments allows the construction of ion intensity chromatograms where the ion intensity (m/z) is plotted against either scan number or GC retention time. Each MS manufacturer has its own computer software for controlling the system and for data analysis. For the identification of eluting compounds, a library based search is necessary either performed manually or automatically with a search program. For example, the HP Chem Station uses a probability-matched algorithm where an unknown spectrum is compared with each reference spectrum using a reverse search technique. A reverse search verifies that the main peaks in the reference spectrum are also present in the unknown spectrum. The commercially available MS libraries are unsuitable for routine urinary organic acid analysis, hence those laboratories offering such a service have their own custom-made libraries.

5. ENANTIOMERIC ANALYSIS

Recently, inherited metabolic diseases have been described in which a particular enantiomeric form of a metabolite is exclusively excreted [42]. Typical examples are D(+)-glyceric aciduria [43] and L-2-hydroxyglutaric aciduria [44]. Using conventional GC–MS for urinary organic acid screening, increased amounts of these

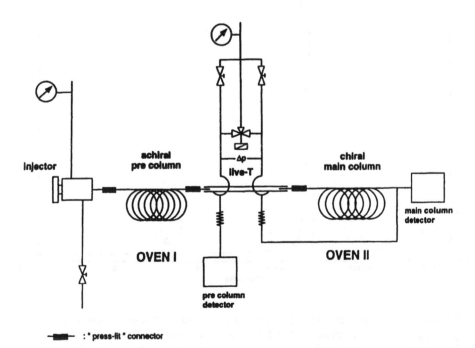

Figure 3 Schematic representation of enantioselective multidimensional GC. (From Ref. 42).

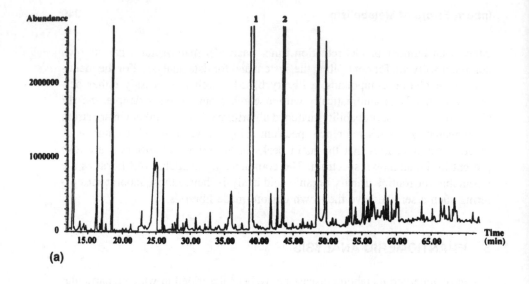

(a)

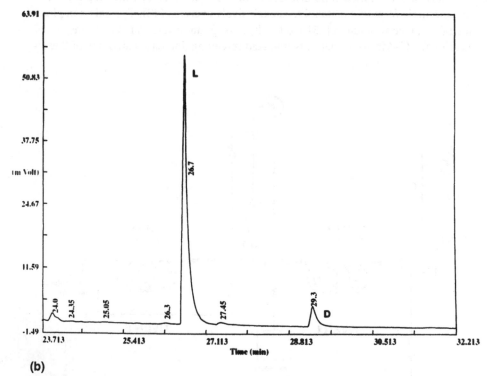

(b)

Figure 4 (a) Urinary organic acid chromatogram from a 17-year-old-patient with 2-hydroxyglutaric aciduria. Peak 1 is 2-hydroxyglutaric acid and peak 2 the internal standard (2-oxocaproic acid). (b) Chiral separation of 2-hydroxyglutaric acid in urine of the same patient using enantioselective multidimensional GC–MS. Peak L corresponds to L(S)-2-hydroxyglutaric acid (92%) and peak D to D(R)-2-hydroxyglutaric acid (8%), confirming the diagnosis of L-2-hydroxyglutaric aciduria, a relatively new neurometabolic disorder.

two metabolites can be seen in certain patients, but the nonchiral column cannot determine which enantiomer is present. This problem has resulted in the introduction of new chiral phases in organic acid analysis. Until very recently, optically active acids were esterified with (−)-menthol and the hydroxyl group acetylated [45] or further esterified as (−)-2-butanol esters [46]. The development of chiral columns for GC–MS using modified cyclodextrin has allowed the simultaneous analysis of different classes of enantiomers [47–49]. A further analytical refinement is the use of enantioselective multidimensional GC–MS (Fig. 3) employing heart cutting techniques from a nonchiral pre-separation column onto a second main chiral column without the need for further sample cleanup or derivatization [50,51]. An example of this analysis is shown in Figure 4 in the case of a patient with L-2-hydroxyglutaric aciduria. Conventional GC–MS of urinary organic acids showed an abnormal peak corresponding to 2-hydroxyglutaric acid (Fig. 4A). Enantioselective multidimensional GC with either quadrupole or time-of-flight MS detection showed this substance to be virtually exclusively L-2-hydroxyglutaric acid (Fig. 4B). This technique has been used in the investigation of urinary metabolites in maple syrup urine disease [52], D-lactic acidosis (unpublished observations), and of 5-oxoproline excretion [53]. Patterns of ketone bodies in patients with β-ketothiolase deficiency and medium-chain acyl-CoA dehydrogenase deficiency [54] have also been investigated. This new development will have an increasingly important role, not only in the diagnosis of inborn errors of metabolism, but also in the extension of our understanding of normal physiological processes.

6. PERSPECTIVES

The introduction of MS has revolutionized and broadened the horizons in the field of selective screening for inborn errors of metabolism. The first use of MS for the diagnosis of an organic aciduria in 1966 [30] has led to the finding of new diseases and their consequent therapy, and in turn has increased our knowledge of biochemical pathways. Although MS has played a pivotal role in elucidating disorders of organic acid metabolism, it should be remembered that physiological processes are exceedingly complex and MS has been used with increasing frequency over the last few years to detect defects in other metabolic pathways. For example, GC–MS has also proven valuable in the diagnosis of disorders of steroid metabolism [55], bile acid catabolism [56], and glycoprotein degradation [57]. In the last group of diseases, the identification and structural analysis by GC–MS of urinary oligosaccharides excreted in these disorders led to the characterization of the specific enzyme deficiencies.

Selective screening is based upon clinical impression, i.e., the symptoms presented by the patient should alert the physician to the possibility of an inborn error of metabolism. This is in contrast to neonatal screening, where samples of all newborns should be routinely screened for rather frequent inborn disorders, the detection of which prior to the onset of clinical symptoms is known to be of great advantage. With the advent of tandem MS in laboratories involved in the diagnosis of inborn errors of metabolism, a wide variety of metabolites can be detected in neonatal blood using minute sample volumes (2 µl), opening the possibility of early detection of more metabolic diseases with subsequent rapid initiation of appropriate therapy [58,59]. This method, however, cannot replace the aforementioned GC–MS methods because not all metabolites can be detected by tandem MS. Thus MS still has an important role to play in the field of selective screening for inborn errors of metabolism and development of suitable technologies, i.e., liquid chromatography–mass spectrometry (LC–MS) for disorders of purine and pyrimidine metabolism and neurotransmitter analysis will no doubt add to our diagnostic repertoire.

REFERENCES

1. J.-M. Saudubray, H. Ogier de Balny, and C. Charpentier, in J. Fernandes, J.-M. Saudubray, and G. van den Berghe (eds.), Inborn Metabolic Diseases: Diagnosis and Treatment, 1996, Springer-Verlag, Berlin, pp. 3–39.
2. M.R. Seashore, Curr. Opin. Pediatr. 10 (1998) 609–614.
3. M. Duran, L. Dorland, S.K. Wadman, and R. Berger, Eur. J. Pediatr. 153 (Suppl. 1) (1994) S27–S32.
4. P.T. Ozand and G.G. Gascon, J. Child Neurol, 6 (1991) 196–219.
5. K. Bartlett and D. Gompertz, Biochem. Med., 10 (1974) 15–23.
6. M. Duran, D.Ketting, R. Van Vossen, T.E. Beckeringh, L. Dorland, L. Bruinvis, and S.K. Wadman, Clin. Chim. Acta, 152 (1985) 253–260.
7. S.I. Goodman and S.P. Markey, Diagnosis of Organic Acidemias by Gas Chromatography–Mass Spectrometry, 1981, Alan R. Liss, New York.
8. R.A. Chalmers and R.W. Watts, Analyst, 97 (1972) 958–967.
9. J.A. Thompson and S.P. Markey, Anal. Chem., 47 (1975) 1313-1321.
10. W.L. Fitch, P.J. Anderson, and D.H. Smith, J. Chromatogr., 162 (1979) 249–259.
11. B.J. Verhaeghe, M. F. Lefevere, and A.P. De Leenheer, Clin. Chem., 34 (1988) 1077–1083.
12. Y. Mardens, A. Kumps, C. Planchon, and C. Wurth, J. Chromatogr. Biomed. Appl., 577 (1992) 341–346.
13. C. Bachmann, R. Bühlmann, and J.P. Colombo, J. Inher. Metab. Dis., 7 (Suppl. 2) (1984) 126.
14. P. B. Mortensen and N. Gregersen, Clin. Chim. Acta, 103 (1980) 33–37.
15. K.M. Gibson, S.I. Goodman, F.E. Frerman, and A.M. Glasgow, J. Pediatr. 114 (1989) 607–610.
16. R. Matalon, K. Michals, D. Sebasta, M. Deanching, P. Gashkoff, and J. Casanova, Am. J. Med. Genet., 29 (1988) 463–471.

17. P. Divry, M.O. Rolland, J. Tessier, M. Desage, and J. Cotte, J. Inher. Metab. Dis., 5 (1982) 41.

18. I.M. Bengtsson and D.C. Lehotay, J. Chromatogr. B, 685 (1996) 1–7.

19. D.C. Lehotay, Biomed. Chromatogr., 5 (1991) 113–121.

20. W. Lehnert, Eur. J. Pediatr., 153 (Suppl. 1) (1994) S9-S13.

21. R.M. Chalmers and A.M. Lawson, Organic Acids in Man: The Analytical Chemistry, Biochemistry and Diagnosis of the Organic Acidurias, 1982, Chapman and Hall, London.

22. G. Hoffmann, S. Aramaki, E. Blum-Hoffmann, W.L. Nyhan, and L. Sweetman, Clin. Chem., 35 (1989) 587–595

23. K. Tanaka, D.G. Hine, A. West-Dull, and T.B. Lynn, Clin. Chem., 26 (1980) 1839–1846.

24. P. Divry, C. Vianey-Liaud, and J. Cotte, Biomed. Environ. Mass Spectrom., 14 (1987) 663–668.

25. J.A. Thompson, B.S. Miles, and P.V. Fennessey, Clin. Chem., 23 (1977) 1734–1738.

26. C. Jakobs, F. Stellard, E. Kvittingen, M. Henderson, and R. Lilford, Prenat. Diagn. 10 (1990) 133–134.

27. C. Jakobs, H.J. ten Brink, S.A. Langelaar, T. Zee, F. Stellard, M. Macek, K. Srsnova, S. Srsen, and W.J. Kleijer, J. Inher. Metab. Dis., 14 (1991) 653–660.

28. C. Weycamp, J. Willems, A. Van Gennip, M. Duran, F. Trijbels, and C. Doelman, J. Inher. Metab. Dis., 21 (Suppl. 2) (1998) 146.

29. E. Klenk and W. Kahlke, Hoppe Seyler's Z. Physiol. Chem., 333 (1963) 133–143.

30. K. Tanaka, M.A. Budd, M.L. Efron, and K.J. Isselbacher, Proc. Natl. Acad. Sci. USA, 56 (1966) 236–242.

31. E.C. Horning and M.G. Horning, Clin. Chem., 17 (1971) 802–809.

32. E.C. Horning and M.G. Horning, J. Chromatogr. Sci., 9 (1971) 129–140.

33. R.A. Hites and K.Biemann, Anal. Chem., 40 (1968) 127–132.

34. R.A. Hites and K. Biemann, Anal. Chem., 42 (1970) 855–860.

35. E. Jellum, O. Stokke, and L. Eldjarn, Clin. Chem., 18 (1972) 800–809.

36. E. Jellum, O. Stokke, and L.Eldjarn, Anal. Chem., 45 (1973) 1099–1106.

37. O.A. Mamer, J.C. Crawhall, and S.S. Tjoa, Clin. Chim. Acta, 32 (1971) 171–184.

38. R.A. Chalmers, P.Purkiss, R.W.E. Watts, and A.M. Lawson, J. Inher. Metab. Dis., 3 (1980) 27–43.

39. R.A. Chalmers, in A. Lawson (ed.), Mass Spectrometry, 1989, Walter de Gruyter, Berlin, pp. 355–403.

40. L. Sweetman, in F.A.Hommes (ed.), Techniques in Diagnostic Human Biochemical Genetics: A Laboratory Manual, 1991, Wiley-Liss, New York, pp. 143–176.

41. A.C. Sewell and H.J. Böhles, Clin. Chem., 37 (1991) 1301– 1302.

42. A.C. Sewell, M. Heil, F. Podebrad, and A. Mosandl, Eur. J. Pediatr., 157 (1998) 185–191.

43. J.R. Bonham, T.J. Stephenson, K.H. Carpenter, J.M. Rattenbury, C.H. Crombie, R.J. Pollitt, and D. Hull, Pediatr., Res., 28 (1990) 38–41.

44. M. Duran, J.P. Kamerling, H.D. Bakker, A.H. Van Gennip, and S.K. Wadman, J. Inher. Metab. Dis., 3 (1980) 109–112.

45. J.P. Kamerling, G.J. Gerwig, J.F.G. Vliegenhart, M. Duran, D. Ketting, and S.K. Wadman, J. Chromatogr. Biomed. Appl., 143 (1977) 117–123.

46. J.P. Kamerling, M. Duran, J. Gerwig, J.F.G. Vliegenhart, D. Ketting, and S.K. Wadman, J. Chromatogr., 222 (1981) 276–283.

47. A. Dietrich, B. Maas, V. Karl, P. Kreis, D. Lehmann, B. Weber, and A. Mosandl, J. High Resolut. Chromatogr., 15 (1992) 176–179.

48. A. Dietrich, B. Maas, W. Messer, G. Bruche, V. Karl, A. Kaunzinger, and A. Mosandl, J. High Resolut. Chromatogr., 15 (1992) 590–593.

49. H-G. Schmarr, A. Mosandl, and A. Kaunzinger, J. Microcol. Sep., 3 (1991) 395–402.

50. A. Kaunzinger, F. Podebrad, R. Liske, B. Maas, A. Dietrich, and A. Mosandl, J. High Resolut. Chromatogr., 18 (1995) 49–53.

51. A. Kaunzinger, A. Rechner, T. Beck, A. Mosandl, A.C. Sewell, and H.J. Böhles, Enantiomer, 1 (1996) 177–182.

52. F. Podebrad, M. Heil, S. Leib, B. Geier, T. Beck, A. Mosandl, A.C. Sewell, and A. Mosandl, J. High Resolut. Chromatogr., 20 (1997) 355–362.

53. M. Heil, F. Podebrad, T. Beck, A. Mosandl, A.C. Sewell, and H.J. Böhles, J. Chromatogr., B, 714 (1998) 119–126.

54. M. Heil, F. Podebrad, E. Prado, T. Beck, A. Mosandl, A.C. Sewell, H.J. Böhles, and W. Lehnert, J. Chromatogr B Biomed. Sci. Appl., 739 (2000) 313–324.

55. J. Honour, Ann. Clin. Biochem., 34 (1997) 32–44.

56. P.T. Clayton, J. Inher. Metab. Dis., 14 (1991) 478–496.

57. J.P. Kamerling and J.F.G. Vliegenhart, in A.M. Lawson (ed.), Mass Spectrometry, 1989, Walter de Gruyter, Berlin, pp. 177–263.

58. L. Sweetman, Clin. Chem., 42 (1996) 345.

59. K. Bartlett, S.J. Eaton, and M. Pourfarzam, Arch. Dis. Child., 77 (1997) F151–F154.

14

Applications of Gas Chromatography–Mass Spectrometry in Clinical and Forensic Toxicology and Doping Control

Hans H. Maurer
University of Saarland, Homburg/Saar, Germany

1. INTRODUCTION

In clinical toxicology, the diagnosis or the definite exclusion of an acute or chronic intoxication is of great importance. Furthermore, patients addicted to alcohol, medicaments, or illegal drugs have to be monitored. For determination of clinical death as a prerequisite for explantation of organs, the presence of drugs, which may depress the central nervous system, must be analytically excluded. The compliance of patients can be monitored by determination of the prescribed drugs. Finally, monitoring of drugs with a narrow margin of therapeutic safety can be performed by the clinical toxicologist. Similar problems arise in forensic toxicology.

In forensic toxicology, proof of an abuse of illegal drugs or of a murder by poisoning are important tasks. Furthermore, drugs, that may reduce the penal responsibility of a criminal, or that may reduce a person's fitness to drive a car must be monitored in body fluids or tissues.

In doping control, the use or abuse of drugs, that may stimulate the buildup of muscles, enhance endurance during competition, lead to reduction of body weight, or that reduce the pain caused by overexertion must be monitored, typically in urine.

The basis of a competent toxicological judgment and consultation is an efficient toxicological analysis. The choice of the method in analytical toxicology or doping control depends on the problems, that have to be solved. Usually, the compounds, that have to be analyzed are unknown. Therefore, the first step before quantification, e.g., in plasma, is the identification of the interesting compounds. The screening strategy of the systematic toxicological analysis (STA) must be very extensive, because several thousands of drugs or pesticides should be considered. It often includes a screening test and a confirmatory test. If only a single drug or category has to be monitored, immunoassays can be used for screening in order to differentiate between negative and presumptively positive samples. Positive results must be confirmed by a second independent method that is at least as sensitive as the screening test and that provides the highest level of confidence in the result. Without doubt, gas chromatography–mass spectrometry (GC–MS), especially in the full-scan electron ionization (EI) mode, is the reference method for confirmation of positive screening tests [1–8]. This two-step strategy is employed only if the drugs or poisons to be determined are scheduled, e.g., by law or by international organizations, and if immunoassays are commercially available. If these requirements are not met, the screening strategy must be more extensive, because several thousands of drugs or pesticides are on the market worldwide [9]. For these reasons, STA procedures are necessary that allow the simultaneous detection of as many toxicants in biosamples as possible. Most often, GC–MS procedures are used today [1,2, 7,8,10–14]. High-performance–liquid chromatography (HPLC) coupled to diode-array detectors [15–17] was also described for general screening procedures, but the specificity is less than full-scan EI-MS. Liquid chromatography–mass spectrometry (LC–MS) procedures were reviewed by Maurer [18].

Most of the STA procedures cover basic (and neutral) drugs, which are more important toxicants. For example, most of the psychotropic drugs have, like neurotransmitters, basic properties. Nevertheless, some classes of acidic drugs or drugs producing acidic metabolites, such as the cardiovascular drugs angiotensin converting enzyme (ACE) inhibitors and angiotensin receptor II (AT-II) blockers, dihydropyridine calcium channel blockers (metabolites), diuretics, coumarin anticoagulants, hypoglycemic sulfonylureas, barbiturates, or nonsteroidal anti-inflammatory drugs (NSAIDs), are relevant to clinical and forensic toxicology or doping. Therefore, GC–MS screening procedures for detection of basic, neutral, and acidic drugs in biosamples are described here. The GC–MS procedures for testing for drugs of abuse are described Chapter 15 of this book.

After the unequivocal identification, quantification of the drugs can also be performed by GC–MS, especially using stable isotopes as internal standards. Suitable procedures are also described here.

1.1. Choice of Biosamples

Concentrations of drugs are relatively high in urine, so that urine is the sample of choice for a comprehensive screening and identification of unknown drugs or poisons [1,2,8]. However, the metabolites of these drugs must be identified in addition or even exclusively. In horse doping control, urine is also the common sample for screening [19]. Blood (plasma, serum) is the sample of choice for quantification. However, if the blood concentration is high enough, screening can also be performed herein. This may be advantageous, since sometimes only blood samples are available and some procedures allow simultaneous screening and quantification [4,7,9,14]. The GC–MS analysis of drugs in alternative matrices such as hair [6], sweat and saliva [5], meconium [20], or nails [21] has also been described, but the toxicological interpretation of the analytical results may be difficult.

1.2. Sample Preparation

Suitable sample preparation is an important prerequisite for GC–MS analysis in biosamples. It may involve cleavage of conjugates, isolation, and derivatisation preceded or followed by cleanup steps.

Cleavage of conjugates can be performed by gentle but time-consuming enzymatic hydrolysis [8]. However, the enzymatic hydrolysis of acyl glucuronides (ester glucuronides of carboxy derivatives such as NSAIDs) may be hindered due to acyl migration [22]. Acyl migration means intramolecular transesterification at the hydroxy groups of the glucuronic acid, which leads to β-glucuronidase-resistant derivatives. In emergency toxicology, it is preferable to cleave the conjugates by rapid acid hydrolysis [1,23,24]. Alkaline hydrolysis is only suitable for cleavage of ester conjugates. However, the formation of artifacts during chemical hydrolysis must be considered [25]. A compromise of both cleavage techniques is the use of a column packed with immobilized glucuronidase/arylsulfatase. It combines the advantages of both methods, the speed of acid hydrolysis and the gentle cleavage of enzymatic hydrolysis [26, 27]. Acyl glucuronides, e.g., of acidic drugs, were readily cleaved under the conditions of extractive alkylation (alkaline pH, elevated temperature) [11,13,28].

Isolation can be performed by liquid–liquid extraction (LLE) at a pH at which the analyte is nonionized or by solid-phase extraction (SPE) preceded or followed by cleanup steps. Sample pretreatment for SPE depends on the sample type: whole blood and tissue (homogenates) need deproteinization and filtration/centrifugation steps before application to the SPE columns, whereas for urine usually a simple dilution step and/or centrifugation is satisfactory. Whatever SPE column is used, the analyst should keep in mind that there are large differences from batch to batch, and that the same sorbents from different manufacturers also lead to different results [29]. Therefore, use of a suitable internal standard (e.g., deuterated analytes) is recommended. The pros and cons of SPE procedures for STA have been discussed by Franke and de Zeeuw [30].

Solid-phase microextraction (SPME) is becoming a modern alternative to SPE and LLE. Solid-phase microextraction is a solvent-free and concentrating extraction technique especially for rather volatile analytes. It is based on the adsorption of the analyte on a stationary phase coating a fine rod of fused silica. The analytes can be desorbed directly into the GC injector. Fast GC–MS procedures for screening, e.g., for barbiturates [31] or amphetamines [32], have been published.

Extractive alkylation has been proved to be a powerful procedure for simultaneous extraction and derivatization of acidic compounds [11–13,28,33–37]. The acidic compounds were extracted at pH 12 as ion pairs with the phase-transfer catalyst (tetrahexyl ammonium iodide, THA$^+$ I$^-$) into the organic phase (toluene). Reaching the organic phase, the phase-transfer catalyst could easily be solvated due to its lipophilic hexyl groups. The principle of the procedure is summarized in Figure 1. The poor solvation of the anionic analytes leads to a high reactivity against the alkylation (most often methylation) reagent alkyl (methyl) iodide. Part of the phase-transfer catalyst could also reach the organic phase as an ion pair with the iodide anion formed during the alkylation reaction or with anions of the urine matrix. Therefore, the remaining part had to be removed to maintain the GC column's separation power and to exclude interactions with analytes in the GC injection port. We tested several SPE sorbents and different

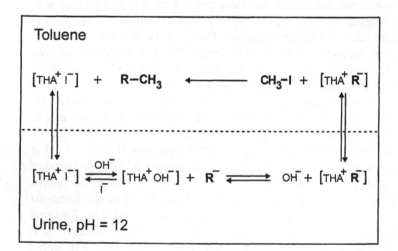

Figure 1   Principle of extractive alkylation.

eluents for efficient separation of the vestige of the phase-transfer catalyst from the analytes. The diol sorbent yielded best reproducibility and recovery under the described conditions. Further advantages of such SPE columns were, that they could easily be handled, were commercially available, and did not need to be manually prepared as described by Lisi et al. [33].

Derivatization steps are necessary if relatively polar compounds containing, e.g., carboxylic, hydroxy, or primary or secondary amino groups are to be determined by GC or GC–MS. The following procedures are typically used for basic compounds: acetylation (AC), trifluoroacetylation (TFA), pentafluoropropiony-lation, heptafluorobutyration (HFB), or trimethylsilylation (TMS), or for acidic compounds: methylation, extractive methylation, pentafluoropropylation, tri-methylsilylation or *tert*-butyldimethylsilylation. Further details and the pros and cons of derivatization procedures can be found in a review by Segura et al. [38].

2. SYSTEMATIC TOXICOLOGICAL ANALYSIS PROCEDURES FOR THE DETECTION OF BASIC OR NEUTRAL DRUGS AND/OR THEIR METABOLITES

Three typical STA procedures are presented here, one for the detection of most of the basic and neutral drugs in urine after acid hydrolysis, LLE, and acetylation [1,2,10,23,25,39–59]; one for the detection of doping-relevant stimulants, beta-blockers, beta-agonists, and narcotics after enzymatic hydrolysis, SPE, and com-bined TFA and TMS derivatization [8] and one for automated screening of barbi-turates, benzodiazepines, antidepressants, morphine, and cocaine in blood after SPE and trimethylsilylation [7,14].

2.1. Screening for Basic and Neutral Drugs

A screening method for detection of most of the basic and neutral drugs in urine after acid hydrolysis, LLE and acetylation was developed and has been improved during the last years. Acid hydrolysis has proved to be very fast and efficient when performed under microwave radiation [24]. Acetylation was suitable for derivatization of numerous drugs and their metabolites [1,2]. It leads to stable derivatives with good gas chromatographic properties. The acetylation mixture can be evaporated before analysis so that the resolution power of capillary col-umns does not decrease in contrast to other derivatization reagents, e.g., for TMS.

This comprehensive full-scan GC–MS screening procedure allows within one run the simultaneous screening and confirmation of the following categories of drugs: amphetamines [60–62], designer drugs [57,59,63], barbiturates and other sedatives-hypnotics [54], benzodiazepines [46], opiates, opioids, and other potent analgesics [41,56], anticonvulsants [52], antidepressants [42,58], phenothiazine and butyrophenone neuroleptics [40,43], non-opioid analgesics [39], antihistamines [47–49,51], antiparkinsonian drugs [44], beta-blockers [45], antiarrhythmics [50,53], diphenol laxatives [55] and their metabolites. The screening is performed using reconstructed mass chromatograms, which may indicate the presence of suspected mass spectra in the full mass spectra stored during GC separation. Positive signals can be confirmed by visual [9] or computerised [64] comparison of the peaks underlying full mass spectra with reference spectra. Eight ions per category were individually selected from the mass spectra of the corresponding drugs and their metabolites identified in authentic urine samples [1,23]. Generation of the mass chromatograms can be started by clicking the corresponding pull-down menu, which executes the user-defined macros [10]. (The macros can be obtained from the author: e-mail: Hans.Maurer@med-rz.uni-sb.de.) The procedure is illustrated in Figures 2 and 3. Figure 2 shows selective

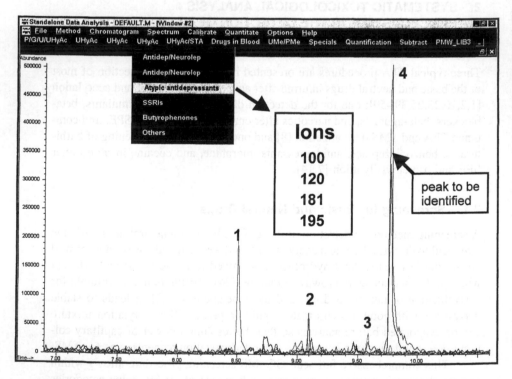

Figure 2   Selective mass chromatograms indicating atypical antidepressants in an urine extract after acid hydrolysis and acetylation. Peak 1 indicates mirtazapine, peak 2 the bis-demethyl metabolite of the opioid tramadol, peak 3 the hydroxy metabolite of mirtazapine, and peak 4 the nor metabolite of mirtazapine.

mass chromatograms indicating atypical antidepressants in a urine extract after acid hydrolysis and acetylation. Peak 1 indicate mirtazapine, peak 2 the bis-de-methyl metabolite of the opioid tramadol, peak 3 the hydroxy metabolite of mirtazapine, and peak 4 the nor metabolite of mirtazapine. The detection of the opioid metabolite in the screening for atypical antidepressants elucidates that mass chromatography is only the screening procedure. Unequivocal identification is achieved only by comparison of the full mass spectrum with the reference spectrum. Such identification is exemplified for peak 4. Figure 3 shows the mass spectrum underlying peak 4 in Figure 2, the reference spectrum, the structure and the hit list found by library search [64].

This procedure allows simultaneous, fast, and specific detection of most of the toxicologically relevant drugs in urine samples after therapeutic doses. Therefore, it has proved to be suitable also for screening of abused medicaments in psychiatry. After unequivocal identification, quantification of the drugs can be performed by GC–MS. The limit of detection (LOD) of parent compounds in

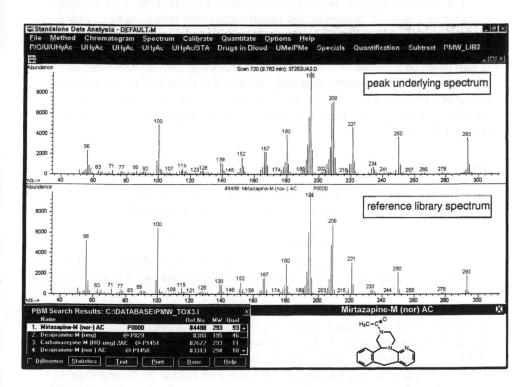

Figure 3 Mass spectrum underlying peak 4 in Figure 2, the reference spectrum, the structure and the hit list found by library search [64].

urine is of minor value if these drugs are mainly or even exclusively excreted in metabolized form. In these cases, what should be studied is whether and how long the intake of a therapeutic drug dose can be monitored by the procedure [13,28].

2.2. Screening Procedure for Doping-Relevant Drugs

Another screening procedure was published for the detection of doping-relevant stimulants, beta-blockers, beta-agonists, and narcotics after enzymatic hydrolysis, solid-phase extraction and combined TFA and TMS derivatization [8]. The time-consuming enzymatic cleavage of conjugates is acceptable for doping analysis, because results do not have to be available as fast as in emergency toxicology. The authors did not focus their procedure on the detection of the metabolites, even if they could be detected in most cases for a longer time and more sensitively than their parent compounds. The chemical properties of the analytes allow the use of SPE with acceptable recoveries. However, the large differences from batch to batch should be kept in mind [29]. The combined TFA and TMS derivatization provides very good GC properties, but underivatized samples cannot be analyzed on the same GC–MS apparatus without changing the column.

This screening procedure, which is limited to some doping-relevant drugs, is also based on full-scan GC–MS and mass chromatography for documentation of the absence of the corresponding drug. Such exclusion procedures are suitable for doping analysis, since the occurrence of positives is low in contrast to clinical toxicology, where the occurrence is high and where many more drugs must be detected or excluded.

2.3. Automated Screening Procedure

An automated screening procedure for barbiturates, benzodiazepines, antidepressants, morphine, and cocaine in blood after SPE and trimethylsilylation was developed [7,14]. The sample preparation consists of SPE and TMS derivatization, both automated using an HP PrepStation.

The samples directly injected by the PrepStation were analyzed by full-scan GC–MS. Using macros, peak identification and the reporting of results were also automated. This fully automated procedure requires about 2 hr, which is unacceptable for emergency toxicology, but acceptable for forensic drug testing or doping control. Automation of the data evaluation is a compromise between selectivity and universality. If the exclusion criteria are chosen too narrowly, peaks can be overlooked. If the window is too large, a series of proposals is given by the computer and the toxicologist needs a lot of time for differentiation. Nevertheless, this procedure is an interesting alternative for drug screening in blood.

3. SYSTEMATIC TOXICOLOGICAL ANALYSIS PROCEDURES FOR DETECTION OF ACIDIC DRUGS

Some classes of acidic drugs or drugs that are metabolized to acidic compounds, such as the cardiovascular drugs ACE inhibitors and AT-II blockers, dihydropyridine calcium channel blockers, diuretics, coumarin anticoagulants, hypoglycemic sulfonylureas, barbiturates, or NSAIDs, are relevant to clinical and forensic toxicology or doping. Therefore, these acidic drugs should also be monitored, ideally in one procedure.

Extractive alkylation has been proved to be a powerful procedure for simultaneous extraction and derivatization of acidic compounds [11-13,28,33-37]. The methyl derivatives are stable and show good gas chromatographic properties.

A comprehensive GC–MS screening procedure for the detection of acidic drugs, poisons and/or their metabolites in urine after extractive methylation was developed [11-13,28,34-37]. The analytes were separated by capillary GC and identified by computerized MS in the full-scan mode. As already described in section 2.1, the possible presence of acidic drugs and/or their metabolites could be indicated using mass chromatography with selective ions. The identity of positive signals in such mass chromatograms was confirmed by comparison of the peaks underlying full mass spectra with the reference spectra [9,64] recorded during the corresponding studies. This STA method allowed the detection in urine of most of the ACE inhibitors and AT II antagonists [28], of coumarin anticoagulants of the first generation [11], of dihydropyridine calcium channel blockers [13], barbiturates [35], diuretics [37], hypoglycemic sulfonylureas (sulfonamide part) [36], NSAIDs [12], and of various other acidic compounds [35]. Mass spectra of all these drugs and metabolites (methylated and silylated) are included in reference libraries [9,64] for specific detection by library search. The higher-dosed drugs, at least, could also be detected in plasma samples after extractive methylation.

Again, the LOD of parent compounds in urine is of minor value if these drugs are mainly or even exclusively excreted in metabolized form. In these cases, what should be studied is whether and how long the intake of a therapeutic drug dose can be monitored by the procedure [13,28].

4. QUANTIFICATION OF DRUGS IN PLASMA

4.1. Quantification Using a Universal Procedure

Therapeutic drug monitoring (TDM) is commonly performed in clinical chemistry using commercial assays. For TDM of rarely used or new drugs, special procedures using combined techniques must be developed. Such combined techniques are common in clinical and forensic toxicology, because several thousand (unknown) drugs or poisons must be analyzed, ideally using one or a few standard procedures. For example, determination of the antiepileptic lamotrigine (3,5-diamino-6-(2,3 dichlorophenyl)-1,2,4-triazine) is described based on such a universal GC–MS procedure [23,25].

A 1-ml portion of plasma was extracted twice (pH 7.4 and 12] using a 1: 1 mixture of diethyl ether and ethyl acetate after addition of methaqualone as internal standard. The extract was analyzed by GC–MS in the selective-ion monitoring (SIM) mode (m/z 185, 187, 255, 257 for lamotrigine and 235, 250 for methaqualone). A five-point calibration graph was established using spiked plasma samples [65]. The procedure was linear from 0.5 to 20 mg/L (r^2 = 0.991) with coefficients of variation of less than 15%. The LOD was 0.1 mg/L. As we have seen, therapeutic concentrations ranged from 1 to 6 mg/L. The presented method has proved to be suitable for TDM as well as for clinical toxicology.

4.2. Quantification Using Special Procedures

For precise quantification of lower dosed drugs in plasma, another workup was necessary to improve the signal-to-noise ratio (S/N). Since only a limited number of quite similar compounds had to be isolated, SPE was preferred, resulting in rather clean extracts. However, to compensate for the known batch-to-batch differences of SPE columns [29], deuterated analytes were used as internal standards. Heptafluorobutyric anhydride derivatization was preferred for primary and secondary amines to reach the sensitivity necessary for determination of low blood levels.

A typical procedure for determination of the designer drugs is given here. Plasma samples, 1 ml, were worked up after addition of deuterated standards by solid phase extraction (pH 6, ICT [Bad Homburg, Germany] Isolute Confirm HCX, ethyl acetate–ammonia, 98:2) [66] followed by derivatization using heptafluorobutyric anhydride (30 min, 56°C).

Quantification was performed in the SIM mode. The concentrations were calculated from the relation of the peak areas to those of the deuterated internal standards. Figure 4 shows mass fragmentograms with the ions m/z 135, 162, 176, 240, and 254 indicating d5-MDA (IS), methylenedioxyamphetamine (MDA), d5-MDMA (IS), and methylenedioxymethamphetamine (MDMA) in plasma after SPE and HFB derivatization. The method was validated for the most important designer drug MDMA. It was linear from 10 to 500 ng/ml (r^2 = 0.9991) with a recovery of better than 90%, an intraday precision of better than 5%, an interday precision of better than 10%, and a detection limit of 3 ng/ml (S/N 3) [59].

The quantification procedures described above in sections 4.1 and 4.2 exemplify the potential of GC–MS for quantitative determination in analytical toxicology. Further quantification procedures are described in Chapter 15 of this book and they were recently reviewed [3,4,67].

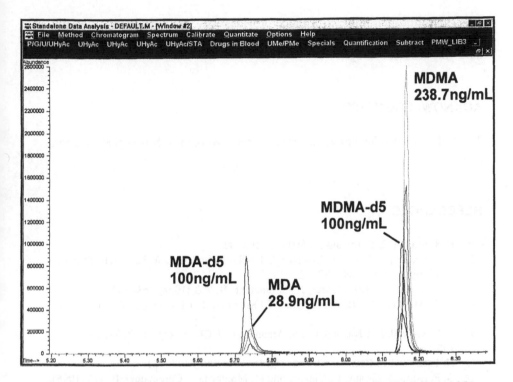

Figure 4 Mass fragmentograms with the selected ions m/z 135, 162, 176, 240, and 254 indicating d5-MDA (IS), MDA, d5-MDMA (IS) and MDMA in plasma after SPE and HFB derivatization.

5. CONCLUSIONS

In the last 10 years many papers have appeared concerning GC–MS detection of unknown drugs and their metabolites in biosamples relevant to clinical toxicology, forensic toxicology, and doping control. They describe procedures either for confirmation of chromatographic or immunological results or for STA. Confirmation was usually performed in the SIM mode, because only a particular compound had to be identified. Since low-priced mass spectrometers are widely used today, many papers have appeared in the last few years in this field, mostly improving previous papers. Today, GC–MS is the method of choice for STA in clinical and forensic toxicology as well as in doping control. If the drug is unknown, full-scan mode is the method of choice, since comparison of the full mass spectra with reference spectra is necessary. The screening can be performed using mass chromatography followed by library search.

Quantification in the SIM mode provides very good precision, especially using stable isotopes as internal standards. However, they are commercially available only for a few drugs.

ACKNOWLEDGMENTS

The author thanks Thomas Kraemer and Armin Weber for their suggestions and help.

REFERENCES

1. H.H. Maurer, J. Chromatogr., 580 (1992) 3–41.
2. H.H. Maurer, J.W. Arlt, T. Kraemer, C.J. Schmitt, and A.A. Weber, Arch. Toxicol., 19 (Suppl.) (1997) 189–197.
3. T. Kraemer and H.H. Maurer, J. Chromatogr. B, 713 (1998) 163–187.
4. M.R. Moeller, S. Steinmeyer, and T. Kraemer, J. Chromatogr. B, 713 (1998) 91.
5. D.A. Kidwell, J.C. Holland, and S. Athanaselis, J. Chromatogr. B, 713 (1998) 111–135.
6. H. Sachs and P. Kintz, J. Chromatogr. B, 713 (1998) 147–161.
7. A. Polettini, A. Groppi, C. Vignali, and M. Montagna, J. Chromatogr. B, 713 (1998) 265.
8. A. Solans, M. Carnicero, R. de la Torre, and J. Segura, J. Anal. Toxicol., 19 (1995) 104–114.
9. K. Pfleger, H.H. Maurer, and A. Weber, Mass Spectral and GC Data of Drugs, Poisons, Pesticides, Pollutants and their Metabolites, 3rd. ed., 2000, Wiley-VCH, Weinheim.
10. H.H. Maurer, Spectroscopy Europe, 6 (1994) 21–23.
11. H.H. Maurer and J.W. Arlt, J. Chromatogr. B, 714 (1998) 181–195.
12. H.H. Maurer, T. Kraemer, and F.X. Tauvel, J. Anal. Toxicol. (2000) submitted.
13. H.H. Maurer and J.W. Arlt, J. Anal. Toxicol., 23 (1999) 73–80.
14. A. Polettini, J. Anal. Toxicol., 20 (1996) 579–586.
15. O.H. Drummer, A. Kotsos, and I.M. McIntyre, J. Anal. Toxicol., 17 (1993) 225–229.
16. M. Balikova, Sb. Lek., 95 (1994) 339–345.
17. Y. Gaillard and G. Pepin, J. Chromatogr. A, 763 (1997) 149–163.
18. H.H. Maurer, J. Chromatogr. B, 713 (1998) 3–25.
19. G. Gonzalez, R. Ventura, A.K. Smith, R. de la Torre, and J. Segura, J. Chromatogr. A, 719 (1996) 251–264.
20. C. Moore, A. Negrusz, and D. Lewis, J. Chromatogr. B, 713 (1998) 137–146.
21. D.A. Engelhart, E.S. Lavins, and C.A. Sutheimer, J. Anal. Toxicol., 22 (1998) 314–318.

22. L.H. Spahn and L.Z. Benet, Drug Metab. Rev., 24 (1992) 5–47.
23. H.H. Maurer, in K. Pfleger, H.H. Maurer, and A. Weber (eds.), Mass spectral and GC data of drugs, poisons, pesticides, pollutants and their metabolites, 3rd ed., 2000, Wiley-VCH, Weinheim.
24. T. Kraemer, A.A. Weber, and H.H. Maurer, in F. Pragst, ed., Proceedings of the Xth GTFCh Symposium in Mosbach, 1997, Helm-Verlag, Heppenheim, pp. 200–204.
25. H.H. Maurer, in K. Pfleger, H.H. Maurer, and A. Weber (Eds.), Mass spectral and GC data of drugs, poisons, pesticides, pollutants and their metabolites, 2d ed, 1992, VCH, Weinheim, pp. 3–32.
26. S.W.H. Toennes and H.H. Maurer, in S.D. Ferrara (ed.) Proceedings of the 35th International TIAFT Meeting, 1997, Centre of Behavioural and Forensic Toxicology, Padova, pp. 227–240.
27. S.W. Toennes and H.H. Maurer, Clin. Chem., 45 (1999) 2173–2182.
28. H.H. Maurer, T. Kraemer, and J.W. Arlt, Ther. Drug Monit., 20 (1998) 706–713.
29. M.J. Bogusz, R.D. Maier, B.K. Schiwy, and U. Kohls, J. Chromatogr. B, 683 (1996) 177–188.
30. J.P. Franke and R.A. de Zeeuw, J. Chromatogr. B, 713 (1998) 51–59.
31. B.J. Hall and J.S. Brodbelt, J. Chromatogr. A, 777 (1997) 275–282.
32. C. Battu, P. Marquet, A.L. Fauconnet, E. Lacassie, and G. Lachatre, J. Chromatogr. Sci., 36 (1998) 1–7.
33. A.M. Lisi, R. Kazlauskas, and G.J. Trout, J. Chromatogr., 581 (1992) 57–63.
34. H.H. Maurer and T. Kraemer, J. Chromatogr. B, (2001) in preparation.
35. H.H. Maurer and T. Kraemer, Ther. Drug Monit., (2001) in preparation.
36. H.H. Maurer, T. Kraemer, and J.W. Arlt, J. Chromatogr. B, (2001) in preparation.
37. H.H. Maurer and T. Kraemer, Ther. Drug Monit., (2001), in preparation.
38. J. Segura, R. Ventura, and C. Jurado, J. Chromatogr. B, 713 (1998) 61–90.
39. H. Maurer and K. Pfleger, Fresenius. Z. Anal. Chem., 314 (1983) 586–594.
40. H. Maurer and K. Pfleger, J. Chromatogr., 272 (1983) 75–85.
41. H. Maurer and K. Pfleger, Fres. Z. Anal. Chem., 317 (1984) 42–52.
42. H. Maurer and K. Pfleger, J. Chromatogr., 305 (1984) 309–323.
43. H. Maurer and K. Pfleger, J. Chromatogr., 306 (1984) 125–145.
44. H. Maurer and K. Pfleger, Fres. Z. Anal. Chem., 321 (1985) 363–370.
45. H. Maurer and K. Pfleger, J. Chromatogr., 382 (1986) 147–165.
46. H. Maurer and K. Pfleger, J. Chromatogr., 422 (1987) 85–101.
47. H. Maurer and K. Pfleger, J. Chromatogr., 430 (1988) 31–41.
48. H. Maurer and K. Pfleger, J. Chromatogr., 428 (1988) 43–60.
49. H. Maurer and K. Pfleger, Fres. Z. Anal. Chem., 331 (1988) 744–756.
50. H. Maurer and K. Pfleger, Fres. Z. Anal. Chem., 330 (1988) 459–460.
51. H. Maurer and K. Pfleger, Arch. Toxicol., 62 (1988) 185–191.
52. H.H. Maurer, Arch. Toxicol., 64 (1990) 554–561.
53. H.H. Maurer, Arch. Toxicol., 64 (1990) 218–230.
54. H.H. Maurer, J. Chromatogr., 530 (1990) 307–326.
55. H.H. Maurer, Fres. J. Anal. Chem., 337 (1990) 144.
56. H.H. Maurer and C.F. Fritz, Int. J. Legal. Med., 104 (1990) 43–46.
57. H.H. Maurer, Ther. Drug Monit., 18 (1996) 465–470.

58. H.H. Maurer and J. Bickeboeller-Friedrich, J. Anal. Toxicol., 24 (2000) 340–347.
59. H.H. Maurer, J. Bickeboeller-Friedrich, T. Kraemer, and F. Peters, Toxicol. Lett., (1999) in press.
60. T. Kraemer, I. Vernaleken, and H.H. Maurer, J. Chromatogr. B, 702 (1997) 93–102.
61. H.H. Maurer and T. Kraemer, Arch. Toxicol., 66 (1992) 675–678.
62. H.H. Maurer, T. Kraemer, O. Ledvinka, C.J. Schmitt, and A.A. Weber, J. Chromatogr. B, 689 (1997) 81–89.
63. H.K. Ensslin, K.A. Kovar, and H.H. Maurer, J. Chromatogr. B, 683 (1996) 189–197.
64. K. Pfleger, H.H. Maurer, and A. Weber, Mass Spectral Library of Drugs, Poisons, Pesticides, Pollutants and their Metabolites, 3rd ed., 2000, Hewlett Packard, Palo Alto, CA.
65. T. Kraemer, J. Bickeboeller-Friedrich, and H.H. Maurer, Arch. Pharm. Pharm. Med. Chem., 331 (1998) 5.
66. M.R. Moeller and M. Hartung, J. Anal. Toxicol., 21 (1997) 591.
67. O.H. Drummer, J. Chromatogr. B, 713 (1998) 201–225.

15

Detection of Drugs of Abuse by Gas Chromatography–Mass Spectrometry

Jennifer S. Brodbelt, Michelle Reyzer, and Mary Satterfield
University of Texas, Austin, Texas

1. INTRODUCTION

Gas chromatography–mass spectrometry (GC–MS) provides one of the most powerful, versatile, and sensitive tools for detection and quantitation of drugs of abuse (Fig. 1), especially in complex mixtures such as urine, blood, and saliva. The growth in the applications of GC–MS in this area is attributed to the development of novel methods of derivatization of compounds that were previously too involatile for gas chromatographic separation, new methods of extraction that also assist in pre-concentration of the targeted analytes, and improvements in the mass spectrometers that give enhanced sensitivities and more elegant data acquisition, often at lower cost than the previous era of GC–MS instruments. Such developments have contributed to the current ability to detect drugs of abuse at levels as low as ppb in many complex mixtures. This chapter provides an overview of many of the recent applications of GC–MS for detection of drugs of abuse with a focus on biological matrices. Derivatization and sample preparation methods are also reviewed in this context.

2. BIOLOGICAL MATRICES

The presence of drugs of abuse and their metabolites may be detected in a number of biological matrices, including blood or plasma, urine, saliva, sweat, hair, and

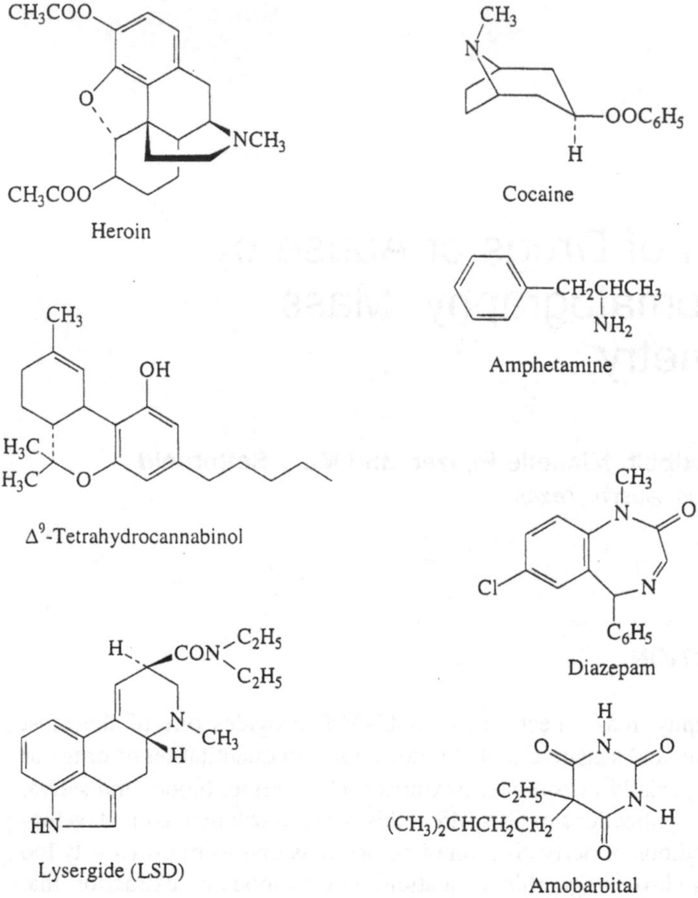

Figure 1 Structures of common drugs of abuse.

meconium. Each of these diverse matrices provides a different perspective on exposure to and usage of illicit drugs and presents a different challenge for analysis [1–8]. Metabolism and enzymatic processing that occur predominantly in the liver and kidneys mean that derivatized forms of the drugs of abuse, such as methylated analogs, glucuronides, sulfates, and other conjugates, may represent the key signature for detection of illicit drug use based on urine or blood samples. Although blood is one of the most commonly used biological fluids for detection of drugs of abuse, it is also one of the most complex mixtures, containing proteins, fats, and cellular material [2]. One notable advantage of the use of blood samples for detection of illicit drugs is that often the drugs may be detected in their native form prior to metabolism in the kidneys or liver. The whole blood matrix may be simplified by removal of the red blood cells by centrifugation, with care taken to prevent rupture of the cells. The resulting plasma or serum may then be analyzed. Serum and plasma are nearly identical with the exception of the presence of the soluble blood-clotting factors that are present only in plasma. However, both serum and plasma still contain high levels of proteins that may interfere with certain types of analyses due to the naturally high affinity between many proteins and polar substances, such as many drugs of abuse. Often, acidic solutions or organic solvents are used to alleviate this problem by promoting precipitation of the proteins.

Urine has a naturally low protein and lipid content, making it a less complex matrix than blood or plasma; however, the specific composition of urine varies on a nearly hourly basis [1]. Thus, analytical procedures developed for urine must not be strongly influenced by such changes. Because most natural components in urine are water soluble, extraction procedures can be developed to maximize the differences in solubility between lipophilic drugs of abuse and water-soluble components. Other sampling problems inherent to urine include potentially large fluctuations in concentration of illicit drugs depending upon metabolism, the presence of masking agents, and the urinary volume, in addition to moderate changes in pH. Two advantages to the use of urine samples for detection of illicit drugs are the plentiful amount available, and the relatively high concentrations of drugs and metabolites that may be found in the urine.

The use of saliva, sweat, and hair as matrices offers less invasive sampling procedures compared with the use of blood and urine [3–7]. Both sweat and saliva offer relatively cleaner matrices than urine or blood due to the lower content of proteins, lipids, and sugars [3–5], but not all drugs of abuse are evident in these matrices. Saliva is generally over 99% water with a pH from 5 to 7, and collection of analyzable amounts can typically be obtained by various saliva stimulation methods, such as chewing inert substances. The concentrations of illicit drugs are often lower than the concentrations present in blood or urine, making the use of saliva for drug detection a greater analytical challenge in terms of detection limits. Sweat offers the possibility of relatively noninvasive sampling, and a fair number of illicit drugs emerge at high concentrations in sweat. Sweat collection based on passive wiping methods affords only microliter sample sizes, but several milliliters can be obtained after exercise-induced perspiration. Sweat suffers from the shortcoming of being susceptible to environmental contamination from drug usage external to the subject, such as in smoke-filled environments. In addition, both saliva and sweat offer relatively short windows of opportunity for detection of drugs of abuse after ingestion of the drugs. Hair offers an unusual matrix in that the concentrations of drugs of abuse along the length of hair may vary as a function of the drug usage at particular points in time, thus giving access to the history of drug use [4,6–7]. Moreover, the acquisition of hair samples is a noninvasive procedure, and hair requires minimal storage requirements, unlike blood, urine, and saliva which need refrigeration. The analysis of meconium, the first fecal matter eliminated from an infant, provides a way to estimate fetal exposure to drugs of abuse in the latter half of pregnancy [8]. The use of meconium samples gives an excellent way to predict subsequent developmental problems in infants, to diagnose health problems related to drug exposure, and to ensure proper treatment at an early stage.

Applications involving the detection of drugs of abuse have been reported in all of these diverse matrices, and usually the matrix selection depends on both the availability and access to specific types of samples and the quantitative objective.

3. SAMPLE PREPARATION

A variety of sample preparation methods are used depending on the matrix of interest. A typical sequence of steps includes addition of an internal standard to the sample, addition of buffer and centrifugation if necessary, and then either conventional liquid–liquid extraction (LLE), solid-phase extraction (SPE) [9,10], or solid-phase microextraction (SPME) [11]. The latter two methods are newer techniques that enhance pre-concentration of the analytes and reduce the amount of organic solvents used. Solid-phase extraction typically involves the extraction of analytes onto a short column or polymeric disk, followed by removal of the analytes from the polymer by a subsequent solvent wash [9,10]. Solid-phase microextraction developed in the late 1980s, involves absorbing analytes onto a modified polymeric solid support, typically a coated fused-silica fiber, and then placing the fiber directly in the GC injection port [11]. The SPME fibers are usually placed directly in an aqueous solution containing the analytes or are used to extract the analytes from the headspace above the sample. For both SPE and SPME, the nature of the polymeric material influences the extraction capabilities, and there are numerous commercially available SPE and SPME devices. The SPME devices are reuseable and minimize the production of organic waste compared with LLE methods. Finally, derivatization methods either prior to or after the extraction step are often used to improve the chromatographic properties of the analytes, as described in the next section.

4. DERIVATIZATION PROCEDURES

Derivatization methods are commonly used in developing analytical strategies for detection of drugs of abuse by GC–MS to enhance the chromatographic properties of the analytes [12]. Many illicit drugs contain polar functional groups, such as hydroxyl groups, ketones, carboxylic acids, and amine groups, making them involatile, thermally unstable, and difficult to separate. Derivatization of the polar groups improves the volatility and thermal stability, thus allowing GC–MS methods to remain at the forefront of analytical methods for detection of drugs of abuse. The three most common derivatization procedures include silylation, acylation, and alkylation (Fig. 2) [12].

Silylation is used to replace reactive hydrogens, such as hydroxyl, thiol, and amine protons, by alkylsilyl groups, resulting in silyl ethers and silyl esters. Perhaps the most popular silylation reaction involves use of *N,O*-bis-trimethylsilyl-trifluoroacetamide (BSTFA) with trimethylchlorosilane (TMCS) as a catalyst. Acylation procedures result in replacement of reactive hydrogens by acyl groups. This process is typically used to derivatize analytes containing hydroxyl, amine, amide, or phenolic moieties by reaction with acyl halides or acid anhydrides. The many variations of trifluoroacetylation and propionylation reactions fall into this category. Popular derivatizing agents include pentafluoro-1-propanol (PFP), *N*-methyl-bis(trifluoroacetamide) (MBTFA), and heptafluorobutyric anhydride

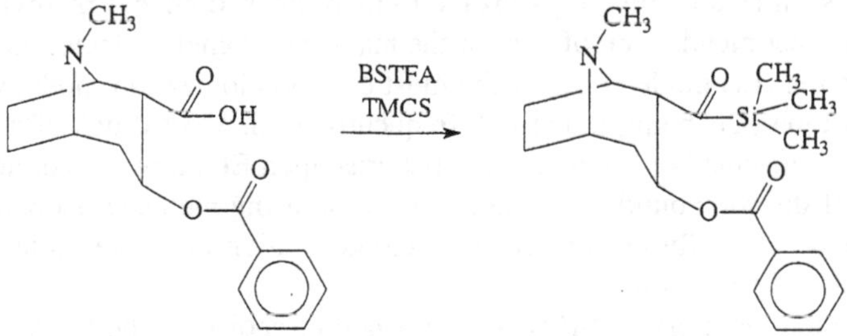

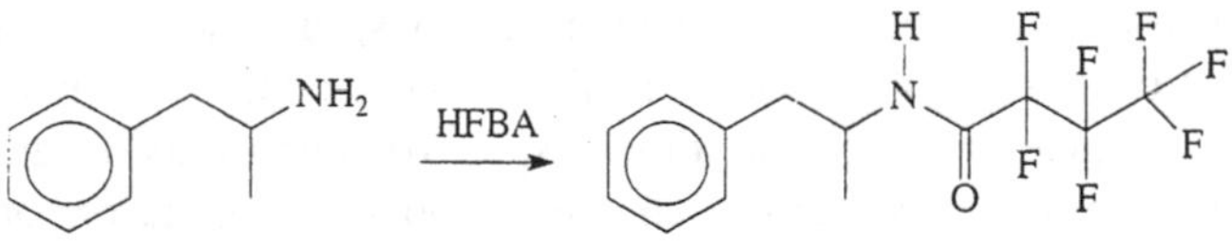

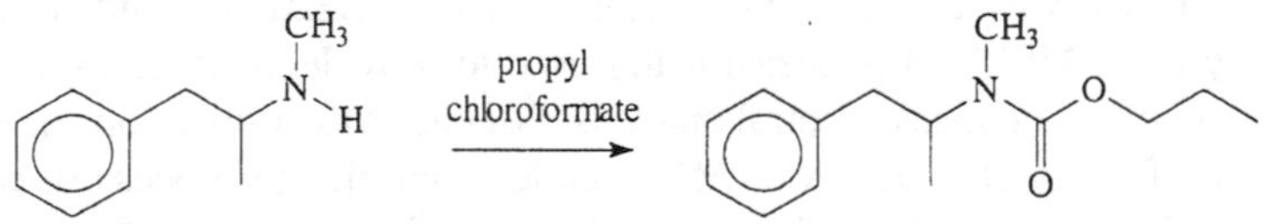

Figure 2 Derivatization reactions for drugs of abuse. (A) Benzoylecgonine reacting with BSTFA/TMCS to form the trimethylsilyl (TMS) derivative. (B) Amphetamine reacting with HFBA to form the heptafluorobutyl (HFB) derivative. (C) Methamphetamine reacting with propyl chloroformate to form the propyl carbamate derivative.

(HFBA). Alkylation is used to replace reactive hydrogens associated with carboxylic acids, hydroxyls, thiols, phenols, and both primary and secondary amines with alkyl groups by using alkyl halide reagents. Esterification of acids by using a simple alcohol and acidic catalyst is a well-known example of this derivatization procedure.

Other new derivatization methods are also being developed [13,14]. For example, the use of alkylchloroformates as derivatizing agents allows conversion of certain amines to carbamates and carboxylic acids to esters.

5. GAS CHROMATOGRAPHY–MASS SPECTROMETRY INSTRUMENTATION

The numerous types of benchtop gas chromatograph–mass spectrometers are clearly the instruments of choice for detection of drugs of abuse, offering the advantages of low cost, high throughput, and unparalleled sensitivity [15]. The GC columns most commonly used for separation of drugs of abuse are fused-silica capillary columns with alkylpolysiloxane stationary phases, ranging from nonpolar ones such as dimethylpolysiloxane to more polar diphenyl methylpolysiloxane, and other blends thereof. As for the mass spectrometric detection, the electron ionization (EI) mode is by far the most common ionization mode, with chemical ionization (CI) being much less frequently used. Most drugs of abuse give structurally informative fragmentation patterns upon EI, allowing confident identification of the compounds. Chemical ionization using methane, isobutane, or ammonia is occasionally used to enhance the generation of stable molecular species, such as $[M+H]^+$ ions.

Selective ion monitoring (SIM) is the method of choice for both quantitation and identification of drugs of abuse. Selective ion monitoring involves rapid switching of the mass analyzer between specific target ions, such as several characteristic fragment ions or fragment ions plus the molecular ion. The enhancement in sensitivity is great because more time is spent detecting ions of interest rather than scanning broad mass ranges containing no pertinent information. Since the mass spectra of most drugs of abuse are well documented, SIM is easily implemented with a minimum of analyte characterization. Tandem mass spectrometry (MS–MS) [16] is occasionally used to provide an even higher degree of specificity in the mass spectral analysis. For GC–MS applications involving detection of drugs of abuse, MS–MS is undertaken via the process of collision-induced dissociation (CID) in which the dissociation of specific ions of interest is promoted by energetic collisions between analyte ions and neutral target molecules, typically an inert gas such as helium. The CID MS–MS mode offers enhanced specificity because only the fragment ions occurring from a selected precursor ion are analyzed and detected, thus requiring two consecutive stages of mass analysis and reducing the interference of undesirable background ions.

6. DRUGS OF ABUSE

6.1. Cocaine

Cocaine is routinely analyzed by GC–MS, typically to confirm positive drug screening tests done by immunoassay [3,17–27]. Advanced applications include using hair samples [17,27], evaluating different metabolites [18], investigating cocaine metabolism postmortem [20], automating analyses [22,24], and improving both extraction [14,22,25] and derivatization [14,23] methodologies. Typical cocaine drug test confirmation by GC–MS involves analysis of benzoylecgonine in urine, but concentrations of another metabolite, ecgonine, are usually higher; thus, techniques have been developed to detect ecgonine in urine. For example, one strategy involved two derivatization steps, one nonylation and one propionylation. The ecgonine and benzoylecgonine derivatives were analyzed by EI and SIM, and limits of detection (LODs) were determined to be about 10 ng/ml. Ecgonine concentrations were found to be about five times those of benzoylecgonine for 104 urine specimens that had previously tested negative for cocaine based on analyzing benzoylecgonine alone [18].

Improvements in the analysis of benzoylecgonine from urine have been reported recently from three groups [14,23,24]. All techniques used EI in conjunction with SIM, and all employed deuterated benzoylecgonine as an internal standard. Solid-phase extraction in conjunction with one-step esterification of benzoylecgonine into either its propyl or isopropyl esters resulted in LODs of at least 10 ng/ml [23]. An automated technique was developed utilizing SPE and trimethylsilyl derivatization that has the capability of analyzing over 100 samples in 2 hr. No LODs were reported [24]. Finally, SPME was utilized to extract the hexylchloroformate derivative (hexyl ester) of benzoylecgonine from urine (Fig. 3). The LOD was reported to be 30 ng/ml for this rapid, solvent-free extraction technique [14].

Analysis of cocaine and several metabolites in hair samples was undertaken by two groups [17,27]. In both analyses, the hair samples were first washed, pulverized, and digested. The analytes were removed from the matrix via LLE or SPE and derivatized into either their trimethylsilyl derivatives using BSTFA/TMCS or into their heptafluoryl or hexafluoryl derivatives using HFBA and hexafluoroisopropanol (HFIP). Mass spectrometry utilized EI and the SIM mode, and deuterated standards were employed for quantitation. The LODs ranged from 0.1 to 0.8 ng/mg in one study [17]. This technique was used to analyse the hair of 2 tenth-century Peruvian mummies, where one mummy tested positive for both cocaine and benzoylecgonine [17]. The other study used this analysis scheme to

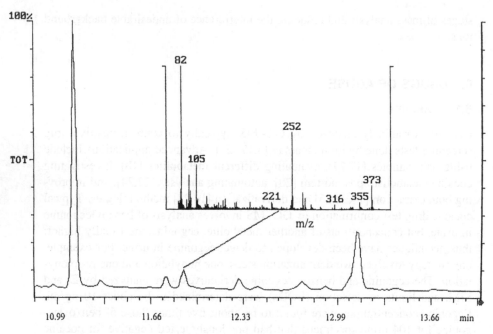

Figure 3 Total ion chromatogram for a urine specimen after performing SPME after hexylchloroformate derivatization. The mass spectrum inset indicates the presence of benzoylecgonine as its benzoylecgonine hexyl ester. (From Ref. 14.)

confirm positive tests by immunoassay, even below the 0.5 ng/mg threshold level [27].

The postmortem concentrations of cocaine and cocaethylene in the blood and tissues (liver, brain, muscle) of rats given both cocaine and ethanol were measured by GC–MS [20]. The blood or tissue homogenate was subjected to LLE in the presence of propylbenzoylecgonine added as an internal standard. The underivatized extract was analyzed by GC–MS. Characteristic ions were analyzed in the SIM mode and resulted in LODs of 10 ng/ml for blood and 30 ng/g for tissue. It was shown that cocaine was metabolized into cocaethylene in the liver up to 1 hr postmortem [20]. Human postmortem blood was analyzed for cocaine and five metabolites using a protein precipitation–sequential derivatization method [22]. First, blood protein was precipitated with methanol, then the analytes were derivatized into their *n*-propyl or *p*-nitro analogs. A final LLE removed the analytes and an aliquot was analyzed via GC–MS. Deuterated analogs of cocaine and three metabolites were used as internal standards. Using EI and SIM, LODs of better than 10 ng/ml were obtained for all compounds [22].

Three metabolites, benzoylecgonine, ecgonine, and ecgonine methyl ester, in addition to cocaine, were analyzed in urine using a "continuous method" of sample preparation in order to minimize analysis time [21]. Urine was mixed with buffer containing codeine as the internal standard and passed through a reversed-phase C18 column to remove matrix, then the analytes were removed, derivatized into their trimethylsilyl (TMS) derivatives, and injected into the GC–MS. Electron ionization was used with SIM. Limits of detection of 1 ng/ml (cocaine and ecgonine methyl ester), 20 ng/ml (benzoylecgonine), and 200 ng/ml (ecgonine) were determined in the presence of potentially interfering drugs [21].

Since unmetabolized cocaine in urine can be an indicator of recent cocaine use, an improved method for the analysis of cocaine to the exclusion of any metabolites was reported [25]. A single-step LLE with petroleum ether of buffered urine removed cocaine while leaving benzoylecgonine unextracted. The underivatized extract was analyzed by EI with SIM. In conjunction with deuterated cocaine as an internal standard, a LOD of about 5 ng/ml was determined, even in the presence of a large excess of benzoylecgonine [25]. Unmetabolized cocaine was also analyzed preferentially from urine using SPME because benzoylecgonine was not effectively extracted with the polydimethylsiloxane fiber used [26]. This method employed EI and SIM, and resulted in a LOD of 50 ng/ml [26].

6.2. Opiates

Numerous methods of detection of opiates exist utilizing all types of biological samples, including hair, sweat, and tissue. Jenkins et al. [19] compared concentrations of heroin and its metabolites in saliva to concentrations in blood and plasma. After SPE of the saliva, blood, and plasma, derivatization with MBTFA of the metabolites was carried out. (Heroin requires no derivatization for detection by GC–MS.) A calibration curve for quantitation was used with LOD at 1 ng/ml using EI and SIM.

Screening tests of hair and urine for simultaneous detection of drugs of abuse, including heroin and its metabolites have been developed. Solans et al. [28] used SPE with N-methyl-N-trimethylsilyl-trifluoroacetamide (MSTFA) and MBTFA derivatization of heroin metabolites to detect and identify more than 100 compounds in urine, including opiates, with a method designed to screen large numbers of samples in venues such as the Olympic Games [28]. Deuterated internal standards with EI and scanning mode mass spectrometry were used. Kintz and Mangin [17] developed a method to detect simultaneously opiates, cocaine, and major metabolites in human hair using LLE and BSTFA derivatization. Using EI and SIM for MS detection, LODs were 0.1 to 0.8 ng/ml. Another screening method of human hair involved initial detection by immunological screening [27]. After SPE, positive samples were derivatized with a mixture of HFBA and HFIP and analyzed using EI and SIM.

Detection of 1-α-acetylmethadol (LAAM) and methadone, opiate derivatives that are used to treat heroin addiction, were both detected with positive-ion CI-MS using a mixture of ammonia and methane as the reagent gases. Moody et al. [29] used LLE and derivatized the metabolites of LAAM with trifluoroacetic anhydride (TFAA) prior to quantitation with deuterated internal standards and calibration curves to determine limits of quantitation of 5 to 10 ng/ml. In contrast, Alburges et al. [30] carried out SPE of plasma, urine, and tissue samples, but did not derivatize the analytes because methadone and its metabolites have no easily derivatized functional groups. Quantitation with deuterated internal standards showed a linear dynamic range of 10 to 600 ng/ml when positive-ion CI and SIM were used for detection. Detection of heroin and its metabolites in sweat has been reported by Kintz et al. [31]. They used BSTFA for derivatization of the metabolites of heroin extracted by LLE with use of a deuterated internal standard and calibration curves for quantitation. The LODs were 0.5 to 1 ng/ml using EI and SIM for MS detection.

6.3. Amphetamines

Amphetamines have also been extensively analyzed by GC–MS [2,32–36]. Recent work has focused on improvements in derivatization procedures [13], extraction techniques [1,33,35,36], and optimizing ionization parameters [14]. Solid-phase microextraction was utilized for headspace analysis of urine to determine both amphetamine and methamphetamine [32]. In conjunction with the use of pentadeuterated methamphetamine as an internal standard, LODs were obtained as low as 0.1 μg/ml for both compounds monitored in the SIM mode and using isobutane CI. This method was found to be 20 times more sensitive than traditional headspace analysis of urine without SPME extraction [32]. Even lower limits of detection were obtained utilizing derivatization of amphetamine and methamphetamine with propylchloroformate and LLE of the propylcarbamate derivatives from urine [13]. When deuterated amphetamine and methamphetamine were used as internal standards, the LODs were 25 ng/ml for both drugs; the LOD improved to 5 ng/ml when N-propylamphetamine was used as the internal standard. In this case, ions were formed via EI and SIM was used (Fig. 4) [13]. This method was shown to be rugged and relatively free from potential interferences from other related drugs.

Solid-phase microextraction was utilized for headspace analysis of blood in conjunction with ''on-column'' derivatization of methamphetamine, amphetamine, and the internal standard pentadeuterated methamphetamine with HFBA [33]. The HFB-derivatives were ionized by EI and analyzed by SIM. The LODs were 0.01 μg/g for both drugs of interest [33]. The HFB-derivatives of amphet-

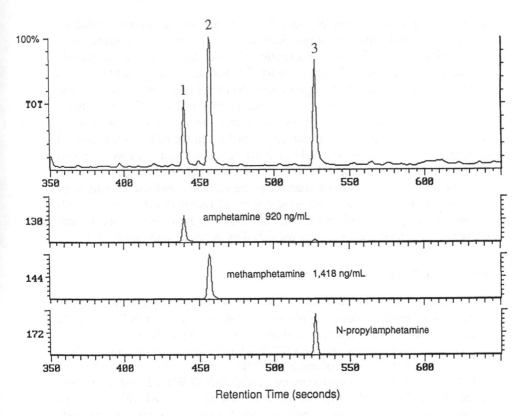

Figure 4   Total ion and extracted ion chromatograms from a urine specimen that screened positive by immunoassay. Peaks are: (1) amphetamine, (2) methamphetamine, and (3) *N*-propylamphetamine. (From Ref. 13.)

amine and methamphetamine extracted from urine by SPE were compared with the nonderivatized drugs using EI and ammonia CI [34]. Using calibration curves and SIM, LODs for ammonia CI of the HFB derivatives were found to be approximately 90 ng/ml for both amphetamine and methamphetamine, while EI gave LODs of about 10 ng/ml for both drugs. Chemical ionization of the underivatized drugs led to higher LODs due to poor chromatography. However, the intense structure-specific ions formed by CI of both derivatized and underivatized amphetamines could be utilized in conjunction with traditional EI methods for rapid screening and identification of amphetamine and methamphetamine [34].

A SPME headspace procedure was recently described for the screening of urine for 21 amphetamine-related compounds, including amphetamine, methamphetamine, and related "designer" drugs [35]. Three deuterated amphetamines were added to the urine samples and used as internal standards for quantitation. A special nonpolar capillary column designed for the separation of amino compounds (Supelco PTA-5) was used and allowed good separation without derivatization. Ions formed from EI were monitored in the SIM mode and LODs were obtained ranging from 1 to 50 ng/ml, with 10 ng/ml being the most common [35]. Another SPME procedure was described by Myung et al. [37] for the determination of amphetamine, methamphetamine, and dimethamphetamine in urine. In this case, the SPME fiber was directly immersed in the urine sample instead of in the headspace and no internal standards were employed. Ions were formed by EI and monitored in the SIM mode. Calibration curves were generated for quantitation, and LODs were comparable to the screening method, ranging from 1 to 10 ng/ml [37]. A comprehensive review of recent (1991 to early 1997) analysis techniques, including GC–MS methods, for the determination of amphetamines in blood and urine was published by Kraemer and Maurer [36].

6.4. Cannabinoids

Because marijuana is the most widely used illegal drug of abuse in this country there is great interest in detection of cannabinoids and their metabolites [38,39]. Initial extraction from blood, hair, and body tissues is typically by LLE [38,40–44]. Goodall and Basteyns [40] used BSTFA for derivatization and found LODs of 0.2 ng/ml for Δ^9-tetrahydrocannabinol (THC) and 11-hydroxy-THC (11-OH-THC) and 2 ng/ml for 11-nor-9-carboxy-THC (THCCOOH) in blood, using deuterated internal standards [40]. For MS detection, EI was used with SIM. This method of analysis is proposed as an alternative to more costly detection methods involving negative-ion CI GC–MS or tandem MS.

Several groups recently have reported analysis of human hair for THC and related compounds [41–43]. Cirimele et al. [41] in "a rapid, simple, and direct screening method" were able to detect THC, cannabidiol (CBD), and cannabinol (CBN) without derivatization in hair using an internal standard. Their LODs ranged from 0.01 to 0.1 ng/mg. In this study, EI with SIM was used. Using PFPA and PFP-OH as derivatizing agents, French researchers were able to detect as little as 5 pg/mg of THCCOOH in hair using internal standards for quantitation [42]. This dramatic drop in LODs is attributed to use of negative-ion CI with methane (CH_4) in conjunction with SIM. Jurado et al. [43] compared methods for testing hair for THC and THCCOOH in Spain vs. France, the aforementioned method used in France while HFBA and HFP-OH were used for derivatization in the Spain study. For detection, the Spanish laboratory used EI with SIM. Interlaboratory results were similar in assessment of identical samples.

Detection and quantitation of THC in human tissue has been accomplished by Kudo et al. [44] using methylation derivatization and deuterated standards. The LODs were 1 ng/g using EI and SIM. A unique method has been developed for detection of Δ^8-THC, Δ^9-THC, CBD, and CBN in saliva requiring minimal sample preparation [45]. Hall et al. (45) used acetic acid to coagulate the protein in saliva that is removed prior to extraction by SPME. Deuterated standards were used to determine a LOD of 1 ng/ml using EI and SIM [45] (Fig. 5). This method was compared with LLE and found to be more specific and accurate.

6.5. Benzodiazepines

Although clinically prescribed as sedatives and anticonvulsants, benzodiazepines can be abused and therefore must be monitored in drug tests. Many GC–MS methods are currently available for the analysis of benzodiazepines [46–50]. Recent advances have been made with regards to preanalysis enzyme hydrolysis [46,48], analysis of the drugs in human tissue [47], and the investigation of new sample preparation techniques [49]. A good review of numerous techniques currently used to analyze benzodiazepines was published by Drummer [50].

Benzodiazepines are frequently found in urine as their β-glucuronide metabolites. Two groups recently reacted the glucuronides found in urine with glucuronidase to recover the parent compounds, removed the matrix by either LLE or SPE, and then used EI along with SIM for analysis [46,48]. Needleman and Porvaznik [46] analyzed seven benzodiazepines prescribed by physicians in this manner and found a low limit of linearity of about 80 ng/ml while using deuterated diazepam as an internal standard [46]. Alternatively, King and King [48] used on-disk derivatization of eight prescribed benzodiazepines with N-methyl-N-(t-butyldimethylsilyl)trifluoroacetamide (MTBSTFA) in conjunction with three deuterated benzodiazepine internal standards to obtain LODs lower than 10 ng/ml.

Analysis of bromazepam in blood and human tissues, including liver, brain, lung, kidney, and muscle, was undertaken by GC–MS [47]. Whole blood or tissue homogenates were extracted via a series of LLEs and analyzed directly, using EI and SIM. N-desmethyldiazepam was used as an internal standard, and LODs of about 2 to 5 ng/g were obtained [47].

Five benzodiazepines, diazepam, nordiazepam, oxazepam, temazepam, and lorazepam, were determined in urine and serum using SPME [49]. The SPME parameters were optimized in aqueous solutions using calibration curves, and LODs ranged from 20 to 70 ng/ml. Serum and urine samples were successfully qualitatively analyzed for the five benzodiazepines studied using EI GC–MS, however it was noted that the extraction of oxazepam and lorazepam was suppressed in the biological matrices compared with aqueous solution [49].

6.6. Others

There are numerous other illicit drugs that have been detected in biological matrices by GC–MS, including barbiturates, lysergide (LSD), phencyclidine (PCP),

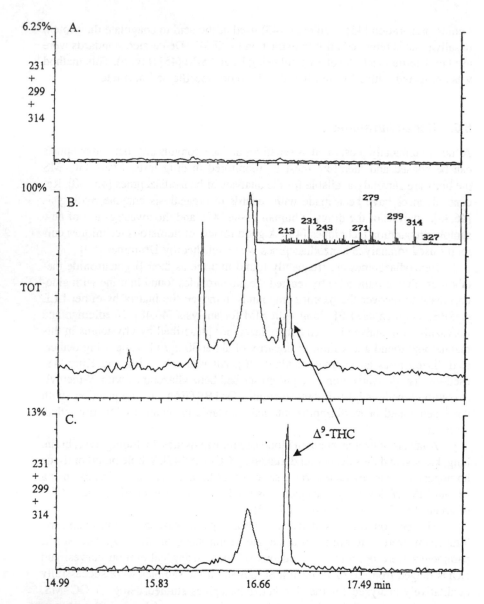

Figure 5 Chromatograms after performing SPME on human saliva samples (A) prior to and (B,C) after marijuana smoking. A and C show the reconstructed selected ion chromatograms obtained by monitoring three characteristic ions of Δ^9-tetrahydrocannabinol. The insert on B shows the mass spectrum for the peak at 16.99 minutes, identified as Δ^9-tetrahydrocannabinol. (From Ref. 45.)

and fenfluaramine. The GC–MS applications involving these miscellaneous drugs of abuse are summarized below.

Detection of barbiturates in urine without sample preparation is possible using SPME as reported by Hall et al. [51]. Eight different barbiturates were successfully detected and quantified in a mixture using calibration curves and deuterated internal standards. A LOD of 1 ng/ml was determined using EI and SIM.

A comprehensive review of the detection of lysergide (LSD) in biosamples was published in 1998 [52]. According to Schneider et al. [52] detection of LSD in biological samples is difficult due to the extremely small amounts ingested for hallucinogenic effects. Most often, immunoassays are used for initial screenings followed up by confirmation with other methods including GC–MS. Analysis using GC requires derivatization by silylation of the sample. With derivatization of blood serum a linearity range for TMS-LSD of 0.1 to 10 ng/ml has been reported. Detection involved EI with SIM. Positive-ion or negative-ion CI using ammonia and methane for reagent gases has been used to detect LSD and its metabolites, yielding a LOD of 10 pg/ml in spiked urine samples. Detection of LSD in rat hair and in human hair of self-reported users has been reported. Derivatized LSD was detected in the hair of rats given the lowest dose of 0.05 mg/ kg. Of the self-reported users, LSD was detected in 2 of the 17 samples. For quantitation, internal standards were used. The SIM mode was used for detection following EI.

Developed for use as an anesthetic, phencyclidine (PCP) is no longer used by the medical community due to its hallucinogenic effects. Schneider et al. [52] carried out a review of the determination of phencyclidine in biosamples and reported on methods developed for a wide range of biological samples from hair to meconium [52]. Most of the methods involved SPE. Many of the methods involved derivatization of the PCP and metabolites via methylation or trimethylsilylation after extraction. Many of the methods used deuterated internal standards and calibration curves to detect and quantify PCP and its metabolites. Most methods included use of EI with SIM although one method employed surface ionization detection and another used tandem MS. The LODs ranged from 25 pg in hair to 0.47 ng/ml in urine.

One unique method of detection of PCP in urine and blood involved SPME of headspace above blood and plasma [53]. Sodium hydroxide and potassium carbonate were added to the sample and extraction performed at 90°C. Both an internal standard and calibration curve were used to determine a LOD of 1 ng/ ml in blood and 0.25 ng/ml in urine. Surface ionization detection was used after GC separation.

Two drugs once widely prescribed together for weight loss, phendimetrazine and fenfluramine (phen-fen), were detected and quantitated in urine using SPME (54). The only pretreatment of the urine necessary was a 1 : 1 dilution of the urine prior to extraction with a 30 μm polydimethylsiloxane (PDMS) fiber. An internal standard with calibration curves was used in calculations of LOD: 40 ng/ml for phendimetrazine and 30 ng/ml for fenfluramine. Metabolites of these two drugs were also detected but in smaller amounts than their precursors. For detection EI was used in conjunction with SIM.

7. CONCLUSIONS

Gas chromatography–mass spectrometry has proven to be a powerful analytical tool for detection of drugs of abuse in biological matrices. In particular, the use of the SIM mode has enhanced the sensitivity of the GC–MS method for detection of drugs of abuse. The versatility of the method has been extended by the use and development of novel derivatization procedures that have allowed a more diverse array of illicit drugs and their metabolites to be converted to thermally stable, volatile compounds. In addition, SPE and SPME methods also have improved the sensitivities of detection due to their capabilities for preconcentration of the targeted analytes.

ACKNOWLEDGMENTS

Financial support from the National Science Foundation and the Welch Foundation is gratefully acknowledged.

REFERENCES

1. J. Chamberlain, The Analysis of Drugs in Biological Fluids, 2d Ed., 1995, CRC Press, Boca Raton, FL.
2. M.R. Moeller, S. Steinmeyer, and T. Kraemer, J. Chromatogr. B, 713 (1998) 91.
3. D.A. Kidwell, J.C. Holland, and S. Athanaselis, J. Chromatogr. B, 713 (1998) 111.
4. T. Inoue, and S. Seta, Forensic Sci. Rev., 4 (1992) 90.
5. W. Schramm, R. H. Smith, P.A. Craig, and A. Kidwell, J. Anal. Toxicol. 16 (1992) 1.
6. H. Sachs and P. Kintz, J. Chromatogr. B, 713 (1998) 147.
7. M.R. Moeller, J. Chromatogr., 580 (1992) 125.
8. C. Moore, A. Negrusz, and D. Lewis, J. Chromatogr. B, 713 (1998) 137.
9. P.D. McDonald and E.S.P. Bouvier (eds.), Solid Phase Extraction Applications Guide and Bibliography: A Resource for Sample Preparation Methods Development, 6th ed., 1995, Waters Corp., Milford, MA.
10. J.P. Franke, and R.A. de Zeeuw, J. Chromatogr. B, 713 (1998) 51.
11. Z. Zhang, M.J. Yang, and J. Pawlizyn, Anal. Chem., 66 (1994) 844A.
12. J. Segura, R. Ventura, and C. Jurado, J. Chromatogr. B, 713 (1998) 61.

13. R. Meatherall, J. Anal. Toxicol., 19 (1995) 316–322.
14. B. Hall, A. Parikh, and J. Brodbelt, J. Forensic Sci., 44 (1999) 519–526.
15. C.Henry, Anal. Chem., 71 (1999) 401A–406A.
16. K.L. Busch, G.L. Glish, and S.A. McLuckey, Mass Spectrometry/Mass Spectrometry: Technique and Applications of Tandem Mass Spectrometry, 1988, VCH Publishers, New York.
17. P. Kintz and P. Mangin, Forensic Sci. Int., 73 (1995) 93–100.
18. C. Hornbeck, K. Barton, and R. Czarny, J. Anal. Toxicol., 19 (1995) 133–138.
19. A. Jenkins, J. Oyler, and E. Cone, J. Anal. Toxicol., 19 (1995) 359–374.
20. F. Moriya and Y. Hashimoto, J. Forensic Sci., 41 (1996) 129–133.
21. S. Cardenas, M. Gallego, and M. Valcarcel, Rapid Commun. Mass Spectrom., 10 (1996) 631–636.
22. D. Smirnow and B. Logan, J. Anal. Toxicol., 20 (1996) 463–467.
23. B. Paul, C. Dreka, J. Summers, and M. Smith, J. Anal. Toxicol., 20 (1996) 506–508.
24. F. Diamond, W. Vickery, and J. de Kanel, J. Anal. Toxicol., 20 (1996) 587–591.
25. D. Garside, B. Goldberger, K. Preston, and E. Cone, J. Chromatogr. B, 692 (1997) 61–65.
26. Y. Makino, T. Takagi, S. Ohta, and M. Hirobe, Chromatography, 18 (1997) 185–188.
27. J. Segura, C. Stramesi, A. Redon, M. Ventura, C. Sanchez, G. Gonzalez, L. San, and M. Montagna, J. Chromatogr. B, 724 (1999) 9–21.
28. A. Solans, M. Carnicero, R. de la Torre, and J. Segura, J. Anal. Toxicol., 19 (1995) 104–114.
29. D.E. Moody, D.J. Crouch, C.O. Sakashita, M.E. Alburges, K. Minear, J.E. Schulthies, and R.L. Foltz, J. Anal. Toxicol., 19 (1995) 343–351.
30. M.E. Alburges, W. Huang, R.L. Foltz, and D.E. Moody, J. Anal. Toxicol., 20 (1996) 362–368.
31. P. Kintz, R. Brenneisen, P. Bundeli, and P. Mangin, Clin. Chem., 43 (1997) 736–739.
32. M. Yashiki, T. Kojima, T. Miyazaki, N. Nagasawa, Y. Iwasaki, and K. Hara, Forensic Sci. Int., 76 (1995) 169–177.
33. N. Nagasawa, M. Yashiki, Y. Iwasaki, K. Hara, and T. Kojima, Forensic Sci. Int., 78 (1996) 95–102.
34. P. Dallakian, H. Budzikiewicz, and H. Brzezinka, J. Anal. Toxicol., 20 (1996) 255–261.
35. C. Battu, P. Marquet, A. Fauconnet, E. Lacassie, and G. Lachatre, J. Chromatogr. Sci., 36 (1998) 1–7.
36. T. Kraemer and H. Maurer, J. Chromatogr. B, 713 (1998) 163-187.
37. S.-W. Myung, H.-K. Min, S. Kim, M. Kim, J.-B. Cho, and T.-J. Kim, J. Chromatogr. B, 716 (1998) 359–365.
38. M.A. Huestis, J.M. Mitchell, and E.J. Cone, J. Anal. Toxicol., 20 (1996) 441–452.
39. B.J. Rowland, J. Irving, and E.S.Keith, Clin. Chem., 40 (1994) 2114–2115.
40. C.R. Goodall, B.J. Basteyns, J. Anal. Toxicol., 19 (1995) 419–426.
41. V. Cirimele, H. Sachs, P. Kintz, and P. Mangin, J. Anal. Toxicol., 20 (1996) 13–16.
42. P. Kintz, V. Cirimele, and P. Mangin, J. Forensic Sci., 40 (1995) 619–622.

43. C. Jurado, M. Menendez, M. Repetto, P. Kintz, V. Cirimele, and P. Mangin, J. Anal. Toxicol., 20 (1996) 111–115.
44. K. Kudo, T. Nagata, K. Kimura, T. Imamura, and N. Jitsufuchi, J. Anal. Toxicol., 19 (1995) 87–90.
45. B. Hall, M. Satterfield-Doerr, A.R. Parikh, and J. Brodbelt, Anal. Chem., 70 (1998) 1788–1796.
46. S. Needleman and M. Porvaznik, Forensic Sci. Int., 73 (1995) 49–60.
47. X. Zhang, K. Kudo, T. Imamura, N. Jitsufuchi, and T. Nagata, J. Chromatogr. B, 677 (1996) 111–116.
48. J. King and L. King, J. Anal. Toxicol., 20 (1996) 262–265.
49. Y. Luo, L. Pan, and J. Pawliszyn, J. Microcolumn Sep., 10 (1998) 193–201.
50. O. Drummer, J. Chromatogr. B, 713 (1998) 201–225.
51. B. Hall and J.S. Brodbelt, J. Chromatogr. A, 777 (1997) 275–282.
52. S. Schneider, P. Kuffer, and R. Wennig, J. Chromatogr. B, 713 (1998) 189–200.
53. A. Ishii, H. Seno, T. Kumazawa, K. Watanabe, H. Hattori, and O. Suzuki, Chromatographia, 43 (1996) 331–333.
54. M. Chiarotti, S. Strano-Rossi, and R. Marsili, J. Microcolumn Sep., 9 (1997) 249–252.

16

Gas Chromatography–Mass Spectrometry Analysis of Explosives

Shmuel Zitrin
Division of Identification and Forensic Science, Jerusalem, Israel

1. INTRODUCTION

Gas chromatography-mass spectrometry (GC–MS) is considered the method of choice in forensic analysis, being both highly reliable and sensitive. As such, it has achieved a leading role in the analysis of explosives, both before and after explosion. Before reaching this status, it had been assumed that GC–MS might not be suitable for the analysis of explosives, due to thermal instability of some explosives. In practice, however, it seems that most explosives and related materials can be successfully analyzed by GC–MS [1]. This is especially true for nonexploded explosives, but successful results have been obtained also in postexplosion analysis. Problems are encountered in the GC–MS of nonvolatile explosives, such as cellulose nitrate (nitrocellulose, NC), which does not elute through the GC column, and in thermally labile explosives such as tetryl or some nitrate esters, which may decompose or hydrolyze in the GC injector.

An important issue in forensic science is the criteria required for a positive identification of a single organic compound. An opinion held by some forensic scientists is that such identification cannot be based on chromatographic methods only, even when several such methods are applied. It is argued that other methods, such as infrared (IR) spectrometry, nuclear magnetic resonance (NMR) spectrometry, or mass spectrometry (MS), should be included.

While adhering to these criteria when unexploded explosives are analyzed is usually straightforward, it is often very difficult in postexplosion analysis. The samples collected from the debris often contain only trace amounts of the original explosives, mixed with large amounts of contaminants. Infrared and NMR spectrometry are usually incompatible with this situation, mainly because they are not generally preceded by on-line chromatographic separation. In many postexplosion analyses they are too insensitive to detect the explosive traces within the large amounts of interfering materials. Although commercially available, gas chromatography–Fourier transform infrared spectrometry (GC–FTIR) has not been routinely used in postexplosion analysis, partly due to sensitivity limitations. This leaves the hyphenated MS methods GC–MS and liquid chromatography-mass spectrometry (LC–MS) as the remaining methods for confirmation. In practice, GC–MS is more widely used than LC–MS in the analysis of explosives. Possible reasons are the lower prices of benchtop GC–MS instruments and their multifunctional use in many forensic laboratories. In addition, unlike mass spectra obtained in LC–MS, the electron ionization (EI) spectra obtained in GC–MS are easily comparable to existing libraries.

It should be emphasized that the routine use of GC–MS for post-explosion analysis is not without problems. Highly contaminated extracts from the debris often decrease the efficiency of the GC–MS analysis. This is especially true for some nitrate esters and nitramines, where a significant decrease in sensitivity has been observed.

The introduction of the chemiluminescence detector, known also as thermal-energy analyzer (TEA), into GC and LC instruments has affected the use of GC–MS in postexplosion analysis. The TEA is very specific: it will not give a positive response unless nitro (or nitroso) groups are present in the analyte molecules. Therefore, it is often used as a basis for positive identification of an explosive even without MS confirmation. It is true that in contrast to a complete mass spectrum, a chromatographic peak in GC–TEA analysis does not constitute an unequivocal identification, but GC–TEA may succeed in some post-explosion analyses where GC–MS fails.

2. GAS CHROMATOGRAPHY–MASS SPECTROMETRY OF EXPLOSIVES CONTAINING NITRO GROUPS

According to the common classification [2], explosives containing nitro groups may be divided as (1) nitroaromatic compounds, which contain a C-NO$_2$ bond (where the carbon atom is part of an aromatic system); (2) nitrate esters, which contain a C-O-NO$_2$ bond; and (3) nitramines, which contain a C-N-NO$_2$ bond.

2,4,6-Trinitrotoluene (TNT) is the most important nitroaromatic explosive. Other examples are 2,4-dinitrotoluene (2,4-DNT), 2,6-DNT, 1,3,5-trinitrobenzene (TNB), and 2,4,6-trinitrophenol (picric acid). Common nitrate ester explosives are glycerin trinitrate (nitroglycerin, NG), ethylene glycol dinitrate (nitroglycol, EGDN), pentaerythritol tetranitrate (PETN), and NC. The most important nitramines are the heterocyclic nitramines 1,3,5-trinitro-1,3,5-triaza-cyclohexane (RDX, hexogen, cyclonite) and 1,3,5,7-tetranitro-1,3,5,7-tetraza-cyclooctane (HMX, octogen). Another nitramine, which has a trinitroaromatic ring, is 2,4,6-*N*-tetranitro-*N*-methylaniline (tetryl). Figure 1 shows the structural formulas of some common explosives containing nitro groups.

2.1. Nitroaromatic Compounds

Most nitroaromatic explosives are quite stable under GC conditions [3], so their GC–MS analysis is easily carried out. This explains why many GC–MS analyses deal only with nitroaromatic compounds.

An early GC–MS analysis of crude TNT was carried out by combining a packed-column GC to a high-resolution (HR) mass spectrometer [4]. Most of the

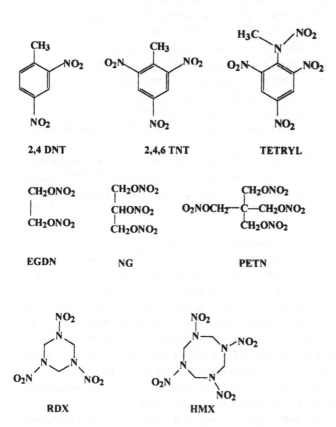

Figure 1 Structural formulas of some common explosives.

2,4,6-TNT was removed prior to the analysis by fractional crystallization. The impurities, identified by their EI mass spectra, included isomers of dinitrobenzene (DNB), DNT and TNT, as well as TNB, 2,4,6-trinitroxylene (TNX), and 1-nitro-naphthalene.

Gas chromatography–mass spectrometry was employed in the analysis of wastewater generated in the production of TNT [5]. Compounds were analyzed and quantified by GC with flame ionization detection (FID) but their identification was carried out by GC–MS, using a packed-column GC instrument. The identi-fied components included (in addition to 2,4,6-TNT) toluene, 2- and 4-nitrotolu-enes, all six DNT isomers, aminonitrotoluene isomers, amino-DNT isomers, 2,3,6-TNT, 1,3-DNB, 1,3,5-TNB, some nitrocresol isomers, and 3,5-dinitroani-line. Several unexpected compounds were also identified, including N-nitroso-morpholine, N-morpholinoacetonitrile and nitrobenzonitrile isomers. The mor-pholine derivatives are not directly related to the nitration of toluene, and are possibly the result of the use of morpholine as an algaecide in water-cooling devices at munitions facilities. The major component in the wastewater was 1,3-DNB. Its formation is attributed to oxidation followed by decarboxylation of the methyl group in 2,4-DNT rather than to direct nitration of benzene.

Gas chromatography–mass spectrometry, along with LC–MS, was used to identify organic compounds present in explosives-related industrial wastewater [6]. Solid-phase extraction techniques were employed to remove the compounds from their aqueous solutions. Compounds identified by GC–MS included 2,4-DNT and its degradation product 2-amino-4-nitrotoluene, TNT, and diphenyl-amine (DPA).

2,4-Dintrotoluene, TNT, and two degradation products of TNT were identi-fied by packed column GC–MS in groundwater contaminated by TNT-related wastes [7]. The two TNT degradation products, 4-amino-2,6-DNT and 2-amino-4,6-DNT, were identified by comparing their EI mass spectra to those of authentic samples.

Gas chromatography–mass spectrometry of nine nitroaromatic compounds was carried out in order to evaluate its potential as an analytical method for the determination of TNT and possible related impurities in environmental samples such as water and soil [8]. Nitrobenzene, the three mononitrotoluenes (MNTs), 1,3-DNB, 2,4-DNT, 2,6-DNT, and TNB were analyzed in addition to TNT. The MS ionization techniques included EI, positive-ion chemical ionization (CI) and negative-ion CI (NCI), all recorded in both full-scan and selective ion monitoring (SIM) modes. They were compared with GC analysis with electron capture detec-tion (ECD). The comparison included sensitivity (detection limits), linear re-sponse range, and precision. The GC–ECD was recommended for those analyses where there was no uncertainty in the identity of the analytes.

Gas chromatography–mass spectrometry was used for the determination of microgram amounts of some mononitroaromatic compounds in samples of river water, sediment, and fish [9]. The analytes included nitrobenzene, nitrotoluenes, nitroanisoles, and chloronitrobenzenes. The mass spectrometric ionization was EI, using both full-scan and SIM nodes. Deuterated nitrobenzene was the internal standard for the quantitative determination.

Nitroaromatic compounds, including nitro derivatives of benzene, toluene, phenol, and aniline were extracted from their aqueous solutions and analyzed by GC (ECD and TEA) and by GC–MS, using EI, CI, and NCI [10]. Methane and isobutane were the reagent gases in CI, and methane and argon in NCI. Best sensitivity (detection limits of 1 to 3 pg for a full-scan) was obtained in NCI. Sensitivity in EI was better than in CI. The GC–ECD was the most sensitive detector (more sensitive than GC–MS) while the GC–TEA was the least sensitive. However, the TEA showed better selectivity than the ECD. Various extraction methods, including liquid–liquid extraction and solid-phase extraction were also evaluated. An actual sample of polluted water was analyzed by GC–MS (EI and NCI). Two degradation products of TNT were identified: 4-amino-2,6-DNT and 2,6-diamino-4-nitrotoluene. A later work by the same group [11] used GC–TEA, LC, and NCI GC–MS to analyze surface water and soil from sites of former ammunition plants in Germany. Mononitro, dinitro, and trinitro derivatives of toluene were found, as well as monoamino and diamino reduction products of TNT.

Nitroaromatic explosives and related compounds were identified by GC–MS in ammunition wastewater and aqueous samples from old ammunition plants and military sites [12]. The nitroaromatic compounds included nitro derivatives of toluene and aniline. The best sensitivity (in the sub-pg range) was obtained by using NCI, with methane as reagent gas, in the SIM mode.

Soils from several military installations in the United States were extracted by acetonitrile and the extracts analyzed by GC–MS [13]. The purpose of the study was to find the major transformation products of TNT, which had been buried in the soils. The most common transformation products were 2-amino-4,6-DNT and 4-amino-2,6-DNT, formed from TNT by microbiological reduction. 2,4,6-Trinitrobenzaldehyde, an oxidation product of TNT, and TNB (formed by further oxidation to 2,4,6-trinitrobenzoic acid followed by decarboxylation) were also found. 3,5-Dinitroaniline, the reduction product of TNB was also identified. 2,4-Dinitrotoluene and 2,6-DNT were found in relatively large amounts. Their presence was attributed to disposal of propellants in the studied locations, rather than to the disposal of TNT.

Gas chromatography–mass spectrometry of TNT, using a short capillary column for sample introduction, was carried out in both CI and NCI, using methane and isobutane as reagent gases [14]. The aim was to determine the detection limits using SIM and to compare them with those obtained by using a solid probe for sample introduction. Lowest detection limits (sub-ng) were obtained in the NCI mode, monitoring the M$^{-\cdot}$ ion at m/z 227.

A mixture of 2,4-DNT and 2,6-DNT was identified by GC–MS in an old, possibly pre–World War I explosive found in Ireland [15]. The DNT isomers were mixed with potassium chlorate.

Some synthetic musk fragrances, widely used in soaps, detergents, and lotions, are polynitroaromatic compounds. As such, they may give false positive results when tested by field kits based on color reactions of explosives.

The GC–MS analysis of four "nitro musks," musk ambrette, musk xylol, musk tibetene and musk ketone was studied [16]. Full EI mass spectra of the four compounds were given. In an actual case, a powder that gave false positive results with an explosive testing kit was identified as musk ambrette [17]. Gas chromatography–mass spectrometry, using both EI and CI, was one of several methods used for the identification.

2.2. Nitrate Esters and Nitramines

Gas chromatography–mass spectrometry analysis of nitrate and nitrite esters was carried out by using photoionization MS and comparing the resulting spectra with the corresponding EI spectra [18]. A short packed column was employed for the GC, and an argon lamp with a lithium window (LiF) window (11.83 and 11.62 eV) was used to ionize the molecules. In addition to simple alkyl nitrates and alkyl nitrites, the explosives EGDN and propylene-1,2-glycol dinitrate (PGDN) were analyzed.

Gas chromatography–mass spectrometry was used for the quantitative analysis of a series of nitrate ester explosives: 1,2,4-butanetriol trinitrate (BTTN), triethylene glycol dinitrate (TEGN), metriol trinitrate (MTN, 1,1,1-tri(hydroxymethyl)ethane trinitrate), PGDN, NG and PETN [19]. The MS ionization techniques included EI, CI and NCI. Methane was the usual reagent gas for CI and NCI, but isobutane and ammonia were also studied. Quantitation was carried out using SIM.

Some nitrate esters are widely used in medicine as coronary vasodilators. Gas chromatography–mass spectrometry was often used to determine these nitrate esters and their metabolites in human plasma. Nitroglycerin and its two metabolites, glycerol-1,2-dinitrate and glycerol-1,3-dinitrate, were most frequently analyzed in human plasma. Other pharmaceutically employed nitrate esters, such as isosorbide dinitrate and PETN, were also determined in blood by GC–MS. The preferred ionization method for the determination of these nitrate esters in blood was NCI in the SIM mode, often monitoring the NO_3^- at m/z 62. The analytical procedures are not discussed here but are reviewed in detail elsewhere [20].

Semtex, a plastic explosive made in former Czechoslovakia has been frequently used by terrorists. Its explosive components are PETN and RDX. It was analyzed by supercritical fluid extraction followed by off-line GC–ECD and GC–MS [21]. Except for its two major components and some hydrocarbons, a minor and rather unexpected impurity was identified as EGDN, which had not been listed by the manufacturer as a Semtex ingredient. The EGDN impurity, having a much higher vapor pressure than PETN and RDX, could be responsible for the positive response to Semtex by devices that detect vapors of explosives ("sniffers").

Tetryl, a nitramine with a trinitroaromatic nucleus, gives a discrete chromatographic peak when analyzed by GC–MS so it was only natural to assume that the mass spectrum of this peak represented the mass spectrum of tetryl [22,23]. However, it was found [24,25] that the mass spectrum was not that of tetryl but of N-methylpicramide. The latter is the hydrolysis product of tetryl, as shown in Figure 2. This hydrolysis, which takes place in the gas chromatograph, most probably in the injector, is an example of an artifact which, if overlooked, may lead to an erroneous attribution of the mass spectrum of one compound (N-methylpicramide) to another compound (tetryl). It demonstrates the caution that should be exercised in interpreting GC–MS results, especially of thermally labile compounds.

While the decomposition of tetryl under GC conditions is well established, it is still uncertain whether the highly nonvolatile heterocyclic nitramine HMX elutes intact from the GC column. While in some studies [12,22] decomposition was noticed, which led to difficulties in its GC (or GC–MS) analysis, it was claimed by others [23,26–29] that by carefully controlling the GC conditions, HMX may elute through the GC column without decomposition. However, a convincing mass spectrum proving the identity of the eluted compound was either absent [26,29] or it contained ions that could not identify HMX unequivocally [23,27,28].

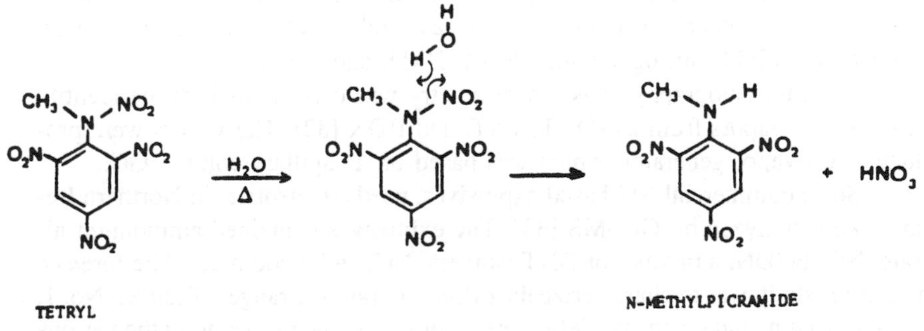

Figure 2 Hydrolysis of tetryl to N-methylpicramide.

2.3.   Nitro-Containing Explosives from the Various Groups

Most GC–MS analyses of explosives were carried out on mixtures of explosives
from the different classes (nitroaromatic, nitrate esters, and nitramines).

In an early work [30], CI-MS of explosives was studied, using methane as
reagent gas. Some of the explosives were introduced into the mass spectrometer
by a packed-column GC and some via a solid probe. The compounds analyzed
by GC–MS included simple nitroaromatic derivatives of benzene, toluene, aniline, and phenol and the explosives TNT, EGDN and NG. The less volatile explosives PETN and RDX as well as the thermally labile tetryl, were introduced
via a solid probe. The military plastic explosive "Composition C-3" and the
commercial dynamite "Hi-drive" were extracted with acetone and the extracts
were analyzed by GC–CI-MS. Mononitrotoluenes, 2,4-DNT, 2,6-DNT, TNT and
RDX were identified in the plastic explosive, while NG and EGDN were identified in the dynamite.

In another early work [22], a packed column GC–MS was used for the
analysis of different explosives, including the nitroaromatic 2,4-DNT and 2,4,6-
TNT, the nitrate ester NG, and the nitramines RDX and tetryl. Ionization modes
were EI, CI, and NCI, using methane, isobutane and ammonia as reagent gases.
Other explosives, which did not elute from the GC column and were introduced
into the mass spectrometer by solid probe, included picric acid, dipicrylamine,
PETN and HMX.

Gas chromatography–mass spectrometry of TNT, EGDN, NG, PETN, and
RDX was studied in both EI and NCI [31]. The NCI mode used methane as a
moderator but its pressure was low, so the formation of negative ions was a
resonance-capture process rather than NCI. The sensitivity was similar in the two
modes. It was in the 100 to 150-ng range for full scan but could be improved
by using SIM.

Thirty-four explosives and related materials were analyzed by both GC–
MS and LC [23]. The analyzed compounds included nitro derivatives of benzene,
toluene, xylene, phenol and naphthalene, as well as TNB, TNT, picric acid,
EGDN, NG, PETN, nitroguanidine, RDX, HMX and tetryl.

Gas chromatography–mass spectrometry was used to confirm the identity
of explosive vapors from 2,4-DNT, TNT, and RDX [32]. The vapors were produced by a vapor generator, which was based on a capillary column GC.

Some commercial NG-based explosives, used by terrorists in Northern Ireland, were analyzed by GC–MS [33]. The explosives contained ammonium nitrate, NG, EGDN, a mixture of DNT isomers, NC, and wood meal. The forensic
aim of the study was to characterize the different brands (Frangex, Frangex No. 1,
Plaster Gelatin, and Opencast Gelignite) according to the ratio among the various
DNT isomers.

Some of the recent GC–MS analyses of explosives were carried out in an ion-trap mass spectrometer [27,28,34–36]. Ion-trap GC–MS with a cooled temperature-programmable injector was used to analyze picogram amounts of explosives in water [27,28]. The analyzed explosives were TNT, PETN, RDX, HMX, tetryl, and five DNT isomers (all except 3,5-DNT). The use of temperature-programmable injector enabled satisfactory analysis of most of the listed explosives, including RDX and PETN, which sometimes pose problems in their GC analysis. The resulting EI spectra of most explosives matched their published spectra, except for tetryl (which decomposed to N-methylpicramide) and HMX (whose spectrum contained interference from a possible plasticizer).

Ion-trap GC–MS was used to analyze 2,4-DNT, 2,6-DNT, TNT, and RDX in sea water [34,35]. Solid-phase microextraction (SPME) was used to concentrate the explosives prior to their GC–MS analysis. Concentrations of pg/ml could be detected.

Ion-trap GC–MS–MS, using a temperature-programmable injector, was used to analyze 2,4-DNT, TNT, EGDN, NG, PETN, and RDX [36]. The sensitivity, which was better than that achieved in quadrupole instruments, was in the picogram range. It was better in the NCI mode than in the CI mode, but the NCI spectra of the nitrate esters (EGDN, NG, and PETN) contained only low-mass ions at m/z 46 (NO_2^-) and 62 (NO_3^-) and therefore could not serve as a basis for identification.

2.4. The Mass Spectrometry of Explosives Containing Nitro Groups

This topic is discussed here only briefly. For a more extensive discussion, the reader is referred elsewhere [37,38]. The EI mass spectra of nitroaromatic compounds usually contain molecular ions or related ions in the high-mass region. The EI spectra (including those of positional isomers) usually differ from each other to an extent that enables reliable identification. In addition to $M^{+\cdot}$, the loss of NO_2 from the molecular ion is characteristic to nitroaromatic compounds. The spectra of trinitroaromatic explosives, such as TNB, TNT, or picric acid, include diagnostic $[M - 3NO_2]^+$ ions. The abundance of the $M^{+\cdot}$ ions in nitroaromatic compounds decreases when a nitro group is in an adjacent ("ortho") position to a hydrogen-containing substituent such as methyl. The major process in these compounds, known as "ortho" effect, is the loss of hydroxyl radical from $M^{+\cdot}$, leading to an abundant $[M - OH]^+$ ion. In 2,6-DNT and 2,4,6-TNT, where two nitro groups are positioned "ortho" to the methyl group, the abundance of the molecular ions becomes negligible, and the base peaks are the $[M - OH]^+$ ions at m/z 165 and 210, respectively.

The CI mass spectra of nitroaromatic explosives usually contain highly abundant $[M + H]^+$ ions and little fragmentation. An important feature in these spectra is the existence of ions at m/z values that are lower by 30 than those of the corresponding $[M + H]^+$ ions. Their appearance and abundance seemed to be dependent on experimental conditions, such as the type and pressure of the reagent gas, ion source design and temperature and the presence of water in the ion source. When first noticed, they had been mistakenly assumed to be formed by the unusual loss of the NO radical from the even-electron $[M + H]^+$ ion [39, 40]. It was later proved [41–44] that their formation was the result of reduction of the nitroaromatic compound to an aromatic amine ($ArNO_2 \rightarrow ArNH_2$), which could take place prior to the ionization. Thus, the ion of the protonated amine, formed in CI, is observed. It is important for the analyst to be aware of this reduction process, which can explain some unusual ions in the mass spectra of nitroaromatic compounds. Thus, the ion at m/z 198 in some GC–CI-MS analyzes of TNT [8,22] was probably the $[M + H]^+$ ion of the reduction product (amino-DNT).

NCI mass spectra of nitroaromatic compounds usually contain $M^{-\bullet}$ and $[M - H]^-$ ions, whose relative abundance depends on the specific compounds and the reactant ions. In compounds such as TNT, where the "ortho" effect may operate, an $[M - OH]^-$ is observed.

Electron-ionization mass spectra of nitrate esters are very similar to each other and usually contain ions in the low-mass region: at m/z 30 (NO^+), 46 (NO_2^+), and 76 ($CH_2ONO_2^+$). No ions in the molecular weight region are observed. While the combination of the three ions (at m/z 30, 46, and 76) is very helpful as group-diagnostic for the presence of a nitrate ester, an identification of a specific nitrate ester should not be made only on the basis of its EI spectrum.

Unlike their EI mass spectra, CI mass spectra of nitrate esters contain ions at the molecular weight region: $[M + H]^+$ and $[M + H-HONO_2]^+$. This makes CI an excellent complementary method to EI for the analysis of individual nitrate esters.

Figure 3 shows the EI and CI mass spectra of EGDN and NG [24]. The similarity between the EI spectra, in contrast to the difference between the CI spectra, can be clearly observed.

The NCI spectra of nitrate esters are characterized by two major ions in the low-mass region: at m/z 46 (NO_2^-) and 62 (ONO_2^-). Ions observed in the molecular weight region are $M^{-\bullet}$ and $[M + ONO_2]^-$.

Electron-ionization mass spectra of the two heterocyclic nitramines RDX and HMX are similar, with their most abundant ions mainly at the low-mass region: at m/z 30 (NO^+), 42 ($CH_2NCH_2^+$), and 46 (NO_2^+). More diagnostic ions appear at m/z 120, 128, and 148 but the spectra lack molecular ions.

When CI spectra of RDX are taken with different reagent gases, they differ from each other not only in the degree of fragmentation (which is usual in CI), but also in the type of the fragment ions [45]. With reactant ions, which are weak Brönsted acids, such as $C_4H_9^+$ (in isobutane) or NH_4^+ (in ammonia), a set of fragment ions at m/z 84, 131, and 176 is observed. With strong Brönsted acids, such as H_3^+ (in hydrogen) or $C_2H_5^+$ (in methane) other major ions are observed

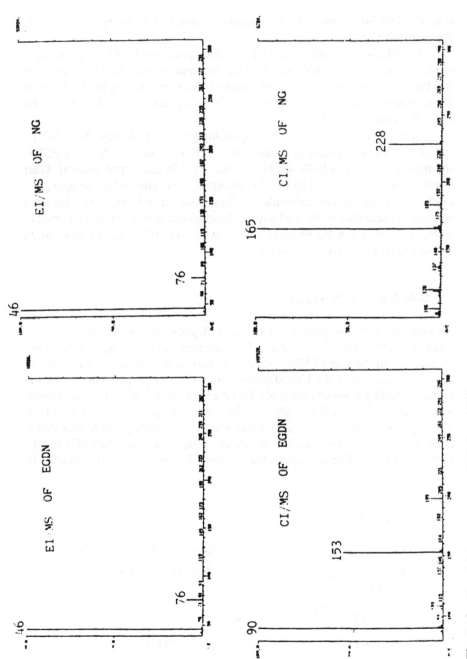

Figure 3 EI and CI mass spectra of EGDN and NG.

at m/z 75 and 149. A mechanistic explanation, based on the energetics involved, was suggested [45].

The EI mass spectrum of tetryl has its base peak at m/z 241, corresponding to a loss of NO_2 from the molecular ion. This ion appears when tetryl is introduced into the mass spectrometer via a solid probe [46,47] but is completely absent in the spectrum of tetryl obtained by GC–MS, where tetryl is hydrolyzed to N-methylpicramide [24,25].

Some unusual adduct ions, corresponding to $[M + NO]^+$ and $[M + NO_2]^+$, appear in the mass spectra of nitrate ester explosives, such as NG and PETN, and nitramines, such as RDX and HMX [38, and references cited therein]. Their formation, which is favored by CI-MS conditions, especially in high sample pressures, is attributed to ion–molecule reactions between NO^+ and NO_2^+ ions and the neutral molecules of the explosives. These adduct ions were also reported to be highly abundant in the EI spectra of nitrate esters [19], when EI was carried out in a tight, dual EI/CI ion source.

3. ORGANIC PEROXIDES

Two important organic peroxides have been illegally used by terrorists in recent years. The most common is 3,3,6,6,9,9-hexamethyl-1,2,4,5,7,8-hexacyclononane or triacetonetriperoxide (TATP), which was first identified in a terrorist case in Israel [48] and later in the United States [49]. Its simple preparation from easily available starting materials has made TATP a high-priority issue for law enforcement agencies throughout the world. Another peroxide, 3,4,8,9,12,13-hexaoxa-1,6-diazabicyclo [4,4,4]tetradecane or hexamethylenetriperoxidediamine (HMTD), was also identified in terrorist-related cases in Israel [48] and the United States [50]. Figure 4 shows the structural formulas of TATP and HMTD.

TATP   HMTD

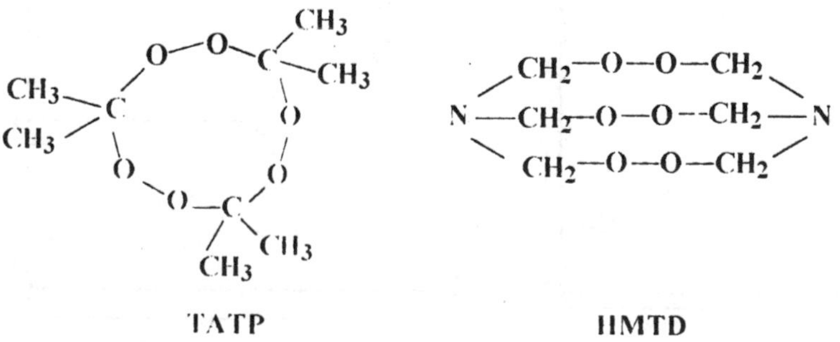

Figure 4 Structural formulas of TATP and HMTD.

The first mass spectra of TATP were recorded by introduction of the sample with a direct-insertion probe [48,49]. The EI mass spectrum was very uninformative and contained its major ions in the low-mass region, with a very low-abundance molecular ion. It had some resemblance to the spectrum of acetone. The CI mass spectra, using methane [49] and isobutane [48,49] contained a distinct [M + H]$^+$ ion at m/z 223. A GC–MS analysis of TATP was later performed, using EI [51–53] and CI-methane [51]. The resulting mass spectra were similar to those obtained by the solid-probe technique.

The EI mass spectrum of HMTD, obtained by solid probe [48], contained a distinct molecular ion at m/z 208 and major fragment ions at m/z 88 and 176. The CI mass spectrum, using isobutane as reagent, had its base peak at m/z 209, corresponding to the [M + H]$^+$ ion. The GC–MS analysis [50] produced similar spectra, but the CI-methane spectrum contained, in addition to the [M + H]$^+$ ion, many fragment ions.

4. GAS CHROMATOGRAPHY–MASS SPECTROMETRY OF INORGANIC ANIONS

While GC–MS has found its place as a leading analytical tool for the analysis of organic explosives and related materials, inorganic ions have been analyzed by chemical spot tests [54], ion chromatography [55], and capillary electrophoresis [55], without a confirmatory MS method. To upgrade the analysis of inorganic ions, the incorporation of inorganic anions into organic molecules and subsequent analysis of the products by GC–MS was successfully attempted.

An attractive idea for nitrate ions (NO$_3^-$) is to convert them (with the aid of sulfuric acid) into nitronium ions (NO$_2^+$), which are the electrophilic agents in the nitration of aromatic rings. To facilitate the electrophilic attack, an activated aromatic ring, containing electron-donating substituents, should be employed.

An early attempt used 1,3,5-trimethoxybenzene as the activated aromatic compound [56]. Surprisingly, the GC–MS analysis of the product failed to detect the expected nitro derivative. Instead, nitrobenzene was identified, using both EI and CI.

An aromatic substrate, which was successfully used to identify nitrate ions in real-life postexplosion cases, was *tert*-butylbenzene [57]. The dried aqueous extract was acidified with concentrated sulfuric acid and reacted with the tert-butylbenzene, producing five products: all three mononitro isomers and two dinitro isomers (2,4- and 2,6-). The reaction products (sometimes not all the five products) were easily identified by GC–MS, using EI. An idea to improve the sensitivity of this method (P. Pigou, personal communication, 1998) is based on the assumption that the second nitration step may lower the sensitivity. Hence, the possibility of using a penta-substituted benzene as the aromatic substrate is being explored.

A different approach is to use a reagent with a functional group that may be replaced by the inorganic anion via a nucleophilic substitution. Thus, the analyzed anion is incorporated into the organic molecule, which can then be analyzed by GC–MS. Based on this approach, a method for the GC analysis of anions was developed, using pentafluorobenzyl bromide [58], pentafluorobenzyl p-toluenesulfonate [59] or pentafluorobenzyl methanesulfonate [60] as organic substrates. The products of the nucleophilic substitution were then analyzed by GC. As the anions to be analyzed are in aqueous solutions, and the organic reaction is carried out in an organic solvent, a phase-transfer catalyst must be employed in order to transfer the anions from the aqueous solution to a nonaqueous one. Quaternary ammonium salts were used for this purpose. Cyanide, nitrite, nitrate, sulfide, and thiocyanate were among the analytes [58–60].

The nucleophilic substitution described above, using tetra-n-octylammonium bromide as phase-transfer catalyst, was successfully applied to explosives-related anions such as nitrate and thiocyanate [57]. The thiocyanate ion (CNS^-) is a highly diagnostic product of the burning of black powder. Sulfide (S^{2-}), hydrosulfide (HS^-) and disulfide (S_2^{2-}), which are also present among the burning products of black powder, are identified by the same method. The products of the reaction of pentafluorobenzyl bromide ($C_6F_5CH_2Br$) with NO_3^- and CNS^- are $C_6F_5CH_2ONO_2$ and $C_6F_5CH_2CNS$, respectively. Their EI mass spectra include distinct molecular ions at m/z 243 and 239, respectively. The sulfur-related products are $C_6F_5CH_2SH$, $C_6F_5CH_2—S—CH_2C_6F_5$, and $C_6F_5CH_2—S—S—CH_2C_6F_5$. They all show molecular ions in their EI mass spectra. Chlorate, an important ion in improvized explosive mixtures, could not be analyzed by this method; its reaction with $C_6F_5CH_2Br$ did not give the expected substitution product [57].

A large variety of anions, including nitrate and azide, were analyzed by GC-MS as their ethyl derivatives [61]. Derivatization was carried out with ethyl-p-toluenesulfonate, using 18-Crown-6 (1,4,7,10,13-hexaoxacyclooctadecane) as a catalyst.

5. GAS CHROMATOGRAPHY–MASS SPECTROMETRY IN POSTEXPLOSION ANALYSIS

Postexplosion analysis is one of the most difficult areas in forensic science [62,63]. It is usually based on the assumption that even in a "complete" explosion, there are some residues of the original explosives in concentrations compatible with the available analytical methods. The first problem is to choose the "right" exhibits, i.e., those with the best chances to contain the residues. No clear rules exist as to what exhibit should be collected (there are sometimes hundreds of them), and collection is usually based on experience and luck. The exhibit may often contain only traces of the original explosive, mixed with large amounts of contaminants. This decreases the efficiency of the GC–MS, a decrease that may be more pronounced with certain explosives. For example, RDX and PETN are more difficult to analyze than TNT. The key to a successful GC–MS analysis of debris is often an efficient cleaning of the extract prior to the analysis.

Gas chromatography–mass spectrometry was part of overall schemes for postexplosion analysis in several laboratories [64–66]. In addition, GC–MS played an important role in the postexplosion analysis of real-life cases [24,50,53,57,64,67].

Gas chromatography–mass spectrometry helped to identify the explosives used in a terrorist bombing [50]. Vapors from the debris were adsorbed on charcoal, which was then extracted with CH_2Cl_2 and the extract was analyzed by GC–MS. The two peaks in the total ion chromatogram (TIC) gave typical EI spectra of nitrate esters, with ions at m/z 30, 46, and 76. Chemical-ionization–mass spectrometry, using methane as reagent gas, identified the two esters as diethyleneglycol dinitrate (DEGDN) and MTN. The $[M + H]^+$ and $[M + H-HONO_2]^+$ ions appeared at m/z 197 and 134, respectively, in the CI spectrum of DEGDN and at m/z 256 and 193, respectively, in the CI spectrum of MTN.

Gas chromatography–mass spectrometry has been routinely used in Israel for postexplosion analysis [24,64]. A typical postexplosion TIC of an acetone extract from debris [24] is shown in the lower part of Figure 5. The TNT peak, emerging after 321 seconds, was first located by reconstructing a mass chromatogram for the ion at m/z 210, as shown in the upper part of Figure 5. Based on its full EI mass spectrum, the peak was attributed to TNT. The use of a mass chromatogram to locate the position of a suspected explosive in the TIC is especially helpful when the peak in the TIC is negligible or appears as a "shoulder"—a situation that occurs often in postexplosion analysis.

The elution of PETN under normal GC conditions is known to occur with loss of sensitivity [68]. The problem was especially serious in postexplosion analysis [24], often resulting in poor peak shapes and rapid deterioration of the GC column. To improve the elution of PETN, the use of very short (1.5 m) capillary columns was suggested [67].

The postexplosion analysis of some nitrate esters, mainly PETN and NG, is further complicated by the presence of hydrolysis products, which are lower nitrate esters of the parent alcohol [67,69,70]. Figure 6 shows the structural formulas of the hydrolysis products of PETN. These compounds appear also in thin-layer chromatography (TLC) [69] and LC–MS [71] of extracts containing PETN and NG. If unaware of their presence, they may lead to erroneous interpretation of the GC–MS results. As the lower nitrate esters have free hydroxyl groups, conversion to trimethylsilyl (TMS) derivatives was carried out prior to the GC–MS analysis [67]. In an actual case, the TLC analysis revealed several Griess-

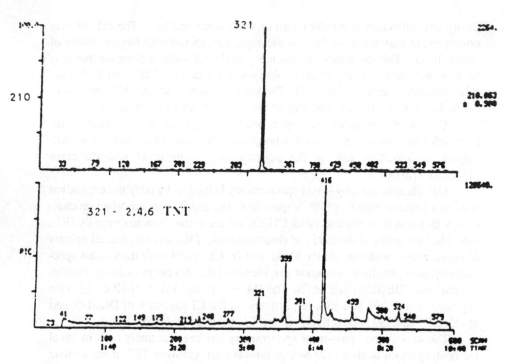

Figure 5 TIC (lower part) and mass chromatogram at m/z 210 (upper part) of an acetone extract from debris. The peak emerging after 321 seconds was identified as TNT.

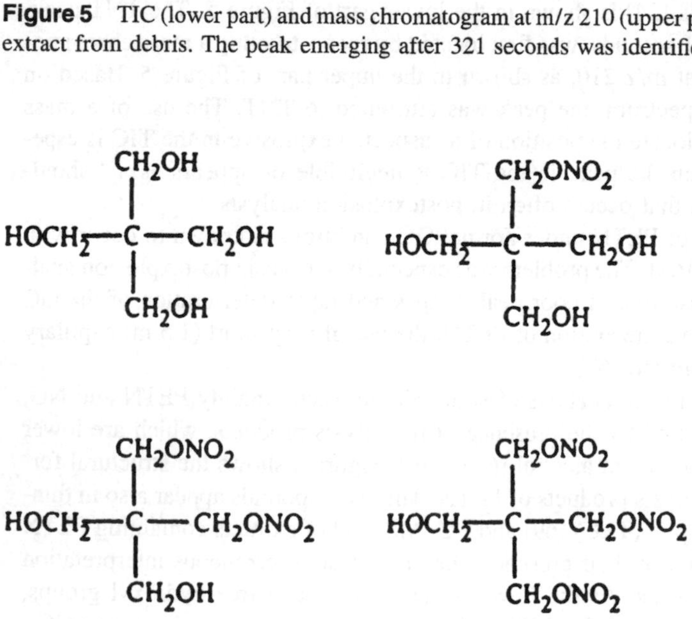

Figure 6 Structural formulas of hydrolysis products of PETN.

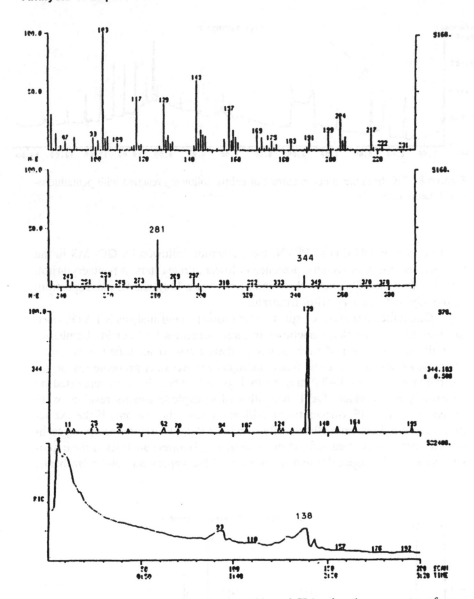

Figure 7 TIC, mass chromatogram (at m/z 344), and CI (methane) mass spectrum from a silylated extract of debris. The peak emerging after 138 seconds was identified as the TMS derivaive of pentaerythritol trinitrate.

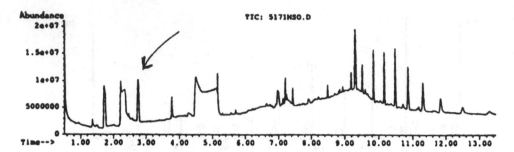

Figure 8 TIC from the aqueous extract of debris, following reaction with pentafluoro-bezyl methanesulfonate.

positive spots in addition to PETN. Derivatization followed by GC–MS in the CI (methane) mode proved the presence of lower nitrate esters of pentaerythritol. Figure 7 shows the TIC, mass chromatogram (at m/z 344), and CI mass spectrum of the silylated pentaerythritol trinitrate.

One of the most elusive explosives in postexplosion analysis is TATP. Used in many terrorist attacks, sometimes in large quantities by ''suicide bombers'' in public buses or crowded markets, it nevertheless was not identified in the debris (its presence was later learned from intelligence sources). A probable reason for the failure to identify TATP may be its high volatility. This was supported by an experiment in which TATP was allowed to explode and its residues were analyzed by GC–MS. Successful identification was achieved only if the extraction was carried out immediately after the blast [52]. Therefore, sampling techniques other than classical extraction were tried [53]: direct analysis of the explosive vapors (''headspace'') and adsorption of the vapors on solid adsorbents,

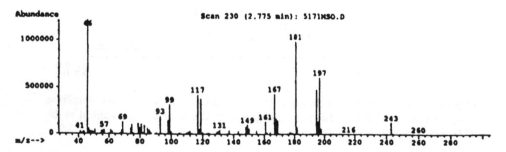

Figure 9 EI mass spectrum of the peak emerging after 2.7 minutes, corresponding to $C_6F_5CH_2ONO_2$.

followed by elution with organic solvents. Using these techniques, TATP was identified in debris by GC–MS [53].

The GC–MS of anions, following their incorporation into organic molecules, was applied to postexplosion analysis [57]. Figure 8 shows the TIC from the aqueous extract of debris, following reaction with pentafluorobenzyl methanesulfonate, using tetra-n-octylammonium bromide as a phase-transfer catalyst. Figure 9 shows the EI mass spectrum of the peak emerging after 2.7 minutes, which corresponds to $C_6F_5CH_2ONO_2$. Thus, the original presence of NO_3^- ions could be deduced.

REFERENCES

1. J. Yinon and S. Zitrin, Modern Methods and Applications in Analysis of Explosives, 1993, Wiley, New York, pp. 122–139.
2. J. Yinon and S. Zitrin, The Analysis of Explosives, 1981, Pergamon Press, Oxford, pp. 1–28.
3. M. Kaplan and S. Zitrin, J. Assoc. Off. Anal. Chem., 60 (1977) 619–624.
4. T.L. Chang, Anal. Chim. Acta, 53 (1971), 445–448.
5. R.J. Spanggord, B.W. Gibson, R.G. Keck, D.W. Thomas, and J.J. Barkley, Environ. Sci. Technol., 16 (1982) 229–232.
6. A.J. Danzig, A.M. Dietrich, C.E. Parker, and K.B. Tomer, Proceedings of ASMS 38th Annual Conference on Mass Spectrometry and Allied Topics, Tuscon, AZ, 1990, pp. 1075–1076.
7. W.E. Pereira, D.L. Short, D.B. Manigold, and P.K. Roscio, Bull. Environ. Contam. Toxicol., 21 (1979) 554–562.
8. D.S. Weinberg and J.P. Hsu, J. High Resolut. Chromatogr. Chromatogr. Commun., 6 (1983) 404–418.
9. Y. Nishikawa and T. Okumura, Anal. Chim. Acta, 312 (1995) 45–55.
10. J. Feltes, K. Levsen, D. Volmer, and M. Spiekermann, J. Chromatogr., 518 (1990) 21–40.
11. J. Feltes and J. Koll, LC-GC Int., 7 (1994) 698–701.
12. K. Levsen, P. Mussmann, E. Berger-Preiss, A. Preiss, D. Volmer and G. Wünsch, Acta Hydrochim. Hydrobiol., 21 (1993) 153–166.
13. M.E. Walsh, T.F. Jenkins, and P.G. Thorne, J. Energetic Mater., 13 (1995) 165–183.
14. M.R. Lee, S.C. Chang, T.S. Kao, and C.P. Tang, J. Res. Natl. Bur. Stand., 93 (1988) 428–431.
15. S.D. McDermott, J. Forensic Sci., 39 (1994) 1103–1106.
16. M.P. Yurawecz and B.J. Puma, J. Assoc. Off. Anal. Chem., 66 (1983) 241–247.
17. Y. Bamberger, S. Levy, T. Tamiri, and S. Zitrin, Proceedings of 3rd International Symposium on Analysis and Detection of Explosives, 1989, Mannheim-Neuostheim, Germany, pp. 26.1–26. 24.
18. H. Tagaki, N. Washida, H. Akimoto, and M. Okuda, Anal. Chem., 53 (1981) 175–179.

19. M R. Lee, D.G. Hwang, and C. P. Tang, Proceedings of 3rd International Symposium on Analysis and Detection of Explosives, 1989, Mannheim-Neuostheim, Germany, pp. 5.1–5. 12.

20. J. Yinon, Toxicity and Metabolism of Explosives, 1990, CRC Press, Boca Raton, FL, pp. 110–113.

21. G.C. Slack, H.M. McNair, and L. Wasserzug, J. High Resolut. Chromatogr., 15 (1992) 102–104.

22. W. Gielsdorf, Fresenius Z. Anal. Chem., 308 (1981) 123–128.

23. Y. Inoue, S. Arakawa, N. Ueda, J. Yamamoto, and R. Nakashima, Tottori Daigaku, 20 (1989) 97–104.

24. T. Tamiri and S. Zitrin, J. Energetic Mater., 4 (1986) 215–237.

25. J. Yinon, S. Zitrin, and T. Tamiri, Rapid Commun. Mass Spectrom., 7 (1993) 1051–1054.

26. M. Hable, C. Stern, C. Asowata, and K. Williams, J. Chromatogr. Sci., 29 (1991) 131–135.

27. J. Yinon, Proceedings of 5th International Symposium on Analysis and Detection of Explosives, 1995, Washington, DC, p. 6.1.

28. J. Yinon, J. Chromatogr. A, 742 (1996) 205–209.

29. M.E. Walsh and T. Ranney, J. Chromatogr. Sci., 36 (1998) 406– 416.

30. C.T. Pate and M.H. Mach, Int. J. Mass Spectrom. Ion Phys., 26 (1978) 267–277.

31. A.S. Cummings and K.P. Park, Proceedings of 1st International Symposium on Analysis and Detection of Explosives, 1983, FBI Academy, Quantico, VA, pp. 259–265.

32. G.A. Reiner, C.L. Heisey, and H.M. McNair, J. Energetic Mater., 9 (1991) 173–190.

33. D.T. Burns and R J. Lewis, Anal. Chim. Acta, 307 (1995) 89–95.

34. S.A. Barshick, J.E. Caton, and W.H. Griest, Proceedings of ASMS 45th Annual Conference on Mass Spectrometry and Allied Topics, Palm Springs, CA, 1997, p. 222.

35. S.A. Barshick and W.H. Griest, Anal. Chem., 70 (1998) 3015–3020.

36. P.A. Dreifuss and T.S. McConnel, Proceedings of ASMS 46th Annual Conference on Mass Spectrometry and Allied Topics, Orlando, FL, 1998, p. 1475.

37. J. Yinon and S. Zitrin, Modern Methods and Applications in Analysis of Explosives, 1993, Wiley, New York, pp. 178–215.

38. S. Zitrin, in A. Beveridge, ed., Forensic Investigation of Explosions, 1998, Taylor and Francis, London, pp. 287–296.

39. R.G. Gillis, M.J. Lacey, and J.S. Shannon, Org. Mass Spectrom., 9 (1974) 359–364.

40. S. Zitrin and J. Yinon, Org. Mass Spectrom., 11 (1976) 388–393.

41. A. Maquestiau, Y. Van Haverbeke, R. Flammang, H. Mispreuve, and J. Elguero, Org. Mass Spectrom., 14 (1979) 117–118.

42. J.J. Brophy, V. Diakiw, R.J. Goldsack, D. Nelson, and J.S. Shannon, Org. Mass Spectrom., 14 (1979) 201–203.

43. A.G. Harrison and R.K.M.R. Kallury, Org. Mass Spectrom., 15 (1980) 284–288.

44. J. Yinon and M. Laschever, Org. Mass Spectrom., 16 (1981) 264-266.

45. S. Zitrin, Org. Mass Spectrom., 17 (1982) 74–78.
46. F. Volk and H. Schubert, Explosivstoffe, 16 (1968) 2–10.
47. S. Zitrin and J. Yinon, Adv. Mass Spectrom., 7 (1978) 1457–1464.
48. S. Zitrin, S. Kraus, and B. Glattstein, Proceedings of 1st International Symposium on Analysis and Detection of Explosives, 1983, FBI Academy, Quantico, VA, pp. 137–141.
49. H.K. Evans, F A.J. Tulleners, B.L. Sanchez, and C.A. Rasmussen, J. Forensic Sci., 31 (1986) 1119–1125.
50. D.J. Reutter, E.C. Bender, and T.L. Rudulph, Proceedings of 1st International Symposium on Analysis and Detection of Explosives, 1983, FBI Academy, Quantico, VA, pp. 149–157.
51. G.M. White, J. Forensic Sci., 37 (1992) 652–656.
52. H. Arai and J. Nakamura, in Current Topics in Forensic Science, vol. 4, 1997 Shunderson Communications, Canada, pp. 209–211. (Paper presented at 14th Meeting of the International Association of Forensic Sciences, 1996, Tokyo.)
53. T. Tamiri, S. Abramovich-Bar, D. Sonenfeld, S. Tsaroom, A. Levy, D. Muller, and S. Zitrin, Paper presented at the 6th International Symposium on Analysis and Detection of Explosives, 1998, Prague, Czech Republic.
54. C.R. Midkiff and W.D. Washington, J. Assoc. Off. Anal. Chem., 57 (1974) 1092–1097.
55. B. McCord and E.C. Bender, in A. Beveridge, ed., Forensic Investigation of Explosions, 1998, Taylor and Francis, London, pp. 251–258.
56. Y.L. Tan, J. Chromatogr., 140 (1977) 41–46.
57. B. Glattstein, S. Abramovich-Bar, T. Tamiri, and S. Zitrin, Proceedings of 5th International Symposium on Analysis and Detection of Explosives, 1995, Washington, DC, p. 33.1
58. S.H. Chen, H.L. Wu, M. Tanaka, T. Shono, and K. Funazo, J. Chromatogr., 396 (1987) 129–137.
59. K. Funazo, M. Tanaka, K. Morita, M. Kamino, M. Shono, and H.L. Wu, J. Chromatogr., 346 (1985) 215–225.
60. M. Tanaka, H. Takigawa, Y. Yasaka, T. Shono, K. Funazo, and H.L. Wu, J. Chromatogr., 404 (1987) 175–182.
61. K.J. Mulligan, J. Microcolumn Separations, 7 (1995) 567–573.
62. C. R. Midkiff, Proceedings of 5th International Symposium on Analysis and Detection of Explosives, 1995, Washington, DC, 1.1.
63. A.D. Beveridge, Forensic Sci. Rev., 4 (1992) 17–49.
64. S. Zitrin, J. Energetic Mater., 4 (1986) 199–214.
65. J.S. Deak, H. Clark, C. Dagenais, S. Jones, D. McClure, and B.W. Richardson, Proceedings of 3rd International Symposium on Analysis and Detection of Explosives, 1989, Mannheim-Neuostheim, Germany, pp. 18.1–18.19.
66. T. Kishi, J. Nakamura, Y. Komo-oka, and H. Fukuda, in J. Yinon, ed., Advances in Analysis and Detection of Explosives, 1993, Kluwer Academic Publishers, Dordrecht, pp. 11–17. (Paper presented at 4th International Symposium on Analysis and Detection of Explosives, 1992, Jerusalem, Israel.)
67. T. Tamiri, S. Zitrin, S. Abramovich-Bar, Y. Bamberger, and J. Sterling, , in J. Yinon, Ed., Advances in Analysis and Detection of Explosives, 1993, Kluwer Academic,

Dordrecht, The Netherlands, pp. 323–324 (Paper presented at 4th International Symposium on Analysis and Detection of Explosives, 1992, Jerusalem, Israel.)

68. P. Kolla, J. Forensic Sci., 36 (1991) 1342–1359.
69. J. Helie-Calmet and H. Forestier, Int. Crim. Police Rev, 1979, 38–47.
70. A. Basch, Y. Margalit, S. Abramovich-Bar, Y. Bamberger, D. Daphna, T. Tamiri, and S. Zitrin, J. Energetic Mater., 4 (1986) 77–91.
71. D.W. Berberich, R.A. Yost, and D.D. Fetterolf, J. Forensic Sci., 33 (1988) 946–959.

17
Gas Chromatography–Mass Spectrometry Analysis of Flavors and Fragrances

M. Careri and A. Mangia
University of Parma, Parma, Italy

1. INTRODUCTION

Gas chromatography (GC) offers high sensitivity and the ability to separate the complex mixtures of food aroma compounds. The direct coupling of GC and mass spectrometry (MS) has aroused considerable interest in flavor research, this technique becoming the major instrumental technique in volatile component identification. In order to be useful in combination with GC, the mass spectrometer should have scan capabilities with sufficient sensitivity to yield a mass spectrum with roughly 1 to 10 ng of component when operated in the electron ionization (EI) mode. Newer mass spectrometers designed specifically for GC–MS use may be of either magnetic sector or quadrupole design. Nowadays, commercial GC–MS instruments are also based on ion-trap technology; multiple-stages of MS–MS experiments can be performed with this mass analyzer, making ion-trap MS particularly attractive in structure elucidation in flavor-related studies. Time-of-flight (TOF) systems are also available, but have not received as much developmental effort as the other two major designs. A number of limitations in the identification exist when using the mass spectrometer as a GC detector in flavor research. As a consequence of the very high data output rate, the production of several hundred mass spectra can be obtained during a single GC–MS analysis, which necessitates a suitable data handling system. Generally, most of the volatile compounds in a food can be readily identified by comparing individual spectra with those in a reference spectrum library. A difficulty that can be encountered is certainly the differentiation of positional isomers and Z or E double-bond isomers on the basis of their mass spectra. In this case, if all possible isomers having similar spectra are available, univocal identification can be performed by using GC relative retention time measurements.

In the last decades, the analysis and identification of volatile compounds in a large variety of foods have been greatly assisted by the development of powerful analytical techniques, such as GC–MS. As a result, a great deal of information on flavor components has been obtained in recent years for a number of food products.

Gas chromatography–mass spectrometry has also been demonstrated to be a powerful tool in both the qualitative and quantitative analyses of essential oil components. Literature on capillary and enantioselective capillary GC on-line coupled to isotope-ratio MS (IRMS) has been reviewed by Mosandl [1]. These techniques, which are valuable tools in control of authenticity of flavors and essential oils, are discussed in relation to various aspects, such as sample preparation and cleanup, chromatographic behavior of enantiomers, detection systems, and the use of internal isotope standards for quantitation.

Very recently, Bicchi et al. [2] compiled an overview on the use of derivatized cyclodextrins as chiral selectors for GC separation of enantiomers in essential oils and flavors. Analytical techniques such as capillary GC and GC–MS, two-dimensional GC, GC–IRMS, two-dimensional GC–IRMS, and liquid chromatography–capillary GC (LC–GC) were reviewed with respect to their applications to the enantiomeric separation.

A comprehensive overview of the techniques most commonly used for instrumental analysis of flavor compounds in food has been recently reported [3]. Several methods used for sample treatment are described, as well as the following techniques for extraction prior to GC analysis: solvent extraction and distillation techniques, headspace methods, and solid-phase microextraction. The use of GC–olfactometry and of ion-trap MS in food aroma analysis is also described.

Among the various GC detection systems used for analyzing aroma compounds, the use of MS has been reported for qualitative and quantitative analyses of flavor compounds in milk and dairy products [4].

As for lipid foods, reviews of the methodology for the isolation and concentration of odor/flavor components in these products have been published [5,6]. Gas chromatography–mass spectrometry analysis was used to identify and quantify a number of volatiles known to relate to good-quality olive oil flavor [6], such as aldehydes, esters, hydrocarbons, ketones, furans and other substances.

This chapter presents an overview of selected publications, that appeared between 1992 and 1999 dealing with the application of GC–MS to the analysis of flavors and fragrances with emphasis on the aroma sampling techniques. In addition, some examples of the role of this coupled technique in this field are more deeply discussed.

2. SURVEY ON GAS CHROMATOGRAPHY–MASS SPECTROMETRY APPLICATIONS TO THE CHARACTERIZATION OF FLAVORS AND FRAGRANCES

2.1. Gas Chromatography–Mass Spectrometry of Flavors

Several techniques for the removal of volatile compounds associated with flavor from foods have been reported: dynamic headspace (DHS) and purge and trap (PT) [7–9], solvent extraction [10], high-vacuum distillation and steam distillation [11], simultaneous steam distillation–extraction (SDE) [8,11,12], supercritical fluid extraction (SFE) [13–15], and solid-phase microextraction (SPME) [16,17].

The aroma compounds of dry-cured Parma ham have been analyzed by thermal-desorption GC–MS after extraction by means of the DHS technique [7]. Using GC–MS, 122 substances, including hydrocarbons, aldehydes, alcohols and esters, were identified in the volatile fraction. The same research group used DHS and SDE techniques to isolate aroma components from Parmesan cheese [8]. By means of GC–MS, more than 100 substances were characterized in the extracts. This application is discussed later in this chapter in section 3.

Volatile compounds present in 14 commercial olive oils and cooking oils (corn, vegetable, and rapeseed oils) have been analyzed by GC–MS [9]; aroma sampling was carried out by PT. This method proved to be effective for the quality control of these oils during production and may give a chromatographic pattern as a useful fingerprint for determination of origin.

Using GC–MS, interesting data on the chemistry of Australian honey volatiles have been obtained by D'Arcy et al. [10]. The aroma compounds of Australian blue gum (*Eucalyptus leucoxylon*) and yellow box (*Eucalyptus melliodora*) honey samples were isolated by solvent (ethyl acetate) extraction. This chemical fingerprint procedure was demonstrated to be promising for identifying compounds that are useful for authenticating the floral origin of honeys. In fact, a variety of distinctive nor-isoprenoids, monoterpenes, benzene derivatives, aliphatic compounds, and Maillard reaction products were identified among the natural honey volatiles, some of these being floral source descriptors for Australian honeys.

Three different well-established sampling techniques, i.e., high-vacuum distillation steam distillation and SDE, have been investigated for the isolation of volatile compounds derived from raw and roasted earth-almond [11]. Using GC–MS, it was possible to identify the indicator compounds for evaluation of the degree of earth-almond roasting. The main volatiles characterizing raw earth-almond were alcohols, whereas in the roasted earth-almond aroma, furans, pyrazines and pyrrols prevailed, as formed by the Maillard reaction during the roasting process.

In a study to obtain information on the effect of γ-irradiation on aroma of fresh mushrooms, the SDE technique followed by GC and GC–MS was applied to qualitatively and quantitatively analyze volatile compounds of nonirradiated mushrooms and samples that had been γ-irradiated with doses of 1, 2, and 5 kGy [12]. The important finding of this investigation was the large reduction of the content of total volatiles, primarily of eight-carbon compounds, due to the irradiation process, resulting in a flavor loss.

Supercritical fluid extraction as a sample preparation technique in food analysis has become increasingly popular in recent years. The usefulness of using supercritical fluids to investigate wine aroma has been demonstrated by both off-line SFE and coupled SFE–GC [13].

A fast and efficient method based on the use of SFE and GC–MS for the analysis of flavor compounds from roasted peanuts has been recently developed [14]. A group of substances known to be related to roasted flavor were identified and quantified using the full-scan mode. Total ion chromatograms of extracts of two peanut samples roasted under different conditions are shown in Figure 1. The GC–MS analysis confirmed the presence of pyrazine compounds and other flavor substances, such as methylpyrrole, that are associated with the roasting conditions and sensory perceptions of a taste panel.

Supercritical fluid extraction combined with high-resolution GC–MS proved to be a powerful tool for the analysis of the virgin olive oil aroma [15]. The volatiles identified were compared with those obtained by using the DHS method. Different aromatic profiles were obtained by applying the two extraction procedures. The profiles obtained by DHS–GC–MS corresponded to a genuine extra-virgin olive oil sample in accordance with previous findings [18]. The presence of off-flavors was not detected. In the SFE extracts, however, markers of oxidation processes were identified, since this technique is also suitable for the extraction of semivolatile compounds. These were volatile compounds related to oxidation of linoleic, linolenic, and oleic acids, and in particular aldehydes and acids, which had been previously found in oxidized olive oil samples [19].

Solid-phase microextraction has been investigated for the analysis of 2,4,6-trichloroanisole, a cork taint compound, in wine samples [16]. This solvent-free procedure was coupled to GC–MS under selective ion monitoring (SIM) conditions using a fully deuterated internal standard ([^{2}H$_5$]trichloroanisole) for quantitative purposes. The SPME–GC–MS method was demonstrated to be selective, precise, and sensitive with a 5 ng L^{-1} limit of quantification.

Solid-phase microextraction has been applied to the analysis of volatile substances in apples. Compounds were separated by time-compressed GC using time-of-flight MS for detection [17]. Time-compressed GC was proposed for reduction of the time required for separation without loss in analytical performance. In fact, unknown compounds, even when not well separated chromatographically, could be characterized by their mass spectra. Time-compressed GC combined

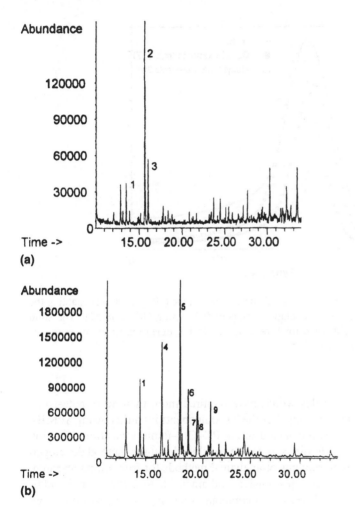

Figure 1 Peak profile for extract from roasted peanuts: (a) mild conditions: 145°C, 3 minutes, and (b) severe conditions: 170°C, 17 minutes. Peak notation: 1, methylpyrrole; 2, hexanol; 3, hexanal; 4, methylpyrazine; 5, 2,6-dimethylpyrazine; 6, furancarboxalde-hyde; 7, 2,3,5-trimethylpyrazine; 8, 2-ethyl-5-methylpyrazine and 2-ethyl-6-methylpyrazine; 9, 3-ethyl-2,5-dimethylpyrazine. (From Ref. 14.)

with time-of-flight MS allows analysis of dozens of components in very few minutes owing to the extremely fast spectral generation rates (up to 500 spectra/sec) [20]. The potential of this technique is illustrated in Figure 2. Using time-compressed GC, apple aroma compounds containing 1 to 15 carbons were eluted in about 2 minutes (Table 1). Performing traditional PT–GC analysis, the separa-

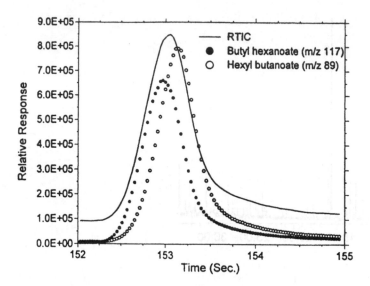

Figure 2 Demonstration of high speed spectral generation (40 spectra/sec) enabling the detection and quantification of coeluting compounds by using GC–TOF-MS. The solid line represents the reconstructed total ion current (RTIC). Retention times differ by approximately 0.2 sec. (From Ref. 17.)

tion of the collected volatiles would have required more than 40 minutes, i.e., 20 times longer. In addition, only 5 of the 29 components identified using SPME–GC–time-of-flight-MS were detected in the PT–GC–flame ionization detection (FID) profile. From these findings, it can be concluded that the method developed proved to satisfy the requirements of analytical methods for fruit aroma research, since it is rapid, solvent-free, inexpensive, and amenable to automation. In addition, SPME provided good linearity for compounds ranging in concentration from ppb to ppm without influence from matrix changes.

Regarding quantitation of aroma substances by GC–MS, a quantitative assay of ultratrace compounds is particularly demanding. Gas chromatography–mass spectrometry using deuterium-labeled methoxypyrazines allowed identification and quantitation of naturally occurring grape methoxypyrazines (Scheme 1) in a variety of red wines, at the low-ng level [21]. These compounds are potent flavorants contributing to the aroma of natural products such as peas and bell peppers [22], and are responsible for the herbaceous/vegetative aroma of wine. Stable isotope dilution GC–MS under chemical ionization (CI) conditions was used to analyze 2-methoxy-3-(2-methylpropyl)pyrazine in nine Australian and three New Zealand Cabernet Sauvignon wines (mean concentration 19.4 ng/L) and in six blended Bordeaux wines (mean concentration 9.8 ng/L). In all of the

Table 1 Volatile Compounds in Golden Delicious Apples, Sampled with SPME (PDMS, 100 μm) and Identified by Time-of-Flight MS

		Retention time (min)	
Peak	Volatile compound	SPME	Purge and trap
1	Pentane	1.082	
2	Acetone	1.132	
3	1-Butanol	1.322	
4	Propyl acetate	1.494	
5	Propyl propanoate	1.524	
6	Butyl acetate	1.619	13.45
7	Ethyl 2-methylbutanoate	1.639	13.80
8	2-Methylbutyl acetate	1.753	15.80
9	Propyl butanoate	1.802	
10	Butyl propyrate	1.932	
11	Pentyl acetate	2.005	
12	Butyl 2-methyl butanoate (?)	2.105	
13	Butyl butanoate	2.164	
14	Hexyl acetate	2.396	31.34
15	Butyl 2-methyl butanoate (?)	2.437	
16	Pentyl butanoate	2.460	
17	Hexyl propyrate	2.492	
18	Propyl 2-methyl-2-butenoate	2.650	
19	Hexyl 2-methylpropyrate	2.855	
20	Not identified	2.893	
21	Hexyl butanoate	3.183	
22	Butyl hexanoate	3.202	
23	4-Methoxyallylbenzene	3.223	
24	Hexyl 2-methylbutanoate	3.264	32.76
25	2-Methylbutyl hexanoate	3.395	
26	Hexyl pentanoate	3.633	
27	2-Methylpropyl 2-methylbutanoate	3.754	
28	Hexyl hexanoate	4.002	38.82
29	α-farnesene	4.453	41.75

Source: Modified from Ref. 17.

wines considered, the determined concentrations of this pyrazine were above the sensory detection threshold in water (about 2 ng/L), thus proving its contribution to wine flavor. As for its isomer, 2-methoxy-3-(1-methylpropyl)pyrazine, its level was less than 2 ng/L, the sensory detection threshold being about 1 ng/L in water, suggesting that this component will seldom significantly influence the wine aroma.

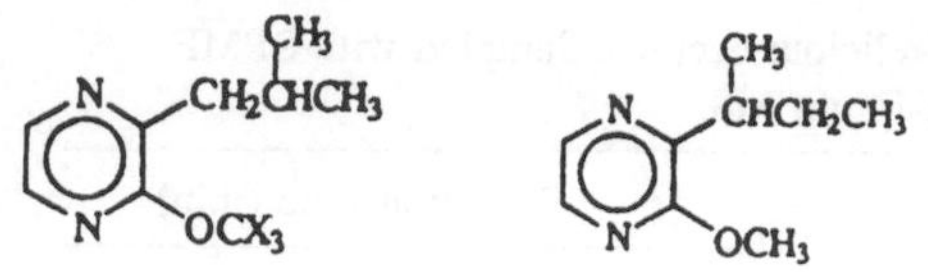

Scheme 1

2.2. Gas Chromatography–Mass Spectrometry of Fragrances

Essential oils are complex mixtures of flavor and fragrance compounds originating from plants. The analysis of essential oils, which consist mainly of mixtures of monoterpene and sesquiterpene hydrocarbons and their oxygenated derivatives, often requires a prefractionation before the chromatographic analysis. Otherwise, identification is difficult even with the use of GC–MS, since mass spectra of the components of the same class present very similar patterns. Consequently, in order to unequivocally identify each essential oil constituent, it is necessary to obtain the spectrum of an extremely pure substance. Among the on-line methods of fractionation, LC–GC allows the separation and analysis of complex mixtures in fully automated mode. Applications of this technique to citrus oil were recently reported [23,24]. The LC–GC–MS technique was proposed for the analysis of monoterpenes, sesquiterpenes, and their oxygenated derivatives in bergamot essential oil [23], and mono- and sesquiterpenes in the volatile fraction of eight essential oils: bergamot, lemon, mandarin, sweet orange, bitter orange, grapefruit, clementine, and Mexican lime [24]. After on-line preseparation of the oil by LC into compound classes and further separation by capillary GC, component identification was carried out by MS, using ion-trap detection (ITD). The advantage of using this on-line technique lies in the ability to obtain mass spectra of the components of a fraction eliminating the interferences that occur when the whole essential oil is analyzed.

In the field of essential oils, separation of individual enantiomers and determination of enantiomer excess play an important role in the characterization of plant material, in the investigation of the origin of the essential oil, and in the search of possible adulterations. Reliable assessment of the genuineness of essential oils is a difficult task, since synthetic analogs of essential oil components are commercially available. Therefore, suitable specific methods in the authenticity control of essential oils are of fundamental interest. Two-dimensional GC is a useful technique for both analytical and preparative purposes. After a separation of aroma components on a first column, they are on-line transferred to a second column coated with a different stationary phase for further separation. Enantioselective two-dimensional GC is based on the combination of a nonchiral precolumn and a chiral column, thus proving useful in evaluating origin-specific enantiomeric ratios and for differentiating natural flavor and fragrance substances from those of synthetic origin. By using this system (Fig. 3), the enantiomeric distribution of monoterpene hydrocarbons (β-pinene, sabinene, limonene) and monoterpene alcohols (linalool, terpinen-4-ol, α-terpineol) has been determined in order to evaluate the genuineness of mandarin essential oils [25]. Besides the characterization of mandarin oil, the method developed could afford the determination of extraneous oils added to or contaminating the oils.

Other authors determined the enantiomeric excess of limonene and limonene-1,2-epoxide by two-dimensional GC and GC–MS with trimethyl-γ-cyclodextrin as chiral selector [26]. They found (R)-(+)-limonene with an enantiomeric purity between 97.1 and 97.4% and (1S,2R,4R)-(+)-limonene-1,2-epoxide with an enantiomeric purity between 88.0 and 91.9%.

On-line two-dimensional GC–MS using a combination of J&W DB-Wax and heptakis(2,6-di-O-methyl-3-O-pentyl)-β-cyclodextrin columns has been proposed in a study of enantiodifferentiation of edulans I and II, the key flavor components of the purple passion fruit [27]. The investigation was applied to a variety of purple passion fruit extracts and distillates of different origins (Kenya, Chile and Ivory Coast). In all samples analyzed, the 2S enantiomers prevailed, as shown in Figure 4.

The results of enantio-two-dimensional GC–MS in authenticity analysis of geranium oil have been recently described [28]. This essential oil, which is obtained by distillation of leaves of *Pelargonium* species, is used in the fragrance industry and also as a flavoring material in foods and beverages. An important olfactive ingredient of the oil is rose oxide (*cis-/trans*-2-(2-methyl-1-propenyl)-4-methyltetrahydropyran). Using SPME and enantio-two-dimensional GC–MS, the mechanistic aspects of rose oxide formation using mixed specifically labeled precursors were studied. These findings proved helpful for authenticity assessments of geranium oils.

Enantioselective GC–IRMS and two-dimensional GC–IRMS represent other powerful techniques in the authenticity control of natural flavors and fragrances; using IRMS, identical δ^{13}C ratios are expected for enantiomers from genuine substances. Interesting investigations have been recently accomplished by using GC–IRMS in the authenticity control of essential oils [29–31].

Authenticity assessment of genuine substances of coriander oil (*Coriander sativum* L.) has been the object of a work dealing with the analysis of 10 authentic coriander essential oil samples of different origins [29]. The techniques used were GC–IRMS and enantioselective two-dimensional GC using a chiral cyclodextrin derivative as the stationary phase. The enantiomer ratio and the δ^{13}C values of 12 characteristic components were compared with those of commercially available

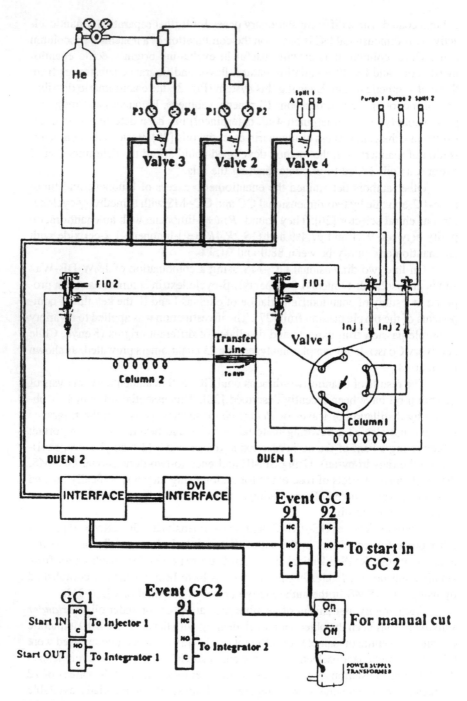

Figure 3 Pneumatic and electronic scheme of the GC–GC system. (From Ref. 25.)

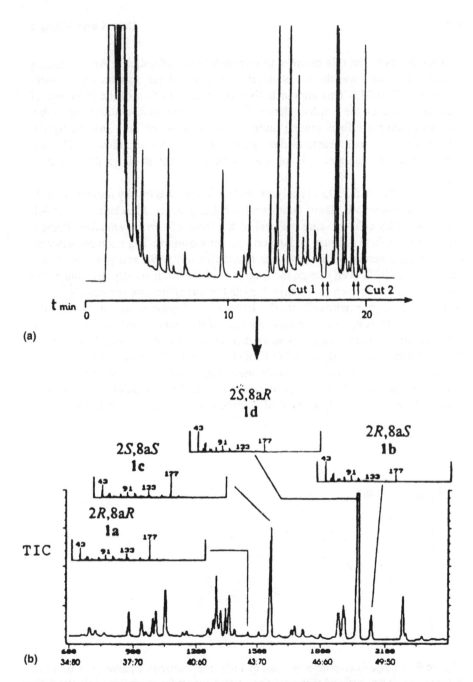

Figure 4 Two-dimensional GC–MS enantiodifferentiation of edulans 1a to d in an extract of purple passion fruit from Chile: (a) preseparation on achiral column (J&W DB-Wax), cut 1, 16.8 to 17.1 minutes, cut 2, 19.0 to 19.3 minutes; (b) separation on a chiral column [heptakis(2,6-di-O-methyl-3-O-pentyl)-β-cyclodextrin]. (From Ref. 27.)

species and essential oils in order to establish their authenticity. An interesting aspect of this work was the use of suitable internal isotopic standards to perform capillary GC–IRMS. This approach allowed the elimination of the influence of isotopic effects on the δ values during CO_2 fixation by photosynthesis in such a way that isotopic effects among authentic substances are only limited by the influence of enzymatic reactions during secondary biogenetic pathways. The isotopic fingerprint characteristic of the coriander essential oil is illustrated in Figure 5.

Faulhaber et al. [30,31] analyzed mandarin essential oils by measuring $\delta^{13}C$ ratios of characteristic flavor compounds of this product. Mandarin oils, which are obtained by cold-pressing the peel of the fruits of *Citrus reticulata* Blanco, are used in the food industry and in perfume compositions. The main constituents of mandarin essential oil are limonene (approximately 69%) and γ-terpinene (approximately 20%), even though the fragrance of this oil is mostly determined by minor components such as methyl *N*-methylanthranilate (approximately 0.4%) and α-sinensal (approximately 0.3%). Evaluation of genuineness of this product is of special interest, since synthetic analogs of the essential oil components are commercially available. In this context, measurements of $\delta^{13}C$ and $\delta^{15}N$ of methyl *N*-methylanthranilate [30] and of $\delta^{13}C$ of characteristic flavor components in this oil proved helpful in the authenticity assessment of cold-pressed mandarin oil [31]. The characteristic profile of mandarin essential oil so established could be applied to the authenticity control of commercially available mandarin oils.

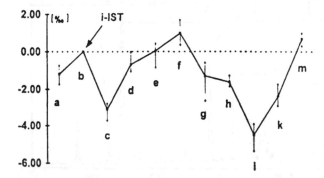

Figure 5 Isotope fingerprint of the essential oils from authentic samples ($n = 10$) of *C. sativum* L. with maximal and minimal values; γ-terpinene is the internal isotopic standard: limonene (a), γ-terpinene (b), *p*-cymene (c), linalol (d), geraniol (e), myrcene (f), geranyl acetate (g), β-pinene (h), camphene (i), terpinolene (k), and sabinene (m). (From Ref. 29.)

3. SELECTED APPLICATION EXAMPLES OF GAS CHROMATOGRAPHY–MASS SPECTROMETRY IN FLAVOR RESEARCH

3.1. Aroma Profile of Parmigiano-Reggiano Cheese

The flavor profile is one of the most significant parameters for the characterization of a food product. A representative example is the definition of the aroma profile of a typical Italian product such as the Parmigiano-Reggiano cheese. Characteristic features of this product are defined area of origin, strictly and traditionally defined production technology, and rules for the feeding of the animals.

A comprehensive study of the volatile fraction of this food has been carried out on a significant number of samples of certified origin and aging, coming from different geographical zones of the typical production area and representative of all seasonal productions [8,32]. For the characterization of an aromatic fingerprint, sampling is a crucial step, which has to be accurately evaluated. To obtain a complete chemical definition of Parmigiano-Reggiano cheese aroma, two different sampling techniques were used, namely DHS and SDE techniques. With the DHS procedure, the most volatile compounds are preferentially collected, whereas the use of SDE also provides interesting information about long-chain components of the cheese volatile fraction. These compounds greatly contribute to the flavor of the product.

Twenty-one samples of aged (24 mo) Parmigiano-Reggiano cheese were obtained from producers in different zones of the production area. Six samples were produced during the winter and five during each other season. The analyses were carried over a period of 1 yr, in such a way that all samples were at the same aging level. For each determination, samples were taken at the center of 1-kg cheese block.

The SDE procedure was performed using a micro-version of a Likens-Nickerson apparatus [33] in the configuration for heavier-than-water solvents. The aroma components were extracted by steam distillation and the aqueous distillate was simultaneously extracted with dichloromethane. The organic extracts were analyzed by GC–FID and GC–MS. The DHS technique was carried out with adsorption on Tenax traps and thermal desorption with cryofocusing of the volatile substances into the GC capillary column.

For the chromatographic separations, a J&W DB-Wax fused-silica capillary column was used. The identification was carried out by GC–MS in the EI mode. The MS identifications were confirmed by comparison of the mass spectra obtained to mass spectra in the National Bureau of Standards (NBS) library and to those of authentic substances, where possible. The compounds identified in the DHS and SDE extracts are listed in Tables 2 and 3, respectively [8]. From the comparison of the GC patterns, noticeable differences in composition can be observed in the vola-

Table 2 Volatile Constituents of Parmigiano-Reggiano Cheese (Headspace)

Peak no.[a]	Compound	ID[b]	Occurrence[c]
1	n-Hexane	MS, RT	21
2	2-Methylhexane	MS	21
3	n-Heptane	MS, RT	21
4	Methyl cyclohexane	MS	18
5	Acetone	MS, RT	21
6	2-Methylpropanal	MS	21
7	n-Octane	MS, RT	10
8	Methyl acetate	MS, RT	8
9	1-Octene	MS, RT	4
10	Tetrahydrofuran	MS, RT	21
11	Tetrachloromethane	MS	15
12	(?)-Octene	MS	21
13	Ethyl acetate	MS, RT	21
14	2-Butanone	MS, RT	21
15	2-Methylbutanal	MS	21
16	3-Methylbutanal	MS, RT	21
17	3-Methyl-2-butanone	MS, RT	21
18	Benzene	MS, RT	21
19	Ethanol	MS, RT	21
20	2-Ethylfuran	MS, RT	21
21	Ethyl propanoate	MS, RT	5
22	2-Pentanone	MS, RT	21
23	Pentanal	MS	1
24	Methyl butanoate	MS, RT	21
25	2,2-Dimethyldecane	MS	5
26	Chloroform	MS	21
27	Isobutylacetate	MS, RT	1
28	Toluene	MS, RT	21
29	(R)-2-butanol	MS, RT	21
30	Ethyl butanoate	MS, RT	21
31	1-Propanol	MS, RT	18
32	Dimethyl disulfide	MS	21
33	Isopropenyl acetate	MS, RT	1
34	Butyl acetate	MS	1
35	Hexanal	MS, RT	16
36	2-Hexanone	MS, RT	21
37	2-Methyl-1-propanol	MS, RT	21
38	Ethylbenzene	MS, RT	12
39	Propyl butanoate	MS, RT	17
40	Ethyl pentanoate	MS, RT	20
41	Isopropylbenzene	MS	13
42	p-Xylene	MS, RT	19

Table 2 Continued

Peak no.[a]	Compound	ID[b]	Occurrence[c]
43	2-Pentanol	MS, RT	21
44	m-Xylene	MS, RT	17
45	1-Butanol	MS, RT	21
46	o-Xylene	MS, RT	5
47	2-Heptanone	MS, RT	21
48	Methyl hexanoate	MS, RT	19
49	Limonene	MS, RT	16
50	4-Methyl-1-hexene	MS	7
51	3-Methyl-1-butanol	MS, RT	21
52	Butyl butanoate	MS, RT	13
53	Ethyl hexanoate	MS, RT	21
54	3-Methyl-3-buten-1-ol	MS, RT	21
55	1-Pentanol	MS, RT	18
56	4-Hydroxy-3-propyl-2-hexanone	MS	3
57	Acetoin	MS, RT	21
58	1-Hydroxy-2-propanone	MS, RT	16
59	1-Methylvinylbenzene	MS	19
60	Propyl hexanoate	MS	13
61	2,6-Dimethylpyrazine	MS, RT	21
62	2-Heptanol	MS, RT	19
63	Ethyl heptanoate	MS, RT	18
64	1,3-Butanediol	MS	10
65	1-Hexanol	MS, RT	20
66	Dimethyl trisulfide	MS	21
67	2-Nonanone	MS, RT	21
68	Nonanal	MS, RT	9
69	Methyl octanoate	MS, RT	5
70	2-Butoxyethanol	MS	20
71	Ethyl octanoate	MS, RT	21
72	Acetic acid	MS, RT	21
73	2-Ethyl-1-hexanol	MS	12
74	Benzaldehyde	MS, RT	21
75	Tetramethylurea	MS	4
76	Propanoic acid	MS, RT	17
77	1-Octanol	MS	2
78	Ethyl nonanoate	MS	1
79	2-Methylpropanoic acid	MS	8
80	2,3-Butanediol	MS	6
81	Benzonitrile	MS	10
82	2-Propoxyethanol	MS	2
83	1-Methoxy-2-propanol	MS	6
84	Methyl decanoate	MS	3

Table 2 Continued

Peak no.[a]	Compound	ID[b]	Occurrence[c]
85	2-Undecanone	MS, RT	21
86	1,2-Propanediol	MS	3
87	Butanoic acid	MS, RT	21
88	2-Hydroxybenzaldehyde	MS	3
89	Acetophenone	MS, RT	20
90	Ethyl decanoate	MS, RT	21
91	Furfuryl alcohol	MS, RT	21
92	3-Methylbutanoic acid	MS	21
93	Dimethyl tetrasulfide	MS	16
94	Naphthalene	MS	18
95	3-Methyl-2-(5H)-furanone	MS	6
96	Pentanoic acid	MS, RT	21
97	2-Phenyl-2-propanol	MS	4
98	Acetamide	MS	11
99	2-(2-Butoxy)-ethoxyethanol	MS	9
100	1-methylnaphthalene	MS	3
101	Hexanoic acid	MS, RT	21
102	Geranyl acetone	MS	9
103	Tetramethyl thiourea	MS	9
104	2-Ethylhexanoic acid	MS	5
105	Benzothiazole	MS	15
106	Heptanoic acid	MS, RT	21
107	1-Dodecanol	MS, RT	10
108	Phenol	MS	19
109	Octanoic acid	MS, RT	17
110	Nonanoic acid	MS, RT	18

[a] For peak sequences see Ref. 8.
[b] MS, Mass spectrum of unknown identical to that in literature; RT, agreement of retention time with authentic compound.
[c] Occurrence of identification in 21 samples.
Source: Modified from Ref. 8.

tile fraction collected using the two techniques [8]. As expected, DHS sampling enables the detection of the most volatile compounds, such as acetone, ethyl acetate, ethanol, and ethyl propionate. In contrast, the chromatograms of the SDE extracts contained abundant signals corresponding to the less volatile compounds. Among the 110 compounds identified in the DHS fraction, hydrocarbons, alcohols, esters, and ketones predominated, while free fatty acids were the most abundant constituents of the aroma SDE extracts, in which 105 components were identified.

Table 3 Volatile Constituents of Parmigiano-Reggiano Cheese (SDE)

Peak no.[a]	Compound	ID[b]	Occurrence[c]
1	2-Pentanone	MS, RT	21
2	Diacetyl	MS	21
3	Methyl butanoate	MS, RT	21
4	Chloroform	MS	21
5	Toluene	MS, RT	21
6	(E)-2-butenal	MS, RT	21
7	Ethyl butanoate	MS, RT	21
8	(Z)-2-butenal	MS	19
9	1-Propanol	MS, RT	18
10	3-Hexanone	MS, RT	20
11	Dimethyl disulfide	MS	21
12	Hexanal	MS, RT	16
13	2-Hexanone	MS, RT	21
14	2-Methyl-1-propanol	MS, RT	21
15	3-Penten-2-one	MS, RT	16
16	Ethyl benzene	MS, RT	12
17	Propyl butanoate	MS, RT	17
18	2-Pentanol	MS, RT	21
19	m-Xylene	MS, RT	17
20	1-Butanol	MS, RT	21
21	3-Penten-2-ol	MS, RT	15
22	2-Heptanone	MS, RT	21
23	Methyl hexanoate	MS, RT	19
24	2-Pentenal	MS, RT	19
25	Limonene	MS, RT	16
26	3-Methyl-1-butanol	MS, RT	21
27	Butyl butanoate	MS, RT	13
28	Ethyl hexanoate	MS, RT	21
29	3-Methyl-3-buten-1-ol	MS, RT	21
30	1-Pentanol	MS, RT	18
31	2-Vinyl-2-butenal	MS	10
32	Acetoin	MS, RT	21
33	2-Octanone	MS, RT	21
34	Methyl heptanoate	MS, RT	6
35	1-Hydroxy-2-propanone	MS, RT	21
36	(Z)-2-heptenal	MS, RT	6
37	Propyl hexanoate	MS	13
38	2,6-Dimethylpyrazine	MS, RT	21
39	2-Heptanol	MS, RT	19
40	Ethyl heptanoate	MS, RT	18
41	1,3-Butanediol	MS	17
42	2-(Chloromethyl)furan	MS	5

Table 3 Continued

Peak no.[a]	Compound	ID[b]	Occurrence[c]
43	1-Hexanol	MS, RT	20
44	Dimethyl trisulfide	MS	21
45	2-Nonanone	MS, RT	21
46	Nonanal	MS, RT	9
47	2,4-Hexandienal	MS, RT	6
48	Isobutyl hexanoate	MS	6
49	Ethyl octanoate	MS, RT	21
50	3-Ethyl-2,5-dimethylpyrazine	MS	6
51	3-(Methyltiopropanal)	MS	21
52	Acetic acid	MS, RT	21
53	Furfural	MS, RT	20
54	Benzaldehyde	MS, RT	21
55	Propanoic acid	MS, RT	17
56	1-Hepten-4-ol	MS	14
57	2-Undecanone	MS, RT	21
58	Butanoic acid	MS, RT	21
59	Phenylacetaldehyde	MS, RT	21
60	Ethyl decanoate	MS, RT	21
61	Furfuryl alcohol	MS, RT	21
62	3-Methyl butanoic acid	MS	21
63	n-Heptadecane	MS, RT	21
64	(?-Dimethylbutanoic acid)	MS	21
65	Pentanoic acid	MS, RT	21
66	Unknown	MS	3
67	(2,6,10,14-Tetramethyl hexadecane)	MS	10
68	1-Decanal	MS	19
69	n-Octadecane	MS, RT	18
70	2-Tridecanone	MS, RT	21
71	Hexanoic acid	MS, RT	21
72	Long-chain carbonyl compound	MS	21
73	1-Undecanol	MS, RT	21
74	2,6-Bis(t-butyl)-4-methylphenol	MS	12
75	Tetradecanal	M, RT	17
76	Heptanoic acid	MS, RT	21
77	δ-Octalactone	MS	11
78	Long-chain carbonyl compound	MS	15
79	Long-chain carbonyl compound	MS	17
80	2-Pentadecanone	MS	21
81	Pentadecanal	MS	14
82	Ethyl tetradecanoate	MS, RT	18
83	Octanoic acid	MS, RT	21
84	Hexadecanal	MS	14

Table 3 Continued

Peak no.[a]	Compound	ID[b]	Occurrence[c]
85	γ-Decalactone	MS, RT	11
86	Nonanoic acid	MS, RT	21
87	δ-Decalactone	MS, RT	21
88	Ethyl hexadecanoate	MS	18
89	Decanoic acid	MS, RT	21
90	9-Decenoic acid	MS	21
91	γ-Dodecalactone	MS, RT	21
92	1-Hexadecanol	MS, RT	12
93	Undecanoic acid	MS	20
94	δ-Dodecalactone	MS, RT	21
95	Dodecanoic acid	MS, RT	21
96	Tridecanoic acid	MS	21
97	(?-Methyltridecanoic acid)	MS	18
98	δ-Tetradecalactone	MS	21
99	Tetradecanoic acid	MS, RT	21
100	9-Tetradecenoic acid	MS	21
101	(?-Methyltetradecanoic acid)	MS	17
102	(?-Methyltetradecanoic acid)	MS	21
103	Pentadecanoic acid	MS, RT	21
104	Hexadecanoic acid	MS, RT	21
105	9-Hexadecenoic acid	MS	20

[a] For peak sequence see Ref. 8.
[b] MS, Mass spectrum of unknown identical to that in literature; RT, agreement of retention time with authentic compound.
[c] Occurrence of identification in 21 samples. Tentative identification in parentheses.
Source: Modified from Ref. 8.

With regard to the different classes of substances, in the DHS extracts, hydrocarbons, which are secondary products of lipid auto-oxidation, ranging from *n*-hexane to 1-methylnaphthalene were identified. Linear long-chain C_{17} and C_{18} hydrocarbons were found in the samples obtained using the SDE method.

Aldehydes account for about 2% of the relative chromatographic peak area in both series of extracts [32]. Short-chain aldehydes can be formed from amino acids during cheese ripening via Strecker degradation [34]. In DHS samples, linear and branched-chain saturated aldehydes as well as two aromatic derivatives were identified. Unsaturated aldehydes were isolated in the SDE samples, together with long-chain compounds, such as tetradecanal, pentadecanal, and hexadecanal. Long-chain aldehydes derive from fatty acids by an α-oxidation mechanism [35]. In the SDE extracts of all 21 samples analyzed, mass spectral data evidenced the presence of a sulfur-containing aldehyde, 3-(methylthio)propanal, which had not previously been reported as a component of Parmigiano-Reggiano cheese aroma.

Ketones are the most abundant constituents of the Parmigiano-Reggiano aroma, accounting for 26% of the total chromatographic peak area of the DHS fraction [32]. 2-Alkanones are the most frequently found. Among them, 2-pentanone and 2-heptanone are the most abundant. Methyl ketones were isolated with both sampling methods used. These compounds, which can be formed as secondary products of the β-oxidation of fatty acids during lipolysis [36], play an important role in defining the flavor of this product.

Fatty acids were the most important components of the SDE extracts, accounting for 81.4% of the total chromatographic peak area (10.7% in the DHS extracts) [32]. In particular, long-chain fatty acids ($C_{10:0}$ to $C_{16:1}$) were isolated. Free fatty acids can derive from three biochemical pathways during the ripening of cheese: lipolysis, proteolysis, and lactose fermentation [37]. Free fatty acids deriving from lipolysis are the linear-chain homologues from C_4 to C_{18}. In addition to the even-numbered straight-chain acids, which are by far the most abundant acids in the fatty acid class, small amounts of odd-numbered branched-chain acids, such as isobutanoic and isovaleric acids, were found. These compounds derive from deamination of amino acids caused by proteolytic enzymes. Short-chain organic acids, e.g., acetic, propionic, and butanoic acids, found in different amounts using the two sampling techniques, derive from lactose fermentation. Free fatty acids are considered to greatly contribute to the flavor of aged cheese. In the samples examined, significant differences in the content of some acids were observed according to the season of production [32].

Among esters, ethyl acetate was the most abundant component. Ethyl esters of odd-numbered acids from C_2 to C_{16} predominated with respect to the methyl and butyl derivatives. Ethyl esters are considered to contribute to the typical fruity note of the cheese.

A number of lactones were isolated in the SDE samples, whereas, owing to their scarce volatility, they were not found using the DHS method. These compounds are known to contribute to the cheese flavor, being recognized as contributors to the pleasant aroma of butter [38]. Among lactones, the long-chain homologues (δ-decalactone and δ-dodecalactone) are quantitatively predominant. Long-chain lactones derive from the corresponding hydroxy acids (C_8 to C_{16}) by the loss of water [39].

Some sulfur compounds were also found, in particular dimethyl disulfide in the DHS fraction. These compounds, which derive from sulfur-containing amino acids by Strecker degradation, can be considered important contributors to the aroma of the product.

Finally, some heterocyclic compounds, such as alkylpyrazines and furans, were identified. These substances are important trace components of flavor of unprocessed and heated foods, as natural components or deriving from nonenzymatic Maillard reaction [40,41].

From the results obtained in this study, it can be inferred that the use of GC–MS in combination with adequate aroma sampling techniques can solve analytical problems of aroma characterization, even in the case of very complex matrices. The combined use of the two extraction procedures proved helpful in providing a complete fingerprint of the aged Parmigiano-Reggiano cheese aroma, which can be used as a useful reference for the characterization of the product.

Further, the results obtained enabled the correlation of the composition of the volatile fraction of Parmigiano-Reggiano cheese with sensory attributes [42]. This subsequent investigation evidences the significance of the flavor in defining the quality of a food product and, in addition, the contribution of the different volatile compound classes or of the individual substances to the sensory attributes. For example, esters, particularly methyl butanoate, ethyl hexanoate, and isobutyl acetate, are found to be positively related to the "fragrant" and "fruity" notes of aged cheese. Short-chain methyl-branched alcohols, secondary products of proteolysis, contribute to a positive "maturation" component of the aroma, in opposition to the sharp stimuli of free fatty acids (Fig. 6).

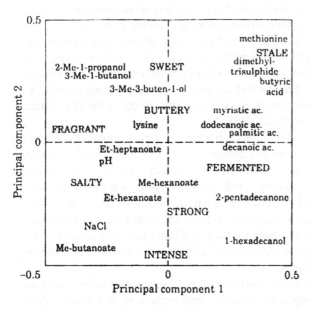

Figure 6 Principal components analysis of volatile, nonvolatile, and sensory data of Parmigiano-Reggiano cheese. Variable loadings on the plane of the first two components. (From Ref. 42.)

3.2. Characterization of Volatile Compounds in Wine-Stopper Cork

A different example of the potential of GC–MS for the analysis of flavors and off-flavors is represented by a series of studies dealing with the characterization of volatile compounds in cork used for the production of wine stoppers [43–45]. The approach chosen for these investigations was the application of DHS for aroma sampling and the use of GC–MS for the analysis of the volatile components of different cork samples in various analytical problems: (1) characterization of the volatile fraction in raw material (bark) and new and used cork wine stoppers [43], (2) characterization of volatile compounds produced by microorganisms isolated from cork [44], and (3) study of the effects of electron-beam irradiation on cork volatile compounds in a sterilization process [45].

Cork is a plant tissue produced from the cork cambium of the cork oak (*Quercus suber* L.). The use of cork for sealing wine bottles is very old and still widespread, particularly for quality wines and those intended for long-term storage. The quality of the cork can influence the organoleptic features of the wine. In fact, volatile substances deriving from microbiological activity can give rise to undesirable off-flavors (musty and moldy smells). In particular, processing and storage of corks under favorable conditions, e.g., high moisture levels, promote the growth of microflora (molds and bacteria) and consequently the production of metabolites having negative effects on wines coming in contact with this material. The main metabolites considered as causes of corkiness in wine are 2,4,6-trichloroanisole, 2-methylisoborneol, 1-octen-3-ol, 1-octen-3-one, and guaiacol [46]. Even the commonly used sterilization treatment with chlorinated solutions can produce chlorophenols and chloroanisoles. Considering these problems, analyses of volatile substances in cork and cork stoppers under different situations were carried out using GC–MS.

In a preliminary study, the volatile fraction of cork samples at different stages of processing (raw material and finished stoppers) was analyzed [43]. One hundred seven components were identified by GC–MS, several of which had not been previously reported in cork. Predominant volatiles were hydrocarbons, alcohols, acids, and carbonyl compounds. A natural origin, such as chemical or enzymatic degradation of the cork, could explain the occurrence of many substances. All the identified compounds with a phenyl ring and with a linear structure, e.g., vinylbenzene, benzyl alcohol, and 2-hydroxybenzaldehyde, could be derived from suberin and lignin.

The correlation between microbial growth and formation of volatile components in the cork was the object of a further investigation by the same authors [44]. The spores of 10 micro-organism strains were cultivated on malt extract (Oxoid) and in sterile water containing sterilized cork powder. The behaviors of the following 10 strains were studied: *Penicillium* (three strains), *Trichoderma* (two strains), *Aspergillus* (two strains), *Mucor*, *Monilia*, and *Streptomyces* (one strain each). In addition, one strain of *Botrytis cinerea* isolated from grapes was considered, taking into account that stoppers could be contaminated during storage in the cellar. The odors deriving from the 11 cultures on cork and on malt extract were defined by a sensorial panel. Correspondingly, the volatile components generated by the microbiological activity were identified by means of GC–MS. Compound-rich chromatographic profiles were obtained from the headspace of the strains cultured on malt. More than 60 compounds were identified, including several sesquiterpenes, 16 carbonyl compounds, 16 alcohols, some esters, and a few hydrocarbons. No carboxylic acids were detected. The cultures on cork gave rise to different aromatic profiles, only 12 compounds being present after microbiological growth. Among these, sesquiterpenes, exhibiting balsamic, fruity, or floral aromas, were synthesized by all microbial strains (Table 4). All strains grown on cork produced chlorobenzenes, whereas the same strains cultured on malt were less able to produce aromatic chlorinated compounds. 1-Octen-3-ol, an alcohol having a typical mushroom odor already noted in tainted cork stoppers [47], was produced by five strains cultured on cork.

To avoid the negative effects of microbiological activity on the quality of cork stoppers, electron-beam irradiation treatment is a valid alternative to the chemical sterilization of cork stoppers, based mainly on the use of chlorine or hydrogen peroxide. The effects of this process on the volatile fraction of

Table 4 Volatile Compounds Produced by Growth on Cork of the Strains Considered

Peak no.	Compounds	Frequency/11 strains
1	Chlorobenzene[a]	11
2	Octan-3-one[b]	8
3	Anisole[a]	11
4	1,3-Dichlorobenzene[a]	11
5	Oct-1-en-3-ol[a]	5
6	α-Santalene[b]	1
7	Caryophyllene[a]	8
8	α-Bergamotene[a]	11
9	β-Bergamotene[b]	10
10	Isocaryophyllene[b]	11
11	β-Bisabolene[b]	10
12	α-Curcumene[b]	8

[a] Identification based on the mass spectrum and confirmed by GC of authentic substances.
[b] Tentatively identified by the mass spectrum.
Source: Ref. 44.

the cork have been studied [45] by qualitative and quantitative comparisons of the composition of the volatile fraction of the nonirradiated cork with that of the same material submitted to three different doses of radiation (25, 100, and 1000 kGy). Figure 7 shows the chromatogram of a nonirradiated cork sample as an example of the aromatic profile of cork. The corresponding compound identifications are listed in Table 5. By comparing the nonirradiated cork samples with those irradiated at 25 kGy (Table 6), it can be noted that the differences involve only 10 substances. The content of light-sensitive terpenes (β-pinene, β-phellandrene, fenchone, isocamphone) greatly decreases, whereas a remarkable increase of some aliphatic hydrocarbons (1-octene, n-nonane, 1-decene, 1-undecene, 1-dodecene, 1-tridecene) is observed. These compounds likely derive from the radiolysis of fatty acids esterified with phenolic fragments present in the suberin [48]. The treatment at 100 kGy leads to an increase of almost all the volatiles identified. The largest increases are observed for aliphatic hydrocarbons and for carbonyl compounds. The increase of the content of carbonyl compounds is explained by radiation-induced oxidation in the presence of oxygen. In the samples treated at the highest dose (1000 kGy), the variation of the composition of the volatile fraction shows the same trend with generally more remarkable quantitative effects. Minor variations are observed for some terpenic substances, such as

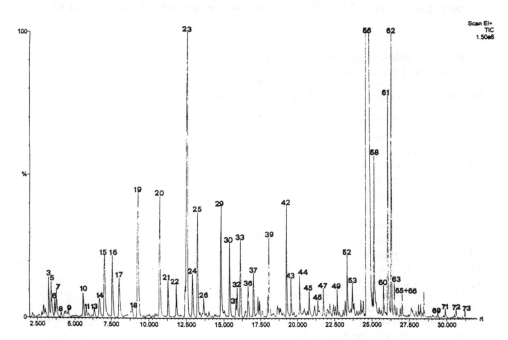

Figure 7 Total ion current gas chromatogram of nonirradiated cork sample. (From Ref. 45.)

Table 5   Volatile Compounds Identified in the Nonirradiated Cork Stoppers

Peak no.[a]	Compound	ID[b]	Calculated KI
1	Methyl acetate	MS, RT, KI	822
2	1-Octene	MS, RT	833
3	Ethyl acetate	MS, RT, KI	836
4	2-Butanone	MS, RT	841
5	n-Nonane	MS, RT	900
6	2-Methylbutanal	MS, KI	906
7	3-Methylbutanal	MS, RT, KI	909
8	3-Methyl-2-butanone	MS, RT	918
9	1-Nonene	MS, RT	930
10	Pentanal	MS, RT, KI	964
11	2-Pentanone	MS, RT	970
12	2-Methylpentanal	MS	975
13	α-Pinene	MS, KI	988
14	n-Decane	MS, RT	1000
15	α-Thujene	MS, KI	1007
16	Trichloromethane	MS, RT	1018
17	Toluene	MS, RT, KI	1026
18	1-Decene	MS	1043
19	Camphene	MS, RT, KI	1051
20	Hexanal	MS, RT, KI	1079
21	β-Pinene	MS, RT, KI	1089
22	n-Undecane	MS, RT	1100
23	Ethylbenzene	MS, RT, KI	1113
24	p-Xylene	MS, RT	1120
25	m-Xylene	MS, RT	1127
26	1-Methoxy-2-propanol	MS	1135
27	1-Chloroheptane	MS, RT	1138
28	1-Undecene	MS	1143
29	1-Butanol	MS, RT, KI	1157
30	o-Xylene	MS, RT	1169
31	2-Heptanone	MS, RT	1177
32	Heptanal	MS, RT, KI	1178
33	Limonene	MS, RT, KI	1183
34	β-Phellandrene	MS, KI	1189
35	(Z)-2-hexenal	MS	1189
36	Eucalyptol	MS	1193
37	n-Dodecane	MS, RT	1200
38	3-Methyl-1-butanol	MS, RT	1215
39	2-Pentylfuran	MS	1228
40	1-Chlorooctane	MS, RT	1243
41	1-Dodecene	MS, RT	1244
42	(?)-Methyl isopropylbenzene	MS	1259

Table 5   Continued

Peak no.[a]	Compound	ID[b]	Calculated KI
43	1,2,4-Trimethylbenzene	MS, RT, KI	1267
44	Octanal	MS, RT, KI	1283
45	n-Tridecane	MS, RT	1300
46	1,2,3-Trimethylbenzene	MS, KI	1321
47	6-Methyl-5-hepten-2-one	MS, RT, KI	1333
48	1-Tridecene	MS, RT	1344
49	1-Hexanol	MS, RT, KI	1347
50	(?)-Methyl isopropylbenzene	MS	1348
51	Fenchone	MS, KI	1382
52	Nonanal	MS, RT, KI	1389
52	n-Tetradecane	MS, RT	1400
54	(E)-2-octenal	MS, Ki	1422
55	(?)-Methyl isopropenylbenzene	MS	1429
56	Acetic acid	MS, RT, KI	1449
57	1-Octen-2-ol	MS, RT, KI	1454
58	Furfural	MS, RT, KI	1457
59	1-Heptanol	MS, RT, KI	1462
60	Copaene	MS	1487
61	2-Ethyl-1-hexanol	MS, RT, KI	1494
62	Camphor	MS, RT, KI	1510
63	Benzaldehyde	MS, RT, KI	1525
64	(E)-2-nonenal	MS, KI	1531
65	Isocamphone	MS	1542
66	Propanoic acid	MS, RT	1542
67	1-Octanol	MS, RT, KI	1557
68	Butanoic acid	MS, RT	1630
69	Acetophenone	MS, RT, KI	1647
70	2-, 3-Methylbutanoic acid	MS	1667
71	Isoborneol	MS, KI	1669
72	Borneol	MS, RT, KI	>1700
73	Naphthalene	MS, RT	>1700

[a] For peak sequence see Figure 7.

[b] MS, identification by comparison with mass spectrum stored in the NIST library; RT, identification by comparison with retention time of authentic reference compounds; KI, agreement with literature data.

Source: Ref. 45.

Table 6 Mean Peak Areas (Arbitrary Units) or Volatiles Identified in the Nonirradiated and Irradiated Cork Stoppers[a,b]

Compound	Nonirradiated stoppers	Irradiated stoppers		
		25 kGy	100 kGy	1000 kGy
Aliphatic hydrocarbons				
1-Octene	92 (26) a	200 (5) a	581 (19) a	3904 (23) b
n-Nonane	197 (36) a	580 (5) a	2018 (1) b	22832 (21) c
1-Nonene	341 (10) a	447 (31) a	1585 (8) a	9859 (9) b
α-Pinene	564 (15) a	466 (11) a	1326 (10) b	n.d.[c]
n-Decane	1240 (8) a	1225 (9) a	3243 (9) a	28245 (18) b
α-Thujene	4132 (12) a	3537 (8) a	8943 (6) b	3957 (20) a
1-Decene	44 (12) a	108 (8) a	535 (16) b	3940 (16) c
Camphene	8174 (15) a	6660. (11) a	16448 (6) b	7004 (13) a
β-Pinene	1837 (17) a	1096 (9) b	2140 (9) a	n.d.
n-Undecane	1375 (16) a	1645 (9) a	4414 (6) ab	45146 (14) b
1-Undecene	331 (18) a	864 (21) a	3685 (9) b	31429 (16) c
Limonene	3490 (15) a	2581 (12) a	10592 (9) b	n.d.
β-Phellandrene	119 (4) a	42 (8) b	78 (13) b	53 (18) b
n-Dodecane	2150 (17) a	1964 (17) a	7217 (29) b	34543 (17) b
1-Dodecene	381 (19) a	817 (17) a	3520 (12) b	26801 (14) c
n-Tridecane	1342 (18) a	1157 (17) a	3679 (13) a	42992 (13) b
1-Tridecene	264 (16) a	519 (15) a	2900 (9) b	23618 (11) c
n-Tetradecane	2318 (30) a	1384 (11) a	4715 (15) b	n.q.[d]
Copaene	1957 (36) a	1469 (19) a	5016 (15) b	2113 (33) a
Aromatic hydrocarbons				
Toluene	2326 (11) a	1421 (7) a	3111 (14) a	15441 (17) b
Ethylbenzene	17238 (17) a	4017 (15) b	21715 (17) a	23226 (21) a
p-Xylene	2486 (18) a	834 (12) a	3148 (9) ab	5611 (13) b
m-Xylene	5811 (28) ab	1926 (12) b	6649 (9) a	7523 (10) a
o-Xylene	3579 (17) a	1333 (13) a	5670 (11) ab	12416 (15) b
(?)-Methyl Isopropylbenzene	4548 (20) a	2587 (13) a	13130 (4) b	n.q.

Table 6 Continued

Compound	Nonirradiated stoppers	Irradiated stoppers		
		25 kGy	100 kGy	1000 kGy
1,2,4-Trimethylbenzene	1562 (18) a	1082 (15) a	2912 (12) a	15242 (14) b
1,2,3-Trimethylbenzene	727 (17) a	373 (17) a	991 (5) a	8144 (8) b
(?)-Methyl isopropylbenzene	747 (14) a	394 (17) a	974 (10) a	n.d.
(?)-Methyl isopropenylbenzene	761 (26) a	490 (18) a	23187 (16) b	n.q.
Naphthalene	315 (32) ab	162 (27) a	558 (13) b	590 (21) b
Alcohols				
1-Methoxy-2-propanol	1151 (30) ab	415 (11) b	1929 (22) a	n.d.
1-Butanol	2656 (42) a	1273 (9) a	3445 (7) ab	6696 (26) b
3-Methyl-1-butanol	454 (23) a	363 (6) a	1556 (4) b	n.d.
1-Hexanol	581 (27) a	292 (14) a	1400 (9) a	13354 (16) b
1-Octen-3-ol	1116 (33) a	867 (30) a	3312 (8) b	2192 (27) b
1-Heptanol	1041 (21) a	561 (19) a	2006 (8) a	7734 (30) b
2-Ethyl-1-hexanol	9959 (30) a	5203 (24) b	13341 (10) a	12165 (20) a
1-Octanol	806 (50) a	365 (6) a	2064 (18) a	5888 (33) b
Isoborneol	616 (40) a	298 (28) a	1143 (10) b	n.q.
Borneol	538 (36) ab	352 (23) a	837 (10) b	507 (27) a
Carboxylic acids				
Acetic acid	87300 (36) a	63858 (21) a	316000 (8) b	1750000 (12) c
Propanoic acid	123 (22) a	83 (13) a	529 (2) a	1616 (25) b
Butanoic acid	n.d.	n.d.	n.d.	1913 (30)
2-, 3-Methylbutanoic acid	n.d.	n.d.	n.d.	4868 (24)
Carbonyl compounds				
Aldehydes:				
2-Methylbutanal	660 (13) a	478 (3) a	1442 (14) b	10317 (23) c
3-Methylbutanal	1218 (12) a	976 (5) a	3126 (6) b	29013 (17) c
Pentanal	1237 (13) a	1352 (7) a	4067 (8) b	19767 (16) c
2-Methylpentanal	249 (20) a	529 (10) a	1007 (11) b	4997 (19) c
Hexanal	6469 (20) a	6618 (8) a	24200 (7) ab	58760 (15) b
Heptanal	1512 (17) a	919 (9) a	3738 (14) a	30363 (21) b

	Nonirradiated	25 kGy	100 kGy	1000 kGy
(Z)-2-hexenal	64 (18) a	61 (6) a	310 (9) b	1057 (10) c
Octanal	1837 (18) a	1252 (7) a	5048 (12) a	57511 (22) b
Nonanal	3638 (31) a	2828 (11) a	10706 (6) b	n.q.
(E)-2-octenal	797 (30) a	412 (11) a	3453 (15) b	n.q.
Benzaldehyde	1668 (30) a	1167 (19) a	4291 (12) ab	8126 (44) b
(E)-2-nonenal	501 (41) a	325 (15) a	2364 (14) b	2708 (34) b
Ketones:				
2-Butanone	1623 (21) a	779 (9) a	2589 (22) a	16056 (16) b
3-Methyl-2-butanone	200 (14) a	301 (13) a	599 (8) b	4707 (22) c
2-Pentanone	448 (19) a	278 (14) b	570 (4) a	n.q.
2-Heptanone	646 (24) a	304 (12) a	1911 (11) ab	n.q.
6-Methyl-5-hepten-2-one	1076 (24) a	2509 (9) a	10064 (11) b	63563 (13) c
Fenchone	859 (22) a	418 (17) b	1262 (13) a	n.q.
Camphor	10853 (31) a	6591 (19) a	18863 (9) a	9491 (33) a
Isocamphone	154 (21) a	84 (29) b	276 (6) c	89 (19) b
Acetophenone	272 (30) a	346 (12) ab	336 (5) ab	707 (22) b
Esters				
Methyl acetate	n.d.	n.d.	n.d.	7227 (33)
Ethyl acetate	1608 (19) a	1962 (6) a	12000 (15) b	7957 (15) c
Ethers				
Eucalyptol	1865 (20) a	1565 (11) a	5895 (12) b	n.d.
Furans				
2-Pentylfuran	3624 (23) ab	1685 (13) b	7412 (7) ab	12715 (48) b
Furfural	7363 (36) ab	2265 (16) a	4213 (12) a	13320 (37) b
Chlorinated hydrocarbons				
Trichloromethane	4330 (11) a	3610 (2) a	1478 (10) a	4208 (17) a
1-Chloroheptane	247 (17) a	194 (13) a	492 (9) b	4001 (16) b
1-Chlorooctane	451 (22) a	314 (7) a	1300 (14) a	24613 (15) b

[a] Means calculated on four samples of nonirradiated stoppers and on three samples of stoppers irradiated respectively at 25, 100, and 1000 kGy; numbers in parentheses, RSD, %.
[b] Means followed by the same letter in a line are not significantly different at the 0.05 level of significance.
[c] n.d., not detected.
[d] n.q., not quantified because of overlapping with the adjacent peak.
Source: Ref. 45.

α-thujene, camphene, copaene, borneol, and camphor, which appear to be very stable toward radiation. By contrast, other terpenes (α-pinene, β-pinene, limonene, eucalyptol) proved to be highly radiation-sensitive and were completely decomposed. While the content of the other volatiles increases regularly with increasing irradiation level, the content of all terpenes identified has a maximum increase at the 100-kGy level and a falloff at the 1000-kGy dose. Likely, the energy corresponding to the 100-kGy level is effective in releasing terpenes from the matrix, whereas at the highest doses, radiation-induced decomposition is prevalent.

4. CONCLUSIONS

Gas chromatography–mass spectrometry is a well-established technique in the field of analysis of very complex mixtures, as flavors and fragrances are. This is confirmed by the continuously growing literature examples and by a wide spectrum of applications dealing not only with the definition of the aromatic profile of a food or an essential oil, but also with technological aspects, such as shelf-life of food products, migration of volatile compounds from packaging materials, and compounds deriving from microbial activity.

On the other hand, some problems in the analysis of flavors and fragrances still exist. The most crucial aspects, in our opinion, are as follows:

1. Effective and reliable sampling procedures are required in terms of yields and reproducibility, using solvent extraction or solvent-free techniques.
2. It is necessary to correlate the composition of the volatile fraction of a product with sensory properties. In fact, less abundant components can be relevant to the aroma of a food. This problem can be solved by performing an adequate statistical treatment to correlate the GC–MS data to experimental sensory evaluations. The success of this approach strongly depends on the reliability of the sampling procedure used. In this context, a combined analytical-sensory technique, such as GC–olfactometry, can help to solve these problems.
3. Quantitative determinations are still an unresolved problem, particularly in the case of solid matrices.

These problems require and are worth further research, and GC–MS will be the basic analytical tool for their solutions.

REFERENCES

1. A. Mosandl, Food Rev. Int., 11 (1995) 597–664.
2. C. Bicchi, A. D'Amato, and P. Rubiolo, J. Chromatogr. A, 843 (1999) 99–121.

3. R. Marsili, Techniques for Analyzing Food Aroma, 1997, Marcel Dekker, New York, pp. 209–235.
4. R. Mariaca and J.O. Bosset, Lait, 77 (1997) 13–40.
5. E.G. Perkins, Proceedings of 79th Annual Meeting of the American Oil Chemists' Society, May 8–12 1988, Phoenix, AZ, 1988, AOCS, Champaign, IL, pp. 35–56.
6. A.K. Kiritsakis, J. Am. Oil Chem. Soc., 75 (1998) 673–681.
7. G. Barbieri, L. Bolzoni, G. Parolari, R. Virgili, R. Buttini, M. Careri, and A. Mangia, J. Agric. Food Chem., 40 (1992) 2389–2394.
8. M. Careri, P. Manini, S. Spagnoli, G. Barbieri, and L. Bolzoni, Chromatographia, 38 (1994) 386–394.
9. S.V. Overton and J.J. Manura, J. Agric. Food Chem., 43 (1995) 1314–1320.
10. B.R. D'Arcy, G.B. Rintoul, C.Y. Rowland, and A.J. Blackman, J. Agric. Food Chem., 45 (1997) 1834–1843.
11. M.J. Cantalejo, J. Agric. Food Chem., 45 (1997) 1853–1860.
12. J.-L. Mau and S.-J. Hwang, J. Agric. Food Chem., 45 (1997) 1849–1852.
13. G.P. Blanch, G. Reglero, and M. Herraiz, J. Agric. Food Chem., 43 (1995) 1251–1258.
14. M. Leunissen, V.J. Davidson, and Y. Kakuda, J. Agric. Food Chem., 44 (1996) 2694–2699.
15. M.T. Morales, A.J. Berry, P.S. McIntyre, and R. Aparicio, J. Chromatogr. A, 819 (1998) 267–275.
16. T.J. Evans, C.E. Butzke, and S.E. Ebeler, J. Chromatogr. A, 786 (1997) 293–298.
17. J. Song, B.D. Gardner, J.F. Holland, and R.M. Beaudry, J. Agric. Food Chem., 45 (1997) 1801–1807.
18. R. Aparicio, M.T. Morales, and V. Alonso, J. Agric. Food Chem., 45 (1997) 1076–1083.
19. M.T. Morales, J.J. Rios, and R. Aparicio, J. Agric. Food Chem., 45 (1997) 2666–2673.
20. B.D. Gardner, J.A. Johnson, V.B. Artaev, O.D. Sparkman, and J.F. Holland, Paper presented at the 43rd ASMS Conference on Mass Spectrometry and Allied Topics, May 21–26, 1995, Atlanta, GA.
21. M.S. Allen, M.J. Lacey and S. Boyd, J. Agric. Food Chem., 42 (1994) 1734–1738.
22. K.E. Murray and F.B. Whitfield, J. Sci. Food Agric., 26 (1975) 973–986.
23. L. Mondello, K.D. Bartle, P. Dugo, P. Gans, and G. Dugo, J. Microcol. Sep., 6 (1994) 237–244.
24. L. Mondello, P. Dugo, G. Dugo, and K.D. Bartle, J. Chromatogr. Sci., 34 (1996) 174–181.
25. L. Mondello, M. Catalfamo, A.R. Proteggente, I. Bonaccorsi, and G. Dugo, J. Agric. Food Chem., 46 (1998) 54–61.
26. G.P. Blanch and G.J. Nicholson, J. Chromatogr. Sci., 36 (1998) 37.
27. G. Schmidt, G. Full, P. Winterhalter and P. Schreier, J. Agric. Food Chem., 43 (1995) 185–188.
28. M. Wüst, T. Beck and A. Mosandl, J. Agric. Food Chem., 46 (1998) 3225–3229.
29. C. Frank, A. Dietrich, U. Kremer, and A. Mosandl, J. Agric. Food Chem., 43 (1995) 1634–1637.
30. S. Faulhaber, U. Hener, and A. Mosandl, J. Agric. Food Chem., 45 (1997) 2579–2583.

31. S. Faulhaber, U. Hener, and A. Mosandl, J. Agric. Food Chem., 45 (1997) 4719–4725.

32. G. Barbieri, L. Bolzoni, M. Careri, A. Mangia, G. Parolari, S. Spagnoli, and R. Virgili, J. Agric. Food Chem., 42 (1994) 1170–1176.

33. M. Godefroot, P. Sandra, and M. Verzele, J. Chromatogr., 203 (1981) 325.

34. H.D. Belitz and W. Grosch, in Food Chemistry, 1987, Springer-Verlag, Berlin, p. 399.

35. H.D. Belitz and W. Grosch, in Food Chemistry, 1987, Springer-Verlag, Berlin, p. 280.

36. H.D. Belitz and W. Grosch, in Food Chemistry, 1987, Springer-Verlag, Berlin, p. 181.

37. S. Kuzdzal-Savoie, Int. Dairy Fed. Bull., 118 (1980) 53–66.

38. E.A. Day, in I. Hornstein, ed., Flavor Chemistry, 1966, American Chemical Society Advances in Chemistry Series 56, American Chemical Society, Washington, DC.

39. H.D. Belitz and W. Grosch, in Food Chemistry, 1987, Springer-Verlag, Berlin, p. 131.

40. J.N. de Man, in Principles of Food Chemistry, 1990, Van Nostrand Reinhold, New York, pp. 100–112.

41. J.E. Hodge, J. Agric. Food Chem., 1 (1953) 928–953.

42. R. Virgili, G. Parolari, L. Bolzoni, G. Barbieri, A. Mangia, M. Careri, S. Spagnoli, G. Panari, and M. Zannoni, Lebensm. Wiss. Technol., 27 (1994) 491–495.

43. V. Mazzoleni, P. Caldentey, M. Careri, A. Mangia, and O. Colagrande, Am. J. Enol. Vit., 45 (1994) 401–406.

44. P. Caldentey, M.D. Fumi, V. Mazzoleni, and M. Careri, Flavour Fragr. J., 13 (1998) 185–188.

45. M. Careri, V. Mazzoleni, M. Musci, and R. Molteni, Chromatographia, 49 (1999) 166–172.

46. J.M. Amon, J.M. Vandepeer, and R.F. Simpson, Austral. N.Z. Wine Ind. J., 4 (1989) 62–69.

47. T.H. Lee and R.F. Simpson, in G.H. Fleet, ed., Wine Microbiology and Biotechnology, 1994, Harwood Academic, Chur, Switzerland, p. 553.

48. P.J. Holloway, Chem. Phys. Lipids, 9 (1972) 158.

18

Gas Chromatography–Mass Spectrometry for Residue Analysis: Some Basic Concepts

H. F. De Brabander, K. De Wasch, and S. Impens
University of Gent, Merelbeke, Belgium

R. Schilt
TNO Nutrition and Food Research, Zeist, The Netherlands

M. S. Leloux
State Institute for Quality Control of Agricultural Products, Wageningen, The Netherlands

1. INTRODUCTION

A residue may be defined as a trace of a component, that is present in a matrix after some kind of administration. The matrix may be anything in which a residue may be present, trapped, or concentrated (meat, urine, feces, liver, etc.). There is no general agreement upon the concentration level of a trace. However, the ppb level (μg kg^{-1}) in which residues are present may certainly be considered as a trace or even an ultratrace level. The impact of results in the field of residue analysis in veterinary food inspection is increasing. This is mainly due to the more severe legislation for some residues (class A or forbidden components). Therefore, the highest standard is required from the control methods used. One of the aspects is the specificity of the method. Specificity is defined as the ability of the method to distinguish between the analyte being measured and other substances. The method must have the ability to distinguish between the analyte being measured at trace or ultra trace concentrations and other substances (possibly) present at 10 to 10.000 times higher concentrations. For the control of registered veterinary drugs, next to the determination of the identity of the analyte (qualification), a good quantification is necessary.

441

A residue analysis procedure consists of three distinct steps: First, the analyte has to be extracted from the matrix. Then, the extract is freed from as many interfering products as possible. The third step is the identification and eventually the quantification of the analyte. In modern residue analysis chromatographic techniques are very important. Next to thin-layer chromatography (TLC) and liquid chromatography (LC), gas chromatography–mass spectrometry (GC–MS) is the technique most frequently used. The choice of a very specific detection technique (GC–MS, MS–MS, MS^n, high-resolution [HR] GC–MS) could result in a reduction of sample pretreatment, which is, at first sight, an advantage. However, it is not always conceivable for a control laboratory to use all these techniques in routine analyses (e.g., due to high instrument costs). Moreover, as the number of substances of different parts to screen is high and not limited (in the future), one detection technique will never be as specific for all compounds. Another option is high-performance liquid chromatography (HPLC) fractionation. This procedure results in several purified fractions, each containing a limited number of analytes and matrix components. Each HPLC fraction may be analyzed with a specific technique, e.g., GC–MS, and if necessary with different techniques, e.g., TLC, LC–MS.

The choice of the analytical strategy must always be seen in the light of the interpretation of analytical results. Inspection services are interested mainly in a YES/NO answer: Has this animal been illegally treated? Is the concentration of the residue higher than a certain value, e.g., maximum residue limit (MRL), maximum permitted limit (MPL)? In fact, all questions may be subsumed in one major question: Is the law violated? When the answer to that question is YES, actions are taken. Different kinds of actions (rejection of animals, inspection at the farm, etc.) may be performed.

Laboratories, on the other hand, evaluate crude analytical data on the basis of predefined criteria for all relevant parameters, e.g., signal-to-noise ratios (S/N ratios), deviations from observed and target values for reference materials. For residue analysis, such minimum quality criteria are part of EC legislation [1]. Based on these criteria, analytical data are transformed into YES/NO answers. Moreover, in most countries, a system of first and second analysis is used. When the final answer of the first analysis is YES, the legal action is suspended until the results of the second analysis, if any, are known.

2. GAS CHROMATOGRAPHY–MASS SPECTROMETRY APPARATUS IN RELATION TO RESIDUE ANALYSIS

For the routine analysis of residues in meat-producing animals there is an increasing use of coupled techniques. In most cases, low-resolution MS coupled to a chromatographic separation is used. In low-resolution GC–MS, two approaches may be distinguished: (1) The use of instruments, that are able to monitor a whole chromatogram in the full-scan mode, e.g., with 1 spectrum per second, without loosing detection power, e.g., instruments based on ion-trap technology, and (2) linear quadrupole instruments using the full-scan mode for high amounts of analyte, and selective ion monitoring (SIM) for detecting low concentrations (<1 to 10 ng).

Both techniques have their pros and cons, and also their own fans. In full-scan GC–MS, a complete spectrum of each point of the chromatogram as well as all kinds of ion chromatograms can be generated afterwards with a data system. Identification of the analyte by library search may be performed. The result of this search is expressed as a figure, which reflects the fit between the standard and the sample spectrum.

With SIM, a limited number of ions (2 to 4 ions) are monitored during a selected time interval. The presence of the analyte is determined by the presence of these "diagnostic" ions in the correct retention time window and within the correct abundance ratio between certain limits, e.g., $\pm20\%$ in chemical ionization (CI) and $\pm10\%$ in electron ionization (EI). In the EU guidelines, the monitoring of 4 ions is mandatory. In practice, the monitoring of 4 ions at low concentrations (<1 ppb) does not always give satisfactory results: some ions disappear at lower concentrations or the ratio between the ions is not reproducible. Moreover, not all analytes in a multiresidue method give enough diagnostic ions with one derivatization method.

Gas chromatography–tandem mass spectrometry (GC–MS–MS) has been available for some time on the larger instruments. However, in most cases, these machines are too expensive for use in field laboratories performing residue analysis. In the mid-1990s, benchtop GC–MS–MS-based ion-trap technology was introduced. In this way, smaller and also less expensive instruments could be constructed. Instead of the classical MS–MS in space, using an instrument consisting of three quadrupoles in series, the MS–MS experiment takes place in one ion trap, in function of time. One ion (the precursor ion, also called parent ion) is isolated and stored in the ion trap. Subsequently, the precursor ion is fragmented by increasing the speed of the ion and by inducing collisions with the helium molecules present in the trap. The product ions (also called daughter ions) are scanned out of the trap resulting in a product (or a daughter) ion mass spectrum. However, it must be noted that the instrument is not limited to only one stage of MS–MS; a target product ion may act as a new precursor ion with formation of secondary product ions (also called granddaughter ions), and so on. Therefore, the nomenclature MS^n is preferred to MS–MS throughout this chapter. In practice, the absolute number of ions decreases with each MS–MS step, and therefore MS^5 is a practical limit in GC–MS^n.

3. POSSIBLE INTERFERENCES IN GAS CHROMATOGRAPHY–MASS SPECTROMETRY ANALYSIS OF RESIDUES

Gas chromatography–mass spectrometry is often considered as a technique with which no identification mistakes could be made, apart from mistakes due to cross contamination. This may be true or nearly true in major and minor component analysis, but this is certainly not the case in residue analysis.

In extracts of biological material (e.g., urine, meat, feces), a large variety of components in a large variety of concentrations are present. An unknown and variable amount of these matrix components are coextracted with the analyte and introduced into the chromatographic system and the mass spectrometer. Interference between these matrix components, possibly present at relatively high concentrations (ppm range, mg kg^{-1} or higher), and analytes, present at very low concentrations (ppb range, $\mu g \cdot kg^{-1}$), is possible and should be avoided.

Interferences mostly result from coeluting peaks or from background noise. The mass spectrum obtained is a mixture of two mass spectra and false interpretation is possible. Isotope interference is another possibility [2]. This phenomenon may occur with any isotope; ^{13}C is used here as an example. Carbon has two natural isotopes: ^{12}C and ^{13}C with a ratio of 98.9 to 1.1 (the exact figures are rounded for simplicity). In residue analyzes, three important parameters should be taken into account:

1. The very large difference in concentration between the analytes and the matrix components.
2. Analytes and/or interferences are mostly organic molecules containing a relative high amount of carbon atoms.
3. Interferences (most likely) may have analogous structures as the analytes.

In a quadrupole instrument using the SIM mode, many interferences are not observed by the highly selective use of the detector. In an ion-trap instrument, high concentrations of coeluting molecules may influence the ionization time of analytes and so the detection limit. This may cause the following phenomena: false negative and/or positive results, and wrong quantification.

3.1. False Negative Results

In SIM, the diagnostic ions of the analyte must be present in the correct relative intensities ($\pm 20\%$ for CI and $\pm 10\%$ for EI) (EC criteria 93/256) [1,3,4]. The higher the number of diagnostic ions monitored, the higher the specificity of the method (fewer false positives) but also the higher the chance of false negative results when the identification criteria are applied (too) strictly. The relative intensity of the ions may be disturbed by background noise and coeluting substances. In Figure 1, the relative abundance of the ion m/z 440 in the sample spectrum is partly due to the analyte (see spectrum standard) and partly due to an interference. The ratio between m/z 440 and 425 is out of range (normal range between 41 and 51). According to the quality criteria, the sample has to be declared negative although the analyte is clearly present. The conclusion "negative" (NO answer) will be the same using SIM or full-scan mode. At full-scan identification of components, the same parameters will influence the (reverse) fit of the spectrum and also disturb the visual comparison. However, in full-scan mode, the presence of interferences (a possible coeluting peak) will be noted more readily.

The interference may be due to a molecule having a molecular or fragment ion equal to one of the diagnostic ions of the analyte. However, interference is also possible with molecules present in a much higher concentration containing ions, that are 1 or 2 amu lower than the diagnostic ions. The latter is of particular interest on shoulder peaks.

3.2. False Positive Results

These may occur when SIM signals are generated by interfering molecules. Nortestosterone (NT), for example, is a well-known anabolic steroid used in cattle fattening. The same component is also endogenous in various animal species [5–7]. For determination of its ditrimethylsilyl (di-TMS) derivative of the β-form, three ions are monitored: m/z 418 (100%), 403 (20%), and 328 (35%). In the urine of pregnant cows, estradiol (E2 in the α-form) is present in concentrations 10^4 to 10^5 times higher than the concentration of NT found after illegal application of the drug. It was demonstrated that the di-TMS derivatives of α-E2 and β-NT are not well separated under the chromatographic conditions used in most laboratories. The structures of the di-TMS derivatives of both components are

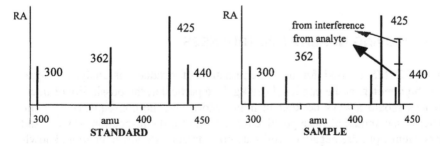

Figure 1 Formation of false negative results by SIM by interferences that disturb the relative ratios of the diagnostic ions.

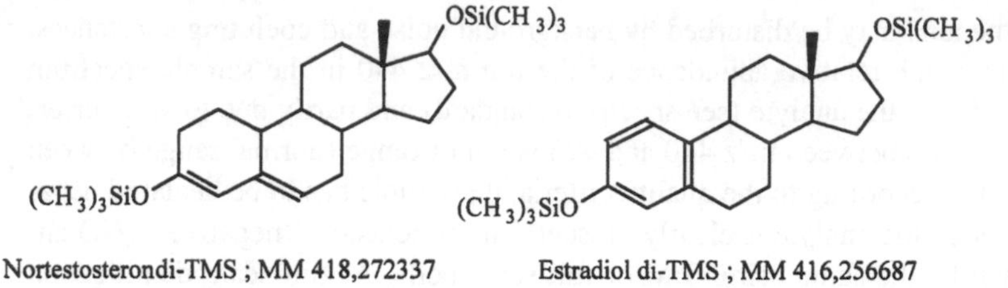

Nortestosterondi-TMS ; MM 418,272337   Estradiol di-TMS ; MM 416,256687

Figure 2 Formulas and molecular masses of nortestosterone-(di-TMS) and estradiol-(di-TMS) showing the similarity of the molecules.

presented in Figure 2. The molecular masses of these two components differ by only 2 amu. Since the structures are similar, fragmentation is also similar.

The mass spectrum of estradiol contains ions at m/z 416 due to the molecular ion and fragment ions at m/z 402 and 326. It was calculated and demonstrated that the isotope peaks of estradiol may generate a correct SIM signal for nortestosterone [2]. Using the SIM method according to the book, the analyst will therefore conclude abuse of NT: the three ions are present within the correct retention time windows and with the correct ratios. The interfering ions may also be generated by several interferences simultaneously or by stable isotopes of other elements. The same reasoning may apply for a set of four ions using HR–GC–MS.

3.3. False Quantification

Quantification of residues is of increasing importance, especially close to the MRL, MPL, or any other decision limit, e.g., the CCα (critical concentration alpha) [8,9]. In a manner analogous to that described above, interferences may influence the abundance of the analyte ions or the internal standard ions resulting in a false quantification.

4. IMPROVING QUALITATIVE TRUENESS

When GC–MS is used for the determination of residues of analytes, illegal growth promoters at the ppb level (μg kg^{-1}) in particular, the possibility of interferences should always be kept in mind. Moreover, the consequences for the owner of the animal of false positive results, and for the inspection services and the consumer of false negative results are considerable. However, caution, knowledge of the background of residue analysis, and investing time (and money) into the analysis may prevent the analyst from making a wrong decision.

4.1. Avoiding False Negatives

False negatives by the loss of the analyte during the cleanup, derivatization, or injection should be monitored by using internal standards. Deuterated analogues of the target components are most suited for this purpose because the behavior of the deuterated component in the extraction procedure, cleanup, and detection is very like that of the analyte. However, their availability, in number as well as in quantity, is limited. Deuterated components are also very useful for quantification and to prove the absence of a certain analyte in a certain sample (real negatives): the signal of the deuterated (heavy) component is present and that of the light component is not. Moreover, deuterated components may be used for balancing the ion ratios.

False negative results due to not fulfilling the quality criteria, e.g., disturbance of the normal peak ratios of the ions from the analyte by one or more interferences or missing peaks, should be dealt with in another way. The analyst should be aware that the statistical possibility of its occurrence is high. Instead of declaring a sample immediately negative (NO answer) because the ratio of one of the ions is not within the range proposed by the quality criteria, other elements should be added to the analysis. Possible solutions are reinjecting the same derivative on another column, using other derivatization reagents or techniques, using GC–MS^n, performing a second analysis with a different method, etc.

Although each technique on its own may not completely fulfill the quality criteria (e.g., by disturbed ion ratios) the combination of these techniques may give enough analytical evidence to prove the presence of the analyte "beyond any reasonable doubt." In the new EC criteria, to be published in 2000 [10], a system of identification points is proposed to interpret the analytical data. For the confirmation of substances listed in Groups A and B of Annex I of Council Directive 96/23/EC, minimums of 4 and 3 identification points, respectively, are required. In Table 1, the number of identification points that each of the basic mass spectrometric techniques can earn is given.

However, in order to qualify for the identification points required for confirmation:

1. A minimum of at least one ion ratio must be measured.
2. All measured ion ratios must meet the criteria described above.
3. A maximum of three separate techniques can be combined to achieve the minimum number of identification points.

Of course, each ion may only be counted once.

High-performance LC coupled with full-scan UV diode-array detection (DAD), with fluorescence detection, or to an immunogram, or two-dimensional TLC coupled to spectrometric detection are techniques that may contribute a

Table 1   Number of Identification Points That Each of the Mass Spectrometric Techniques Can Earn

MS technique	Identification points earned per ion
Low-resolution mass spectrometry (LR)	1.0
LR-MS^n precursor ion	1.0
LR-MS^n transition products	1.5
High-resolution mass spectrometry (HR)	2.0
HR-MS^n porecursor ion	2.0
HR-MS^n transition products	2.5

maximum of one identification point (for substances in Group A) provided that the relevant criteria for these techniques are fulfilled.

This approach allows the use of the classical low-resolution GC–MS in the SIM mode (four ions and three correct ratios) for laboratories using older equipment as described in 93/256 [1]. Combinations of modern methods, e.g., GC–MS yielding two ions (and one ratio) in combination with LC–MS yielding two ions (and one ratio), are also allowed. New instruments brought on the market may be included very easily into the system in the future. In some cases, earning four identification points without the power of MS^n or HRMS may pose a problem. Therefore, an extra point can be earned with HPLC and HPTLC. Some molecules, e.g., trenbolone and methyltestosterone, yield a very typical fluorescence spectrum with identification power certainly equivalent to an ion in mass spectrometry [11]. On the other hand, the analyst should be restricted to four identification points only. Depending upon the analyst's analytical experience and skill, the equipment available, and the impact of the analytical result, many more points (evidence) can be gathered (see below).

4.2. Avoiding False Positives

As demonstrated above, false positives may result from the presence of diagnostic ions that do not originate from the analyte but are generated by one or more interferences present at high concentration in the final extract. The fact that the correct ions (with correct ratios) can be produced from the interfering (endogenous) compounds is transparent to the analyst when using the GC–MS in the SIM mode.

With instruments that are not able to take a full-scan run at low concentration, the following strategy can be recommended. In the case of a positive signal, a second full-scan run on the same sample is performed in order to exclude the presence of isotope-generating peaks at the retention time of the analyte. The absence of substantial concentrations of isotope peak generators in the full-scan mass spectrum has to be considered as a quality criterion. The SIM mode could also be used for screening purposes only and suspect samples rechromatographed and fully identified with the other system.

Isotope interferences may be avoided by using instruments capable of operating in the full-scan mode at low concentration levels. The quality criteria (3 to 4 ions) may be extended by using full-scan spectrum matches between the sample spectrum and a (home made) library spectrum. However, in order to obtain a good fit, the sample spectrum should be (reasonably) free of interfering peaks. At lower concentrations (how low: depending upon the cleanup used) the diagnostic ions become less abundant in comparison with the background ions (Fig. 3).

The use of MS–MS is a well-known way to reduce background. If a positive signal is obtained, the sample is rechromatographed (eventually several times) in the MS–MS mode in order to obtain more information. In most cases, there is no need for an extra extraction and cleanup: usually only 1 μl out of 25 μl is injected in the GC and there is enough sample left for the extra injections. Full MS^2 spectra on each diagnostic ion, e.g., three ions, could be taken during

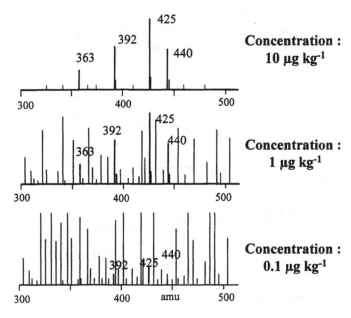

Figure 3 Diagnostic ions in relation to background ions at various concentrations (theoretical example).

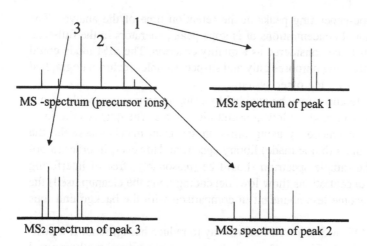

Figure 4 Precursor ion spectrum and the resulting product ion spectra taken during four successive GC–MS and GC–MS² runs (theoretical example).

each run. In Figure 4, a precursor ion spectrum and the resulting product ion spectra are shown (theoretical example).

As can be seen from this figure, at least 10 identification points are earned with this technique (three precursor ions and several product ions, presuming that correct ratios are obtained). Such a number of identification points may be

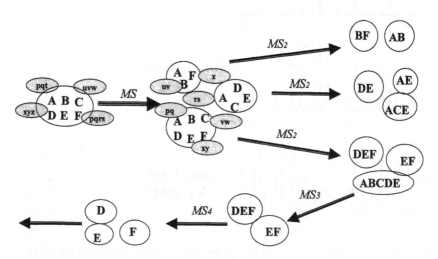

Figure 5 Difference between MSⁿ and multiple MS² runs on different precursor ions.

earned in only one extra GC run using MS^n. In this technique, a product MS^n ion serves as precursor ion of the MS^{n+1} run. In Figure 5, the difference between MS^n and different MS^2 experiments is shown. The latest versions of ion-trap software enable the upgrade from MS–MS to MS^n (up to MS^5) on the basis of software changes only.

Another approach to obtaining more analytical evidence (and more identification points) is higher mass accuracy. High-resolution GC–MS with magnetic sector instruments has been available for a long time. However, for most laboratories, these instruments are too expensive. Recently, higher mass accuracy may be obtained with more affordable benchtop instruments, based on time-of-flight technology.

5. QUALITATIVE AND/OR QUANTITATIVE ANALYSIS

The classical difference between qualitative and quantitative methods is described below. In quantitative analysis, the result is expressed as a figure (e.g., 75 µg kg^{-1}). Quantitative analysis is necessary for the determination of residues of components that may be present in food at maximum allowable concentration, e.g., at the MRL or MPL. The method must be able to establish whether the concentration of the analyte is lower or higher than that limit. The methods must have a limit of quantification (LOQ) that is lower than the MRL. Recently, consensus was reached on the fact that analytical methods to be used for controlling an MRL should have a LOQ of (at least) 0.5 MRL. When the result obtained is higher than the MRL (and action may follow), quality criteria must be used for the qualification of the residue. For values much lower than the MRL, qualitative errors play a less important role unless the (screening) method should miss the analyte completely.

Qualitative methods do not produce figures: the results are expressed as YES/NO answers. These methods could be used for residues of forbidden substances (e.g., products with an estrogenic, androgenic or gestagenic action). However, qualitative methods always have a semiquantitative character: the minimum amount (a quantitative figure) of analyte to discriminate a signal from the background. In practice, a sort of action limit is applied for the determination of the presence or absence of the analyte.

The difference between qualitative and quantitative methods in residue analysis is not as simple as described above. A method is not necessarily quantitative because a figure is produced, but only when that figure fulfills certain criteria of accuracy (trueness and precision). However, too many people neglect that item and consider a figure produced by any instrument automatically as a quantitative result. An illustration of that (normal) human behavior is the common negligence of the rounding of figures: the number of significant figures must reflect the precision of the analysis. In residue analysis, the coefficient of variation increases with decreasing concentration according to the so-called Horwitch curve [12]. In most cases, according to the rounding rules, only the magnitude of a result could be given, e.g., 2.10^1, which is something nobody likes.

For registered veterinary drugs, quantification is only necessary in a small concentration range. The analytical method is validated in the small range close to the MRL, e.g., with an MRL of 50, the range could be from 25 to 75 (Fig. 6). Therefore, the fact that a result (not a method) is qualitative or quantitative does not depend upon the method, but upon the result of the analysis. In Figure 6, the concentration axis is divided into three parts: (1) the quantitative part around the MRL ($25 < x < 75$; the validation range), (2) a qualitative part in the range of $x < 25$, and (3) a qualitative part in the range $x > 75$.

The final result of this quantitative method is not expressed as a figure, but again as a YES/NO answer, i.e., above or below the MRL. In the context of residue analysis, a YES/NO answer is not to be interpreted in terms of "positive" or "negative," but in more general terms of "violation" or "nonviolation." The expressions "positive" and "negative" should in principle be avoided: a negative result does not necessarily mean that the residue is absent, because the residue level may be under the MRL or the component may be endogenous in a certain species. The expression "positive" can also be confusing: Recently, a politician declared that the results of the analysis were positive because no residues were found.

In addition, the method of quantification is very important. When MS is used, the quantitative result may be calculated in various ways. When the full signal of the sample versus the signal of the internal standard (calibrated against a series of standards) is used (Figure 7), the quantification is not very reliable because the sample spectrum could be different from the standard spectrum: some ions could be missing and/or some ratios between ions may be disturbed.

Alternatively, the sum of a number of diagnostic ions (including MS^n ions), the most important ion or an algorithm including the correct ratios of the ions

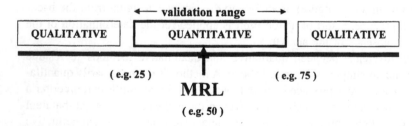

Figure 6 Representation of the terms qualitative method or quantitative method upon the result obtained.

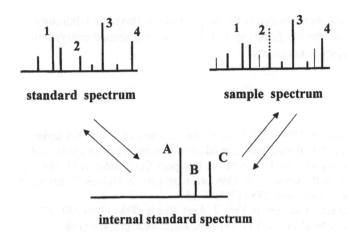

Figure 7 Quantification in MS.

may be used. All these methods of quantification will give different results and can be the cause of contradictions. In Figure 7, for example, peak 2 of the sample is distorted by an interference: taking the blind sum of all peaks will result in false quantification. A correction for the correct peak ratios should be made. The best method of quantification is the use of deuterated internal standards. However, as mentioned before, their availability in number as well as in quantity is limited.

6. CONCLUSIONS

Gas chromatography–mass spectrometry is a very powerful technique for the analysis of residues in veterinary products. During the 1990s, the classical SIM mode was increasingly complemented by other techniques offering full-scan mode at low concentrations, e.g., GC–MS–MS, GC–MS^n and HR-GC–MS. However, it is still very dangerous to consider GC–MS as an absolute error-less technique. As in any other analytical technique, false positive and false negative results as well as false quantification can be obtained. However, when the analyst is aware of the possible causes of these errors, the application of some simple rules and the investment of a little more time in analyzes may prevent most of these mistakes. A very good strategy is to consider a first "violation" result just as a "suspect" result and to repeat the complete analysis immediately within the laboratory. In that second analysis, deuterated internal standards preferably should be used. Only when the two successive results match, qualitatively as well as in magnitude of concentration, is the result ready to leave the laboratory. Even then, an open mind for the followup of the results of a second analysis in an independent laboratory is necessary.

REFERENCES

1. EEC (1993) document 93/256. EEC Commission decision of 14/4/1993 laying down the methods to be used for detecting residues of substances having a hormonal or a thyreostatic action. Official Journal of the European Community, L 118/64.
2. H.F. De Brabander, P. Batjoens, C. Vanden Braembussche, P. Dirinck, F. Smets, and G. Pottie, Anal. Chim. Acta, 275 (1993) 9–15.
3. R.W. Stephany and L.G. Van Ginkel, Fres. J. Anal. Chem., 338 (1990) 370–377.
4. W.G. De Ruig, R.W. Stephany, and G. Dijkstra, J. Chromatogr., 489 (1989) 89–93.
5. H.F. De Brabander, J. Van Hende, P. Batjoens, L. Hendriks, J. Raus, F. Smets, G. Pottie, L. van Ginkel, R. W. Stephany, Analyst, 119 (1994) 2581–2586.
6. M. Vandenbroeck, G. Van Vyncht, P. Gaspar, C. Dasnois, P. Delahaut, G. Pelzer, J. De Graeve, and G. Maghuin-Rogister. J. Chromatogr., 564 (1991) 405–412.
7. G. Debruykere, C. Van Peteghem, H.F. De Brabander, and M. Debackere. Vet. Quart., 12 (1990) 247–250.
8. P. Gowik, B. Julicher, and S. Uhlig, J. Chromatogr. B, 716 (1998) 221–232.
9. B. Juhlicher, P. Gowik, and S. Uhlig, Analyst, 123 (1998) 173–179.
10. EEC document. EEC Commission decision laying down the methods to be used for detecting residues of substances having a hormonal or a thyreostatic action. Official Journal of the European Community, to be published in 2000.
11. R. Verbeke, J. Chromatogr., 177 (1979) 69–84.
12. W. Horwitz, L.R. Kamps, and K.W. Boyer, JAOAC, 63 (1980) 1344–1354.

19

Applications of Gas Chromatography–Mass Spectrometry in Residue Analysis of Veterinary Hormonal Substances and Endocrine Disruptors

R. Schilt
TNO Nutrition and Food Research, Zeist, The Netherlands

K. De Wasch, S. Impens, and H. F. De Brabander
University of Gent, Merelbeke, Belgium

M. S. Leloux
State Institute for Quality Control of Agricultural Products, Wageningen, The Netherlands

1. INTRODUCTION

1.1. Anabolic Agents and Endocrine Disruptors

In Europe, the word ''hormones'' has a very bad reputation because of the possible danger for public health of residues of some of these products in foodstuffs of animal origin. Toxicologists have demonstrated that DES (diethylstilbestrol, a synthetic estrogen) is a potential carcinogen [1–3]. In human medicine, analogous experiences with DES were found (the so-called DES-daughters) [4]. Recently, several cases of poisoning have occurred in Spain and France due to the consumption of liver from animals treated with clenbuterol [5,6]. Moreover, some environmentally persistent alkyl-phenolic compounds (such as nonylphenol) and perhaps other chemicals show estrogenic activity [7]. These ''environmental'' estrogens are brought up in relation to the decreasing quality of human sperm and regarded as an assault on the male.

455

The fields of growth promoters and endocrine disruptors are very large and present many challenges. Treatment of cattle with anabolic components may be detected by the residues present in plasma, excreta, meat, or organs of the animals. In regulatory control at the farm, plasma, urine, and/or feces of the animals may be sampled. At the slaughterhouse, tissue as well as excreta are available for sampling. At the retail level (butcher, supermarket) or in case of import/export, sampling is restricted to tissue only. Finally, all kinds of matrices (powders, cocktails, fodder) circulating on the (black) market must be analyzed for the presence of illegal substances.

The issue of the presence of endocrine disruptors in the environment has initiated a large research effort [8]. The variety of sample materials for estrogens is very large and ranges from pure substances to extremely low levels in surface water.

1.2. Analytical Aspects

For studies unraveling the endocrinology of humans, steroids, especially cortisol and its metabolites and precursors, have been extensively analyzed. Many studies were published on the hydrolysis of glucuronic and sulfate conjugates, the extraction of the analytes, and the subsequent derivatization before gas chromatography (GC) [9]. The coupling of GC with flame ionization detection has played a major role in this field. However, with the availability of reasonably priced benchtop mass spectrometers in the 1980s, the analysis of anabolic steroids using GC–mass spectrometry (MS) within the fields of doping analysis and veterinary residue analysis started to gain popularity quite rapidly. Automated GC–MS instruments enabled high-throughput analysis under routine conditions, for example, for doping analysis during the Olympic Games and for veterinary residue control in cattle, when GC–MS was generally regarded as a very expensive technique.

In this chapter, the application of GC–MS or GC–tandem mass spectrometry (MS–MS) for residue analysis for different types of hormonally active substances is discussed. In addition, special attention is focused upon the forensic use of GC–MS.

2. ANABOLIC STEROIDS

2.1. Introduction

Many analogs of the endogenous steroids 17β-estradiol, 17β-testosterone, and progesterone (Fig. 1) have been developed by the pharmaceutical industry for therapeutic purposes. Functional groups have been modified or added, by small or larger chemical modifications, to alter the strength and mode of action, to enable oral administration, and to influence other pharmacokinetic properties.

In general, a straightforward sample pretreatment is carried out (Table 1). The number of steps depends on the sample matrix and the levels of the substances to be analyzed. For the analysis of anabolic steroids in human urine (doping control), the degree of cleanup needed is generally lower than the corresponding analysis of anabolic steroids in the urine of cattle. In the latter case, high-performance liquid chromatography (HPLC) is often routinely used to further purify the extracts obtained.

2.2. Deconjugation

The deconjugation of steroids is an essential step in the analysis. In general, an enzymatic deconjugation is carried out using commercially available Helix pomatia juice (glucuronidase and sulfatase) or bacterial glucuronidase. Earlier studies describe the use of acid hydrolysis instead of enzymatic hydrolysis. A serious drawback of acid hydrolysis is the greater chance of unwanted reactions such as decomposition and degradation of the target analytes. Because of a lack of suitable conjugated standard analytes, the yield of the deconjugation is difficult to verify. Only a few studies have been published, and some data are available from nonpublished studies. In summary, a 16-hr deconjugation with Helix pomatia juice at 37°C has proven to be the best generally applicable approach, although unwanted side effects may occur.

2.3. Sample Cleanup

Although GC–MS provides a specific method of detection, the sample cleanup plays an important role in the final result. Often the procedures are used for multianalyte analysis destined to detect a number of chemically and/or structurally related analytes in one run. Especially with anabolic steroids, this implies the coextraction of many endogenous steroids and steroid metabolites that may be present at levels far higher than the target levels of the analytes (0.5 to 50 µg/L or kg). Furthermore, other sample constituents can interfere with the analysis and influence the final result both in a qualitative way (false positive or false negative result) as well in a quantitative way. Therefore, it is important that a particular method, if possible, is compared with other methods slightly differing in the extraction procedure [10].

Although limited in the range of analytes to be extracted in a single analysis, immunoaffinity chromatography (IAC) has clear advantages in the selectivity and purity of the isolated fraction [11]. For the confirmation of stanozolol with high-resolution GC–MS, IAC was successfully applied [12]. By combining several types of antibodies raised against different steroids, columns can be prepared to selectively extract mixtures of steroids [13–15].

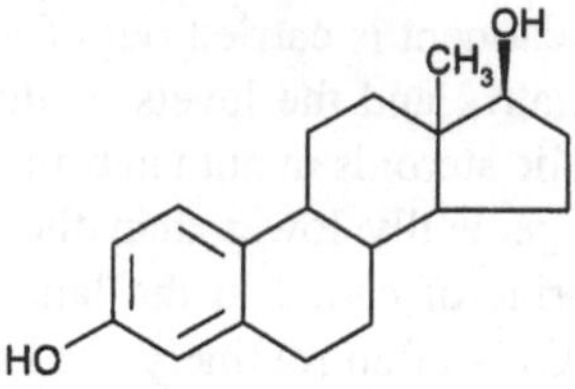

(a)

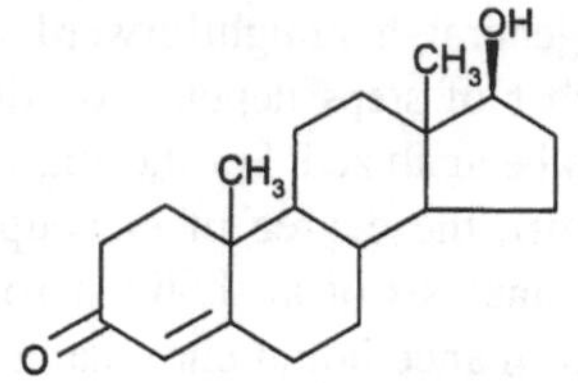

(b)

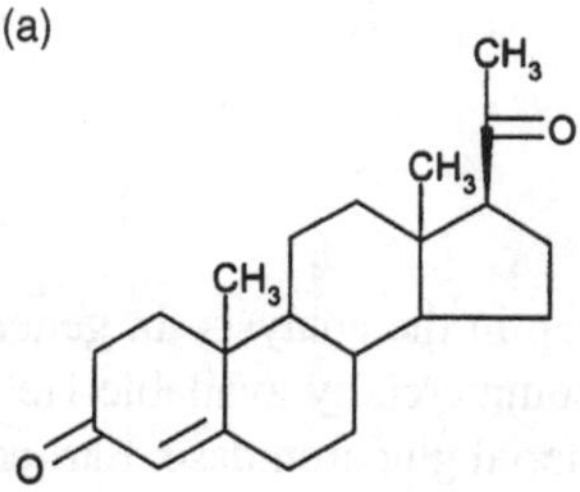

(c)

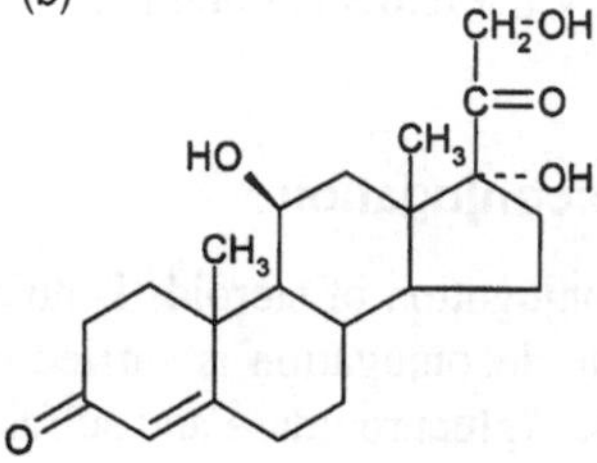

(d)

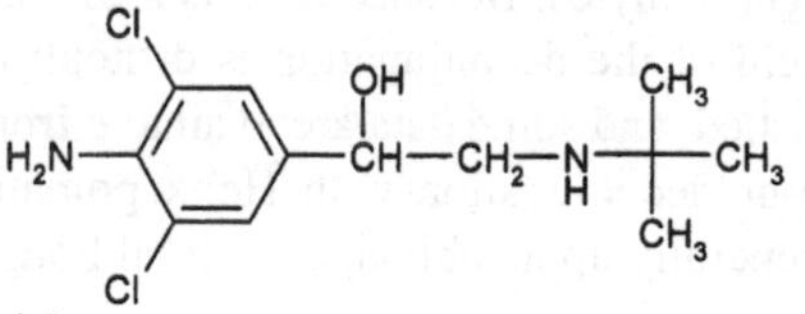

(e)

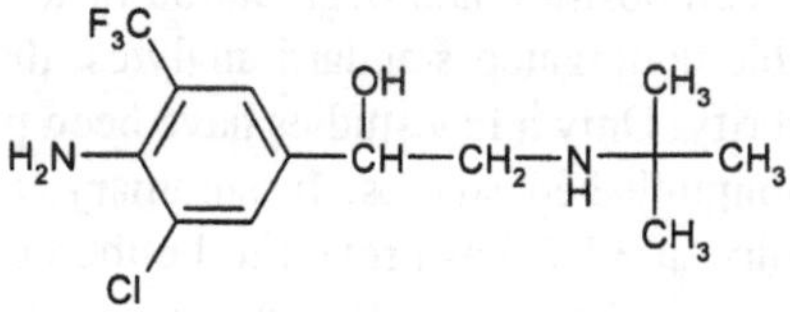

(f)

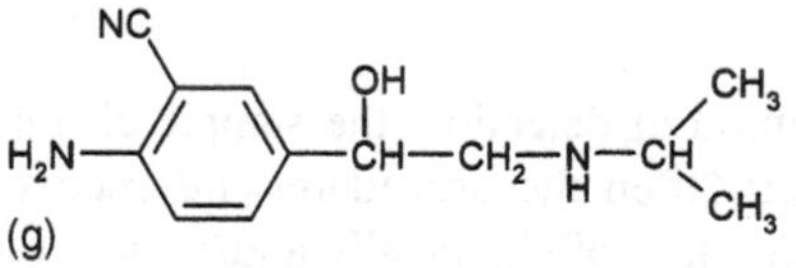

(g)

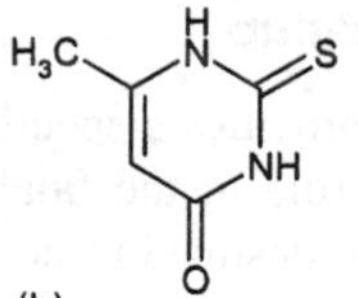

(h)

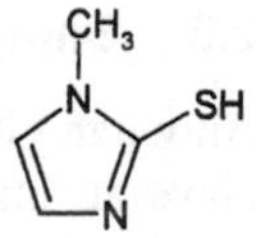

(i)

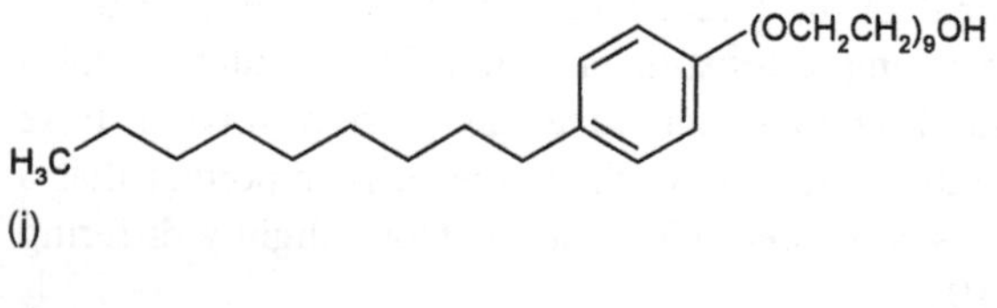

(j)

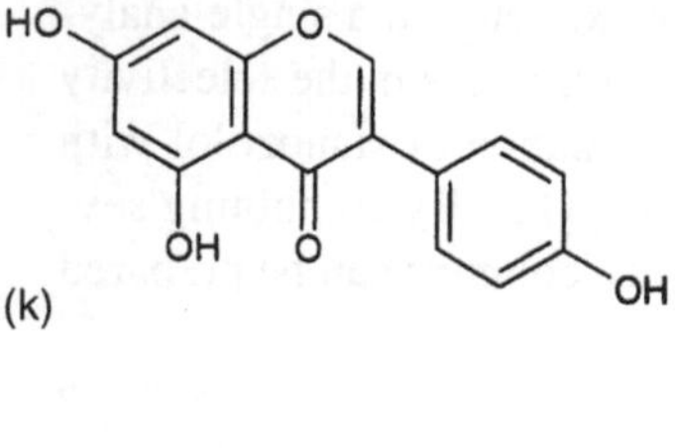

(k)

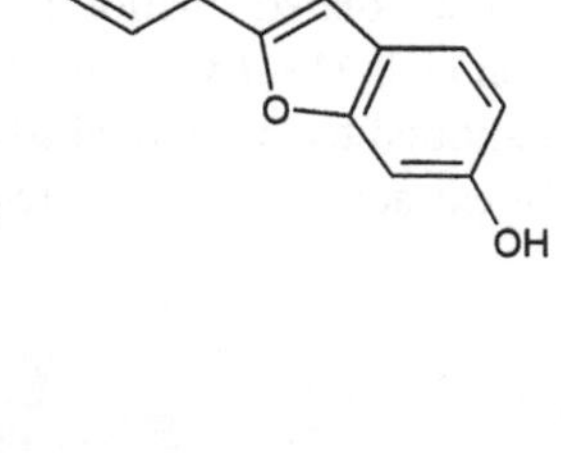

(l)

Table 1 Scheme of Analytical Procedures

Urine, bile, liver	Tissue (meat)
Deconjugation	Homogenization
Extraction and purification (L–L, SPE, IAC)	Extraction and purification (L–L, SPE, IAC)
HPLC fractionation	HPLC fractionation
Derivatization	Derivatization
GC–MS analysis (SIM or full scan)	GC–MS analysis (SIM or full scan)

2.4. Derivatization

The need for derivatization procedures applicable for large ranges of analytes to be analyzed within a single method especially necessitates the continuous optimization of this relatively small but very important part of the analytical procedure [16]. The use of internal (deuterated) or external standards is important to check the derivatization yield, which occasionally can show large variations. Because an extensive description of the derivatization itself, despite its importance, is not within the scope of this chapter, the major derivatization procedures are briefly discussed [16,17].

2.4.1. Silyl Derivatives

The majority of analytical procedures employ silylation as the derivatization reaction. Many reagents and combinations have been described. Depending on the catalyst present in a silylation mixture, keto functional groups can be converted into enol-trimethylsilyl (TMS) derivatives. Iodine ions (I^-) in the well-known N-methyl-N-trimethylsilyltrifluoroacetamide (MSTFA) with ammonium iodide and dithioerythritol (1000:2:4 v/w/w) mixture are often used for this purpose [10, 18,19].

2.4.2. Acylated Derivatives

The most commonly used acylated derivative is the heptafluorobutyric acid derivative. An advantage is the high-mass ions generation of high mass ions increasing

---

Figure 1 Basic chemical structures of growth promoters and endocrine disruptors. (a) 17 β-estradiol, (b) 17 β-testosterone, (c) progesterone, (d) cortisol (hydrocortisone), (e) clenbuterol, (f) mabuterol, (g) cimaterol, (h) methylthiouracil, (i) tapazole (methimazole), (j) nonoxynol-9, (k) genistein (isoflavone), and (l) coumestrol (coumestane).

the specificity. Furthermore, the response of the signals obtained is high, enabling sub-nanogram quantities to be analyzed. A nice example is the determination of 17β-estradiol in bovine plasma [20].

A special use of a combination of acylation and silylation for the GC–MS analysis of stanozolol in calf urine has been described [21,22]. A similar approach has been used for indolalkylamines [23].

The growing interest in GC–combustion–isotope-ratio MS (GC–C–IRMS) has intensified the use of acylation using acetic acid anhydride [24–33]. The advantage is the introduction of a limited number of carbon atoms in order to minimize the effect on the $^{12}C/^{13}C$ ratio.

2.4.3. Methoxylated Derivatives

A specific reagent used for the derivatization of keto groups is methoxylamine or ethoxylamine. The resulting derivative is a stable oxime. This derivatization can be used alone or in combination with silylation. Examples are described for boldenone [34] and chlorotestosterone (clostebol) [35–37].

2.5. Gas Chromatography–Mass Spectrometry Analyses

2.5.1. Gas Chromatography–Mass Spectrometry

As was also discussed in Chapter 18, the analysis of anabolic steroids, other growth promoters, and estrogens has long been closely linked to GC–MS. The low levels in complicated sample materials containing large amounts of closely related substances that are prone to interfere with the analysis, require the use of capillary GC for separation. Remarkably, in (human) doping analysis, split injection has been the method of choice for many years. In the field of veterinary anabolics, this injection technique is not appropriate for this purpose. The target levels are often below 10 μg/L or μg/kg, necessitating the use of splitless injection. Careful maintenance of the GC injector and the column are essential for successful measurements. In most laboratories, fused-silica nonpolar or low-polarity columns, 25 to 30 m in length are used. In particular cases, columns of up to 60 m in length have been used to optimize the separation of different isomers [38].

An additional cleanup of the extracts can be achieved by using GC–GC with column switching. Rozijn et al. [39] have shown that a HPLC cleanup step can be omitted and replaced by GC column switching under routine conditions.

The use of deuterated standards is important for both qualitative and quantitative analysis. In the last decade, many substances have been synthesized by laboratories or chemical companies.

2.5.2. Gas Chromatography–Mass Spectrometry in Multianalyte Analysis

The GC–MS technique is especially suited for multianalyte analysis. Within our laboratories, a large number of anabolic steroids, stilbenes, and resorcylic acid lactones (RALs) such as zeranol are analyzed within one run. After an initial screening with two ions per analyte, suspected samples are reanalyzed using four ions per analyte in selective ion monitoring (SIM) mode, full-scan mode (ion trap) or MS–MS mode (minimum two daughter ions). Using a quadrupole mass spectrometer, the preparation of the instrument settings, i.e., the acquisition parameters, is time consuming. After maintenance of the gas chromatograph (removal of a small polluted part of the column) or column changes, the acquisition parameters of the mass spectrometer must be checked and readjusted.

In Table 2 the mixture of standard analytes and deuterated internal standards used for the calibration of the GC–MS is shown. The analytes were derivatized with MSTFA+ + resulting in the formation of TMS derivatives of both hydroxyl and keto groups. The number of derivatization products formed (often di-TMS, tri-TMS, or isomers) is also included in Table 2. The presence of more than one product has the disadvantage that the minimum detectable level is increased. An advantage is that a combination of products enhances the selectivity and the certainty of the identification.

2.5.3. Gas Chromatography–Tandem Mass Spectrometry

After the development of triple quadrupole mass spectrometers and, moreover, the availability of benchtop ion-trap mass spectrometers, MS–MS has greatly extended the possibilities in the laboratory.

Advantages of MS–MS are the decreased background noise due to the isolation of the parent ion and the subsequent fragmentation. In this way, similar fragments not originating from interferences (sample constituents or column bleed) with a mass comparable to the parent ion cannot be present and will not interfere with the analysis [40]. Recent (unpublished) work in our laboratory demonstrated the power of GC–MS–MS (ion trap) as an important tool for the identification of chloroandrostenedione (CLAD), a metabolite of chlorotestosterone, at a level of 0.2 μg/L in urine samples of cattle.

2.6. Use of Mass Spectrometry as a Confirmation or Identification Tool

The use of MS is often regarded as the ultimate tool for confirmation or identification. Confirmation is generally regarded as identical to identification, with respect to the unambiguous determination of the identity of the analyte. However,

Table 2 Selection of Ions for Multianalyte Steroid Screening Using a Mass Selective Detector

Standard analyte	Remark	RT (min.)	Ions selected for SIM acquisition:	Analyte number
Hexestrol		9.939	207	1
Diethylstilbestrol-d6 (*cis*)	istd	9.354	418; 386	2
PCB-138	GC standard	9.81	360	3
Diethylstilbestrol (*cis*)		9.833	412; 383	4
Hexestrol-d4	istd	9.939	209	5
Dienestrol		10.012	410; 395	6
Dienestrol-d2	istd	10.016	412	7
Diethylstilbestrol-d6 (*trans*)	istd	10.019	418; 386	8
Diethylstilbestrol (*trans*)		10.039	412; 383	9
5α-Estrane-3β, 17α-diol		11.204	332; 242	10
Nortestosterone-17α		12.170	418; 194	11
Trenbolone-17α-1-TMS		12.370	342; 211	12
Boldenone-α		12.401	430; 206	13
Estradiol-17α		12.539	416; 285	14
Nortestosterone-17β-d3	istd	12.653	421; 406	15
Nortestosterone-17β		12.691	418; 194	16
Stanolone (dihydrotestosterone)		12.836	434; 405	17
Boldenone-β		13.018	430; 206	18
Estradiol-17β		13.031	416; 285	19
Estradiol-17β-d3		13.067	419; 285	20
Testosterone-17α		13.232	432; 417	21
Trenbolone-17β-d2-1-TMS	istd	13.258	344	22
Testosterone-17β		13.279	432; 417	23
Trenbolone-17β-1-TMS		13.279	342; 211	24
Testosterone-17β-d2	istd	13.289	434; 419	25
Ethynylestradiol-3-methylether-1-TMS (mestranol)		13.491	382; 367	26
Methandriol		13.742	448; 343	27
Norethynodrel (iso-1)		13.799	442; 427	28
Norethindrone		14.109	442; 427	29
Norethynodrel (iso-2)		14.127	442; 427	30
Delta(9)-11-dehydromethyltestosterone	derivatization standard	14.208	444; 339	31

Compound	istd	RT	m/z	No.
Dianabol-d3	istd	14.218	447; 206	32
Dianabol (methylboldenone)		14.263	444; 206	33
Zearalanon-3-TMS (iso-1)		14.439	521; 307	34
Methyltestosterone-17α-d3	istd	14.514	449; 301	35
Methyltestosterone-17α		14.533	446; 301	36
Ethynylestradiol-17α		14.593	440; 425	37
Norethynodrel (iso-3)		14.610	442; 427	38
17α-Ethynyl-testosterone (ethisterone)		14.709	456; 301	39
Zearalanone-3-TMS (iso-2)		14.712	521; 307	40
Zearalanone-3-TMS (iso-3)		14.964	521; 307	41
4-Chloroandrostendione (iso-1)		15.141	464; 449	42
Zearalanol-α-d4-3-TMS		15.174	437; 307	43
Zearalanol-α-3-TMS		15.263	433; 307	44
Zearalanone-3-TMS (iso-4)		15.369	521; 307	45
Zearalanol-β-d4-3-TMS		15.372	437; 307	46
Zearalanol-β-3-TMS		15.449	433; 307	47
Norgestrel		15.449	456; 316	48
Zearalenone-3-TMS		15.490	519; 305	49
4-Chlorotestosterone (clostebol) (sio-1)		15.555	466; 431	50
17α-CH3-androstan-17β-ol-3-one (mestanolone)		15.567	446; 287	51
Norethandrolone		15.643	446; 287	52
Zearalenol-α-3-TMS		15.950	431; 305	53
Chloroandrostenedione (iso-2)		16.029	464; 449	54
Progesterone iso-1		16.098	458; 443	55
Zearalenol-β-3-TMS		16.202	431; 305	56
2-Methoxy-ethynylestradiol		16.244	470; 455	57
Progesterone iso-2		16.451	458; 443	58
Chlorotestosterone (clostebol)		16.510	466; 431	59
Fluoxymesterone (3-TMS)		16.702	552; 462	60
Fluoxymesterone (2-TMS)		16.992	480; 390	61
4-Bromoestradiol		17.001	496; 365	62
Stanozolol-1-TMS		18.712	400; 385	63

Mw = molecular weight of the derivative.
istd = Deuterated internal standard.

there is a difference between the identification and the assurance that no other substance can lead to the same result. Specifically with regard to the presence of isomers, this distinction is of great importance as is shown in the example in section 2.6.3.

2.6.1. Interferences Related to the Presence of Endogenous Metabolites of Steroids

Using a routine GC–MS procedure in SIM mode for the screening and subsequent identification of anabolic steroids in the urine of cattle, relatively high levels (>50 µg/L) of 17 α-hydroxy-19-nortestosterone (epi-nortestosterone) were likely to be present. In animal experiments with calves after intramuscular treatment with a 17β-nortestosterone ester, it was observed that the levels in urine generally vary between 2 and 10 µg/L for 17β-nortestosterone and between 5 and 20 µg/L for the 17α-metabolite [41]. Closer examination revealed that the ions detected (m/z 346, 331, 256, and 215) were generated by the presence of a reduced metabolite of testosterone, i.e., one of the eight possible tetrahydrotestosterone (THT or androstanediol) isomers. The difference in retention time in that particular case was only a few seconds, approximately 2 to 5 scans. By analyzing all THT isomers, the 5β-androstan-3β, 17β-diol (*bbb*THT) was identified as the interfering substance. Remarkably, all four ions selected of 17α-nortestosterone were present in the spectrum of the interference, although the relative intensities were different (Fig. 2). Applying different margins for the relative intensity ratios will lead to the fulfilment of an increasing number of ion ratios (Table 3). Allowing a larger margin than 10%, i.e., 40%, yields three ion ratios and thus a positive identification.

2.6.2. Criteria for Forensic Analysis in Doping, Crime Investigation, and Detection of Anabolics in Cattle

A number of approaches have been used dealing with the difficult question how many ions are sufficient for an unambiguous identification. Sphon [42] used a simple approach using a (per definition) limited mass spectral library. He concluded that three ions were sufficient to distinguish DES from all other library entries. Many laboratories followed this three-ion approach. Within the European Union, a four-ion approach was officially adopted [43,44]. For the development of these criteria, the approach as described by Pesyna et al. [45] was used to estimate the chance for the occurrence of a steroid with that combination of four ions and their relative abundances based on data from mass spectral libraries. For nortestosterone, this chance, regarded as a chance for a false positive result, was estimated as 1 to $2.7*10^8$.

In many publications reviewed, the use of GC–MS as a tool for unambiguous identification is mentioned. When looked at in detail, a more specific descrip-

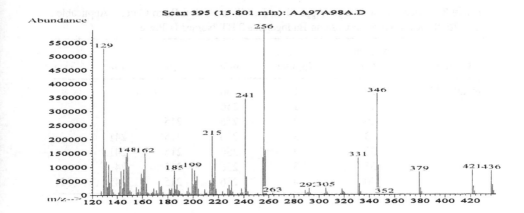

bbb-THT

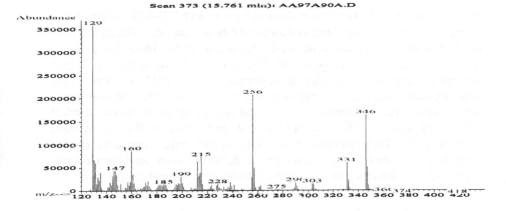

α-NT

Figure 2 Mass spectra of the TMS derivatives of 5β-androstan-3β, 17β-diol (*bbb*THT) and 17α-nortestosterone (αNT)

tion of the data evaluation is missing or only sparsely mentioned. If a quadrupole mass spectrometer was used, the (diagnostic) ion or ions selected for the measurement are mentioned and often shown in chromatograms. However, the precise way in which the complete mass spectrum or the selected ion spectrum was evaluated is infrequently discussed. Infrequently, the relative intensities of the ions measured are compared with the relative intensities obtained for analytical standards.

Table 3 Effect of the Margin Applied on Fulfilling the Ion Ratio Criteria Applicable for 17α-Nortestosterone in Case an Endogenous THT Isomer Is Present

THT isomer	Margin (%)	No. ions	m/e	m/e	m/e	m/e
5bA3b17bD	10	1	256			
	15	1	256			
	20	2	256	215		
	25	3	256	215	241	
	30	3	256	215	241	
	35	3	256	215	241	
	40	4	256	215	241	331

2.6.3. Similarity of Spectra

Special attention is drawn to the analysis of isomers with GC–MS. An illustrative example is the analysis of dexamethasone and betamethasone. The analytes are identical with only one difference in the orientation of the 16-methyl group (α resp. β). Usually in reversed-phase HPLC systems, the retention times are almost identical. Underivatized or after derivatization to (tetra-)TMS derivatives, the resulting mass spectra are virtually indistinguishable. Using the oxidation procedure as described by Courtheyn [46,47], the spectra are again almost identical (Fig. 3). However, a difference is seen between the ratio of the two oxidized isomers formed. Dexamethasone has a higher α to β ratio, whereas betamethasone shows a reversed ratio. If a mixture of dexamethasone and betamethasone is present, the difference in ratios cannot distinguish the analytes anymore.

2.7. Use of Mass Spectrometry to Study Metabolism

The use of GC–MS as tool to study the metabolism and excretion of one or more metabolites has been described in many publications. Although this was not the scope of this chapter to present a complete review, a selection of references has been summarized in Table 4.

3. CORTICOSTEROIDS

3.1. Introduction

Similar to the anabolic steroids, a large number of synthetic corticosteroids were developed by chemical modifications of the endogenous corticosteroid cortisol (hydrocortisone) (see Fig. 1). Corticosteroids are widely used e.g., for the treat-

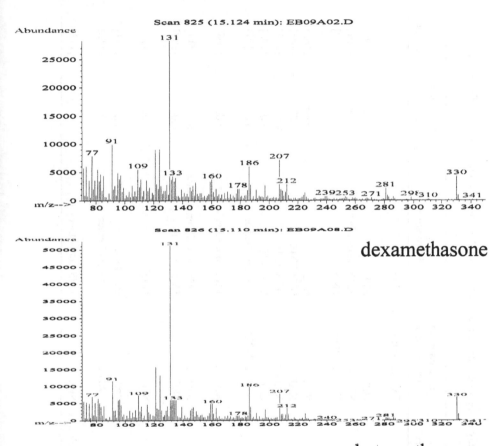

Figure 3 Mass spectra of the oxidized derivatives of dexamethasone and betamethasone.

ment of inflammation or dermal diseases. For several years it has become obvious that corticosteroids have gained popularity in the world of the veterinary anabolics [48]. Often (mixtures of) corticosteroids were detected in combinations with anabolic steroids or β-agonists. The mechanism or mechanisms behind the combined use have not been resolved fully. However, a significant effect on the excretion of, for example, clenbuterol has been noticed [49].

Similar to the anabolic steroids, the group of corticosteroids is large. The variations in chemical structures are larger, which has consequences for the analytical procedures.

Table 4 Selection of References Focused on Study of the Metabolism and Excretion of Anabolic Steroids and Their Endogenous Analogs

Analyte(s)	Sample matrix	Extraction and purification	Derivative	Level or range (µg/L or µg/kg)	Reference
Norethisterone and 6 metabolites	Human plasma	Extrelut and DCM	TMS	0.5–8	[99]
Chlormadinone acetate	Veal calf urine	HPLC	HFB	2–50	[100]
Stanozolol and 2 metabolites	Veal calf urine	SPE, IAC	TMS (enol-ether)	1–100	[101]
Stanozolol and several metabolites	Human urine	XAD-2, ether	TMS	≥1	[102]
Stanozolol and 3 metabolites	Cattle urine	SPE	HFB	≥0.001	[103]
Stanozolol and 1 metabolite	Human urine	XAD-2	HFB-TMS	≥2	[104]
Stanozolol	Rat urine	XAD-2	HFB-TMS	≥50	[105]
Norethisterone and metabolites	Human plasma	L–L	MO-TMS	n.d.	[106]
Methandienone and metabolites	Human urine	XAD-2, HPLC	TMS enol-ether	n.d.	[107]
17α-Methylated anabolic steroids	Human urine	XAD-2	TMS enol-ether	n.d.	[108]
Anabolic steroids	Human urine	XAD-2	TMS or TMS enol-ether	n.d.	[17]
Methandienone and related steroids	Human urine	SPE	TMS enol-ether	n.d.	[109]
Methandienone and metabolites	Horse urine	SPE	TMS	n.d.	[110]
Anabolic steroids	Horse urine	SPE	MO-TMS	n.d.	[111]
Anabolic steroids and 6β-hydroxylated metabolites	Human urine	XAD-2	TMS enol-ether	n.d.	[112]
Testosterone	Horse urine and plasma	SPE	TMS enol-ether		[113]

n.d. = Not determined.

3.2. Sample Cleanup

The sample cleanup for corticosteroids is rather similar to that used for anabolic steroids. The polarity is somewhat higher, allowing the possibility to separate the two groups. Also IAC has been described [50].

3.3. Silyl Derivatives

The presence of a number of hydroxyl groups enables the formation of silyl derivatives. However, from the analysis of corticosteroids, it was observed that sterically hindered functional groups could not be derivatized or only with a low yield. Applications of special mixtures of derivatization regents using the strongest catalysts (TriSil TBT, for example, from Pierce) in combination with prolonged heating have been successful for cortisol metabolites; however, they are less suitable for fast routine analysis. Bagnati et al. [50] have described the use of a TMS derivative of dexamethasone and betamethasone in bovine urine and McLaughlin and Henion [51] described the analysis of a TMS derivative of dexamethasone in tissue.

3.4. Oxidized Derivatives

Due to the relative instability of silylated corticosteroids, alternatives have been developed by Her and Watron [52] and Courtheyn et al. [47]. For dexamethasone and structurally related analytes, an oxidation reaction with pyridinium chlorochromate has proven to be useful in several laboratories. Later, the procedure was improved by changing the oxidizing reagent to dichromate and reducing the reaction time. Furthermore, the application of a hydrolysis step permitted triamcinolone acetonide to be included in the analytical procedure [46]. As described above, special attention should be given to the identification of dexamethasone and betamethasone because of their similar mass spectra and inadequate peak separation.

4. BETA-AGONISTS

4.1. Introduction

Substances with $\beta(2)$-adrenergic action can be used as anabolic agents to increase growth, enhance performance, i.e., carcass quality (less fat and more meat: ''repartitioning''), and increase profits [53]. European Union legislation (EU directive 96/22/EC) prohibits the use of hormonally active anabolics (anabolic steroids, thyrostatic agents, and β-agonists) in animal fattening. In the EU, member-state monitoring and meat inspection programs are carried out to detect illegal use. However, as has been already observed in regulatory control, the number of possibly used β-agonists is large. Small chemical modifications have led to the detection of chemically closely related analogs of clenbuterol, cimaterol, and mabuterol (see Fig. 1). However, the variation of molecular structures that could show a repartitioning effect via β-adrenoceptor activation can be much larger [54–56]. A multitude of GC–MS methods using different derivatives have been described [57,58].

4.2. Sample Cleanup

For current regulatory control for β-agonists in, for example, the Netherlands, a two-stage approach has been developed. This consists of screening methods based

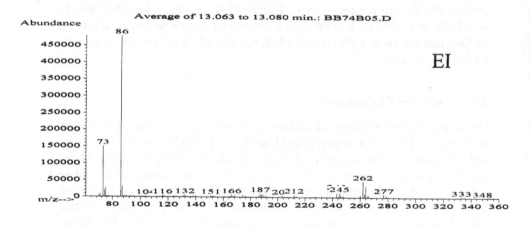

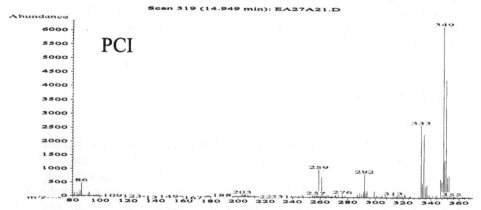

Figure 4 Mass spectra of the TMS derivative of clenbuterol in EI-mode and in PCI-mode (methane).

on enzyme immunoassays (EIA) [59,60], and confirmatory methods using IAC [61] in combination with GC–MS [62].

4.3. Silyl Derivatives

Clenbuterol can be analyzed as a silyl derivative. Unfortunately, the mass spectrum obtained in electron ionization (EI) mode limits the possibilities for an unequivocal identification at lower levels due to the fact that the higher mass ions are only present with very low abundances (Fig. 4). As an alternative, the combination of two analyses, one in EI mode and one in positive chemical ionization (PCI) mode has been successfully used for confirmatory purposes. Furthermore, the combination of two silyl derivatives has been applied [63,64]. The use of alkylboronic acid derivatives has been described, although the gas chromatographic behavior is less favorable [65,66].

5. THYROSTATICS

5.1. Introduction

In contrast to most anabolic agents, there is a general agreement on the ban of thyrostatic drugs: these drugs may be harmful for human health and the meat derived from animals treated with the drugs may be of inferior quality. The weight gain obtained with thyrostatics consists mainly of an increased filling of the gastrointestinal tract and a higher water retention of the animal. Due to the inhibiting effect on the thyroid function by selectively blocking iodine uptake by the thyroid and thus preventing the biosynthesis and excretion of thyroxine (T_4) and triiodothyronine (T_3), the thyrostatics are called antihormonal agents.

The most important and powerful thyrostatic drugs hitherto used are thiouracil and analogous compounds, especially methylthiouracil (MTU) and tapazole (TAP) (see Fig. 1).

5.2. Sample Cleanup

Due to their polar properties, extraction of the thyrostatics is only possible with polar solvents or silica columns. The complexation properties of the thiol compounds offer an elegant alternative method for metal-mediated affinity chromatography [67,68].

Specific procedures for the detection of this group of drugs have been described [67,69,70]. These methods are based on thin-layer chromatography (TLC) in combination with fluorescence induction of the NBD-derivatives (7-chloro-4-nitrobenzo-2-oxa-1,3-diazole) of the drugs with cystein, combined with a rapid and selective extraction procedure, based on a specific complex formation of the drugs with mercury ions. The sample cleanup with the mercurated affinity column was also used by Schilt et al. [68] in a GC–nitrogen-phosphorus detection (NPD) and GC–MS method.

5.3. Derivatization

The thyrostatics can be alkylated to obtain better GC behavior. After extraction, an alkylation with methyl iodide can be made [71]. Alternatively, an extractive alkylation can be performed [68]. Recently, an acylation with pentafluorobenzyl bromide has been successfully applied [72]. After a TLC separation, the thyrostatics can be isolated from the two-dimensional TLC plate and silylated with MSTFA, as was described by Batjoens et al. [73], yielding additional qualitative information.

6. ENDOCRINE DISRUPTORS, SUBSTANCES WITH ESTROGENIC PROPERTIES

Humans are exposed to endocrine disruptors in the workplace, home, community, or during medical care; wildlife are exposed through food and water consumption. Endocrine disruptors are industrial chemicals and environmental pollutants able to disrupt reproductive development in wildlife and humans by mimicking or inhibiting the action of the gonadal steroid hormones estradiol and testosterone, e.g., by binding to estrogen or androgen receptors. The toxicity of these so-called environmental endocrine disruptors is especially insidious during sex differentiation and development due to the crucial role of gonadal steroid hormones in regulating these processes. Published data indicate that chemical exposures may cause alterations in reproductive behavior and contribute to infertility, pregnancy loss, intrauterine fetal demise, birth defects, and ovarian failure in laboratory animals and wildlife.

Data on the association of chemical exposures and adverse reproductive outcomes in humans, however, are equivocal and often controversial. Some studies indicate that chemical exposures are associated with infertility, spontaneous abortion, or reproductive cancer in women. In contrast, other studies indicate that there is no association between chemical exposure and adverse reproductive outcomes. The reasons for such ambiguous findings in human studies are unknown, but likely include the fact that many studies are limited by multiple confounders, inadequate methodology, inappropriate end points, and small sample size. The mechanisms by which chemicals alter reproductive functions in all species are complex and may involve hormonal and/or immune disruption, deoxyribonucleic acid (DNA) adduct formation, altered cellular proliferation, or inappropriate cellular death. There is very little information on the effects of metabolism and intracellular binding proteins on target cell uptake of endocrine disruptors.

Furthermore, individual endocrine disruptors may be agonists/antagonists for more than one endocrine response pathway [74].

A variety of chemicals, ubiquitous in our environment, are suspected reproductive toxicants. Chemicals with known estrogenic or androgenic properties include some herbicides, fungicides, pesticides, e.g., dichlorodiphenyltrichloroethane (DDT), organochlorone pesticides such as methoxychlor and chlordecone (kepone), organophosphate pesticides such as parathion, malathion, and diazanon, phthalates, phytoestrogens, some industrial waste products, solvents such as perchloroethylene, toluene, and styrene, ethylene oxide, 4-vinylcyclohexene, 2,3,7,8-tetrachlorodibenzo-p-dioxin, and polychlorinated biphenyls [75,76]. Analytical chemical aspects of two main groups of endocrine disruptors are discussed below: nonionic surfactants, such as alkylphenol ethoxylates, and phytoestrogens.

7. ALKYLPHENOL ETHOXYLATES

7.1. Introduction

Nonionic surfactants, applied for detergency and (de)foaming processes in industry, consist of alcohol ethoxylates (AE) and alkylphenol ethoxylates (APE). Alkylphenol ethoxylates derived from nonylphenol are called nonylphenol ethoxylates (NPE) and comprise about 80% of the total market volume. Over 60% of these detergents (about 300,000 t/a of alkylphenol ethoxylates in Europe and the United States) end up in an aquatic environment via sewage treatment. Here, environmental biodegradation through anaerobic digestion occurs. The metabolites are nonylphenols, short-chain nonylphenol ethoxylates (typically 1 to 3 ethoxylate units), and acids (such as nonylphenoxy acetic acid and [(nonylphenoxy)ethoxy] acetic acid). These metabolites are even more persistent than their parent compounds and may accumulate in food chains [77–79]. These environmentally stable nonylphenols behave like steroid hormones. The metabolite nonylphenol is about 10 times more toxic than the parent compound [80]. Although the small quantity of nonylphenols may not be directly toxic to aquatic wildlife, an alarming long-term effect of estrogen and progesterone xenobiotics is now evident because of accumulation [81]. The observation that almost every river in the United States has been found to contain biologically active amounts of nonylphenol metabolites has recently led to concerns about their impact on the environment and has prompted an increasing effort to survey Western European environments for the presence of NPE.

7.2. Sample Cleanup

The analytical chemistry of NPE is outlined as follows. The technique most commonly applied to extract nonionic surfactants from aqueous samples has been a procedure called sublation. In this technique, fine bubbles of a inert gas (helium or nitrogen) are dispersed through the aqueous sample, thereby transferring the surface-active materials into an overlaying organic layer. The coextracted ionic surfactants are subsequently removed by ion-exchange chromatography. Centrifugal partition chromatography has been used as an alternative because of its ability to concentrate sample volumes of several liters into organic solvent amounts of several tens of milliliters. Solid-phase extraction (SPE) has been shown to be a very powerful and robust alternative to the traditional methods of extraction. Using this approach, determination of AE and APE in aqueous samples is possible at the ppb level [82].

The determination of alkylphenols and nonylphenol ethoxylates is challenging due to the complexity of the mixtures, comprised of various isomers with different branching of the alkyl moiety and oligomers (with different numbers of ethoxylate units) [82–85].

7.3. Gas Chromatography–Mass Spectrometry

High-performance LC has become the favored method of analysis for these compounds because of its ability to separate and quantitate the various homologues and oligomers by length of the alkyl and ethoxylate chains [86]. The confirmation of the compounds may be done by GC–MS, after the collection of HPLC fractions.

7.4. Derivatization

Derivatization (using N,O-bis(trimethylsilyl)trifluoroacetamide (BSTFA) or hexamethyldisilazane (HMDS)) is only needed for the analysis of oligomers containing more than 5 or 6 ethoxylate units. Alternatively, high-temperature GC may be used, although analyte degradation may occur [87].

8. PHYTOESTROGENS

8.1. Introduction

Phytoestrogens are chemicals naturally occurring in plants, soy products, legumes, and grains, and thought to affect adversely the female reproductive system of laboratory rats, sheep, and humans ([75] and references cited therein). Phytoestrogens appear to have both estrogenic and antiestrogenic activity. The exact mechanisms and effects of phytoestrogens are still unknown. There is evidence that isoflavones exert a protective effect to cancer risk [88,89]. For example, antitumor activity has been demonstrated by inhibition of tumor cell growth. Also, they appear to act in other ways that may inhibit tumor formation and growth, e.g., elevation of sex-hormone-binding globulin and possibly lowering of endogenous estrogen levels [90]. However, concerns about possible adverse effects of isoflavones exist also, especially with children. These are related to knowledge of the role of estrogens at critical stages of development and in mediating reproductive or neuroendocrine disruption in various animal species [91].

The three major chemical types of phytoestrogens that have been identified are flavones, isoflavones, and coumestans. The estrogenic potency of these compounds is variable. The flavones are weak estrogens. The methoxyflavone, tricin, is a constituent of alfalfa and weakly estrogenic in the mouse.

Isoflavones, such as genistein (see Fig. 1) and daidzein, are found in just a few botanical families, e.g., soy. Up to 3 mg/g of genistein and daidzein and their β-glycosides are present in soybeans. Information on isoflavone content in soy products, such as soy milk, tofu, and fermented products such as miso, soy sauce, and tempeh, is given by Coward [92] and Wang [93].

Among the isoflavones, genistein is the most active estrogen with the highest binding affinity for the estrogen receptor. Coumestrol (see Fig. 1), the most potent of the coumestans, has higher binding affinity for the estrogen receptor than genistein. [94].

The metabolic fate of isoflavones, "dietary estrogens," is rather complex. It is of interest that the general pattern of isoflavone conjugates in urine is similar to that of endogenous steroids [94].

8.2. Sample Cleanup

The analysis of isoflavones has been discussed by several authors. The extraction of isoflavones from soy products preferably occurs by acid hydrolysis of the glucoside conjugates followed by a (hot) extraction with acidic methanol, ethanol, or acetonitrile. Extraction of isoflavones from biological fluids, such as serum, plasma, urine, and milk usually occurs using solid-phase extraction methodology [95].

8.3. Gas Chromatography–Mass Spectrometry

Traditionally, GC–MS was applied to determine soy isoflavones and their metabolites in human biological fluids [90,96,97]. Silylation is the most often used method for the derivatization. For quantification, many deuterated standards have been synthesized [97]. Similar to the developments for steroid analysis, LC–MS is more and more frequently used [95,98].

9. GENERAL REMARKS

The widespread application of GC–MS in residue analysis has improved the analytical possibilities enormously and has largely contributed to the quality of the results obtained. Similar to the rapid expansion of GC–MS in routine laboratories in the past decade, LC–MS is following this pattern. In general, GC–MS enables measurement at low residue levels with a high degree of specificity. Unfortunately, no estimations of the error probabilities, especially for false positive identifications, are available yet. Currently, a limited number of studies are being carried out to develop models to predict these possibilities. The application of quality criteria destined to objectively evaluate the mass spectrometric results [44] greatly improves the results obtained. However, their general use is still limited as is observed in many publications reviewed. Usually one or more diagnostic ions are mentioned and depicted, whereas no reference is made to the actual procedure followed to compare spectra of pure standards with spectra obtained from analytes present in the sample.

As has happened with GC–MS, there is a clear trend that the use of LC–MS, because of the increasing availability of robust and sensitive instruments, will be very common in the field of the analysis of growth promoters and endocrine disruptors. In many laboratories, LC–MS is used already to detect and identify anabolic steroids, corticosteroids, and β-agonists at low μg/L or μg/kg levels. Looking at multiresidue analysis, LC–MS will earn its position as a very useful technique. The development of quality criteria for LC–MS, similar to the well-known criteria for GC–MS, for confirmatory analysis will certainly boost the possibilities for more general use.

It is expected, however, that it will not fully replace GC–MS because of the (complicated) behavior of several growth promoters (especially steroids) that possess unfavorable LC–MS properties (either too low or too high degree of fragmentation). Furthermore, the resolution obtained with LC is normally lower than that obtained with capillary GC. The modern laboratory will use both GC–MS and LC–MS to solve their analytical challenges.

ACKNOWLEDGMENTS

The authors thank Mr. H. Hooijerink, Mr. E. O. van Bennekom, and Mr. B. Brouwer for their valuable assistance and contributions.

REFERENCES

1. R. Ferrando, J.P. Valette, N. Henry, R. Boivin, and A. Parodi, C.R. Acad. Sci. Hebd. Seances Acad. Sci. D, 278 (1974) 2067–2070.
2. R. Ferrando, and J.P. Valette, Folia Vet. Lat., 5 (1975) 27–44.
3. A.I. Marcus, Cancer from Beef, DES. Federal Food Registration and Consumer Confidence. 1994, John Hopkins University Press, Baltimore, MD.
4. S. Melnick, P. Cole, A.D., and A. Herbst, N. Engl. J. Med., 316 (1987) 514–516.

5. C. Pulce, D. Lamaison, G. Keck, C. Bostvironnois, J. Nicolas, and J. Descotes, Vet. Hum. Toxicol., 33 (1991) 480–481.
6. J.F. Martinez-Navarro, Lancet, 336 (1990) 1311.
7. R. White, S. Jobling, S.A. Hoare, J.P. Sumpter, and P. MG, Endocrinology, 135 (1994) 175–182.
8. R.J. Kavlock, Chemosphere, 39 (1999) 1227–1236.
9. Shackleton and e. aanvullen, corticosteroids aanvullen. J. Chromatogr., 379 (1986) 91–156.
10. T. Hamoir, D. Courtheyn, H. De Brabander, P. Delahaut, L. Leyssens, and G. Pottie, Analyst, 123 (1998) 2621–2624.
11. L.A. van Ginkel, J. Chromatogr., 564 (1991) 363–384.
12. W. Schanzer, P. Delahaut, H. Geyer, M. Machnik, and S. Horning, J. Chromatogr. B, 687 (1996) 93–108.
13. L.A. van Ginkel, R.W. Stephany, H.J. van Rossum, H.M. Steinbuch, G. Zomer, E. Van de Heeft, and A.P. De Jong, J. Chromatogr., 489 (1989) 111–120.
14. M. Dubois, X. Taillieu, Y. Colemonts, B. Lansival, J. De Graeve, and P. Delahaut, Analyst, 123 (1998) 2611–2616.
15. T.L. Fodey, C.T. Elliott, S.R.H. Crooks, and W.J. McCaughey, Food Agric. Immunol., 8 (1996) 157–167.
16. J. Segura, R. Ventura, and C. Jurado, J. Chromatogr. B, 713 (1998) 61–90.
17. W. Schanzer, and M. Donike, Analytica Chimica Acta, 275 (1993) 23–48.
18. E. Daeseleire, R. Vandeputte, and C. Van Peteghem, Analyst, 123 (1998) 2595–2598.
19. M. Walshe, M. O'Keeffe, and B. Le Bizec, Analyst, 123 (1998) 2687–2691.
20. L.A.v. Ginkel, R.W. Stephany, A. Spaan, S.S. Sterk, and L.A. Van Ginkel. Bovine blood analysis for natural hormones. Analytica Chimica Acta, 275 (1993) 75–80.
21. H.F. De Brabander, et al., Analyst, 123 (1998) 2599–2604.
22. R. Schilt, M.J. Groot, P.L.M. Berende, V. Ramazza, J.S. Ossenkoppele, W. Haasnoot, E.O. Van Bennekom, L. Brouwer, and H. Hooijerink, Analyst, 123 (1998) 2665–2670.
23. M. Donike, R. Gola, and L. Jaenicke, J. Chromatogr., 134 (1977) 385–395.
24. R. Aguilera, M. Becchi, H. Casabianca, C.K. Hatton, D.H. Catlin, B. Starcevic, and H.G. Pope, J. Mass Spectrom., 31 (1996) 169–176.
25. R. Aguilera, M. Becchi, C. Grenot, H. Casabianca, and C.K. Hatton, J. Chromatogr. B., 687 (1996) 43–53.
26. R. Aguilera, D.H. Catlin, M. Becchi, A. Phillips, C. Wang, R.S. Swerdloff, H.G. Pope, and C.K. Hatton, J. Chromatogr. B., 727 (1999) 95–105.
27. M. Bjoroy, J.A. Williams, D.L. Dolcater, M.K. Kemp, and J.C. Winters, Marine Petrol. Geol. 13 (1996) 3–23.
28. V. Ferchaud, B. Le Bizec, F. Monteau, and F. Andre, Analyst, 123 (1998) 2617–2620.
29. P.M. Mason, S.E. Hall, I. Gilmour, E. Houghton, C. Pillinger, and M.A. Seymour, Analyst, 123 (1998) 2405–2408.
30. W. Schanzer, Doping and dope analysis, Chemie Unser. Zeit, 31 (1997) 218–228.
31. C.H.L. Shackleton, A. Phillips, T. Chang, and Y. Li, Steroids, 62 (1997) 379–387.

32. C.H.L. Shackleton, E. Roitman, A. Phillips, and T. Chang, Steroids, 62 (1997) 665–673.
33. M. Ueki, Jpn. J. Toxicol. Environ. Health, 44 (1998) 75–82.
34. M.C. Dumasia, and E. Houghton, Biomed. Environ. Mass Spectrom. 17 (1988) 383–392.
35. L. Leyssens, E. Royackers, B. Gielen, M. Missotten, J. Schoofs, J. Czech, J.P. Noben, L. Hendriks, and J. Raus, J. Chromatogr. B, 654 (1994) 43–54.
36. L. Hendriks, B. Gielen, L. Leyssens, and J. Raus, Vet. Rec. 134 (1994) 192–193.
37. M. Van Puymbroeck, M.E.M. Kuilman, R.F.M. Maas, R.F. Witkamp, L. Leyssens, D. Vanderzande, J. Gelan, and J. Raus, Analyst, 123 (1998) 2681–2686.
38. C.J.M. Arts, R. Schilt, M. Schreurs, and L.A.v. Ginkel, in Proceedings of Euroresidue III (1996), 1996, p. 212.
39. R. Rozijn, B.A. Koops, M. Schreurs, and L.M.H. Frijns. Screening and confirmatory analysis of anabolic steroids, stilbenes and resorcylic acid lactones by two-dimensional gas chromatography coupled with mass spectrometry, paper presented at the Third international symposium on hormone and veterinary residue analysis, 1998, Bruge.
40. D. Courtheyn, H. De Brabander, J. Vercammen, P. Batjoens, M. Logghe, and K. De Wasch, in Proceedings of Euroresidue III (1996) N. Haagsma and A. Ruiter, eds. University of Ulrecht, The Netherlands, 1996, p. 75.
41. W. Haasnoot, R. Schilt, A.R. Hamers, F.A. Huf, A. Farjam, R.W. Frei, and U.A. Brinkman, J. Chromatogr., 489 (1989) 157–171.
42. J.A. Sphon, J. Assoc. Offic. Analyt. Chem. 61 (1978) 1247–1252.
43. 87/410/EEC, E.D., Commission Decision of 14 July 1987 laying down the methods to be used for detecting residues of substances having a hormonal action and of substances having a thyreostatic action (87/410/EEC). Official Journal of the European Community, 1987. L 223 (Aug. 11, 1987) p. 18–36.
44. 93/256/EEC, E.D., Commission Decision of 14 April 1993 laying down the methods to be used for detecting residues of substances having a hormonal or a thyreostatic action (93/256/EEC). Official Journal of the European Community, 1993. L 118 (May 14, 1993) p. 64–74.
45. G.M. Pesyna, F.W. McLafferty, R. Venkataraghavan, and H.E. Dayringer, Analyt. Chem. 47 (1975) 1161–1164.
46. D. Courtheyn, J. Vercammen, M. Logghe, H. Seghers, K. De Wasch, and H. De Brabander, Analyst, 123 (1998) 2409–2414.
47. D. Courtheyn, J. Vercammen, H. De Brabander, I. Vandenreyt, P. Batjoens, K. Vanoosthuyze, and C. Van Peteghem, Analyst, 119 (1994) 2557–2564.
48. D. Courtheyn, et al., in Proceedings of Euroresidue II (1993) N. Haagsma and A. Ruiter, eds. University of Ulrecht, The Netherlands, 1993, p. 251.
49. M.J. Groot, R. Schilt, J.S. Ossenkoppele, P.L.M. Berende, and W. Haasnoot, J. Vet. Med., 45 (1998) 425–440.
50. R. Bagnati, V. Ramazza, M. Zucchi, A. Simonella, F. Leone, A. Bellini, and R. Fanelli, Anal. Biochem., 235 (1996) 119–126.
51. L.G. McLaughlin, and J.D. Henion, J. Chromatogr., 529 (1990) 1–19.
52. G.R. Her, and J.T. Watson, Biomed. Environ. Mass Spectrom., 13 (1986) 57–63.

53. J.P. Hanrahan, ed., Proceedings of the seminar in the CEC programme of Co-ordination on Research in Animal Husbandry, 1987, Brussels, 1987, Elsevier, London.
54. J. Keck, G. Krüger, K. Noll, and H. Machleidt, Arzneimittel Forschung, 22 (1972) 861–869.
55. G. Asato, et al., Agric. Biol. Chem., 48 (1984) 2883–2888.
56. G. Krüger, J. Keck, K. Noll, and H. Pieper, Arzeneimittelforschung 34 (1984) 1612–1624.
57. A. Polettini, J. Chromatogr. B, 687 (1996) 27–42.
58. D. Boyd, O.K. M., and M.R. Smyth, Analyst, 121 (1996) R1–R10.
59. W. Haasnoot, A.R.M. Hamers, R. Schilt, and C.A. Kan in M.R.A. Morgan, C.J. Smith, and P.A. Williams, eds., Food Safety and Quality Assurance: Applications of Immunoassay Systems, Proceedings of an International Conference, March 19–22, 1991, Bowness-on-Windermere, Cumbria, UK, 1992, Elsevier, London; p. 237.
60. M.E. Ploum, W. Haasnoot, R.J.A. Paulussen, G.D.v. Bruchem, A.R.M. Hamers, Schilt, R., and F.A. Huf, J. Chromatogr. 564 (1990) 413–427.
61. W. Haasnoot, M.E. Ploum, R.J.A. Paulussen, R. Schilt, and F.A. Huf, J. Chromatogr. 519 (1990) 323–335.
62. R. Schilt, W. Haasnoot, M.A. Jonker, H. Hooijerink, and R.J.A. Paulussen, in Proceedings of Euroresidue I (1990) N. Haagsma, A. Ruiter, and P. B. Czedik-Eijsenberg, eds., University of Utrecht, The Netherlands, 1990, p. 320.
63. J.A. Van Rhijn, W.A. Traag, and H.H. Heskamp, J. Chromatogr. B, 619 (1993) 243–249.
64. J.A. Van Rhijn, H.H. Heskamp, M.L. Essers, H.J. Van De Wetering, H.C.H. Kleijnen, and A.H. Roos, J. Chromatogr. B, 665 (1995) 395–398.
65. A. Polettini, M.C. Ricossa, A. Groppi, and M. Montagna, J. Chromatogr., 564 (1991) 529–535.
66. W.J. Blanchflower, S.A. Hewitt, A. Cannavan, C.T. Elliott, and D.G. Kennedy, Biol. Mass Spectrom., 22 (1993) 326–330.
67. H.F. De Brabander, The Determination of Thyreostatic Drugs in Biological Material, 1986, Ghent, Belgium, Ghent.
68. R. Schilt, J.M. Weseman, H. Hooijerink, H.J. Korbee, W.A. Traag, M.J. van Steenbergen, and W. Haasnoot, J. Chromatogr., 489 (1989) 127–137.
69. H.F. De Brabander and R. Verbeke, in Proceedings of the 30th European Meeting of the Meat Research Workers, 1984, Bristol, UK.
70. H.F. De Brabander, and R. Verbeke, J. Chromatogr., 108 (1975) 141–151.
71. Yu Gladys, Y.F., E.J. Murby, and J. Wells Robert, J. Chromatogr. B, 703 (1997) 159–166.
72. B. LeBizec, F. Monteau, D. Maume, M.P. Montrade, C. Gade, and F. Andre, Anal. Chim. Acta, 340 (1997) p. 201–208.
73. P. Batjoens, H.F. DeBrabander, and K. DeWasch, J. Chromatogr., 750 (1996) 127–132.
74. S. Safe, K. Connor, K. Ramamoorthy, K. Gaido, and S. Maness, Reg. Toxicol. Pharmacol., 26 (1997) 52–58.
75. F.I. Sharara, D.B. Seifer, and J.A. Flaws, Fertil. Steril., 70 (1998) 613–622.
76. M. Castillo, and D. Barceló, Trends Anal. Chem., 16 (1997) 574–583.

77. C.A. Staples, J. Weeks, J.F. Hall, and C.G. Naylor, Environm. Tox. Chem., 17 (1998) 2470–2480.
78. H. Maki, K. Tokuhiro, Y. Fujiwara, M. Ike, K. Furukawa, and M. Fujita, J. Environm. Sci., 8 (1996) 275–284.
79. H. Maki, H. Okamura, I. Aoyama, and M. Fujita, Environm. Tox. Chem., 17 (1998) 650–654.
80. T.W. Loo, and D.M. Clarke, Biochem. Biophys. Res. Comm., 247 (1998) 478–480.
81. A.S. Bourinbaiar, AIDS, 11 (1997) 1525–1526.
82. A.T. Kiewiet, and P. de Voogt, J. Chromatogr., 733 (1996) 185–192.
83. M.S. Holt, G.C. Mitchell, and R.J. Watkinson, In O. Hutzinger, ed., The Handbook of Environmental Chemistry, volume 3, part F, 1992, Springer-Verlag, Berlin, p. 89.
84. G. Kloster, Detergents and Environment, Surfactant Science Series, vol. 65, 1997, Marcel Dekker, New York.
85. N. Garti, V.R. Kaufman, and A. Aserin, In J. Cross, ed., Nonionic Surfactants: Chemical Analysis. 1987, Marcel Dekker, New York.
86. K.B. Sherrard, P.J. Marriott, R.G. Amiet, and M.J. McCormick, Chemosphere, 33 (1996) 1921–1940.
87. P.d. Voogt, K.d. Beer, and F.v.d. Wielen, Trends Anal. Chem., 16 (1997) 584–595.
88. S. Barnes, Proc. Soc. Exp. Biol. Med., 217 (1998) 386–392.
89. M.J. Messina, V. Persky, K.D.R. Setchell, and S. Barnes, Nutr. Cancer, 21 (1994) 113–131.
90. J.T. Dwyer, B.R. Goldin, N. Saul, L. Gualeiri, S. Barakat, and H. Adlercreutz, J. Am. Diet. Assoc., 94 (1994) 739–743.
91. K.D.R. Setchell, L. Zimmer-Nechemias, J. Cai, and J.E. Heubi, Am. J. Clin. Nutr., 68(suppl) (1998) 1453S–1461S.
92. L. Coward, N.C. Barnes, K.D.R. Setchell, and S. Barnes, J. Agric. Food. Chem., 41 (1993) 1961–1967.
93. H. Wang, and P.A. Murphy, J. Agric. Food Chem., 42 (1994) 1666–1673.
94. M.L. Brandi, Calcif. Tissue Int., 61 (1997) S5–S8.
95. S. Barnes, M. Kirk, and L. Coward, J. Agric. Food Chem., 42 (1994) 2466–2474.
96. G.E. Joannou, G.E. Kelly, A.Y. Reeder, M. Waring, and C. Nelson, J. Steroid Biochem. Molec. Biol., 54 (1995) 167–184.
97. H. Adlercreutz, J. Van Der Wildt, J. Kinzel, H. Attalla, K. Wahala, T. Makela, T. Hase, and T. Fotsis, J. Steroid Biochem. Molec. Biol., 52 (1995) 97–103.
98. S. Barnes, L. Coward, M. Kirk, and J. Sfakianos, Proc. Soc. Exp. Biol. Med., 217 (1998) 254–262.
99. F. Pommier, A. Sioufi, and J. Godbillon, J. Chromatogr. B, 674 (1995) 155–165.
100. H. Hooijerink, R. Schilt, E.O. Van Bennehom, B. Brouwer, and P.L.M. Berende, in Proceedings of Euroresidue III (1996), 1996, p. 505.
101. P. Delahaut, X. Taillieu, M. Dubois, K. DeWasch, H.F. DeBrabander, and D. Courtheyn, Archiv Lebensmittelhygiene, 49 (1998) 3–7.
102. W. Schanzer, G. Opfermann, and M. Donike, J. Steroid Biochem., 36 (1990) 153–174.

103. V. Ferchaud, B. LeBizec, M.P. Montrade, D. Maume, F. Monteau, and F. Andre, J. Chromatogr. B, 695 (1997) 269–277.
104. H.Y.P. Choo, O.S. Kwon, and J. Park, J. Analyt. Toxicol., 14 (1990) 109–112.
105. J.C. Ryu, O.S. Kwon, Y.S. Song, J.S. Yang, and J. Park, J. Appl. Toxicol., 12 (1992) 385–391.
106. M.S. Rizk, N.A. Zakhari, M.I. Walash, S.S. Toubar, C.J. Brooks, and R. Anderson, Acta Pharm. Nord., 3 (1991) 205–210.
107. W. Schanzer, H. Geyer, and M. Donike, J. Steroid Biochem. Mol. Biol., 38 (1991) 441–464.
108. W. Schanzer, G. Opfermann, and M. Donike, Steroids, 57 (1992) 537–550.
109. R. Masse, H.G. Bi, C. Ayotte, P. Du, H. Gelinas, and R. Dugal, J. Chromatogr., 562 (1991) 323–340.
110. H.W. Hagedorn, R. Schulz, and A. Friedrich, J. Chromatogr. Biomed. Applic., 577 (1992) 195–203.
111. P. Teale, and E. Houghton, Biol. Mass Spectrom., 20 (1991) 109–114.
112. W. Schanzer, S. Horning, and M. Donike, Steroids, 60 (1995) 353–366.
113. Y. Bonnaire, L. Dehennin, P. Plou, and P.L. Toutain, J. Analyt. Toxicol., 19 (1995) 175–181.

20

Identification of Terpenes by Gas Chromatography–Mass Spectrometry

A. Orav
Tallinn Technical University, Tallinn, Estonia

1. INTRODUCTION

Monoterpenes, sesquiterpenes, and their oxygen-containing derivatives are an important class of volatile constituents in essential oils of plants that have been found in many types of environmental and geological samples and in other natural or synthetic sources. The identification of terpenes as flavor, fragrance, and curative compounds is important in medicine, veterinary science, and the cosmetics, flavor, and food industries.

The identification of terpenoic compounds is very complicated because these substances usually occur in mixtures of closely related compounds, and they are sensitive to various kinds of isomerization and to the rearrangement of the carbon skeleton. Gas chromatography (GC) has proved to be the most useful and fastest method of separating the terpenes in complex mixtures; substantial progress has been achieved after making use of capillary columns. The very complex nature of many essential oils does not allow a complete chromatographic separation even on high-resolution columns.

The introduction of gas chromatography–mass spectrometry (GC–MS) has been one of the most significant steps toward a reliable identification of components of such mixtures [1–3]. In recent years, the mass spectra of many terpenes have been established. Information about them can be found in databases,in specialized works or atlases, and in original articles or reviews. However, there are some complications in identifying terpenes by GC–MS only. Various compounds (homologues, positional isomers, and stereoisomers) have similar spectra; for example, many monoterpenes have a base peak at 93. Sometimes mass spectra from different sources for the same compound may be different. The errors in identification by GC–MS may also be caused by overlapping of GC peaks.

The reliable identification of terpenes has been achieved using GC–MS together with GC retention indices (RIs) [2,3]. It is recommended that gas chromatography–flame ionization detection (GC–FID) analysis be performed on two columns of different polarity. Usually, terpene analysis is now carried out on capillary columns with nonpolar dimethylsiloxane and polar polyethylene glycol 20M (PEG 20M) stationary phases [4–33]. The GC analysis of terpenes of a wide range of boiling temperatures in complex mixtures requires temperature-programming conditions. The RI values in temperature-programming conditions are highly dependent on operating conditions, including column length and diameter, film thickness, linear velocity of the carrier gas, and the temperature-programming rate. These changes of RIs of terpenes are greater in polar PEG 20M than in nonpolar phases, owing to greater temperature dependence of RIs of terpenes in the polar phase. A more polar stationary phase leads to more variation in the retention data caused by variation in the coating thickness of the stationary phase. The column measurements, film thickness of the stationary phase, and the operating conditions of GC analysis must therefore be known before data from literature about RIs can be used. In recent years, the reproducibility of retention parameters reported by different authors has improved.

Identification precision can be increased by applying additional characteristics, such as the difference in RI values for columns with different polarity, ΔRI, which gives an indication of the functionality of the compound. The large values of ΔRI indicate higher polarities of these components. Zenkevich et al. [4–6] have used an additional identification parameter—the coefficient of partition, K_p, of the analytes between two immiscible liquids (n-hexane and acetonitrile), the values of which ensure the efficiency of group identification.

The quantitative agreement between the two columns gives additional information for the identification. Discrimination is always possible when a sample splitter is used.

In this chapter, the practical results obtained in our laboratory for the identification of terpenoic compounds in volatile natural products are presented using GC–FID on two capillary columns and GC–MS instruments. The difficulties in the identification procedure that arose in this work are also presented.

2. IDENTIFICATION OF TERPENOIC COMPOUNDS

2.1. Experimental

2.1.1. Gas Chromatography - Flame Ionization Detection

The separation of terpenes was carried out on Chrom-5 chromatographs with FID on two fused-silica capillary columns (50 m × 0.2 mm) with bonded stationary phases OV-101 (film thickness 0.5 μm) and PEG 20M (film thickness 0.25 μm). The carrier gas was helium with a splitting ratio of approximately 1:150; flow rates of about 1.3 ml/min for OV-101 and 1.5 ml/min for PEG 20M were applied. The column temperature programming was from 50 to 250°C (OV-101) and from 70 to 230°C (PEG 20M) at the rate of 2°/min. The injector temperature was approximately 160°C. Fused-silica capillary columns and the low injection temperature helped to avoid the formation of artifacts.

A Hewlett-Packard Model 3390A integrator and IBM 286 personal computer were applied in data processing. The temperature-programmed RI was calculated with the following equation:

$$RI = 100\, z + 100\, (t_{Rx} - t_{Rz})/(t_{R(z+1)} - t_{Rz}) \tag{1}$$

where t_{Rx}, t_{Rz}, $t_{R(z+1)}$ are the retention times of the unknown substance x and the n-alkanes z and $z+1$ carbon atoms surrounding the compound x. The reproducibility of RI values expressed in terms of the standard deviation (SD) was 0.1 to 2.5 index units (iu).

The difference in RI values for columns with different polarity, ΔRI, was calculated as follows:

$$\Delta RI = RI(polar) - RI(nonpolar) \tag{2}$$

where RI(polar) and RI(nonpolar) are the retention index values on PEG 20M and OV-101 columns, respectively.

A program written in BASIC was used to calculate the RI values for each peak on two columns and to identify the terpenes using our RI library data and peak areas [37]. For the creation of a RI data bank on two stationary phases in temperature-programming conditions, some standard terpene compounds and some well-characterized mixtures as well as RI literature data on the same columns in the same operating conditions as in our laboratory were used.

2.1.2. Gas Chromatography–Mass Spectrometry

The mass spectrometric analyses were carried out on a Hitachi M-80 B gas chromatograph double-focusing mass spectrometer using OV-1 (50 m × 0.25 mm), PEG 20M (50 m × mm), and RSL 300 (30 m × 0.32 mm) fused-silica capillary columns. The electron energy was 70 eV, the mass range 30 to 300. The temperature programs were 70 to 280°C at 5°/min (OV-1), 60 to 120°C at 5°/min, 120 to 190°C at 10°/min (RSL 300), and 40 min at 70°C, then 70 to 240°C at 3°/min (PEG 20M).

2.2. Identification of Terpene Hydrocarbons

2.2.1. Monoterpenes

The region of RI values of monoterpenes studied was from 877 to 1077 (200 iu) on OV-101 and from 990 to 1283 (293 iu) on PEG 20M (Table 1). All 18 monoterpenes were separated except α-thyjene and α-pinene on PEG 20M and limonene and β-phellandrene on OV-101. The ΔRI values of monoterpenes were from 100 to 210 iu, depending on the number of double bonds in the molecule, and 260 for *p*-cumene (aromatic cycle) (Table 2).

There are many data about RIs for monoterpenes in the literature [2–33]. This is one of the reasons why the reproducibility of RIs for monoterpenes is better than for other terpenes. The reproducibility of RIs is better on OV-101 than on PEG 20M. This is true of all terpenes and is caused by a smaller number of reference RI data and greater temperature dependence of terpenes in the polar phase (Table 3). It is noticeable that the differences in RI values of the same monoterpenes from different references depend mainly on the starting temperature of the program, especially for monoterpenes with two or three double bonds (aliphatic and monocyclic structures) and for the aromatic cycle, which have greater RI temperature dependence than other monoterpenes [36].

The mass spectra of all monoterpenes studied are obtainable from MS reference libraries and from special articles. The majority of mass spectra of monoterpenes have a base peak at 93 [5], but it is possible to identify these compounds by comparison with other mass fragments also. As the number of monoterpenes

Table 1 Regions of Retention Index (RI) Values of Terpenoic Groups on OV-101 and PEG 20M Columns

	RI	
Compound group	OV-101	PEG 20M
Monoterpenes	877–1077	990–1280
Sesquiterpenes	1300–1530	1477–1820
Oxygenated monoterpenes	1020–1584	1210–2186
Oxygenated sesquiterpenes	1550–1730	1960–2400

Table 2 Differences in RI Values of Terpenes on PEG 20M and
OV-101 (ΔRI)

Component groups	No. of double bonds	Structure	Region of ΔRI (iu)
Monoterpenes	0	Tricyclic	100
	1	Bicyclic	100–160
	2	Monocyclic	170–200
	3	Aliphatic	177–210
Sesquiterpenes	1	Tricyclic	112–190
	2	Bicyclic	175–250
	3	Monocyclic	210–260
	3	Aliphatic	250–257
Oxygenated monoterpenes		Oxides	193–336
		Esters	260–380
		Aldehydes	350–486
		Ketones	332–510
		Alcohols	440–620
Oxygenated sesquiterpenes		Oxides	390–500
		Alcohols	480–680
Aromatic hydrocarbons			260

frequently appearing in essential oils is usually below 18 and the elution order
of these compounds on both columns is known, it is not a problem to identify
monoterpenes using RI values. The MS data can confirm the RI identification.

2.2.2. Sesquiterpenes

The number of sesquiterpenes is high (over 150 compounds). There are compounds with aliphatic (farnesenes), monocyclic (humulenes, bisabolenes, elem-
enes), bicyclic (cadinenes, muurolenes, selinenes, caryophyllenes, germacrenes),
tricyclic (copaenes, ylangenes, longifolenes,longicyclenes), and tetracyclic (sati-
vene) structures.

The RI values of sesquiterpenes studied in our laboratory range from 1330
to 1533 (203 iu) on OV-101 and from 1464 to 1766 (302 iu) on PEG 20M (see
Table 1). From these data it is evident that the polar PEG 20M is more selective
for the separation of sesquiterpenes, but some sesquiterpene hydrocarbons do not
separate on PEG 20M either (for example γ- and δ-cadinenes). The ΔRI values
of sesquiterpenes are somewhat higher (112 to 260 iu) than for monoterpenes
and, like ΔRI values of monoterpenes, the values depend mainly on the number
of double bonds in the molecule (see Table 2).

Table 3 Reproducibility of Literature RI Values of Some Terpenes on Two
Stationary Phases

Compound	OV-101			PEG 20M		
	No. of studies	Mean RI	SD	No. of studies	Mean RI	SD
α-Pinene	17	934 ± 2	4	12	1025 ± 6	9
β-Pinene	17	974 ± 3	6	14	1114 ± 6	11
Limonene	20	1025 ± 2	4	19	1202 ± 6	13
Terpinolene	12	1083 ± 4	7	11	1289 ± 6	9
α-Copaene	16	1378 ± 4	8	10	1498 ± 12	19
β-Capyophyllene	17	1420 ± 4	8	15	1602 ± 13	23
α-Humulene	16	1452 ± 5	10	13	1670 ± 11	19
δ-Cadinene	13	1517 ± 3	5	13	1753 ± 12	20
Linalol	19	1085 ± 4	8	13	1542 ± 13	22
α-Terpineol	20	1177 ± 4	8	15	1695 ± 14	25
Geranyl acetate	10	1367 ± 8	10	7	1751 ± 10	10
Spathylenol	8	1566 ± 9	11	4	2124 ± 33	21
α-Cadinol	7	1645 ± 15	16	3	2223 ± 11	6

Source: Determined from Refs. 7–35.

Literature on RI data are available for over 80 sesquiterpene hydrocarbons
[2–35]. However, there are more references for nonpolar phases than for PEG
20M. The RI data of sesquiterpenes in the polar phase are less reproducible also.
The mean value of SDs for four frequently occurring sesquiterpenes on OV-101
was 8 iu, but on PEG 20M it was 20 iu (see Table 3).

For some sesquiterpenes, there are no mass spectra present in our MS libraries, while for some other sesquiterpenes, the mass spectra are very similar. Different sources sometimes give different spectra for the same compound [1,5,28,34].
As the number of sesquiterpenes is high and the elution region of sesquiterpenes
peaks is wider on the polar PEG 20M column, it is necessary to separate the
sesquiterpenes by GC–MS on the polar column, too. For positive identification
of sesquiterpene hydrocarbons, the agreement of at least three identification parameters is necessary (RI data on two columns and GC–MS data on the polar
column).

Since authentic sesquiterpene hydrocarbons are not available for the determination of RI and MS data, the identification of many sesquiterpenes is problem-
atic in our laboratory. Another problem is incorrect spectra of peaks caused by
overlapping of compounds.

2.3. Identification of Oxygenated Terpenes

2.3.1. Oxygenated Monoterpenes

On nonpolar OV-101, the majority of oxygenated monoterpenes eluted between monoterpenoic and sesquiterpenoic hydrocarbons (see Table 1), except some low-boiling oxides and hydrates (cineoles, sabinene hydrate, linalol oxides, fenchon) eluting with monoterpenes and some monoterpenoic esters (neryl and geranyl acetate, geranyl isovalerate, etc.) eluting in the region of sesquiterpenes. The retention data for 1,8-cineole (distributed constituent of essential oils) is very similar to these data for β-phellandrene in both phases studied. The moderately polar RSL 300 column completely separated 1,8-cineole, β-phellandrene, and limonene [38]. The RI region of oxygenated monoterpenes on polar PEG 20M is very wide (about 975 iu) and covered the region of sesquiterpenes on this column (see Table 1). For the identification of oxygenated monoterpenes, it is necessary to consider the values of ΔRI (see Table 2). The ΔRI values of oxygenated monoterpenes (>260 iu) are higher than for terpene hydrocarbons (<260 iu), with the exception of some oxides. Sesquiterpenoic alcohols (440 to 620 iu) have the highest values of ΔRI.

There are literature RI data of oxygenated monoterpenes for over 150 compounds [2–33], but a small number of data for single oxygenated monoterpenes. For this reason, the reproducibility of these data is low [6]. If the number of reference RI data exceeds 10 (n > 10), the mean RI does not depend so much on single data, but the RI reproducibility of oxygenated monoterpenes is still over 10 iu on PEG 20M (see Table 3).

Mass spectra for some oxygenated monoterpenes are missing in our data bank or the correct isomer has not yet been identified. According to the data of Zenkevich et al. [6], $(m/z)_{max}$ data are presented for 168 compounds from the 214 oxygenated monoterpenes studied.

2.3.2. Oxygenated Sesquiterpenes

Oxygenated sesquiterpenes elute mainly after other terpenoic groups (see Table 1) and separate better in the polar phase (region of RI 444 iu) than in the nonpolar phase (RI region 197 iu). The ΔRI region of these compounds is 390 to 680 iu, much higher than for sesquiterpenes. Similarly to oxygenated monoterpenes, ΔRI values for oxygenated sesquiterpenes were lowest for oxides, similar for aldehydes and ketones, and highest for sesquiterpene alcohols as more polar compounds (see Table 2).

Literature data are available for 98 oxygenated sesquiterpenes on OV-101 column, but for only 27 compounds on PEG 20M [3–27]. Only for caryophyllene oxide, nerolidol, cedrol, cadinols, and β-bisabolol do the number of RI references on OV-101 exceed five. The reproducibility of the literature RI data for oxygenated sesquiterpenes is very poor (>10 iu), especially in the polar phase (see Table 3).

There are a few mass spectra of oxygenated sesquiterpenes in our MS library; in many cases the result is sesquiterpene alcohol or oxide, but not a real compound. Some mass spectra for oxygenated sesquiterpenes are very similar (e.g., farnesols and nerolidols).

3. MAIN PROBLEMS OF IDENTIFICATION OF TERPENES

The main problem is to create the data bank of RI values for terpenes on two columns sufficient for the identification of these compounds in complex mixtures. This work is very time consuming because we have only a few standard compounds and the reproducibility of literature RI data is often low, especially for the polar column. During the last five years, we have created a RI database containing RI values of over 100 terpenes in OV-101 and PEG 20M stationary phases in temperature-programming conditions. Mainly, we use natural extracts with known composition and GC–MS data together with literature RI values on two columns in the same analysis conditions as in our laboratory.

The other problem in our laboratory is caused by older GC equipment. The splitting of the injected sample and the lack of a regulator of the carrier gas flow velocity complicate obtaining identical chromatograms in temperature-programming conditions, especially on the polar column. This makes it necessary to make a series of parallel analyses for improving the reproducibility of the results. Our GC–MS apparatus is up to the mark, but as GC–MS works without splitting, it is difficult to obtain similar chromatograms of the same sample by GC–FID and GC–MS. This is very important for the comparison of MS data with GC/RI data.

The final identification criterion is chosen according to the characteristics of both GC–MS and GC–FID, which must compliment each other perfectly. An additional comment concerns the quantitative agreement between the two columns. Unfortunately, this parameter cannot be used for substances occurring in trace concentrations or in cases where two or more substances are overlapping on the column used.

4. IDENTIFICATION OF TERPENES IN ESSENTIAL OIL FROM YARROW (*ACHILLEA MILLEFOLIUM* L.)

The essential oil from the aerial parts of *A. millefolium* L. (yarrow) plants growing wild in Estonia was isolated by steam distillation extraction during 2 h of distillation [38] and analyzed by GC–FID and GC–MS as described above. The results

Table 4 Identification Parameters (RI, m/z) and Relative Peak Areas (%) for Terpenoic Constituents of Essential Oil from *Achillea millefolium* L.

No.	Compound	RI		%		Major peaks of MS
		OV-101	PEG 20M	OV-101	PEG 20M	(m/z)
1	Tricyclene	917	1017	tr	tr	93, 121, 41
2	α-Thyjene	921	1029	0.3	} 7.5	93, 77, 91
3	α-Pinene	930	1029	6.5		93, 92, 91, 77, 79, 121
4	Camphene	940	1074	0.8	0.7	93, 121, 79, 67, 107
5	Sabinene	965	1125	} 28.4	14.2	93, 77, 41, 79, 91
6	β-Pinene	967	1116		13.6	93, 41, 79, 69, 91
7	Myrcene	982	1161	0.2	0.2	93, 69, 41, 79, 53
8	α-Phellandrene	994	1167	tr	0.1	93, 77, 91, 136
9	α-Terpinene	1008	1180	0.7	0.7	121, 93, 136, 91, 77
10	p-Cumene	1011	1270	0.9	0.7	119, 134, 91
11	1,8-Cineole	1019	1211	} 12.3	10.6	43, 81, 71, 108, 84
12	Limonene	1022	1204		0.4	68, 93, 67, 79, 121
13	(Z)-β-Ocimene	1027	1232	0.1	0.1	93, 79, 91.77, 105, 121
14	(E)-β-Ocimene	1038	1250	0.2	0.2	93, 41, 79, 53
15	γ-Terpinene	1050	1246	1.6	1.5	93, 77, 121, 136, 91
16	Terpinolene	1077	1283	0.2	0.2	121, 93, 136, 39, 79, 91
17	(Z)-Sabinene hydrate	1084	1544	0.2	0.1	43, 93, 71, 41, 81
18	Linalol	1086	1547	0.1	0.2	71, 41, 43, 93, 56, 69
19	β-Terpineol	1110	1558	0.3	0.2	43, 71, 93, 136, 69, 58
20	Camphor	1121	1515	1.0	0.9	95, 81, 41, 55, 68, 108
21	(Z)-Sabinol	1124	1648	0.1	0.1	92, 91, 83, 109, 70
22	Myrtenal	1135	1623	0.2	0.4	79, 107, 39, 91, 41, 77
23	Borneol	1152	1693	1.4	1.9	95, 41, 110, 55
24	Terpinen-4-ol	1161	1593	1.3	1.2	71, 111, 93, 43
25	Myrtenol	1168	1783	0.1	0.1	79, 108, 93, 41
26	α-Terpineol	1175	1693	0.8	b	59, 93, 121, 136, 79
27	Piperitone	1225	1720	0.1	0.2	82, 110, 95, 137, 41
28	Bornyl acetate	1267	1574	0.5	0.4	95, 43, 136, 121, 93
29	Geranyl acetate	1365	1755	} 0.5	0.2	69, 43, 41, 93, 68, 121
30	α-Cubebene	1365	1477		0.2	161, 105, 119, 91, 41
31	α-Copaene	1369	1486	0.1	0.2	119, 105, 93, 41, 81
32	(E)-β-Caryophyllene	1412	1589	5.2	4.6	93, 69, 41, 133, 71, 91
33	Aromadendrene	1438	1631	0.1	0.1	41, 69, 91, 93, 55
34	Sesquiterpene	1440	1636	0.3	0.2	69, 41, 93, 55, 91
35	α-Humulene	1443	1660	0.7	0.7	93, 121, 80, 204, 41
36	(Z)-β-Caryophyllene	1449	1667	0.1	0.2	41, 93, 69, 133, 71
37	Allo-aromadendrene	1464	1685	0.5	0.6	93, 91, 41, 69, 105, 161
38	Germacrene D	1471	1700	} 2.6	1.9	161, 91, 105, 119, 133
39	γ-Muurolene	1471	1709		0.4	105, 91, 119, 41, 204
40	Sesquiterpene	1471	1712		0.5	93, 119, 69, 41, 91
41	α-Muurolene	1483	1722		0.1	105, 161, 81, 93, 119

Table 4 Continued

No.	Compound	RI		%		Major peaks of MS (m/z)
		OV-101	PEG 20M	OV-101	PEG 20M	
42	(Z,Z)-α-Farnesene	1483	1733	⎱ 0.3	0.2	41, 63, 69
43	(Z,E)-α-Farnesene	1483	1740	⎰	0.1	41, 63, 69
44	(E,E)-α-Farnesene	1490	1742	0.3	0.3	41, 93, 69, 55, 107
45	β-Bisabolene	1494	1737	⎱ 0.5	0.3	69, 41, 93, 161, 204
46	γ-Cadinene	1497	1747	⎰	⎱ 0.8	161, 105, 91, 119
47	δ-Cadinene	1510	1749	⎱ 0.8	⎰	161, 134, 204, 119
48	(R)-Cuparene	1512	1766	⎰	0.2	132, 145, 119, 105
49	α-Bisabolene	1533	1762	⎱ 0.3	0.1	69, 41, 93
50	(Z)-Nerolidol	1533	2028	⎰	0.2	41, 69, 134, 91, 93, 79
51	(E)-Nerolidol	1549	2035	4.2	4.0	69, 41, 93, 43, 71, 55
52	Unknown	1562	2077	0.4	0.6	69, 41, 43, 81, 93, 55
53	Caryophyllene oxide	1565	1968	2.3	2.4	41, 43, 79, 69, 93
54	Veridiflorol	1575	2069	0.4	0.6	43, 41, 69, 109, 161
55	Ledol[a]	1588	2100	0.3	0.3	43, 69, 41, 122, 81
56	Unknown		2110		0.3	59, 143, 105, 41, 43
57	Guaiol	1608	2126	1.5	1.5	59, 161, 107, 41, 81
58	β-Bisabolol	1614	2142	0.3	0.3	81, 98, 41, 111, 119, 69
59	δ-Cadinol	1624	2156	0.7	0.5	161, 43, 119, 109, 204
60	T-Cadinol	1624	2161		0.5	161, 204, 105, 134
61	T-Muurolol	1630	2172	0.6	0.7	95, 43, 121, 161, 204
62	Unknown	1633	2192	⎱ 0.1	0.5	119, 41, 69, 93, 43
63	α-Cadinol	1633	2218	⎰	0.4	95, 121, 204, 43, 164
64	Caryophyllenol II	1637	2237	2.2	1.7	43, 81, 189, 161, 55
65	Farnesol[a]	1652	2255	⎱ 0.5	0.3	69, 41, 93, 81
66	Farnesol[a]	1652	2275	⎰	0.2	69, 43, 93, 81
67	Unknown		2293		0.9	136, 41, 69, 91
68	Unknown	1659	2360	1.2	1.2	41, 109, 91, 43, 55
70	(Z,E)-Farnesol	1665	2347	⎱ 2.1	1.1	69, 41, 81, 93
71	α-Bisabolol	1665	2205	⎰	1.1	109, 119, 69, 204, 93
72	Chamazulene	1700	2373	2.3	2.0	184, 169, 183, 153

[a] Correct isomer not identified.
[b] Eluates with borneol.
tr = traces (<0.05%).

of the identification of terpenoic constituents of yarrow oil and percentage quantities of peaks are presented in Table 4. The yarrow oil was a complex mixture of approximately 100 compounds, 65 of which were identified as representing up to 90% of the total oil. The main groups of components in this oil were monoterpenes (55%), while oxygenated sesquiterpenes constituted 20%, sesquiterpenes 12%, and oxygenated monoterpenes 6% of the total. The peaks of the main components in oil, sabinene, and β-pinene overlapped on OV-101 column, but were separated on PEG 20M. The peaks of limonene and 1,8-cineole also overlapped on OV-101, but from the PEG 20M data it can be concluded that 1,8-cineole constituted about 10% and limonene only 0.4% in this yarrow sample.

The main difficulties occurred in the identification of sesquiterpenes and oxygenated sesquiterpenes. There are many overlapping peaks (especially on OV-101 column) and insufficient literature on RI data in the polar phase for sesquiterpenes and oxygenated sesquiterpenes in these regions of the chromatograms.

ACKNOWLEDGMENT

We thank Dr. M. Müürisepp for performing GC–MS analyses.

REFERENCES

1. Y. Masada, Analysis of Essential Oils by Gas Chromatography and Mass Spectrometry, 1976, Wiley, New York.
2. P. Sandra and C. Bicchi (eds.), Capillary Gas Chromatography in Essential Oil Analysis, 1987, Huetig, Heidelberg, Germany.
3. R.P. Adams, Identifications of Essential Oil Components by Gas Chromatography/Mass Spectrometry, 1995, Allurd, Carol Stram, IL.
4. V.A. Isidorov, I.G. Zenkevich, E.N. Dubis, A. Slowikowski, and E. Wojciuk, J. Chromatogr. A, 814 (1998) 253–260.
5. I.G. Zenkevich, Rastit. Resur., 32 (1996) 48–58.
6. I.G. Zenkevich, Rastit. Resur., 33 (1997) 16–28.
7. N.W. Davies, J. Chromatogr. A, 503 (1990) 1–24.
8. A. Srinivara Rao, B. Rajanikanth, and R. Seshadri, J. Agric. Food Chem., 37 (1989) 740–743.
9. J.-L. Le Quere and A. Latrasse, J. Agric. Food Chem., 38 (1990) 3–10.
10. J.H. Loughrin, T.R. Hamilton-Kemp, R.A. Andersen, and D.F. Hildebrand, J. Agric. Food Chem., 38 (1990) 455–460.
11. P.R. Venskutonis, A. Dapkevicius, and M. Baranauskiene, in G. Charalambous (ed.), Food Flavours: Generation, Analysis and Process Influence, 1995, Elsevier, Amsterdam, pp. 833–847.

12. P.R. Venskutonis, J. Essent. Oil Res., 8 (1996) 91–95.
13. V.K. Kaul, B. Singh, and R.P. Sood, J. Essent. Oil Res., 8 (1996) 101–103.
14. M.F. Kerslake and R.C. Menary, Perfum. Flavor., 9 (1985) 13–24.
15. K. Yamaguchi and T. Shibamoto, J. Agric. Food Chem., 29 (1981) 366–370.
16. A. Koyasako and R.A. Bernhard, J. Food Sci., 48 (1983) 1807–1812.
17. C. Bicchi, C. Frattini, G.M. Nano, and A.D. Amato, J. High Resolut. Chromatogr., 11 (1988) 56–60.
18. L. Maat, E.J.M. Straver, T.A. van Beek, M.A. Posthumus, and F. Piozzi, J. Essent. Oil Res., 4 (1992) 615–621.
19. N. Chanegriha and A. Baaliouamer, J. Chromatogr. A, 633 (1993) 163–168.
20. I. Nykänen, Z. Lebenm. Unters Forsch., 183 (1986) 267–272.
21. I. Loayza, D. Abujder, R. Aranda, J. Jakupovic, G. Collin, H. Deslauriers, and F.-I. Jean, Phytochemistry, 38 (1995) 381–389.
22. R.G. Binder and R.A. Flath, J. Agric. Food Chem., 37 (1989) 734–736.
23. E. Stashenko, N. Quiroz Prada, and J.R. Martinez, J. High Resolut. Chromatogr., 19 (1996) 353–358.
24. P.R. Venskutonis, A. Dapkevicius, and T.A. van Beek, J. Essent. Oil Res., 8 (1996) 211–213.
25. C. Blanco Tirado, E.E. Stashenko, M.Y. Combariza, and J.R. Martinez, J. Chromatogr. A, 697 (1995) 501–513.
26. R.G. Berger, F. Drewert, and H. Kollmannsberger, Z. Lebensm. Unters Forsch., 188 (1989) 122–126.
27. A.L. Morales, D. Albarracin, J. Rodriquez, and C. Dugue, J. High Resolut. Chromatogr., 19 (1996) 585–587.
28. R.L. Miller, D.D. Bilss, and R.G. Buttery, J. Agric. Food Chem., 37 (1989) 1476–1479.
29. Chu-Chin Chen and Chi-Thang Ho, J. Agric. Food Chem., 36 (1988) 322–328.
30. A. Padrayuttawat and H. Tamura, J. High Resolut. Chromatogr., 19 (1996) 365–369.
31. Y. Holm, R. Hiltunen, and I. Nykänen, Flavour Fragrance J., 3 (1988) 109–112.
32. M.Y. Combariza, C. Blanco Tirado, and E. Stashenko, J. High Resolut. Chromatogr., 17 (1994) 643–646.
33. M. Sakho, J. Crouzet, and S. Seck, J. Food Sci., 50 (1985) 548–550.
34. V.O. Elias, B.R.T. Simoneit, and J.N. Cardoso, J. High Resolut. Chromatogr., 20 (1997) 305–309.
35. E. Stashenko, H. Wiame, S. Dassy, and J.R. Martinez, J. High Resolut. Chromatogr., 18 (1995) 54–58.
36. E. Tudor, J. Chromatogr. A, 779 (1997) 287–297.
37. A. Orav, K. Kuningas, and T. Kailas, J. Chromatogr. A, 697 (1995) 495–499.
38. A. Orav, T. Kailas, and M. Liiv, Chromatographia, 43 (1996) 215–219.

Appendix

AC	acetylation
ACE	angiotensin converting enzyme
AED	atomic emission detector
AGC	automatic gain control
AMS	accelerator mass spectrometry
AP	appearance potential
ASI	associative surface ionization
ASTED	automated sequential trace enrichment of dialysates
AT-II	angiotensin receptor II blocker
beta-BSCD	tert-butyl dimethylsilylated-beta-cyclodextrin
beta-TBDM	heptakis (6)-tert-butyl-dimethylsilyl-2,3-di-O-methyl)-beta-cyclodextrin
BSA	bis-trimethylsilylacetamide
BSTFA	N,O,-bis-trimethylsilyl-trifluoroacetamide
BTEX	benzene, toluene, ethylbenzene and xylenes
CAH	congenital adrenal hyperplasia
CFS	continuous-flow systems
CI	chemical ionization
CID	collision-induced dissociation
CSP	chiral stationary phases
DHS	dynamic headspace
DMA	dimethylamine
DPA	diphenylamine
DSI	dissociative surface ionization
ECD	electron-capture detector
ECNI	electron-capture negative ionization
EI	electron ionization
ELCD	electrolytic conductivity detector
FI	field ionization
FID	flame-ionization detector
FTIR	Fourier-transform infrared
FTR	fractional turnover rate
GC	gas chromatography

GC-C-IRMS   GC-combustion-IRMS
GC-P-IRMS   GC-pyrolysis-IRMS
HFB   heptafluorobutyric ester, heptafluorobutyration
HFBA   heptaflurorbutyric anhydride
HFBI   heptafluorobutyric imidazole
HFIP   hexafluoroisopropranol
HRGC   high-resolution gas chromatography
HRMS   high-resolution mass spectrometry
HSD   hydroxysteroid dehydrogenase
IASPE   immunoaffinity SPE
IE   ionization energy
IR   infrared
IRMS   isotope-ratio mass spectrometry
IS   internal standard
ITD   ion-trap detector
IV   intravenous
LAAM   l-a-acetylmethadol
LC   liquid chromatography
LLE   liquid-liquid extraction
LMCO   low-mass cut-off
LOD   limit of detection
LRMS   low-resolution mass spectrometry
LVI   large-volume injection
MBTFAN   methyl-bis-(trifluoroacetamide)
MFI   multifrequency irradiation
MIDA   mass isotopomer distribution analysis
MMA   monomethylamine
MS   mass spectrometry
MSD   mass-selective detector
MSI   molecular surface ionization
MS-MS   tandem mass spectrometry
MSTFA   N-methyl-N-trimethylsilyl-trifluoroacetamide
MTBSTFA   N-methyl-N-ter-butyl-dimethylsilyl-trifluoroacetamide
NCI   negative-ion chemical ionization
NMR   nuclear magnetic resonance
NPD   nitrogen-phosphorous detector
NSAID   nonsteroidal anti-inflammatory drug
PBM   probability based matching
PCB   polychlorobiphenyl
PCDD   polychlorodibenzo-p-dioxin
PCDF   polychlorodibenzofuran
PCI   positive-ion cemical ionization
PCSE   partially concurrent solvent evaporation
PCT   polychlorinated terphenyls
PDB   Pee Dee Belemnite
PDMS   polydimethylsiloxane

PFB	pentafluorobenzyl
PFB-Cl	pentafluorobenzoyl-chloride
PFP	pentafluoro-1-propanol
PFPA	pentafluoropropionic acid anhydride
PLOT	porous-layer open-tubular column
PT	purge-and-trap
PTV	programmed temperature vaporizer
RF	radiofrequency
RI	(Kovats) retention index
RSI	reaction surface ionization
S/N	signal-to-noise ratio
SCOT	suface-coated open-tubular column
SDE	simultaneous steam distallation–extraction
SFE	supercritical fluid extraction
SFI	single-frequency irradiation
SFM	secular frequency modulation
SI	surface ionization
SIM	selection ion monitoring
SIOMS	surface ionization organic mass spectrometry
SPE	solid-phase extraction
SPETD	thermal-desorption SPE
SPME	solid-phase microextraction
SRM	selective reaction monitoring
STA	systematic toxicological analysis
SVE	solvent vapour exit
SVE	solvent-vapour exit
TBA	tributylamine
TDM	therapeutic drug monitoring
TEA	thermal-energy analyser
TEF	toxicity equivalent factor
TFA	trifluoroacetylation
TFAA	trifluoroacetic anhydride
TIC	total ion chromatogram
TMCS	trimethylchlorosilane
TMS	trimethylsilyl, trimethylsilylation
TNA	trimethylamine
TOF	time-of-flight
TSQMS	triple-stage quadrupole mass spectrometry
WCOT	wall-coated open tubular column
WSCOT	wall-coated superior-capacity open-tubular column

Index

Acetic acid, 428
Acetone, 209, 424
Acetonitrile, N-morpholino-, 390
1α-Acetylmethadol, 378
Acridine, 44
Adenine, N-3-alkyl, 209
Aflatoxins, 219
Alachlor, 185
Aldehydes, 410–412, 427
Alfentanyl, 267, 273, 277
Alkaloids, 45
Alkylamines, 43–44
Alkylphenol ethoxylates, 473–474
Alphaprodine, 262
Ambient air analysis, 200–201,
    202–207, 240–244
Amino acids, 45, 288, 291, 298–299
Aminoalcohols, 44
Ammonia, 209
Amphetamine, 252, 253, 358, 359,
    364, 378–380
  methylenedioxy, 364
Analgesics, 359
Androst-5-ene-3β,17β-16-keto, 314
Androst-5-ene-3β,17β-diol, 312,
    314
5α-Androstane-3α,17α-diol, 315–
    316
Androstanediol, 326, 329, 334, 464
4-Androstenedione, 326, 329, 333

Androstenedione, chloro, 461
Androstenetetrol, 314
Androstenetriol, 312
Androsterone, 312, 323–324
  11-hydroxy, 312, 323, 325
  11-oxo, 312
  dehydro-epi- (DHEA), 312, 321,
    325–326, 329, 334
    15β,16α-dihydroxy, 314
    16-α-hydroxy, 312, 314,
      321
  sulfate, 313, 329, 334–335
Anesthetics, 235–240, 267–281
Angiotensin converting enzyme in-
    hibitors, 356
Angiotensin receptor II blockers,
    356
Aniline, 35, 40, 163
  dimethyl, 160, 171, 176
  dinitro, 390–391
Anions
  azide, 399–400
  chlorate, 399–400
  cyanide, 399–400
  disulphide, 399–400
  hydrosulphide, 399–400
  nitrate, 399–400, 405
  nitrite, 399–400
  sulphide, 399–400
  thiocyanate, 399–400

Anisole
  chloro, 412, 430
  nitro, 391
  trichloro, 412, 430
Antiarrhythmics, 360
Anticonvulsants, 359
Antidepressants, 359, 361–362
Antihistamines, 360
Anti-inflammatory drugs, 252
Antiparkinsonian drugs, 359
Arachidonic acid, 298
Aromatics, 69–71, 76, 79–80, 86–87
Articaine, 267, 269
Atomic emission detection, 27,
      189–191, 279–280
Atrazine, 49–50, 158, 176, 179
Automated cleanup, 252–263

Barbiturates, 170, 257, 356, 358–
      359, 362–363, 381–383
Benzaldehyde, 45, 181
  2-hydroxy, 430
  trinitro, 391
Benzene, 66, 74, 238, 240–244
  alkyl, 70–71, 169
  chloro, 169, 171, 174, 176, 178,
      431
    nitro, 391
  dinitro, 390
  ethyl, 240–244
  hexachloro, 163
  methoxy, 178
  trinitro, 388–391, 394–395
  vinyl, 430
Benzodiazepines, 170, 183, 257,
      261, 359, 362, 381
Benzoic acid, trinitro, 391
Benzoylecgonine, 257, 261, 375–
      377
Benzyl alcohol, 45, 430
Beta-agonists, 356, 362, 363, 469–
      471

Beta-blockers, 359–360, 362
Betamethasone, 466, 469
Bile acid, 293, 296, 299–300
Biomarker analysis, 59–68
Biphenyl
  4-amino, 201, 219
  polychlorinated, 106–114, 129,
      201, 473
Bisabolene, 487
Bisabolol, 445, 489
Boldenone, dehydro, 460
Bornadienes, polychlorinated (see
      Toxaphene)
Bornanes, polychlorinated (see
      Toxaphene)
Bornenes, polychlorinated (see
      Toxaphene)
Borneol, 438
Bromazepam, 381
Bromophos-ethyl, 171
BTEX Analysis, 240–245
Bupivacaine, 267, 274, 277
Butadiene, hexachloro, 163
Butanetriol trinitrate, 392
Butanoate, methyl, 429
Butanoic acid, 428
Butyric acid, 4-hydroxy, 343
Butyrolactone, 275
Butyrophenone, 360

Cadinene, 487
Cadinol, 489
Caffeine, 263
Camphene, 438
  polychlorinated, 127, 136
Camphor, 438
Cannabidiol, 381
Cannabinoids, 380–381
Cannabinol, 380
  delta-9-tetrahydro (THC), 380
Caproic acid, 2-oxo, 345
Carbamates, N-methyl, 163, 170

Carbaryl, 185
Carbazole, 44
Carbofuran, 185
Carbohydrates, 297–298
Caryophyllene, 487
    oxide, 489
Chenodeoxycholic acid, 299–
    300
Chlordecone, 473
Chlorpyrifos, 158
Cholestane, 65–66
Cholesterol, 296, 299–300, 310–
    311
Cholesteryl butyrate, 315
Cholic acid, 293, 299
Cimaterol, 470
1,8-Cineole, 489, 493
Clenbuterol, 455, 467, 470–471
Clinazolam, 275
Clostebol, 460
Cocaethylene, 376
Cocaine, 257, 261, 262, 359, 362,
    375–377
Codeine, 259, 262
Copaene, 438
Corticosteroids, 311–325, 466–
    469
Corticosterone, 311, 313, 322
    11-deoxy, 322
    hydroxy, 322
    tetrahydro, 312
    -11-dehydro-, 312
Cortisol, 311–313, 316–317, 322,
    325, 466, 469
    (20α-dihydro), 325
    20α-hydroxy, 312
    6β-hydroxy, 312, 325
    6-hydroxy, tetrahydro-11-deoxy,
    321
    deoxy, 321
    tetrahydro, 312, 317, 322–325
    -11-deoxy-, 312, 316

Cortisone, 322
    hydroxy-tetrahydro-, 314
    tetrahydro, 312, 314, 322
α-Cortol, 6β-Hydroxycortisol, 312
β-Cortolone, 312, 314, 322
α-Cortolone, 312, 322
    6a-hydroxy, 314
Coumarin anticoagulants, 356, 363
Coumestans, 475
Coumestrol, 475
Cyclohexene, 44
Cyclonite, 389

Daidzein, 475
DDT, 473
Decalactone, 428
Decene, 432
Derivatization, 22–23, 86–87,
    210, 252, 271–273, 315,
    328, 345–346, 358–359,
    372–374, 459–461, 469,
    471, 472, 474
Designer drugs, 253, 359, 364,
    378–380
Dexamethasone, 466, 469
Diazepam, 381
Diazinon, 473
Dibenzofuran, polychlorinated,
    97–106
Dibenzo-p-dioxin, polychlorinated,
    97–106, 473
Dibenzothiophenes, 76, 84–85
Dibucaine, 277
Diesel fuel emission, 79, 202–205
Diethylglycol-dinitrate, 401
Diethylstilbestrol, 455
Difluoromethoxy-2,2-difluoroacetic
    acid, 270, 273
Dihydropyridine calcium channel
    blockers, 356, 363
Dimethamphetamine, 380
Dimethoate, 171

Dimethyl disulphide, 428
Dimethylamine, 46–47
Dimethylenediphenyl diisocyanate, 201
Dioxins, 97–106, 143
Diphenol laxatives, 359
Diphenylamine, 36, 390
Dipicrylamine, 394
Diuretics, 356, 363
DNA adducts, 209, 215–222
Docosahexanoic acid, 298
Dodecalactone, 428
Dodecane, 43
Dodecene, 432
Drugs of abuse, 170, 247–263, 355–365, 369–384
Dynamic headspace, 411

Ecgonine, 261, 375–377
Edulans I and II, 417
Eltoprazine, 158
Enantiomeric analysis, 349–351, 416–417
Enflurane, 269, 273
EPA 8270 method, 75–76
Estradiol, 445–446, 456, 460, 472
Ethanol, 424
Ethion, 171
Ethyl acetate, 424, 428
Ethyl hexanoate, 429
Ethyl propionate, 424
Ethylene glycol dinitrate, 388, 392–396, 401
Ethylene oxide, 473
Etiocholanalone, 312
11-hydroxy, 312
Eucalyptol, 438
Explosives, zie

False negative results, 444–445, 447–448

False positive results, 445–446, 448–451
Farnesene, 487
Farnesol, 490
Fatty acids, 288, 298, 424, 428–429, 432
3-hydroxy, 201
Fenchone, 432
Fenfluramine, 383–384
Fentanyl, 267, 270–272, 276–280
Flavones, 475
Flavours, 409–410, 421–438
Fourier-transform infrared spectroscopy, 27, 280
Fragrances, 409–420
Fructose, 297
Furans, 97–106, 410–411, 428
Fusarium toxins, 201

GC, injection techniques, 3, 20–22, 155–194
GC, on-line LLE, 162–163
GC, on-line SPE, 163–179
Genistein, 475
Geranyl acetate, 489
Germacrene D, 487
Glucose, 288, 297–298
Glutaric acid, L-2-hydroxy, 346, 349–351
Glycerol, 296–297
dinitrate, 392
trinitrate, 388
Glycine, 299, 302
Guaiacol, 430
Guanidine, nitro, 394
Guanine, N-7-alkyl, 209

Haemoglobin adducts, 215–222
Halothane, 235–240, 270
Heptan-2-one, 428
Heroin, 259, 377–378

Hexamethylenetriperoxidediamine, 398–399
Hexane, 207, 427
Hexogen, 389
Human breath analysis, 200–201, 207–209
Humulene, 487
Hydrazines, 32, 44
Hydrocortisone, 466
γ-Hydroxybutyrate, 267
Hypoglycemic sulfonylureas, 356, 363

Imipramine, 49
Immunoaffinity chromatography, 457
Indolalkylamines, 460
Indole, 44
Internal standard, stable isotope, 276, 286, 327
Ionization
    charge-exchange, 81–84
    chemical, 11–13, 84
    electron-capture NCI, 128–136, 142
    field, 13, 84
    surface, 31–54
Ion-trap
    application of, 95–114, 148, 185–187, 199–224, 348
    instruments, 16–17, 96, 289, 416
Isoborneol, 2-methyl, 430
Isobutanoic acid, 428
Isobutyl acetate, 429
Isocamphone, 432
Isoflavones, 475
Isoflurane, 235–240
Isoprene, 57, 209
Isoprenoids, 60–62
Isosorbide dinitrate, 392

Isotope-ratio MS, 67–68, 285– 302, 410, 417–420
Isovaleric acid, 428
    3-hydroxy, 346

Ketamine, 267, 270, 273–274, 277–278

Lactate, 297
Lactones, 159, 428
Lactose, 288, 297, 301
Lactulose, 288
Lamotrigine, 364
Large volume injection, 20–22, 155–194
LC-GC
    heart-cut, 158–162
    techniques, 157–163
Leucine, 298–299
Lidocaine, 47–49, 262, 267, 270, 277
Limonene, 417, 420, 438, 493
    -1,2-epoxide, 417
Linalol, 417, 489
Lindane, 176
Linoleic acid, 288–298, 412
Linolenic acids, 298, 412
Liquid–liquid extraction, 251, 343, 357, 372, 411
Lorazepam, 381
Lysergide, 381–383

Mabuterol, 470
Malathion, 473
Malonic acid, 345–346
Maltose, 288
Mebeverine, 159
Medazepam, 275
Mepivacaine, 277
Mercapturic acids, 209
Metallocenes, 45

Metalochlor, 185
Methadone, 262, 378
Methamphetamine, 252, 261, 262, 378–380
  methylenedioxy, 364
Methanol, 209
Methaqualone, 364
Methoxychlor, 473
Methylamine, 44, 46–47
Methylene dianiline, 201
Methyl-N-methylanthranilate, 428
Methylpyrrole, 412
Methyl-t-butyl ether, 71
4-Methyltetrahydropyran, 2-(2-methyl-1-propenyl), 417
Methylthiouracil, 471
Metriol trinitrate, 392, 401
Mevinphos, 170
Midazolam, 267, 272, 275
Mirtazapine, 361
Monoterpenes, 411, 416, 432, 438, 486–487
  oxygenated, 489
Morphine, 259–260, 359, 362
MS
  analyte ionization, 10–13, 34–37, 80–84
  high-resolution, 15, 26–27, 84–85, 136–142, 273–274
  instrumentation, 10–18, 39–40, 289, 442–443
MS–MS, application of, 63–66, 97–104, 143–148, 374, 461
Multidimensional detection, 27–28, 268–269
Muramic acid, 201
Muurolene, 487

Naphthalene, 66
  1-nitro, 180
  alkyl, 71

[Naphthalene]
  methyl, 427
  nitro, 390
  polychlorinated, 119
Narcotics, 247–263, 355–366, 369–384
Nerolidol, 489–490
Neuroleptics, 360
Nitramines, 388–389, 392–394, 396–398
Nitrate esters, 387–388, 392–398, 401
Nitro musks, 392
Nitrobenzonitril, 390
Nitrocellulose, 387–388
Nitrocresol, 390
Nitrofluoranthenes, 207
Nitroglycerin, 388, 392–398, 401
Nitroglycol, 388
N-Nitrosamines, 219
N-Nitrosomorpholine, 390
Nitrous oxide, 235
Nonane, 432
Non-steroidal anti-inflammatory drugs, 252, 259, 356–357, 363
Nonylphenol ethoxylates, 473–474
Noralfentanil, 277
Nordiazepam, 381
Norisoprenoids, 411
Norketamine, 270–271, 277–278
  5,6-dehydro, 270–271, 277–278
Nortestosterone, 168, 445, 464

Oct-1-en-3-ol, 430–431
Oct-1-en-3-one, 430
Octanoic acid, 298
Octene, 432
Octogen, 389
Olefins, 57, 68–69, 87–91
Oleic acids, 412
Opiates, 256, 257, 359, 377–378

Opioids, 359
Organic acids, 160, 163, 171,
    342–352
Organometallic compounds, 44–45
Oxazepam, 381
Oxoacids, 342, 345, 347

PAH, nitro, 201–202
Palmitic acid, 288, 293, 298
Paraffins, 69–71, 80, 88
Parathion, 473
Pentacaine, 274
Pentaerythritol, tetranitrate, 388,
    392, 394–395, 398, 401
Pentan-2-one, 428
Pentane, 209
Perchloroethylene, 473
Pesticides, 49–50, 159, 163, 171,
    179, 185–187, 191, 473
  organochlorine, 119, 473
  organophosphorous, 28, 163,
    187, 191, 473
Phallandrene, 432, 486, 489
Phenanthrene, 76–78
Phencyclidine, 381–383
Phendimetrazine, 383–384
Phenol, 3-chloro-4-nitro, 180
  chloro, 169, 430
  dimethyl, 176
  trinitro, 388
Phenols, 163, 170, 178, 183
Phenothiazine, 359
Phenylalanine, 299
Phospholipids, 299
Phthalate, 159, 171, 473
  di(2-ethylhexyl), 201
Phytane, 62
Phytanic acid, 346
Phytoestrogens, 473–475
Picramide, N-methyl, 393–395,
    398
Picric acid, 388, 394–395

Pinene, 417, 432, 438, 486, 493
PIONA, 69
Pipecolic acid, 287
Piperidine, 43
Polycyclic aromatic hydrocarbons,
    44, 76–79, 201, 219
Porphyrins, 60–61
Pregn-5-ene,3β,17α,20α-triol,
    312
Pregn-5-ene-3β,15β,17α-triol-20-
    one, 314
Pregn-5-ene-3β,16α,diol-20-one,
    314
Pregn-5-ene-3β,20α,21-triol, 314
Pregn-5-ene-3β,21-diol-20-one,
    314
Pregnanediol, 312, 316
Pregnanetriol, 312, 316–317, 321
  11-oxo, 312, 321
Pregnanolone, 17-hydroxy or 15-
    hydroxy, 311–312, 321,
    326
Pregnenediol, 312
Pregnenetetrol, 314
Pregnenetriol, 17-hydroxy, 321
Pregnenolone, 311
  17α-hydroxy, 327, 329, 334
Prilocaine, 277
Pristane, 62
Process GC-MS, 72–74
Profiling
  plasma steroid, 326–332
  urinary steroid, 313–326
Progesterone, 311, 313, 456, 473
  17-hydroxy, 321, 326, 329,
    332–333
Propanal, 3-methylthio, 428
Propionic acid, 428
Propofol, 267, 270, 273–275, 279
Propylene glycol dinitrate, 392
Purge-and-trap, 411
Purine bases, 45

Pyrazines, 411–415, 428
 methoxy, 411
Pyrene
 amino, 206, 213–215
 hydroxy-amino, 209–215
 hydroxy-N-acetylamino, 209–215
 N-acetyl-1-amino, 214
 nitro, 79, 202–209
 nitro-4,5-dihydrodiol, 222
Pyrethroids, 170
Pyridine, 35, 40, 42–43, 44
Pyrolysis GC-MS, 66
Pyrrole, 44

Quadrupole instruments, 15–16, 289
Quaternary ammonium salts, 44
Quinol, 273, 275, 279
1,4-Quinol, 2,6-di-isopropyl, 270, 273
Quinoline, 44
Quinone, 79

Remifentanil, 267, 277
Residue analysis, 441–454
Resorcyl acid lactones, 461
Ropivacaine, 278
Rose oxide, 417

Sabinene, 417, 493
 hydrate, 489
Sector instruments, 13–15, 289
Semtex, 392–393
Sesquiterpenes, 416, 431, 487–488
 oxygenated, 489–490
Sevoflurane, 235–240
Simazine, 171, 179
Simetryne, 49–50
Sinensal, 428

Sitostanol, 300
Solid-phase extraction, 202, 251, 343, 357, 372
Solid-phase micro-extraction, 229–245, 251, 358, 372, 411
Somatotropins, 469
Soxhlet extraction, 202
Stanozolol, 457, 460
Steam distillation, 411
 and extraction, 411, 490
Stearic acid, 298
Sterane, 60–66
Steroids, 60–66, 296–300, 309–337, 445–448, 456–469
 anabolic, 445, 456–466
 sulphate, 3β-hydroxy-5-ene, 313
Sterol, 61
Stigmasterol, 315–316
Stilbene, 461
Styrene, 473
Sufentanil, 267, 273
Supercritical fluid extraction, 88–89, 202, 411
Surface ionization, 31–54
Systematic toxicological analysis, 359–363

Tapazole, 471
Temazepam, 275, 381
Terpanes, 61–62, 66
Terpenes
 mono-, 411, 416, 432, 438, 483–493
 sesqui-, 416, 431, 483–493
Terpenoids, 45
Terphenyl, polychlorinated, 117–119, 125–127, 137–138, 143–145
Terpinen-4-ol, 417
Terpinene, 420

Terpineol, 417
Testosterone, 326, 328, 334, 456,
    472
  5α-dihydro-, 326, 329, 334
  methyl, 448
  tetrahydro, 464
9-Tetrahydrocannabinol, 380
  11-hydroxy, 380
  11-nor-9-carboxy, 380
N-Tetranitro-N-methylalanine, 389
Tetranitrotetrazacyclooctane, 389,
    393–398
Tetryl, 389, 393–395, 398
Thiouracil, 471
Thujene, 438
Thymol, 274–275, 279
Thyreostatic agents, 471–472
Thyroxine, 471
Time-of-flight instruments, 17–18,
    412–414
Tobacco smoke analysis, 203–204
Toluene, 35, 176, 191, 240–244,
    390, 473
  4-methoxy, 45
  amino-dinitro, 390, 396
  amino-nitro, 390
  diamino-nitro, 391
  diisocyanate, 201
  dinitro, 388, 390–392, 394–395
  nitro, 390, 394
  trinitro, 388–392, 394–396, 401
o-Toluidine, 201
Toxaphene, 119–120, 127–135,
    139–142, 145–148
  nomenclature, 120–121

Toxicity equivalent factors, 107–
    110
Tramadol, 361
Trenbolone, 448
Triacetonetriperoxide, 398–399,
    404–405
Triazine, 49–50, 168, 169, 187
Tributylamine, 42–43
Tricin, 475
Tridecene, 432
Triethylene glycol dinitrate, 392
Triglyceride, 298–299
Tri-iodothyroxine, 471
Trimethylamine, 46–47
Trimethyloethane trinitrate, 392
Trinitrotriazacyclohexane, 389,
    392–398, 401
Trinitroxylene, 390
Triterpanes, 59–64

Undecene, 432
Urea, 299, 302
Urinary metabolites, analysis of,
    209–215
Urinary organic acids, 341–352

Valerolactone, 275
Valine, 294, 299, 302
4-Vinylcyclohexene, 473

Work function of surface, 34

Xylene, 174, 240–244

Zeranol, 461

Milton Keynes UK
Ingram Content Group UK Ltd.
UKHW021920071024
449327UK00022B/1687